Springer Geochemistry

The Springer Geochemistry series seeks to publish a broad portfolio of scientific books, aiming at researchers, students, and everyone interested in geochemistry. The series includes peer-reviewed monographs, edited volumes, textbooks, and conference proceedings. It covers the entire research area including, but not limited to, Isotope Geochemistry, Biogeochemistry, and Environmental Geochemistry.

More information about this series at http://www.springer.com/series/13486

Roberto Cioni · Luigi Marini

A Thermodynamic Approach to Water Geothermometry

Springer

Roberto Cioni
Istituto di Geoscienze e Georisorse
Consiglio Nazionale delle Ricerche
Pisa, Italy

Luigi Marini
Viareggio, Italy

ISSN 2366-6285 ISSN 2366-6293 (electronic)
Springer Geochemistry
ISBN 978-3-030-54317-4 ISBN 978-3-030-54318-1 (eBook)
https://doi.org/10.1007/978-3-030-54318-1

This Springer imprint is published by the registered company Springer Nature Switzerland AG
The registered company address is: Gewerbestrasse 11, 6330 Cham, Switzerland

Preface

Dear reader, I am sure you have seen the beautiful images of Earth taken from space. Among these images, those portraying both an area still lit by the Sun to the west and an already dark area to the east are particularly fascinating to me. As a geologist, I can appreciate the prevalence of natural elements in the lit side of our planet, such as the different continental landscapes and the blue ethereal vastness of the oceans, while the dark area appears to be dominated by the presence of man, especially those continental areas sprinkled with countless man-made light sources.

By now, the anthropogenic impact on climate is plain for everyone to see—apart from a minority of people, who deny this evidence for trivial and despicable economic reasons only. After all, economy is not a science, but a set of processes that can easily be manipulated according to the interests of few people, whose attempt is to make economy look like a scientific discipline. Not surprisingly, economy had not been included among the original disciplines worthy of the Nobel Prize. Allow me a brief digression on this point. Alfred Bernhard Nobel, a Swedish chemist who invented dynamite and ballistite, made a fortune thanks to his patents and industrial activities. In his will, he arranged that his economic resources were to be managed by a foundation, which was to award five annual prizes to exceptional benefactors of humanity in the fields of chemistry, physics, medicine or physiology, literature, and the defence of friendly relations between peoples. The prize was awarded for the first time in 1901. Only many years later, in 1968 (ironically, the year of the idealistic revolution that moved the young people of the time, including Roberto and myself) another prize was established by the Bank of Sweden, but this time it was called "In memory of Nobel," and it was to be awarded to the improperly called "economic sciences"[1].

Today, the exploitation of renewable resources as an alternative to fossil fuels—or a temporary additional support, at least for a certain period of time—is one of the possible ways to satisfy our growing energy demand, while limiting our obvious impact on our planet. One of these renewable resources is geothermal energy, whose exploitation for electricity production started in 1904 in Italy—more

[1]Translated and modified from https://www.focus.it/cultura/storia/come-e-nato-il-premio-nobel.

precisely in Larderello (Tuscany), not far from where Roberto and I worked together for many years—when Prince Piero Ginori Conti managed to turn on five electric bulbs using a dynamo coupled to a piston engine, which was powered by the geothermal steam coming from a fumarole or a shallow well (Marinelli 1990). For the first time ever, the heat produced within our planet by the radioactive decay of the isotopes of uranium, thorium, and potassium was used not directly as heat, but transformed into mechanical energy by a motor and subsequently into electricity by a dynamo. The same energy that moves continental and oceanic plates, controls orogenic cycles, and causes earthquakes and volcanic eruptions was finally used to improve the quality of human life (Marinelli 1990).

With this book, Roberto and I wish to provide our small contribution to the state of the art and possible developments in the use of geothermometers: the most acknowledged and widespread geochemical tools for identifying and exploiting geothermal resources. In this way, we hope to contribute—even though to a small extent—to the oceans of things to be done in order to reduce the enormous anthropogenic impact on our planet, which is the only one we have. Above all, we hope that this book will be useful food for thought for the new generations of geochemists, and that it might encourage a different methodological approach in their future scientific research.

I have now reached a certain age, while Roberto has left us forever on January 31, 2020. However, I believe Roberto will always be part of the scientific community, thanks to his important discoveries in the geochemistry of volcanic and geothermal gases, as well as of thermal waters and crater lakes. In particular, we have to thank him for conceiving the measurement of CO_2 fluxes from soil through accumulation chambers, a method used to date for the surveillance of active volcanoes, geothermal exploration, and environmental control all over the world: It is a creature of Roberto, the most splendid jewel he left us.

Viareggio, Italy
February, 2020

Luigi Marini

Acknowledgements We are indebted to Prof. Niniek Rina Herdianita of the Bandung Institute of Technology, Indonesia, for permission to use the compositional data of prehnite, wairakite, laumontite, garnet, and calcite from the Darajat geothermal field that she studied as part of her Ph.D. thesis, which is a mine of information on hydrothermal mineralogy.

This book has greatly benefitted from the editorial assistance of Annett Buettner and from the critical review of an anonymous reviewer, who are both gratefully acknowledged.

Reference

Marinelli G (1990) Larderello at the origins of geothermic energy. Fratelli Alinari, Florence, p 102

Contents

About the Authors

Roberto Cioni (Leghorn, September 4, 1946–Pisa, January 31, 2020) served at the Institute of Geosciences and Earth Resources of the Italian National Research Council (formerly Institute of Geochronology and Isotopic Geochemistry) in Pisa, Italy.

His scientific activity was centered on the study of fluid chemistry to reconstruct the conceptual models of volcanic and volcanic-hydrothermal systems and understand the ongoing processes especially during periods of enhanced activity. His talents and his unique intuitions have always accompanied and characterized his entire career. He was the first to recognize the role of reactive gas species, such as H_2, CO and COS, as basis of geothermometers and geobarometers.

He also developed different prototypes of scientific instruments and related methodologies, the most important of which is the accumulation chamber for the measurement of diffuse CO_2 fluxes from soils, representing one of the most widely used techniques by the whole international scientific community.

Luigi Marini (Milan, January 22, 1953) has an activity record of more than 40 years in applied geochemistry, both in the industrial sector and in the teaching and research sector, having served at the University of Genoa, Italy, as Researcher and Associate Professor.

He participated in several industrial projects aimed at the exploration and exploitation of geothermal resources and the mitigation of volcanic hazard, in both Italy and abroad.

His scientific activity has been mainly focused on the origin and evolution of fluids in geothermal and volcanic systems, including crater lakes as well as on the terrestrial CO_2 flux and the geological sequestration of CO_2.

He is co-author of more than 100 scientific publications on ISI indexed journals and sole author of the book *Geological Sequestration of Carbon Dioxide: Thermodynamics, Kinetics, and Reaction Path Modeling*.

Chapter 1
Introduction

Abstract The reasons for moving from the traditional geothermometers and f_{CO_2}-indicators to our theoretical, activity based tools are briefly discussed and the main advantages of the theoretical, activity based geothermometers and f_{CO_2}-indicators presented in this book are underscored.

It might seem that there is something mysterious, almost magical, in the capability of geothermometers to estimate the temperature of geothermal aquifers where hot water comes from. The same applies to the potential of f_{CO_2}-indicators to evaluate the CO_2 fugacity of waters hosted in geothermal reservoirs. Actually, there is no mystery and no magic, because geothermometers and f_{CO_2}-indicators are based on two simple and reasonable hypotheses. The first is the occurrence of thermo-chemical equilibrium between the aqueous solution and the hydrothermal minerals in the geothermal aquifer. The second is the lack of disturbing processes, such as mixing with shallow cold waters or re-equilibration upon cooling, during the ascent of the geothermal fluid to the surface.

Quite surprisingly, most "traditional" geothermometers and f_{CO_2}-indicators call for total (analytical) concentrations. On the one hand, the involvement of analytical data together with the mathematical simplicity of most "traditional" geothermometers explains why they are used so frequently. In fact, most people like simple things. On the other hand, this fact contrasts with the rather complicate speciation of several dissolved components, sometimes determining considerable differences between the total concentration of relevant solutes (e.g., SiO_2, Na, K, Ca and Mg) and the activity of the species actually involved in the mineral-solution reactions of interest (e.g., undissociated SiO_2 and the free ions Na^+, K^+, Ca^{2+} and Mg^{2+}). Since the formation of ion-pairs and aqueous complexes depends on temperature, CO_2 fugacity, and total ionic salinity,[1] in a work we carried out in the early '90s together with our colleagues Giovanni Chiodini and Massimo Guidi (Chiodini et al. 1991), we derived functions

[1] Total ionic salinity is defined as $\Sigma_{eq} = \Sigma\ m_i \cdot |z_i|$, where m_i is molality of the ith species and $|z_i|$ is the absolute value of its ionic charge.

R. Cioni and L. Marini, *A Thermodynamic Approach to Water Geothermometry*, Springer Geochemistry, https://doi.org/10.1007/978-3-030-54318-1_1

relating the total concentrations of solutes and the ratios between the total concentrations of solutes to the three afore-mentioned controlling variables. However, in retrospect, it is more correct and convenient to compute the activities of the species of interest for each individual sample, maintaining the thermodynamic equilibrium constants as benchmarks, rather than to derive functions involving total concentrations, manipulating the thermodynamic equilibrium constants, as done by Chiodini et al. (1991).

Other complications affecting the geoindicators are related to the hydrothermal minerals occurring in geothermal aquifers, several of which are not pure solid phases and some of which exhibit highly variable compositions. These aspects were never considered or were not properly considered, in the derivation of traditional geothermometers and f_{CO_2}-indicators.

A surprising fact, already mentioned above, is the mathematical simplicity of traditional geothermometers, resulting from the adoption of the van't Hoff equation integrated under the assumption of constant reaction enthalpy (e.g., Langmuir 1997) to express the temperature dependence of pertinent thermodynamic equilibrium constants, in most previous studies. The use of this form of the van't Hoff equation requires that the isobaric heat capacity of the reactions controlling water geothermometers is close to zero. This is usually a reasonable or relatively reasonable approximation for the dissolution reactions of silica minerals and the ion exchange reactions governing ionic solute geothermometers as shown in Sect. 5.1.3. However, as already noted by some authors (e.g., Arnórsson 2000), there is no need to rely on this approximation and to obtain simple linear functions relating the logarithm of the thermodynamic equilibrium constant to the absolute temperature inverse.

Activity coefficients of individual ions, γ_j, were neglected in several previous studies of water geothermometers, based on the assumption that γ_j ratios are close to unity (e.g., Lindsay 1980). This is true or nearly so for the γ_j ratios involving two ions of the same charge (e.g., the Na/K ratio) but the γ_j ratios involving cations of difference charge, such as the K^2/Mg, K^2/Ca, Na^2/Mg, and Na^2/Ca ratios deviate significantly from unity. In any case, there is no need to use this approximation.

Starting from these premises, we decided to investigate water geothermometers and f_{CO_2}-indicators from the theoretical point of view, adopting the thermodynamic equilibrium constants of suitable mineral-solution reactions as foundation of geothermometric functions and f_{CO_2}-indicators derived in this work. This implies that water geothermometers and f_{CO_2}-indicators involve activity ratios, which must be computed for each sample using a suitable speciation program. Owing to this theoretical approach, the thermodynamic background is summarized in Chap. 2.

Since the geothermometers and f_{CO_2}-indicators are based on the hypothesis of thermo-chemical equilibrium between the aqueous solution and the hydrothermal minerals presumably occurring in the geothermal aquifer, a large effort was devoted to the characterization of both the reservoir liquids and the hydrothermal minerals from active geothermal systems.

For this reason, we have reconstructed over 1000 chemical analyses of reservoir liquids, presumably representative of mineral-solution thermo-chemical equilibrium at aquifer temperatures of 100–350 °C, combining the chemical analyses of the liquid and vapor phases discharged from drilled wells and collected at known separation temperature and pressure. The relevant chemical characteristics of reservoir liquids are presented in Chap. 3. Moreover these reservoir liquids are used to test the geothermometers and f_{CO_2}-indicators, both those traditional and those derived in this work, in subsequent chapters.

For the same reason, over 2200 chemical analyses of hydrothermal minerals from active geothermal systems were compiled and processed, as discussed in Chap. 4. In this way, it was possible to identify the minerals occurring as pure solid phases or nearly so and those present as solid solutions. For the solid solutions, the average activity of the endmembers of interest and other statistical parameters were computed. Average activities were then used in the implementation of water geothermometers and f_{CO_2}-indicators, in subsequent chapters.

Chapter 5 is devoted to the traditional geothermometers and f_{CO_2}-indicators. Their characteristics, strengths and weaknesses are thoroughly treated. Our focus is mainly on the silica, Na–K, Na–K–Ca, Na–Ca, K–Ca, K–Mg and Na–Mg geothermometers as well as on the K–Ca f_{CO_2}-indicator. Nevertheless, other ionic solute geothermometers (e.g., Na–Li, Mg–Li, Ca–Mg, SO_4–F) are briefly recalled, as well as multicomponent chemical geothermometry and the influence of ion complexing on geothermometers and f_{CO_2}-indicators.

The main outcomes of this work are illustrated in Chaps. 6, 7, and 8, which are centered on the theoretical, activity-based Na–K geoindicators, K–Mg and Na–Mg geoindicators, and Ca–K and Ca–Na geoindicators, respectively.

We are aware that the approach we propose here is much more intricate than the simple formulas of traditional geoindicators and that users must have a background in geochemistry and thermodynamics or must acquire it. However, the precision on the calculated geothermal aquifer temperatures and CO_2 fugacities and the additional information on the hydrothermal minerals presumably present in the geothermal reservoir represent significant improvements with respect to the results given by traditional geoindicators and multicomponent chemical geothermometry. We think it is worth a try.

References

Arnórsson S (2000) The quartz and Na/K geothermometers. I. New thermodynamic calibration. In: Proceedings of the world geothermal congress 2000, pp 929–934. Kyushu-Tohoku, Japan

Chiodini G, Cioni R, Guidi M, Marini L (1991) Chemical geothermometry and geobarometry in hydrothermal aqueous solutions: a theoretical investigation based on a mineral-solution equilibrium model. Geochimica et Cosmochimica Acta 55:2709–2727

Langmuir D (1997) Aqueous environmental geochemistry. Prentice Hall, Upper Saddle River, NJ, p 600

Lindsay Jr WT (1980) Estimation of concentration quotients for ionic equilibria in high temperature water: The model substance approach. In: Proceedings of International Water Conference 41:284–294

Chapter 2
Thermodynamics, Geochemical Modeling and Related Considerations

Abstract In this chapter we intend (1) to briefly recall the thermodynamic functions and the relations linking them which are of use in geochemical applications, (2) to present the software package SUPCRT92, which is used to calculate the standard molal thermodynamic properties of minerals, gases, aqueous species, and H_2O, as well as of the reactions involving these entities, as a function of temperature and pressure, from 0 to 1000°C and from 1 bar to 5 kbar, (3) to summarize the main aspects of equilibrium state models and reaction path models, focusing of multicomponent chemical geothermometry, which is a reaction path model of special interest. In particular, the simple equations linking the Gibbs free energy and the thermodynamic equilibrium constant of the dissolution reaction of a given solid solution to the corresponding properties of the dissolution reaction of the key endmember are presented in section 2.2.1.

In the their textbook on geochemical thermodynamics, Nordstrom and Munoz (2006) recall the impact of thermodynamics on the famous physicist Arnold Sommerfeld (1868–1951), who was honored with many distinguished awards for his important contributions to atomic physics. Despite the unbeaten record of eighty one nominations, he never won the Nobel Prize, which was awarded to four PhD students of him, namely Peter Debye, Werner Karl Heisenberg, Wolfgang Pauli, and Hans Bethe. According to Arnold Sommerfeld, the first time he studied thermodynamics, he thought he understood it except for a few minor points. The second time he studied thermodynamics, he thought he did not understand it except for a few minor points. The third time, he knew he did not understand the subject, but by then it did not matter, because he could use thermodynamics effectively. We suspect that Sommerfeld was too strict with himself, but we agree that it is important to be able to use thermodynamics, avoiding excessive insights and details. Therefore, we tried to organize this chapter accordingly.

First, in Sect. 2.1, thermodynamic principles and relations are briefly recalled. The reader is referred to textbooks on thermodynamics (e.g., Lewis and Randall 1961; Denbigh 1981; Anderson and Crerar 1993; Nordstrom and Munoz 2006) for the derivation of relationships from first principles and related discussion. Second,

R. Cioni and L. Marini, *A Thermodynamic Approach to Water Geothermometry*, Springer Geochemistry, https://doi.org/10.1007/978-3-030-54318-1_2

in Sect. 2.2, the software package SUPCRT92 is presented and a related topic is treated in subsection 2.2.1. Third, in Sect. 2.3, emphasis is placed on the aspects of geochemical modeling which are relevant for the development of theoretical, activity-based geothermometers and f_{CO_2} indicators, as we plan to do in this work. The reader is referred to textbooks on geochemical modeling (e.g., Grenthe and Puigdomenech 1997; Bethke 2008) for calculation techniques and examples of application.

2.1 A Synthesis of Thermodynamics for Geothermal Geochemistry

2.1.1 Gibbs Free Energy, Enthalpy, and Entropy

Different functions, known as thermodynamic potentials or heat potentials or fundamental functions, can be utilized to describe the state of the system of interest, that is to understand the proximity to (or the distance from) the condition of chemical equilibrium as well as the direction in which a reversible reaction will proceed to attain the equilibrium condition. Because geochemists generally use temperature as thermal parameter and pressure as mechanical parameter, the most suitable thermodynamic potential is the Gibbs free energy, G. The Gibbs free energy is a state function. This means that G (like the other thermodynamic potentials) depends only on the state of the system and does not depend on the path followed by the system to reach its current state.

The enthalpy, H, which represents the amount of heat exchanged in a process, and the entropy, S, which depends on the state of disorder of the system, are the two components of the Gibbs free energy, as indicated by the fundamental relation:

$$\mathrm{G} = \mathrm{H} - \mathrm{T} \cdot \mathrm{S} \tag{2.1}$$

Like G, both H and S are state functions.

2.1.2 Standard State, Activity and Fugacity

It is possible to measure the variations of state functions resulting from changes in temperature, pressure, and composition, whereas it is impossible to measure the absolute values of state functions. Therefore, the value of a state function at the temperature and pressure of interest is defined with respect to the value that it assumes at an appropriate reference state, which is called standard state. This is a constant that cancels out in the computation of the variation in the state function ensuing from any change in temperature, pressure, and composition. The standard state conventions generally assumed are:

- pure phases at all pressures and temperatures for solids and liquids, including water;
- hypothetical one molal solution referenced to infinite dilution at any pressure and temperature, for aqueous species other than water; the term hypothetical indicates that the adopted standard state convention has nothing to do with the real one molal solution;
- hypothetical ideal gas at 1 bar and any specified temperature for gases.

Incidentally, these standard state conventions imply that gases obey the ideal gas law and solutes obey the Henry's law or the Raoult's law.

The Gibbs free energy of one mole of any substance i (either a solid or a liquid or a gas or a solute) in the standard state is generally indicated by the superscript °, i.e., as G_i^o, whereas G_i indicates the Gibbs free energy of one mole of i in the considered system. The difference between G_i and G_i^o depends on a function called activity, as expressed by the relation:

$$G_i = G_i^o + RT \ln a_i \quad (2.2)$$

where T is the absolute temperature (in Kelvin degrees) and R is the universal gas constant, 8.3144626 J K^{-1} mol^{-1} or 1.9872036 cal K^{-1} mol^{-1}.

From Eq. (2.2), it follows that the activity is one at standard state. Therefore, pure solids and liquids (including water) have unit activity at all pressures and temperatures. Aqueous solutes have unit activity in the hypothetical infinitely diluted, one molal solution at any pressure and temperature. Pure ideal gases have unit activity at 1 bar and any temperature.

Activity coefficients are dimensionless factors introduced to deal with real substances which are not in their standard states and deviate from ideal behavior. The activity of the j-th component in a non-ideal solid or liquid mixture i, $a_{j,i}$, is related to its molar fraction, $x_{j,i}$, by the equation:

$$a_{j,i} = \lambda_{j,i} \cdot x_{j,i}. \quad (2.3)$$

The activity of the j-th component in a non-ideal aqueous solution i, $a_{j,i}$, is related to its molality, $m_{j,i}$, by the equation:

$$a_{j,i} = \gamma_{j,i} \cdot m_{j,i}. \quad (2.4)$$

The distinct approaches adopted to calculate activity coefficients of solute species in dilute and concentrated electrolyte solutions are concisely recalled in Sects. 2.3.3 and 2.3.4, respectively, because of the pivotal role of aqueous solutions.

Although Eq. (2.2) applies to any state of matter, usually fugacity is used instead of activity for pure gases and gas mixtures. Fugacity is related to activity by this simple relation:

$$a_i = \frac{f_i}{f_i^o}. \tag{2.5}$$

If $f_i^o = 1$ bar because of standard state convention, then $a_i = f_i$. This fact cannot be taken for granted because other choices of the standard state are possible. The fugacity of a pure gas i is related to total gas pressure, P, by the relation:

$$f_i = \Gamma_i \cdot P \tag{2.6}$$

where Γ_i is the fugacity coefficient of the considered pure gas. The fugacity of the j-th component of a gas mixture i is defined by the equation:

$$f_{j,i} = \Gamma_{j,i} \cdot P \cdot y_{j,i}. \tag{2.7}$$

where $\Gamma_{j,i}$ and $y_{j,i}$ are the fugacity coefficient and the molar fraction of the j-th component of the gas mixture i.

2.1.3 The Equilibrium Constant

Let us consider a generic reaction:

$$aA + bB = cC + dD. \tag{2.8}$$

The Gibbs free energy of this reaction,ΔG_r, is the sum of the molal Gibbs free energy of reactants and products, each multiplied times the corresponding stoichiometric coefficient, that is:

$$\Delta G_r = c \times G_C + d \times G_D - a \times G_A - b \times G_B. \tag{2.9}$$

To be noted that stoichiometric coefficients are positive for products and negative for reactants. For the considered reaction, it is also possible to compute the standard molal Gibbs free energy of the reaction, ΔG_r^o, in which each product and each reactant is at standard state:

$$\Delta G_r^o = c \times G_C^o + d \times G_D^o - a \times G_A^o - b \times G_B^o \tag{2.10}$$

Because Eq. (2.2) can be written for each product and reactant, it follows that:

$$\Delta G_r = \Delta G_r^o + RT \ln\left(\frac{a_C^c \cdot a_D^d}{a_A^a \cdot a_B^b}\right) \tag{2.11}$$

Since, the Gibbs free energy of the reaction, ΔG_r, is zero at equilibrium, it is possible to write:

$$\Delta G_r^o = -RT \ln\left(\frac{a_C^c \cdot a_D^d}{a_A^a \cdot a_B^b}\right)_{equilibrium} = -RT \ln K, \quad (2.12)$$

where K is the ***thermodynamic equilibrium constant***. Equation (2.12) has been called the most important and useful relation of thermodynamics, because it tells us that the thermodynamic equilibrium constant depends on a difference between standard Gibbs free energy only, and consequently on T and P only, and does not depend on the composition of the system. According to the IUPAC Gold Book (see https://goldbook.iupac.org/terms/view/S05915), the symbol K° should be used instead of K and the adjective "standard" should be used instead of "thermodynamic", although in the IUPAC Gold Book it is written that "*some chemists prefer the name thermodynamic equilibrium constant and the symbol K*". Sorry, I am one of these chemists.

2.1.4 *Thermodynamic Properties of Solids and Gases as a Function of T and P*

To calculate the thermodynamic equilibrium constant at any T and P, it is necessary to compute the changes in standard Gibbs free energy with T and P, as expressed by the following relation:

$$\Delta G^o_{r,T,P} = \Delta G^o_{r,Tr,Pr} + \int_{Tr}^{T} \frac{\partial \Delta G_r^o}{\partial T} dT + \int_{Pr}^{P} \frac{\partial \Delta G_r^o}{\partial P} dP$$
$$= \Delta G^o_{r,Tr,Pr} - \int_{Tr}^{T} \Delta S_r^o dT + \int_{Pr}^{P} \Delta V_r^o dP. \quad (2.13)$$

In Eq. (2.13), $\Delta G^o_{r,Tr,Pr}$ is the standard Gibbs free energy of the considered reaction at the reference temperature Tr and reference pressure Pr, ΔS_r^o is the standard entropy of reaction (i.e., the sum of the standard molal entropies of reactants and products, each multiplied times the corresponding stoichiometric coefficient) and ΔV_r^o is the standard volume of reaction (i.e., the sum of the standard molal volumes of reactants and products, each multiplied times the corresponding stoichiometric coefficient). In other words, ΔS_r^o and ΔV_r^o are defined by relations completely analogous to Eq. (2.10).

If a gas participate to the reaction of interest, it is convenient to split the integral $\int_{Pr}^{P} \Delta V_r^o dP$, in two parts, one for the solids and one for the gas (e.g., CO_2), as follows:

$$\int_{Pr}^{P} \Delta V_r^o dP = \int_{Pr}^{P} \Delta V^o_{solids} dP + \int_{Pr}^{P} V^o_{CO_2} dP, \quad (2.14)$$

in which $V^o_{CO_2}$ is the standard molal volume of gaseous CO_2. The reason for splitting the integral $\int_{Pr}^{P} \Delta V^o_r dP$ is that the volume of solids is practically pressure independent in the range of pressure of geothermal systems (that is from 1 to a few hundred bars), whereas the volume of gases experiences large changes with pressure. Assuming that the reaction volume of solids is constant, and recalling Eqs. (2.2) and (2.5), it is possible to rewrite Eq. (2.14) as follows:

$$\int_{Pr}^{P} \Delta V^o_r dP = \Delta V^o_{solids}(P - 1) + RT \cdot \ln\left(\frac{f_{CO_2,P}}{f_{CO_2,Pr}}\right). \tag{2.15}$$

As shown by Eq. (2.13), the standard entropy of the reaction of interest, ΔS^o_r, is needed to compute the standard Gibbs free energy of the reaction at any temperature $T \neq Tr$. In turn, the standard entropy of reaction is a function of temperature as expressed by the following relation:

$$\Delta S^o_{r,T} - \Delta S^o_{r,Tr} = \int_{Tr}^{T} \frac{\Delta C^o_{P,r}}{T} dT, \tag{2.16}$$

where $\Delta C^o_{P,r}$ is the standard heat capacity at constant pressure of the reaction (i.e., the sum of the standard molal heat capacities of reactants and products, each multiplied times the corresponding stoichiometric coefficient). Isobaric heat capacities of solids and gases are generally defined by simple temperature functions which can be easily integrated, such as the Maier-Kelley equation (Maier and Kelley 1932):

$$C^o_P = a + b \cdot T + c \cdot T^{-2}. \tag{2.17}$$

Adoption of the Maier-Kelley equation for the standard molal heat capacity at constant pressure of solids and gases leads to:

$$\Delta S^o_{r,T} = \Delta S^o_{r,Tr} + \Delta a \cdot \ln\frac{T}{Tr} + \Delta b(T - Tr) + \frac{\Delta c}{2}\left(\frac{1}{T^2} - \frac{1}{Tr^2}\right) \tag{2.18}$$

and

$$\Delta G^o_{r,T} = \Delta G^o_{r,Tr} + \Delta a \cdot \left(T - Tr - T \cdot \ln\frac{T}{Tr}\right) + \frac{\Delta b}{2}\left(-T^2 - Tr^2 + 2 \cdot T \cdot Tr\right) + \Delta c\left(\frac{T^2 + Tr^2 - 2 \cdot T \cdot Tr}{2 \cdot T \cdot Tr^2}\right). \tag{2.19}$$

Equation (2.19) is used to compute the standard Gibbs free energy of the considered reaction at any given temperature $T \neq Tr$.

The standard isobaric heat capacity of the reaction is also involved in the relation expressing the temperature dependence of the standard enthalpy of the reaction:

$$\Delta H^{o}_{r,T} - \Delta H^{o}_{r,Tr} = \int_{Tr}^{T} \Delta C^{o}_{P,r} dT, \quad (2.20)$$

Adopting the Maier-Kelley equation for the standard molal heat capacity at constant pressure of solids and gases, the following relation is obtained from Eq. (2.20):

$$\Delta H^{o}_{r,T} = \Delta H^{o}_{r,Tr} + \Delta a(T - Tr) + \frac{\Delta b}{2}(T^2 - Tr^2) + \Delta c\left(\frac{1}{T} - \frac{1}{Tr}\right) \quad (2.21)$$

The temperature dependence of the thermodynamic equilibrium constant K of any reaction is expressed by the van't Hoff equation:

$$\ln \frac{K_T}{K_{T_r}} = \int_{T_r}^{T} \frac{\Delta H^{o}_{r}}{RT^2} dT, \quad (2.22)$$

Integration of Eq. (2.22) adopting the Maier-Kelley equation for the $\Delta C^{o}_{P,r}$ leads to the following relation:

$$\ln K_T = \ln K_{Tr} - \frac{\Delta H^{o}_{r,Tr}}{R}\left(\frac{1}{T} - \frac{1}{Tr}\right) + \frac{\Delta a}{R}\left(\ln \frac{T}{Tr}\right) + \frac{\Delta b}{2R}(T - Tr) + \frac{\Delta c}{2R}\left(T^2 - Tr^2\right). \quad (2.23)$$

Equation (2.23) is used to compute the thermodynamic equilibrium constant at any given temperature $T \neq Tr$.

The effect of possible phase transitions in solids was not considered, to avoid complicating too much previous equations. For completeness, see Johnson et al. (1992).

2.1.5 *Thermodynamic Properties of Aqueous Species as a Function of T and P*

The thermodynamic properties of electrolyte solutions at elevated temperatures and pressures can be predicted by means of a theoretical model implemented by Harold C. Helgeson and coworkers between 1974 and 1981 (Helgeson and Kirkham 1974a,b; Helgeson and Kirkham 1976; Helgeson et al. 1981) and later revised by Tanger and

Helgeson (1988) and Shock et al. (1992). This theoretical model is known as revised HKF model.

The standard molal volume and isobaric heat capacity of a solute are given by the two expressions:

$$V^o = -\omega \cdot Q + \left(\frac{1}{\varepsilon} - 1\right) \cdot \left(\frac{\partial \omega}{\partial P}\right)_T + a_1 + \frac{a_2}{\Psi + P} + \frac{a_3}{T - \Theta} + \frac{a_4}{(\Psi + P) \cdot (T - \Theta)} \tag{2.24}$$

$$\begin{aligned} C_P^o =& \omega TX + 2TY\left(\frac{\partial \omega}{\partial T}\right)_P - T\left(\frac{1}{\varepsilon} - 1\right)\left(\frac{\partial^2 \omega}{\partial T^2}\right)_P \\ &+ c_1 + \frac{c_2}{(T-\Theta)^2} - \frac{2T}{(T-\Theta)^3}\left[a_3(P - P_r) + a_4 \ln\left(\frac{\Psi + P}{\Psi + P_r}\right)\right] \end{aligned} \tag{2.25}$$

in which:

- ω is the electrostatic Born coefficient which is a property of each solute;
- a_1, a_2, a_3, a_4, c_1, and c_2 are coefficients which assume specific values for each solute;
- Ψ is equal to 2600 bar and Θ is a sort of singular temperature for water, equal to 228 K (−45 °C)
- Q, X, and Y are functions of the dielectric constant of water, ε:

$$Q = \frac{1}{\varepsilon^2} \cdot \left(\frac{\partial \varepsilon}{\partial P}\right)_T \tag{2.26}$$

$$Y = \frac{1}{\varepsilon^2} \cdot \left(\frac{\partial \varepsilon}{\partial T}\right)_P \tag{2.27}$$

$$X = \left(\frac{\partial Y}{\partial T}\right)_P = \frac{1}{\varepsilon^2} \cdot \left(\frac{\partial^2 \varepsilon}{\partial T^2}\right)_P - 2 \cdot \varepsilon \cdot Y^2. \tag{2.28}$$

The standard molal entropy, enthalpy, and Gibbs free energy of each solute are obtained through integration, as shown in the following expressions:

$$\begin{aligned} S^o_{P,T} =& S^o_{Pr,Tr} + \int_{Tr}^{T} \frac{C_P^o}{T} dT - \int_{Pr}^{P} \left[\left(\frac{\partial V^o}{\partial T}\right)_P\right]_T dP \\ =& S^o_{Pr,Tr} + c_1 \cdot \ln\frac{T}{T_r} - \frac{c_2}{\Theta} \cdot \left[\left(\frac{1}{T-\Theta}\right) - \left(\frac{1}{T_r - \Theta}\right) + \frac{1}{\Theta} \cdot \ln\left(\frac{T_r \cdot (T-\Theta)}{T \cdot (T_r - \Theta)}\right)\right] \\ &+ \left(\frac{1}{T-\Theta}\right)^2 \cdot \left[a_3 \cdot (P - P_r) + a_4 \cdot \ln\left(\frac{\Psi + P}{\Psi + P_r}\right)\right] + \omega Y - \left(\frac{1}{\varepsilon} - 1\right)\left(\frac{\partial \omega}{\partial T}\right)_P \\ &- \omega_{Pr,Tr} Y_{Pr,Tr} \end{aligned} \tag{2.29}$$

$$
\begin{aligned}
H^o_{P,T} =& H^o_{Pr,Tr} + \int_{Tr}^{T} C^o_P dT + \int_{Pr}^{P} \left[V^o - T\left(\frac{\partial V^o}{\partial T}\right)_P\right]_T dP \\
=& H^o_{Pr,Tr} + c_1 \cdot (T - T_r) - c_2\left[\left(\frac{1}{T-\Theta}\right) - \left(\frac{1}{T_r-\Theta}\right)\right] + a_1(P - P_r) + a_2 \cdot \ln\left(\frac{\Psi + P}{\Psi + P_r}\right) \\
&+ \left(\frac{2T-\Theta}{(T-\Theta)^2}\right)\left[a_3(P - P_r) + a_4 \cdot \ln\left(\frac{\Psi+P}{\Psi+P_r}\right)\right] + \omega\left(\frac{1}{\varepsilon} - 1\right) + \omega TY \\
&- T\left(\frac{1}{\varepsilon} - 1\right)\left(\frac{\partial \omega}{\partial T}\right)_P - \omega_{Pr,Tr}\left(\frac{1}{\varepsilon_{Pr,Tr}} - 1\right) - \omega_{Pr,Tr} T_r Y_{Pr,Tr}
\end{aligned} \tag{2.30}
$$

$$
\begin{aligned}
G^o_{P,T} =& G^o_{Pr,Tr} - S^o_{Pr,Tr}(T - T_r) + \int_{Tr}^{T} C^o_P dT - \int_{Tr}^{T} \frac{C^o_P}{T} dT + \int_{Pr}^{P} V^o dP \\
=& G^o_{Pr,Tr} - S^o_{Pr,Tr}(T - T_r) - c_1\left[T \cdot \ln\left(\frac{T}{T_r}\right) - T + T_r\right] + a_1(P - P_r) + a_2 \cdot \ln\left(\frac{\Psi+P}{\Psi+P_r}\right) \\
&- c_2\left\{\left[\left(\frac{1}{T-\Theta}\right) - \left(\frac{1}{T_r-\Theta}\right)\right] \cdot \left(\frac{\Theta - T}{\Theta}\right) - \frac{T}{\Theta^2} \cdot \ln\left(\frac{T_r \cdot (T-\Theta)}{T \cdot (T_r - \Theta)}\right)\right\} \\
&+ \left(\frac{1}{T-\Theta}\right) \cdot \left[a_3(P - P_r) + a_4 \cdot \ln\left(\frac{\Psi+P}{\Psi+P_r}\right)\right] + \omega\left(\frac{1}{\varepsilon} - 1\right) - \omega_{Pr,Tr}\left(\frac{1}{\varepsilon_{Pr,Tr}} - 1\right) \\
&- \omega_{Pr,Tr}(T - T_r)Y_{Pr,Tr}
\end{aligned} \tag{2.31}
$$

In spite of the apparent complexity of these equations, they involves seven specific parameters for each solute, namely ω, a_1, a_2, a_3, a_4, c_1, and c_2, which have been defined above.

2.2 Supcrt92

The software package SUPCRT92 (Johnson et al. 1992) is used to compute the standard molal thermodynamic properties of minerals, gases, aqueous species, and H_2O, as well as of reactions involving these entities as a function of temperature and pressure, from 0 to 1000 °C and from 1 bar to 5 kbar. In fact, SUPCRT92 encodes the expressions presented in Sects. 2.1.4 and 2.1.5. An on-line interface to compute the thermodynamic properties of the reactions of interest is available at the following link http://geopig3.la.asu.edu:8080/GEOPIG/pigopt1.html.

Zimmer et al. (2016) expanded the SUPCRT92 thermodynamic dataset and modified the SUPCRT92 program implementing a software package called SUPCRTBL, which is available online at https://models.earth.indiana.edu/applications_index.php. Zimmer et al. (2016) used the mineral database of Holland and Powell (2011) and modified the original computer code to accommodate the different polynomial function for isobaric heat capacity, the molal volume as a function of temperature and pressure, and mineral phase transitions using the Landau model (Holland and Powell 1998). Several species were added to the database including As-bearing minerals and aqueous species (from Nordstrom and Archer 2002; Zhu et al. 2005; Langmuir et al. 2006; and Marini and Accornero 2007, 2010), Al-bearing aqueous species from Tagirov and Schott (2001), aqueous silica from Apps and Spycher (2004) and

Stefánsson (2001), and dawsonite from Benezeth et al. (2007). Although these modifications are certainly needed for studies in metamorphic petrology, environmental geology, and geological sequestration of CO_2, the original SUPCRT92 database is still the most appropriate for this work, in spite of possible problems on the thermodynamic data of Al-bearing minerals (see below).

In the original SUPCRT92 database, the thermodynamic properties of all relevant minerals are from Helgeson et al. (1978), the thermodynamic properties of all pertinent aqueous species are from Helgeson and coworkers (e.g., Shock and Helgeson 1988; Shock et al. 1989, 1997; Sverjensky et al. 1997), the thermodynamic properties of all gases (although CO_2 is the only gas of interest for this work) are from Kelley (1960) and Wagman et al. (1982), whereas the thermodynamic and electrostatic properties of H_2O are calculated by SUPCRT92 using the equations and data given by Helgeson and Kirkham (1974a), Uematsu and Franck (1980), Levelt Sengers et al. (1983), Pitzer (1983), Haar et al. (1984), and Johnson and Norton (1991). Consequently, the results of SUPCRT92 have a high level of internal consistency. This is a good reason to use SUPCRT92 and its original database for computing the log K and other thermodynamic properties of mineral-solution reactions relevant for this work.

However, the reliability of the thermodynamic data of Helgeson et al. (1978) for Al-bearing minerals was questioned by some authors. Hemingway et al. (1982) underscored that there is a systematic difference of about -6.5 kJ (mol Al)$^{-1}$ between the enthalpies of formation reported by Helgeson et al. (1978) for Al-bearing minerals and the enthalpies of formation recommended by Robie et al. (1979), Hemley et al. (1980), and Haas et al. (1981) for the same solid phases. According to Hemingway et al. (1982), this systematic difference in the enthalpies of formation of Al-bearing minerals derives from a questionable procedure used by Helgeson et al. (1978) to obtain the thermochemical data of kaolinite, which was then used as a secondary reference standard for Al in the construction of their thermodynamic database. This error in the thermodynamic data of Helgeson et al. (1978) for Al-bearing minerals was also underscored by Arnórsson and Stefánsson (1999) in a review paper on the thermodynamic properties of feldspars, by Nordstrom and Munoz (2006) in the their textbook on geochemical thermodynamics, and more recently by Tutolo et al. (2014) and Wolery and Jové Colón (2007). The work of Tutolo et al. (2014) is certainly interesting but represents a sort of first step forwards, because revised, internally consistent data are provided for low-albite, maximum-microcline, muscovite, and quartz only, whereas several Al-bearing minerals of interest for this work (i.e., high-albite, high-sanidine, 7Å-clinochlore, clinozoisite, laumontite, prehnite, and wairakite) are disregarded.

In Sect. 4.2.4 it will be shown that the transition temperatures of alkali feldspars are satisfactorily reproduced by using the thermodynamic data of Helgeson et al. (1978), whereas the thermodynamic data proposed more recently by other authors (e.g., Holland and Powell 1998; Arnórsson and Stefánsson 1999; Holland and Powell 2011) lead to computed transition temperatures of alkali feldspars at variance with the values reported in the scientific literature. Therefore, it is better to use the thermodynamic data of Helgeson et al. (1978) rather than those of other thermodynamic compilations mentioned above, at least for the alkali feldspars.

Moreover, it must be noted that even if the enthalpies of formation proposed by Helgeson et al. (1978) for Al-bearing minerals are affected by a systematic error, such an error does not preclude their use in exchange reactions as long as Al is conserved in the solid phases.

The equilibrium constants computed by means of SUPCRT92 are incorporated in the databases of several geochemical modeling codes. This procedure is particularly straightforward for the EQ3/6 software package, because the SUPCRT92 computer code can be used to generate the log K of the reactions of interest at the temperatures and pressures required by EQ3/6.

2.2.1 *The Equilibrium Constant of the Dissolution Reaction of a Solid Solution*

As discussed by Helgeson et al. (1978), if the thermodynamic properties are to be assigned to all the solid solutions present in nature, thermodynamic databases would be flooded by an almost infinite numbers of these solid solutions. Among the hydrothermal minerals occurring in active geothermal systems, illites, chlorites, epidotes, garnets, prehnites and wairakites are solid solutions of varying composition (as detailed in Chap. 4) and would contribute to the flooding of thermodynamic databases. Since this practice leaves much to be desired, it is advisable to take a different approach, considering only the solid solutions in which the key endmembers have average activity or an activity value of interest. The key endmembers are those for which thermodynamic data are available, such as muscovite in illites, 7Å-clinochlore in chlorites, clinozoisite in epidotes, prehnite in prehnite/ferri-prehnite solid solutions, wairakite in wairakite/analcime solid solutions and grossular in garnets, among the hydrothermal minerals occurring in active geothermal systems.

The activity of the key endmember in a given solid solution can be calculated from site mixing approximations, using the relations given in different sections of Chap. 4. Here, we intend to write the simple equations linking the Gibbs free energy and the thermodynamic equilibrium constant of the dissolution reaction of a solid solution to the corresponding properties of the dissolution reaction of the key endmember. Taking illite and muscovite as example, we begin to write the dissolution reaction of muscovite:

$$KAl_3Si_3O_{10}(OH)_2 + 10H^+ = K^+ + 3Al^{3+} + 3SiO_{2(aq)} + 6H_2O. \qquad (2.32)$$

whose standard molal Gibbs free energy is:

$$\Delta G^o_{r,Ms} = G^o_{K^+} + 3 \cdot G^o_{Al^{3+}} + 3 \cdot G^o_{SiO_{2(aq)}} + 6 \cdot G^o_{H_2O} - G^o_{Ms}. \qquad (2.33)$$

To be noted that (i) Eq. (2.33) is a peculiar case of the general Eq. (2.10) and (ii) the Gibbs free energy of hydrogen ion does not appear in Eq. (2.33) being zero by convention. Assuming that the stoichiometry of illite is not very different from that of pure muscovite, the Gibbs free energy of the illite dissolution reaction can be written as:

$$\Delta G^{o}_{r,Ill} = G^{o}_{K^+} + 3 \cdot G^{o}_{Al^{3+}} + 3 \cdot G^{o}_{SiO_{2(aq)}} + 6 \cdot G^{o}_{H_2O} - G^{o}_{Ill}. \tag{2.34}$$

Based on the general Eq. (2.2), the Gibbs free energy of illite can be written as:

$$G^{o}_{Ill} = G^{o}_{Ms} + RT \ln a_{Ms}. \tag{2.35}$$

Inserting Eqs. (2.35) into (2.34) and rearranging we obtain:

$$\Delta G^{o}_{r,Ill} = G^{o}_{K^+} + 3 \cdot G^{o}_{Al^{3+}} + 3 \cdot G^{o}_{SiO_{2(aq)}} + 6 \cdot G^{o}_{H_2O} - G^{o}_{Ms} - RT \ln a_{Ms}. \tag{2.36}$$

Comparing Eqs. (2.33) and (2.36), it turns out that:

$$\Delta G^{o}_{r,Ill} = \Delta G^{o}_{r,Ms} - RT \ln a_{Ms}. \tag{2.37}$$

Eqn (2.37) links the Gibbs free energies of the dissolution reactions of illite and muscovite and is one of the sought relations. Let us now write the general Eq. (2.12) for the dissolution reactions of muscovite:

$$\ln K_{Ms} = -\frac{\Delta G^{o}_{r,Ms}}{RT} \tag{2.38}$$

and illite:

$$\ln K_{Ill} = -\frac{\Delta G^{o}_{r,Ill}}{RT}. \tag{2.39}$$

Plugging Eqs. (2.35) into (2.39) and rearranging leads to:

$$\ln K_{Ill} = \frac{-\Delta G^{o}_{r,Ms} + RT \ln a_{Ms}}{RT} = \ln K_{Ms} + \ln a_{Ms}. \tag{2.40}$$

Equation (2.40) links the thermodynamic equilibrium constant of the dissolution reactions of illite and muscovite and is another of relation of interest. Since it is common practice to use the decimal logarithm of K instead of its natural logarithm, it is convenient to write the following expression:

$$\log K_{Ill} = \log K_{Ms} + \log a_{Ms}. \tag{2.41}$$

It must be kept in mind that Eqs. (2.34) and the relations derived from it are all the more approximate the more the composition of illite deviates from that of muscovite.

Amorphous silica scales formed as a results of polymerization of aqueous monomeric silica often contain significant concentrations of Fe and Al, as reported for several geothermal fields (Gallup 1997, 1998). To test Eq. (2.41), we have used it to predict the log K of the dissolution reaction of Al- and Fe-rich amorphous silica based on the composition of these solid phases and the log K of the dissolution reaction of pure amorphous silica (data from Gallup 1998).

In Fig. 2.1, the predicted log K values are compared with the log K values obtained from the solubilities of Al- and Fe-rich amorphous silica, which were measured by Gallup (1998). The results obtained at 25 °C were disregarded being too low, possibly because the duration of the experiments was insufficient for the attainment of the equilibrium condition. The agreement between measured and predicted log K values is satisfactory taking into account the experimental uncertainties and the approximations affecting Eq. (2.41).

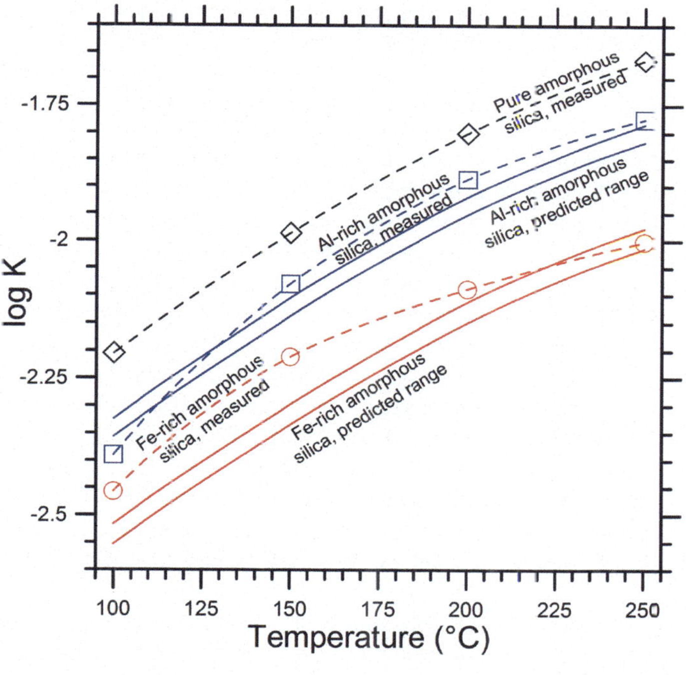

Fig. 2.1 The log K of the dissolution reaction of amorphous silica, pure and rich in Al and Fe, obtained from the solubilities measured at different temperatures (Gallup (1998); dashed lines with symbols) is compared with the predicted log K of Al- and Fe-rich amorphous silica (solid lines)

2.3 Geochemical Modeling

There are two distinct types of geochemical models. The geochemical models of the first type compute the chemical speciation and the saturation state of the aqueous solution of interest. They are often called equilibrium state models. The first calculations of this type were probably performed by Garrels and Thompson (1962), who computed (by hand!) the distribution of aqueous species in seawater starting from its bulk composition, that is from the results of the chemical analysis. Further details are given in Sect. 2.3.1.

The geochemical models of the second type are usually called reaction path models and are used to simulate different processes, such as the dissolution of one or more solid phases in an aqueous solution, mixing of two distinct aqueous solutions, and cooling/heating of an aqueous solution (see Sect. 2.3.2). A geochemical model of the second type was probably implemented for the first time by Garrels and Mackenzie (1967), who simulated the reactions occurring during the gradual evaporation of spring water. First, they calculated aqueous speciation of the initial spring water. Then they removed step-by-step a given quantity of water and recomputed the distribution of aqueous species. Thus, geochemical modelling was extended from a single state to a number of states and a whole process was simulated.

The full-fledged method of reaction path modelling was proposed by Helgeson and coworkers. The equations needed to compute the irreversible water-rock mass transfer processes were presented by Helgeson (1968), whereas the first computer code for reaction path modeling, named PATHI ("path-one"), was used to study several processes, including evaporation, weathering, sediment diagenesis, ore deposition, and hydrothermal alteration (e.g., Helgeson et al. 1969, 1970).

Today, both speciation-saturation calculations and reaction path calculations are performed by several software packages, such as PHREEQC and PhreeqcI (Parkhurst and Appelo 2013; Charlton and Parkhurst 2002), EQ3/6 (Wolery 1992; Wolery and Daveler 1992), SOLVEQ/CHILLER (Reed 1982), and the Geochemist's Workbench (Bethke 2008). Speciation-saturation calculations are carried out by several computer programs, including WATEQ4F (Ball and Nordstrom 1991), MINTEQA2 and Visual MINTEQ (Allison et al. 1991; https://vminteq.lwr.kth.se/), MINEQL and MINEQL+ (Westall et al. 1976; Schecher and McAvoy 1992), and SOLMINEQ.88 (Perkins 1992).

The computer program WATCH is unique, because it was conceived as a tool for modeling the chemistry of geothermal fluids, including the bi-phase fluids discharged from geothermal wells. The program combines the chemical analyses of the liquid and vapor phases separated at specified pressure, temperature conditions and calculates the chemical speciation and the saturation state of reservoir liquids. For excess enthalpy wells, volatile species are partitioned between the liquid and vapor phases coexisting in the geothermal aquifer. Moreover, the effects of boiling and conductive cooling can be simulated. The most recent version of WATCH was implemented by Bjarnason (2010), but the background to the code and the adopted computation methods were previously described by Arnórsson et al. (1982). WATCH, PhreeqcI, and EQ3/6 were used in this work as explained in Sect. 3.1.

2.3.1 Equilibrium State Models

In equilibrium state models, the molalities, activities, and activity coefficients of the aqueous species actually present in an electrolyte solution are computed from measured parameters (e.g., total concentrations, pH, Eh, and temperature) by means of speciation calculations involving:

1. A series of mass balance equations, one for each fundamental chemical component utilized to describe the composition of the electrolyte solution or basis species in EQ3/6 terminology (e.g., Wolery 1992).
2. A series of mass action equations, or thermodynamic equilibrium constants, one for each aqueous complex.
3. The balance of electrical charges, as dictated by the principle electroneutrality:

$$\sum_{i} m_i \cdot Z_i = 0, \qquad (2.42)$$

where the summation includes all charged species (free ions and aqueous complexes) of molality m_i and charge Z_i.

A brief digression is needed on alkalinity, which is defined as the sum of all the bases dissolved in the aqueous solution. To be noted that (i) alkalinity is nearly equal to the concentration of HCO_3^- ion in most natural waters and (ii) the determination of alkalinity by acidimetric titration is straightforward. Owing to these two reasons, alkalinity is frequently included in the analytical routine instead of total dissolved inorganic carbon, DIC, which is rarely analyzed. This explains why DIC or alkalinity are accepted as input by some computer codes, such as PhreeqcI and EQ3 up to a temperature of 50 °C, although only DIC is admitted as input by other computer codes, such as WATCH, which requires DIC as mg/kg of CO_2.

However, in waters of high or relatively high pH, including the geothermal brines that have experienced boiling, a significant proportion of alkalinity is explained by other dissolved species, such as OH^-, $H_3SiO_4^-$, $H_2BO_3^-$, HS^-, and NH_3°. Formation waters usually contain high concentrations of organic acid anions, also contributing to alkalinity. For all these aqueous solutions, it is advisable to determine DIC rather than relying on alkalinity. The measurement of DIC can be performed by infrared detection of the CO_2 released from the acidified aqueous solution, which is our preferred method because it can be carried out directly in the field (Cioni et al. 2007). Alternatively, DIC can be obtained by (i) ion chromatography using a KOH solution as eluent or (ii) an extended alkalinity titration, comprising acidimetric titration followed by removal of CO_2 from the water by bubbling gaseous N_2 and back titration with NaOH or (iii) ICP-AES in the absence of organic carbon (Arnórsson et al. 2006). To be noted, however, that these three methods cannot be performed in the field and the result of the second method is the sum of carbonate alkalinity and sulfide. Further details on alkalinity and DIC are found in textbooks (e.g., Langmuir 1997; Zhu and Anderson 2002; Appelo and Postma 2005) and computer code manuals (e.g., Wolery 1992; Parkhurst and Appelo 2013).

Resuming the main discussion, an example of mass balance equation is:

$$m_{\mathrm{Na,tot}} = m_{\mathrm{Na^+}} + m_{\mathrm{NaCl^o}} + m_{\mathrm{NaSO_4^-}} + m_{\mathrm{NaHCO_3^o}} + m_{\mathrm{NaCO_3^-}} + m_{\mathrm{NaOH^o}} \ldots, \quad (2.43)$$

where the left-hand term is the total analytical molality of Na, resulting from chemical analysis, whereas the right-hand term is the sum of the molalities of all individual Na-bearing aqueous species, both charged and neutral.

An example of mass action equation is the dissociation reaction of the $\mathrm{NaHCO_3^o}$ aqueous complex:

$$\mathrm{NaHCO_3^o} = \mathrm{Na^+} + \mathrm{HCO_3^-}, \quad (2.44)$$

which produces the free ions $\mathrm{Na^+}$ and $\mathrm{HCO_3^-}$, both of which are fundamental chemical components or basis species. The thermodynamic equilibrium constant of the dissociation reaction (2.44) is:

$$\mathrm{K_{NaHCO_3^o}} = \frac{a_{\mathrm{Na^+}} \cdot a_{\mathrm{HCO_3^-}}}{a_{\mathrm{NaHCO_3^o}}}. \quad (2.45)$$

Eqn (2.45) can be rewritten as follows, recalling Eq. (2.4):

$$\mathrm{K_{NaHCO_3^o}} = \frac{m_{\mathrm{Na^+}} \cdot m_{\mathrm{HCO_3^-}} \cdot \gamma_{\mathrm{Na^+}} \cdot \gamma_{\mathrm{HCO_3^-}}}{m_{\mathrm{NaHCO_3^o}} \cdot \gamma_{\mathrm{NaHCO_3^o}}}. \quad (2.46)$$

As shown in Sect. 2.3.3, activity coefficients are a function of the true ionic strength I, which is defined by the equation:

$$\mathrm{I} = \frac{1}{2} \sum_{\mathrm{i}} m_{\mathrm{i}} \cdot \mathrm{Z_i^2}, \quad (2.47)$$

where the summation includes the molal concentrations of all dissolved ionic solutes computed through speciation calculations. Therefore, an iterative approach is needed to compute the equilibrium state. Incidentally, the true ionic strength is different from the stoichiometric ionic strength which is calculated assuming complete dissociation, that is neglecting aqueous complexes.

Upon computation of aqueous speciation, as described above, it is possible to calculate the saturation state of the aqueous solution with respect to the solid phases of interest. The saturation state can be described by two different variables. One is the saturation index, which is a dimensionless variable defined by the following equation:

$$SI_i = \log(Q_i/K_i) \tag{2.48}$$

where K_i is the thermodynamic equilibrium constant of the dissolution reaction of the i-th solid phase, often coinciding with its solubility product, whereas Q_i identifies the corresponding activity product for the considered water sample. The other variable used to describe the saturation state is thermodynamic affinity, A_i, which is defined as follows:

$$A_i = 2.303 \cdot R \cdot T \cdot \log(Q_i/K_i) = 2.303 \cdot R \cdot T \cdot SI_i \tag{2.49}$$

has the magnitude of an energy (i.e., kJ/mole or kcal/mol) and represents the energy driving towards the equilibrium condition in a chemical potential field. For the specific case of the dissolution/precipitation of a given mineral in an aqueous solution, three general situations are possible.

1. the aqueous solution with SI_i and A_i equal to zero is in perfect equilibrium with the i-th mineral;
2. the aqueous solution with SI_i and $A_i < 0$ is undersaturated with respect to the i-th mineral and, therefore, the mineral can dissolve;
3. the aqueous solution with SI_i and $A_i > 0$ is supersaturated with respect to the i-th mineral and, consequently, the mineral can precipitate. However this is a necessary but not sufficient condition, because precipitation can be prevented by extra-thermodynamic (e.g., kinetic) constraints.

2.3.2 *Multicomponent Chemical Geothermometry: A Reaction Path Model of Special Interest*

Reaction path modeling is treated in several textbooks (e.g., Anderson and Crerar 1993; Zhu and Anderson 2002; Marini 2006; Bethke 2008) and computer code manuals (e.g., Wolery 1992; Parkhurst and Appelo 2013). Among the processes that can be simulated by means of reaction path modeling, here we focus on the heating of an aqueous solution, because it represents a popular geothermometric technique, known as multicomponent chemical geothermometry. This geothermometric technique was first proposed by Michard (1977), Michard and Roekens (1983), and Reed and Spycher (1984), and was further developed by Reed, Spycher, and coworkers.

Multicomponent chemical geothermometry implies, first, the calculation of the saturation indices with respect to relevant hydrothermal (secondary) minerals at the temperature of pH and alkalinity measurement. Then, temperature is increased in a series of steps and the saturation indices are recomputed at each step. Finally, the computed saturation indices are plotted against temperature. Since the geothermal liquid is presumably in equilibrium with the considered hydrothermal minerals at the aquifer temperature, all the SI-temperature curves are expected to converge to zero at the aquifer temperature.

This approach has become more and more popular over the years, thanks to the increasing easiness to model the heating of aqueous solutions using different computer programs (see above). Recently, the GeoT software code was developed by Spycher and coworkers (Spycher et al. 2011, 2014, 2016a, b) to make the application of this technique automatic and rigorous, introducing a series of objective criteria to evaluate the aquifer temperature of the water of interest from the computed saturation indices. Indeed, the SI-temperature plot represents an effective technique provided that an internally consistent thermodynamic database is used and possible obstacles are circumvented.

The use of multicomponent chemical geothermometry is complicated by the need to involve aluminum concentration in speciation-saturation calculations since most hydrothermal minerals are Al-silicates. However, Al concentrations are not routinely measured in geothermal liquids. Moreover, Al concentrations of geothermal waters measured at the surface are often not representative of reservoir conditions due to attainment of oversaturation with respect to Al oxyhydroxides (e.g., gibbsite, diaspore and böehmite) and other Al-bearing minerals during the ascent of geothermal waters towards the surface and consequent loss of dissolved Al caused by the precipitation of these solid phases. In addition, finely dispersed Al-oxyhydroxides and Al-bearing minerals may pass through the membrane filters (Kennedy and Zellweger 1974; Laxen and Chandler 1982) and may be dissolved upon acidification leading to overestimate the Al concentration of the aqueous solution of interest. The absence or the poor reliability of Al data can be circumvented assuming that the Al concentration is constrained by equilibrium with a suitable Al-silicate mineral in the geothermal aquifer (Pang and Reed 1998). The choice of this Al-silicate solid phase must be done with caution because different Al-silicate minerals lead to distinct concentrations of dissolved Al and affect the calculation of the saturation state with respect to Al-silicate solid phases. In addition, this assumption implies that the log K (and the corresponding log Q) of the exchange reactions between any Al-bearing solid phase and the Al-bearing mineral chosen to constrain Al concentration are considered instead of the log K (and the corresponding log Q) of the dissolution-precipitation reactions of individual Al-bearing solid phases.

If the deep water has experienced mixing with shallow cold waters or gas loss along its pathway towards the surface, it is necessary to reconstruct the chemical composition of the deep water applying suitable corrections for mixing and gas loss, before calculating the saturation indices. In other terms, the diluting water is subtracted from the analyzed mixture and/or the separated gases are added back to the analyzed mixture (Reed and Spycher 1984; Pang and Reed 1998).

Palandri and Reed (2001) applied multicomponent chemical geothermometry to sedimentary formation waters suggesting how to circumvent typically encountered problems, such as loss of silica, aluminum, and iron, underestimation of organic acid anions and concurrent overestimation of carbonate alkalinity. For instance, if the formation temperature is known and silica loss has occurred, SiO_2 activity under formation conditions can be computed forcing equilibrium with quartz or chalcedony.

If aluminum concentration is not reported, it can be estimated assuming equilibrium with microcline or muscovite as in the case of geothermal waters (see above). Usually, the aqueous solutions result to be supersaturated with calcite if bicarbonate is overestimated because organic acid anions are present but are not considered. If so, alkalinity can be reduced until calcite saturation is attained. However, calcite saturation might be alternatively caused by CO_2 loss. These two possibilities can be distinguished because CO_2 loss determines important pH variations, while lack of organic acid anions causes negligible pH changes.

One of the main limitations of multicomponent chemical geothermometry is the strong dependence of the saturation indices with respect to several minerals (silicates, Al-silicates, carbonates, hydroxides) on the pH of the aqueous solution under aquifer conditions. It must be underscored that the calculation of aquifer pH is affected by strong uncertainties, especially for fluids discharged from excess enthalpy wells (see Sect. 3.1.2) and boiling springs. Corrections are possible, as discussed above, but the reduction or suppression of uncertainties are not guaranteed.

Another essential requirement for the application of multicomponent chemical geothermometry is the use of an internally consistent thermodynamic database in geochemical modeling. This requirement applies not only to multicomponent chemical geothermometry but also to the theoretical activity-based geothermometers and f_{CO_2}-indicators developed in this book. However, if this need is fulfilled it is feasible to move from empirical to theoretical geothermometry.

2.3.3 *Activity Coefficients in Relatively Dilute Solutions*

In relatively dilute aqueous solutions, activity coefficients of individual ions are usually calculated by means of the extended Debye-Hückel equation, also known as B-dot equation:

$$\log \gamma_i = -\frac{A_\gamma \cdot Z_i^2 \cdot \sqrt{I}}{1 + \mathring{a}_i \cdot B_\gamma \cdot \sqrt{I}} + B^\circ \cdot I, \qquad (2.50)$$

where $\mathring{a}_i$ is the ion-size parameter of the i-th ion (which is assumed to be temperature independent), A_γ and B_γ are functions of temperature as well as the density and dielectric constant of the solvent, and I is the true ionic strength of the electrolyte solution, defined by Eq. (2.47). The B-dot equation was obtained by Helgeson (1969) based on the experimental data of NaCl aqueous solutions available for the temperature range 25–300 °C and the ionic strength interval 0–3 mol/kg. The B-dot equation works properly for aqueous solutions of ionic strength up to 0.3–1 mol/kg.

The activity coefficients of $CO_{2(aq)}$, and nonpolar neutral aqueous species, such as $H_{2(aq)}$, $O_{2(aq)}$, and $N_{2(aq)}$, are commonly computed using the expression of Drummond (1981):

$$\ln \gamma_n = \left(C + F \cdot T + \frac{G}{T}\right) \cdot I - (E + H \cdot T) \cdot \left(\frac{I}{I+1}\right) \tag{2.51}$$

where C = −1.0312, F = 0.0012806, G = 255.9, E = 0.4445, and H = −0.001606. The activity coefficients of polar neutral aqueous species, such as $SiO_{2(aq)}$, $B(OH)_{3(aq)}$, and $NaCl_{(aq)}$, are generally set to unity, as recommended by Garrels and Christ (1965).

The activity of water is often calculated using the following expression (Wolery 1992):

$$\ln a_{H_2O} = \frac{18.015}{1000}\left(-\frac{\sum m}{\ln(10)} + \frac{2}{3} A_\gamma I^{1.5} \sigma\left(\mathring{a}_j \cdot B_\gamma \cdot I^{0.5}\right) - \overset{o}{B} I^2\right), \tag{2.52}$$

in which the σ-parameter is:

$$\sigma(x) = \frac{3}{x^3}\left[1 + x - \frac{1}{1+x} - 2 \cdot \ln(1+x)\right] \tag{2.53}$$

with $x = \mathring{a}_j \cdot B_\gamma \cdot I^{0.5}$.

2.3.4 Activity Coefficients in Concentrated Solutions

Pitzer (1973, 1975) proposed a set of semi-empirical equations to compute activity coefficients in aqueous electrolytes. Subsequent researches proven that these equations can be successfully applied to concentrated aqueous solutions, to describe both their speciation (e.g., Pitzer and Kim 1974) and the equilibrium between concentrated aqueous solutions and evaporite minerals (e.g., Harvie and Weare 1980; Harvie, Møller, and Weare 1984).

Without entering into details, it must be noted that the Pitzer's theory involves the so-called Pitzer interaction parameters for cation-anion pairs, for cation-cation pairs and triplets, anion–anion pairs and triplets, neutral species–cation pairs, and neutral species–anion pairs. In the '70s and 80s, the lack of these parameters for $SiO_{2(aq)}$, Al^{3+} ion and its hydroxide complexes, at any temperature, prevented the applicability of the Pitzer's approach to geothermal brines. However, Azaroual et al. (1997) established the interaction parameters of $SiO_{2(aq)}$ with major anions, major cations, and Li^+ ion at temperatures up to 250 °C, the interaction parameters for the system H–Al^{3+}–Na–K–Cl–H_2O up to 100 °C were derived by Christov et al. (2007), and the interaction parameters for cationic Al species with both $SiO_{2(aq)}$ up to 300 °C and $CO_{2(aq)}$ up to 90 °C were evaluated using empirical relations by Accornero and Marini (2009). Therefore, today is possible to apply the Pitzer's equations also to geothermal brines although further experimental data are needed to achieve a higher degree of confidence.

References

Accornero M, Marini L (2009) Empirical prediction of the Pitzer's interaction parameters for cationic Al species with both $SiO_{2(aq)}$ and $CO_{2(aq)}$: implications for the geochemical modelling of very saline solutions. Appl Geochem 24:747–759

Allison JD, Brown DS, Novo-Gradac KJ (1991) MINTEQA2/PRODEFA2, a geochemical assessment model for environmental systems: version 3.0 user's manual. Report EPA/600/3-91/021. U. S. Environmental Protection Agency, Office of Research and Development, Environmental Research Laboratory, Athens, Georgia, p 106

Anderson GM, Crerar DA (1993) Thermodynamics in Geochemistry. The equilibrium model. Oxford University Press, New York, p 588

Appelo CAJ, Postma D (2005) Geochemistry, groundwater and pollution, 2nd edn. A.A. Balkema Publishers, Leiden, p 634

Apps J, Spycher N (2004) Data qualification for thermodynamic data used to support THC calculations. ANL-NBS-HS-000043 REV 00, Bechtel SAIC Company, LLC, Las Vegas, NV. DOC. 20041118.0004, p 172

Arnórsson S, Stefánsson A (1999) Assessment of feldspar solubility constants in water in the range of 0° to 350 °C at vapor saturation pressures. Am J Sci 299:173–209

Arnórsson S, Bjarnason JÖ, Giroud N, Gunnarsson I, Stefánsson A (2006) Sampling and analysis of geothermal fluids. Geofluids 6:203–216

Arnórsson S, Sigurdsson S, Svavarsson H (1982) The chemistry of geothermal waters in Iceland. I. Calculation of aqueous speciation from 0 to 370 °C. Geochim Cosmochim Ac 46:1513–1532

Azaroual M, Fouillac C, Matray JM (1997) Solubility of silica polymorphs in electrolyte solutions. II. Activity of aqueous silica and solid silica polymorphs in deep solutions from the sedimentary Paris Basin. Chem Geol 140:167–179

Ball JW, Nordstrom DK (1991) User's manual for WATEQ4F, with revised thermodynamic data base and test cases for calculating speciation of major, trace, and redox elements in natural waters. U.S. Geological Survey Open File Report, pp 91–183

Bénézeth P, Palmer DA, Anovitz LM, Horita J (2007) Dawsonite synthesis and reevaluation of its thermodynamic properties from solubility measurements: implications for mineral trapping of CO_2. Geochim Cosmochim Ac 71:4438–4455

Bethke CM (2008) Geochemical and biogeochemical reaction modeling, 2nd edn., p 564

Bjarnason JÖ (2010) The chemical speciation program WATCH, version 2.4. ISOR–Iceland GeoSurvey, Reykjavik, Iceland

Charlton SR, Parkhurst DL (2002) PhreeqcI-A graphical user interface to the geochemical model PHREEQC, p 2. U.S. Geological Survey Fact Sheet FS-031-02

Christov C, Dickson AG, Møller N (2007) Thermodynamic modeling of aqueous aluminum chemistry and solid-liquid equilibria to high solution concentration and temperature. I. The acidic $H–Al–Na–K–Cl–H_2O$ system from 0 to 100 °C. J Solution Chem 36:1495–1523

Cioni R, Gambardella B, Marini L (2007) Field determination of total dissolved inorganic carbon (TDIC) in natural waters using an IR analyzer: I. Preliminary laboratory tests. Geothermics 36:47–62

Denbigh KG (1981) The principles of chemical equilibrium, 4th edn. Cambridge University Press, p 494

Drummond SEJr (1981) Boiling and mixing of hydrothermal fluids: chemical effects on mineral precipitation. Ph.D. thesis, p 380. The Pennsylvania State University, University Park

Gallup DL (1997) Aluminum silicate scale formation and inhibition: scale characterization and laboratory experiments. Geothermics 26:483–499

Gallup DL (1998) Aluminum silicate scale formation and inhibition (2): scale solubilities and laboratory and field inhibition tests. Geothermics 27:485–501

Garrels RM, Christ CL (1965) Solutions, minerals, and equilibria. Harper and Row, New York, p 450

Garrels RM, Mackenzie FT (1967) Origin of the chemical composition of some springs and lakes. In: Stumm W (ed) Equilibrium concepts in natural water systems, advances in chemistry series. American Chemical Society, Washington, DC, pp 222–242

Garrels RM, Thompson ME (1962) A chemical model for sea water at 25 °C and one atmosphere total pressure. Am J Sci 260:57–66

Grenthe I, Plyasunov A, Spahiu K (1997) Estimations of medium effects on thermodynamic data. In: Grenthe I, Puigdomenech I (eds.) Modeling in aquatic chemistry. Nuclear Energy Agency, OECD: Paris, France, 325–426

Haar L, Gallagher JS, Kell GS (1984) NBS/NRC steam tables, p 320. Hemisphere, Washington DC

Haas JrJL, Robinson JrGR, Hemingway BS (1981) Thermodynamic tabulations for selected phases in the system $CaO–Al_2O_3–SiO_2–H_2$ at 101.325 kPa (1 atm) between 273.15 and 1800 K. J Phys Chem Ref Data 10:575–670

Harvie CE, Weare JH (1980) The prediction of mineral solubilities in natural waters: the $Na–K–Mg–Ca–Cl–SO_4–H_2O$ system from zero to high concentrations at 25 °C. Geochim Cosmochim Ac 44:981–997

Harvie CE, Møller N, Weare JH (1984) The prediction of mineral solubilities in natural waters: the $Na–K–Mg–Ca–H–Cl–SO_4–OH–HCO_3–CO_3–CO_2–H_2O$ system to high ionic strengths at 25 °C. Geochim Cosmochim Ac 48:723–751

Helgeson HC (1968) Evaluation of irreversible reactions in geochemical processes involving minerals and aqueous solutions: I. Thermodynamic relations. Geochim Cosmochim Ac 32:853–877

Helgeson HC (1969) Thermodynamics of hydrothermal systems at elevated temperatures and pressures. Am J Sci 267:729–804

Helgeson HC, Brown TH, Nigrini A, Jones TA (1970) Calculation of mass transfer in geochemical processes involving aqueous solutions. Geochim Cosmochim Ac 34:569–592

Helgeson HC, Delany JM, Nesbitt HW, Bird DK (1978) Summary and critique of the thermodynamic properties of rock-forming minerals. Am J Sci 278A:229

Helgeson HC, Garrels RM, Mackenzie FT (1969) Evaluation of irreversible reactions in geochemical processes involving minerals and aqueous solutions: II. Applications. Geochim Cosmochim Ac 33:455–481

Helgeson HC, Kirkham DH (1974a) Theoretical prediction of the thermodynamic behavior of aqueous electrolytes at high pressures and temperatures: I. Summary of the thermodynamic/electrostatic properties of the solvent. Am J Sci 274:1089–1198

Helgeson HC, Kirkham DH (1974b) Theoretical prediction of the thermodynamic behavior of aqueous electrolytes at high pressures and temperatures: II. Debye- Hückel parameters for activity coefficients and relative partial molal properties. Am J Sci 274:1199–1261

Helgeson HC, Kirkham DH (1976) Theoretical prediction of the thermodynamic behavior of aqueous electrolytes at high pressures and temperatures: III. Equation of state for aqueous species at infinite dilution. Am J Sci 276:97–240

Helgeson HC, Kirkham DH, Flowers GC (1981) Theoretical prediction of the thermodynamic behavior of aqueous electrolytes at high pressures and temperatures: IV. Calculation of activity coefficients, osmotic coefficients and relative partial molal properties to 600 °C and 5 kb. Am J Sci 281:1249–1516

Hemingway BS, Haas JrJL, Robinson JrGR (1982) Thermodynamic properties of selected minerals in the system $Al_2O_3–CaO–SiO_2–H_2O$ at 298.15 K and 1 bar (10^5 Pascals) pressure and at higher temperatures, vol 1544. U.S. Geological Survey Bulletion, p 70

Hemley JJ, Montoya JW, Marinenko JW, Luce RW (1980) Equilibria in the system $Al_2O_3–SiO_2–H_2O$ and some general implications for alteration/mineralization processes. Econ Geol 75:210–228

Holland TJB, Powell R (1998) An internally consistent thermodynamic data set for phases of petrological interest. J Metamorph Geol 16:309–343

Holland TJB, Powell R (2011) An improved and extended internally consistent thermodynamic dataset for phases of petrological interest, involving a new equation of state for solids. J Metamorp Geol 29:333–383
Johnson JW, Norton D (1991) Critical phenomena in hydrothermal systems: state, thermodynamic, electrostatic, and transport properties of H_2O in the critical region. Am J Sci 291:541–648
Johnson JW, Oelkers EH, Helgeson HC (1992) SUPCRT 92: a software package for calculating the standard molal thermodynamic properties of minerals, gases, aqueous species, and reactions from 1 to 5000 bars and 0–1000 °C. Comput Geosci 18:899–947
Kelley KK (1960) Contributions to the data in theoretical metallurgy XIII: high temperature heat content, heat capacities and entropy data for the elements and inorganic compounds, vol 584. U.S. Bureau of Mines Bulletin, p 232
Kennedy VC, Zellweger GW (1974) Filter pore-size effects on the analysis of Al, Fe, Mn, and Ti in water. Water Resour Res 10:785–790
Langmuir D (1997) Aqueous environmental geochemistry. Prentice Hall, Upper Saddle River, NJ, p 600
Langmuir D, Mahoney J, Rowson J (2006) Solubility products of amorphous ferric arsenate and crystalline scorodite ($FeAsO_4 \cdot 2H_2O$) and their application to arsenic behavior in buried mine tailings. Geochim Cosmochim Ac 70:2942–2956
Laxen DPH, Chandler IM (1982) Comparison of filtration techniques for size distribution in freshwater. Anal Chem 54:1350–1355
Levelt Sengers JMH, Kamgar-Parsi B, Balfour FW, Sengers JV (1983) Thermodynamic properties of steam in the critical region. J Phys Chem Ref Data 12:1–28
Lewis GN, Randall M (1961) Thermodynamics, 2nd edn. In: Pitzer KS, Brewer L, p 723. McGraw-Hill, New York
Maier CG, Kelley KK (1932) An equation for the representation of high temperature heat content data. Am Chem Soc J 54:3243–3246
Marini L (2006) Geological sequestration of carbon dioxide: thermodynamics, kinetics, and reaction path modeling. Developments in geochemistry, 11. Elsevier, Amsterdam, p 453
Marini L, Accornero M (2007) Prediction of the thermodynamic properties of metal–arsenate and metal–arsenite aqueous complexes to high temperatures and pressures and some geological consequences. Environ Geol 52:1343–1363
Marini L, Accornero M (2010) Prediction of the thermodynamic properties of metal–arsenate and metal–arsenite aqueous complexes to high temperatures and pressures and some geological consequences. Environ Earth Sci 59:1601–1606
Michard G (1977) Modification de la répartition des espèces chimiques lors du refroidissement d'une eau thermale. C.R. Acad Sci II 284:949–952
Michard G, Roekens E (1983) Modelling of the chemical composition of alkaline hot waters. Geothermics 12:161–169
Nordstrom DK, Archer DG (2002) Arsenic thermodynamic data and environmental geochemistry. In: Welch AH, Stollenwerk KG (eds) Arsenic in ground water. Springer, US, Boston, MA, pp 1–25
Nordstrom DK, Munoz JL (2006) Geochemical Thermodynamics. Blackburn Press, 2nd Ed, pp 504
Palandri JL, Reed MH (2001) Reconstruction of in situ composition of sedimentary formation waters. Geochim Cosmochim Ac 65:1741–1767
Pang Z-H, Reed M (1998) Theoretical chemical thermometry on geothermal waters: problems and methods. Geochim Cosmochim Ac 62:1083–1091
Parkhurst DL, Appelo CAJ (2013) Description of input and examples for PHREEQC version 3: a computer program for speciation, batch-reaction, one-dimensional transport, and inverse geochemical calculations. Techniques and Methods 6-A43. U.S. Geological Survey
Perkins EH (1992) Integration of intensive variable diagrams and fluid phase equilibrium with solmineq.88 pc.shell. In: Kharaka YK, Maest AS (eds.) Water-rock interaction. Rotterdam, Balkema, pp 1079–1081

Pitzer KS (1973) Thermodynamics of electrolytes. I. Theoretical basis and general equations. J Phys Chem 77:268–277
Pitzer KS (1975) Thermodynamics of electrolytes. V. Effects of higher-order electrostatic terms. J Solution Chem 4:249–265
Pitzer KS (1983) Dielectric constant of water at very high temperature and pressure. P Natl Acad Sci USA 80:4575–4576
Pitzer KS, Kim JJ (1974) Thermodynamics of electrolytes. IV. Activity and osmotic coefficients for mixed electrolytes. J Amer Chem Soc 96:5701–5707
Reed MH (1982) Calculation of multicomponent chemical equilibria and reaction processes in systems involving minerals, gases, and aqueous phase. Geochim Cosmochim Ac 46:513–528
Reed MH, Spycher NF (1984) Calculation of pH and mineral equilibria in hydrothermal waters with application to geothermometry and studies of boiling and dilution. Geochim Cosmochim Ac 48:1479–1492
Robie RA, Hemingway BS, Fisher JR (1979) Thermodynamic properties of minerals and related substances at 298.15 K and 1 bar (10^5 Pascals) pressure and at higher temperatures, vol 1452. U.S. Geological Survey Bulletin, p 456
Schecher WD, McAvoy DC (1992) MINEQL+: a software environment for chemical equilibrium modeling. Comput Environ Urban 16:65–76
Shock EL, Helgeson HC (1988) Calculation of the thermodynamic and transport properties of aqueous species at high pressures and temperatures: correlation algorithms for ionic species and equation of state predictions to 5 kb and 1000 °C. Geochim Cosmochim Ac 52:2009–2036
Shock EL, Helgeson HC, Sverjensky DA (1989) Calculation of the thermodynamic and transport properties of aqueous species at high pressures and temperatures: standard partial molal properties of inorganic neutral species. Geochim Cosmochim Ac 53:2157–2183
Shock EL, Oelkers EH, Johnson JW, Sverjensky DA, Helgeson HC (1992) Calculation of the thermodynamic properties of aqueous species at high pressures and temperatures: Effective electrostatic radii, dissociation constants, and standard partial molal properties to 1000 °C and 5 kb. J Chem Soc (London) Faraday Trans 88:803–826
Shock EL, Sassani DC, Willis M, Sverjensky DA (1997) Inorganic species in geologic fluids: correlations among standard molal thermodynamic properties of aqueous ions and hydroxide complexes. Geochim Cosmochim Ac 61:907–950
Spycher N, Finsterle S, Dobson P (22–24 February, 2016) New developments in multicomponent geothermometry. In: Proceedings, 41st Workshop on Geothermal Reservoir Engineering, p 9. Stanford University, Stanford, California, SGP-TR-209
Spycher N, Peiffer L, Sonnenthal EL, Saldi G, Reed MH, Kennedy BM (2014) Integrated multicomponent solute geothermometry. Geothermics 51:113–123
Spycher N, Peiffer L, Finsterle S, Sonnenthal EL (2016) GeoT user's guide. A computer program for multicomponent geothermometry and geochemical speciation, p 42. Version 2.1. Lawrence Berkeley National Laboratory Report
Spycher N, Sonnenthal E, Kennedy BM (2011) Integrating multicomponent chemical geothermometry with parameter estimation computations for geothermal exploration. Geoth Res T 35:663–666
Stefánsson A (2001) Dissolution of primary minerals of basalt in natural waters: I. Calculation of mineral solubilities from 0 to 350 °C. Chem Geol 172:225–250
Sverjensky DA, Shock EL, Helgeson HC (1997) Prediction of the thermodynamic properties of aqueous metal complexes to 1000°C and 5 kb. Geochim Cosmochim Ac 61:1359–1412
Tagirov B, Schott J (2001) Aluminum speciation in crustal fluids revisited. Geochim Cosmochim Ac 65:3965–3992
Tanger JC, Helgeson HC (1988) Calculation of the thermodynamic and transport properties of aqueous species at high pressures and temperatures: Revised equations of state for the standard partial molal properties of ions and electrolytes. Am J Sci 288:19–98

Tutolo BM, Kong X-Z, Seyfried JrWE, Saar MO (2014) Internal consistency in aqueous geochemical data revisited: applications to the aluminum system. Geochim Cosmochim Ac 133:216–234

Uematsu M, Frank EU (1980) Static dielectric constant of water and steam. J Phys Chem Ref Data 9:1291–1306

Wagman DD, Evans WH, Parker VB, Schumm RH, Halow I, Bailey SM, Churney KL, Nuttall RL (1982) The NBS tables of chemical thermodynamic properties: selected values for inorganic and C_1 and C_2 organic substances in SI units. J Phys Chem Ref Data 11(2):392

Westall J, Zachary JL, Morel F (1976) MINEQL, a computer program for the calculation of chemical equilibrium composition of aqueous systems. Technical Note no. 18, Ralph M. Parsons Lab. MIT, Cambridge

Wolery TJ (1992) EQ3NR, a computer program for geochemical aqueous speciation-solubility calculations: theoretical manual, user's guide and related documentation (version 7.0). Report UCRL-MA-110662 PT III. Lawrence Livermore National Laboratory, Livermore

Wolery TJ, Daveler SA (1992) EQ6, a computer program for reaction path modeling of aqueous geochemical systems: theoretical manual, user's guide, and related documentation (version 7.0). Report UCRL-MA-110662 PT IV. Lawrence Livermore National Laboratory, Livermore

Wolery TW, Jove-Colon C (2007) Qualification of thermodynamic data for geochemical modeling of mineral-water interactions in dilute systems. Sandia National Laboratories Report ANL-WIS-GS-000003 REV 01.

Zhu C, Anderson G (2002) Environmental applications of geochemical modeling, p 284. Cambridge University Press

Zhu Y, Zhang X, Xie Q, Chen Y, Wang D, Liang Y, Lu J (2005) Solubility and stability of barium arsenate and barium hydrogen arsenate at 25 °C. J Hazard Mater 120:37–44

Zimmer K, Zhang Y, Lu P, Chen Y, Zhang G, Dalkilic M, Zhu C (2016) SUPCRTBL: a revised and extended thermodynamic dataset and software package of SUPCRT92. Comput Geosci 90:97–111

Chapter 3
The Reservoir Liquids

Abstract Over 1000 chemical analyses of reservoir liquids coming from different geothermal systems were reconstructed combining the chemical analyses of the liquid and vapor phases discharged from drilled wells and collected at known separation temperature and pressure. All these reservoir liquids, apart from a few exceptions, have probably attained or closely approached the condition of chemical equilibrium with the hydrothermal (secondary) minerals occurring in the geothermal reservoirs of provenance at temperatures of 100-350°C. Therefore, they can be used to test geothermometers and f_{CO_2}-indicators, both the traditional ones and these derived in this work. The chemical characteristics of these reservoir liquids are illustrated through a comprehensive water-classification approach, including not only the triangular diagram of major anions but also the triangular diagram of main cations and suitable salinity plots, all prepared from concentrations in equivalent units. The fractions of free Na^+, K^+, Mg^{2+}, and Ca^{2+} ions and undissociated $SiO_{2(aq)}$ and the activity coefficients of these entities are examined to highlight the effects of relevant association-dissociation reactions as well as of solute-solvent and solute-solute interactions.

The data from the real world used to test the traditional geothermometers and f_{CO_2}-indicators as well as the theoretical, activity-based geoindicators that we intend to propose in this work are 1013 chemical analyses of reservoir liquids, which can be reasonably assumed to be representative of mineral-solution thermochemical equilibrium at aquifer temperatures in the range 100–350 °C. The criteria adopted to select these reservoir liquids are provenance from drilled wells and availability of sufficiently complete chemical and physical data for the calculation of reservoir (aquifer) composition. Upon complete purging of drilling fluids, initial discharge fluids are of particular interest, being representative of the natural, pre-exploitation

Electronic supplementary material The online version of this chapter (https://doi.org/10.1007/978-3-030-54318-1_3) contains supplementary material, which is available to authorized users.

R. Cioni and L. Marini, *A Thermodynamic Approach to Water Geothermometry*, Springer Geochemistry, https://doi.org/10.1007/978-3-030-54318-1_3

state, whereas fluids sampled at later stages may be affected by production-induced processes, such as mixing with re-injected fluids (brines and steam condensates) and inflow of external waters into the geothermal reservoir due to its depressurization.

3.1 Calculation of the Chemical Composition of Reservoir Liquids

In most cases, the chemical composition of reservoir liquids, including pH, was computed by merging the chemical analyses of liquid and vapor phases, separated at known temperature and pressure, and considering both total-discharge enthalpy and aquifer temperature in calculations. These calculations were generally carried out by means of the computer code WATCH, version 2.4 (Arnórsson et al. 1982; Bjarnason 2010). Further details are given in Sect. 2.3 and in the following sections. Only the composition of a few reservoir liquids (some of Salton Sea and those of Asal, Aluto-Langano, and Latera) was taken as reported by the authors due to the unavailability of data for the separated liquid and vapor phases. The boiling spring model was adopted for the Icelandic systems of medium temperature (details in Sect. 3.1.3).

3.1.1 Calculation of Reservoir Liquid Chemistry for Liquid Enthalpy Wells

For the liquid enthalpy wells, whose total discharge enthalpy is close to that of the saturated liquid water at the aquifer temperature, it is reasonable to assume that the system of interest, including the well and the nearby aquifer zones, behaves as an ***isolated system*** and that adiabatic boiling occurs from the reservoir temperature to the vapor-liquid separation temperature. If so, the steam fraction, y, in the two-phase fluid (in mass units)[1] at the separation temperature is computed using the simple relation:

$$y = \frac{H_O - H_L}{H_V - H_L} \tag{3.1}$$

which is obtained by rearranging the enthalpy conservation equation:

$$H_O = H_L \cdot (1 - y) + H_V \cdot y. \tag{3.2}$$

[1]The steam fraction is defined by the ratio $y = Q_V/(Q_V + Q_L)$, where Q_V and Q_L are the mass flow-rates of the vapor and liquid phases, respectively, at vapor-liquid separation conditions.

Please note that the equations have been renumbered to maintain sequential order. I noted that the correct numbers (3.8) and (3.9) were assigned to the first two equations in section 3.8.1. Thank you!

In Eqs. (3.1) and (3.2), H indicates the specific enthalpy of the specified phase and subscripts O, L, and V stand for the reservoir liquid, the separated liquid, and the separated vapor, respectively. Since water is by far the major constituent of geothermal fluids, the specific enthalpies of pure water (tabulated by Lemmon et al. 2017) are generally used in Eqs. (3.1) and (3.2).

The composition of the single liquid phase in the reservoir is then computed through the simple mass balance:

$$C_{i,O} = C_{i,L} \cdot (1 - y) + C_{i,V} \cdot y \tag{3.3}$$

where $C_{i,O}$, $C_{i,L}$, and $C_{i,V}$ represent the concentrations of the i-th chemical component in the reservoir liquid, the separated liquid, and the separated vapor, respectively, in consistent measurement units.

However, the pH of the reservoir liquid can neither be calculated by this simple mass balance [Eq. (3.3)] nor measured in situ, but requires the use of a relatively complex calculation procedure, which takes into account the dissociation constants of weak acids and bases and a suitable balance, either on total ionizable hydrogen ion, or titration alkalinity, or the electrical charge (e.g., Truesdell and Singers 1974; Arnórsson et al. 1982; Reed and Spycher 1984; Henley et al. 1984).

Therefore, a suitable speciation code is needed and the best option is WATCH because it calculates automatically the composition of reservoir fluids, having been specifically developed for this purpose. In contrast, a relatively intricate approach must be followed if other software packages, such as EQ3/6 (Wolery 1992; Wolery and Daveler 1992) and PHREEQC Interactive (Parkhurst and Appelo 2013; Charlton and Parkhurst 2002) are used.

For instance, Marini et al. (2003) used the software package EQ3/6 for reconstructing the chemistry of Miravalles deep waters, assuming that a single saturated liquid phase is present in the reservoir, which is a reasonable hypothesis for this geothermal system. The calculation approach is schematically summarized in Fig. 3.1 and is briefly illustrated here below.

1. First, two distinct EQ3 runs were performed to compute the chemical speciation of the liquid and vapor phases separated at known P, T conditions and cooled at 25 °C. This is the temperature at which the pH and alkalinity of the separated liquid were measured. Since the pH of the condensed vapor phase was not measured, it was assumed to be fixed by the electric charge balance in EQ3 calculations. In general, it is not a good idea to compute the pH by balancing on H^+. In this case, however, this choice is reasonable because H^+ is expected to be a major component and the concentrations of other major components, i.e., CO_2, H_2S and NH_3, are known.

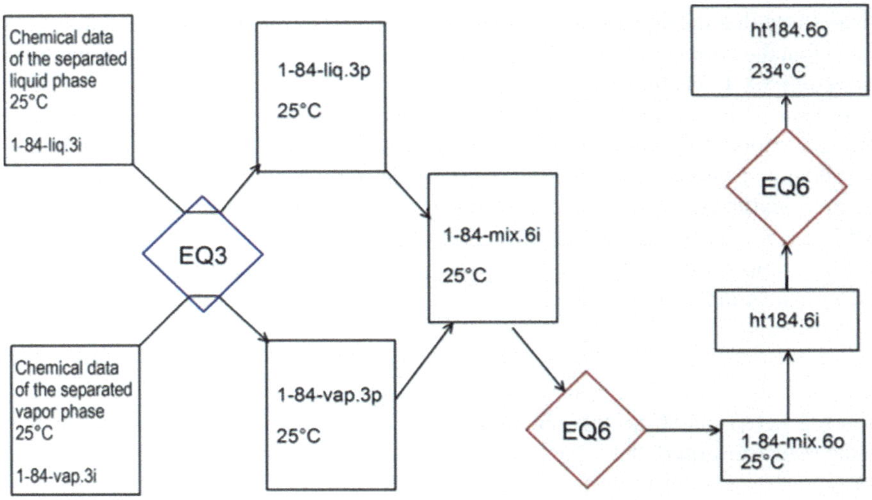

Fig. 3.1 Approach adopted by Marini et al. (2003) to compute the chemical composition of the single liquid phase presumably present in the Miravalles geothermal reservoir using the software package EQ3/6. The fourth step, in which the reservoir liquid is equilibrated with a suitable Al-silicate, is not shown

2. Second, EQ6 was run for mixing the separated, cooled liquid and vapor phases, keeping the temperature at 25 °C, and considering the steam fraction, y, computed for adiabatic boiling from the reservoir temperature to the separation temperature.
3. Third, EQ6 was run for heating the mixture obtained in the previous step from 25 °C to the reservoir temperature.
4. Fourth, EQ3 was run to equilibrate the aqueous solution with a suitable Al-silicate, either muscovite or kaolinite, because analytical Al data are not available.

The same calculations can be carried out by PHREEQC Interactive using the following keyword data blocks (Fig. 3.2):

- SOLUTION_1 to define the temperature and chemical composition of the separated liquid cooled at room temperature and to perform related speciation-saturation calculations (simulation 1);
- SOLUTION_2 to define the temperature and chemical composition of the separated vapor condensed and cooled at room temperature and to perform related speciation-saturation calculations (simulation 2);
- MIX_1 to mix together the separated liquid and condensed vapor at room temperature (simulation 3);
- REACTION_TEMPERATURE_1 to heat the liquid + condensed vapor mixture from room temperature to reservoir temperature (simulation 4);
- EQUILIBRIUM_PHASES_1 to equilibrate the reservoir liquid with the selected Al-silicate (low-albite in this example), which is dissolved or precipitated to attain saturation (simulation 5).

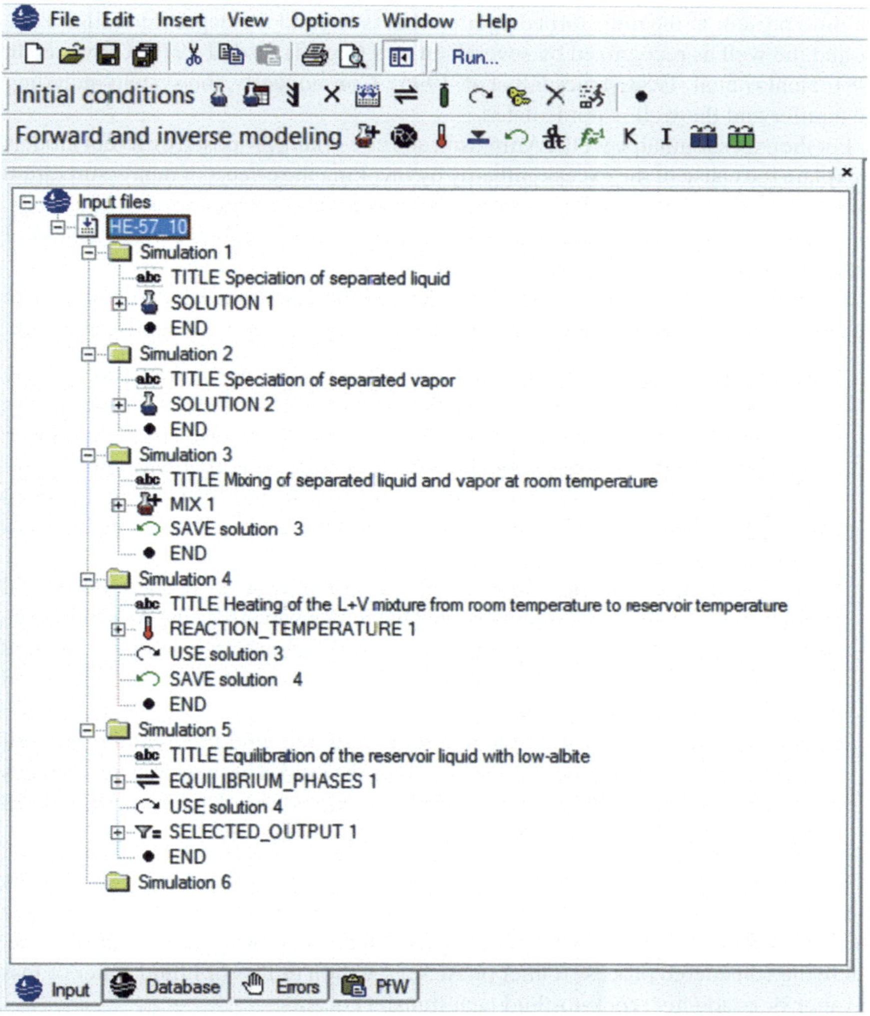

Fig. 3.2 Keyword data blocks used to compute the chemical composition of the liquid phase under reservoir condition for liquid enthalpy wells by means of PHREEQC Interactive

3.1.2 *Calculation of Reservoir Liquid Chemistry for Excess Enthalpy Wells*

The assumptions of isolated system behavior and adiabatic boiling are questionable for the wells with excess enthalpy or excess steam (e.g., Arnórsson and Stefansson 2005b; Arnórsson et al. 2007, 2010). In fact, because of extensive boiling in the producing aquifers of a wet-steam well, its discharge enthalpy is larger than that of the fluid present in the undisturbed aquifer, that is beyond the depressurization zone around the well as recognized by several authors (e.g., Truesdell 1979; Glover et al. 1981; Henley et al. 1984; Truesdell et al. 1995). Consequently, the system including the aquifer and the well is not isolated.

For the excess enthalpy wells, Arnórsson and coworkers proposed different models to explain the cause of the excess enthalpy by invoking the effects of depressurization boiling, liquid phase segregation in the depressurization zones around the wells, conductive heat transfer from the aquifer rock to the flowing fluid as well as loss of steam from the fluid flowing into the wells.

In addition, Arnórsson and coworkers derived the equations needed to calculate the aquifer steam fractions and fluid chemistry for excess enthalpy wells. In detail:

1. Liquid phase segregation means that liquid water is partially or almost completely retained in the aquifer because of its adhesion onto the surfaces of mineral grains by capillary forces, whereas the vapor phase is totally transferred from the aquifer into the well. Liquid phase segregation changes both the enthalpy and the chemical composition of the well discharge compared to those of the undisturbed aquifer fluid. Therefore, the system comprising the aquifer and the well acts as an ***open system***.
2. Loss of steam from the fluid flowing into the well may occur in a sub-horizontal aquifer downstream of the upflow zone. Due to this vapor loss the well discharge suffers a depletion of volatile components and a decrease in enthalpy compared to the undisturbed aquifer. Also in this case, the system including the aquifer and the well acts as an ***open system***.
3. Conductive heat transfer from the aquifer rock to the fluid flowing into the well occurs because depressurization boiling cools the fluid. This heat transfer changes the enthalpy of the fluid but not its chemical composition. Therefore, the system comprising the aquifer and the well acts as a ***closed system***.

For all the models (as in the isolated system case), it is necessary to specify the aquifer temperature and, in general, it is also necessary to assume a suitable value for the temperature (or pressure) in the intermediate zone or depressurization zone, where the considered process (either phase segregation or loss of liquid water or loss of vapor or conductive rock-to-fluid heat transfer) occurs.

Using WATCH, these models were applied to different geothermal systems, such as Hellisheidi, Nesjavellir, and Krafla in Iceland, Olkaria in Kenya, Mahanagdong and Pataan in the Philippines (Arnórsson et al. 1990, 2010; Angcoy 2010; Karingithi et al. 2010; Remoroza 2010; Scott 2011; Scott et al. 2014). The open-system model with liquid retained in the formation provided realistic results in these applications, with concentrations of volatile components and pH values significantly different from those obtained using the isolated-system model. In contrast, concentrations of non-volatile components for the open-system (phase segregation) model resulted to be similar to those for the isolated-system model.

To gain further insight into this matter, the results of the open-system (phase segregation) model for Olkaria well discharges (from Karingithi et al. 2010) are compared, here below, with the results of both the isolated-system (gas partitioning)

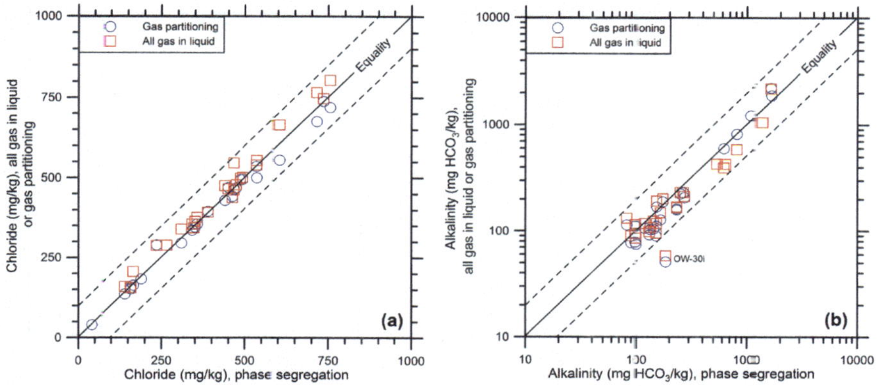

Fig. 3.3 **a** Chloride concentration and **b** total alkalinity of the Olkaria reservoir liquids given by the open-system model with liquid retained in the formation (phase segregation) are compared with those given by the isolated-system model (gas partitioning) and the all-gas-in-liquid model. The dashed lines show deviations from the equality condition

model and those obtained through a still different approach, which is based on the assumption that all gas species are contributed by the liquid phase initially stored in the geothermal aquifer and that the vapor phase is gas-free pure water. This is a realistic hypothesis if vapor is produced through boiling (evaporation) of a gas-poor liquid, as is the case of injection-derived steam (e.g., Panichi 2004). WATCH calculations were performed by merging the chemical analyses of the separated liquid and vapor phases, specifying the reservoir temperature, but omitting total-discharge enthalpy. Since enthalpy is not given to WATCH, the code computes it, assuming that the enthalpy corresponds to that of pure liquid water saturated with steam at the chosen reservoir temperature.

As expected, the three models provide similar results for the non-volatile components, such as chloride (Fig. 3.3a), and relatively similar outcomes for total alkalinity (Fig. 3.3b). In fact, for chloride, the absolute average deviation from the open-system (phase segregation) model is 6.1 $\pm$ 5.9% for the isolated-system (gas partitioning) model and 3.4 $\pm$ 4.2% for the all-gas-in-liquid model. For total alkalinity, the absolute average deviation from the open-system (phase segregation) model is 26 $\pm$ 21% for the isolated-system (gas partitioning) model and 26 $\pm$ 22% for the all-gas-in-liquid model.

Part of the discrepancies between the three models are due to the different reservoir temperatures, with an absolute average deviation from the open-system (phase segregation) model of 8.8 $\pm$ 7.6 °C for the isolated-system (gas partitioning) model and of 8.1 $\pm$ 7.6 °C for the all-gas-in-liquid model.

For the volatile components, such as total CO_2 (Fig. 3.4a) and total H_2S (Fig. 3.4b), the results of the isolated-system model (gas partitioning) are significantly lower than those of the open-system (phase segregation) model and of the all-gas-in-liquid model, whereas the latter two models give relatively similar outcomes.

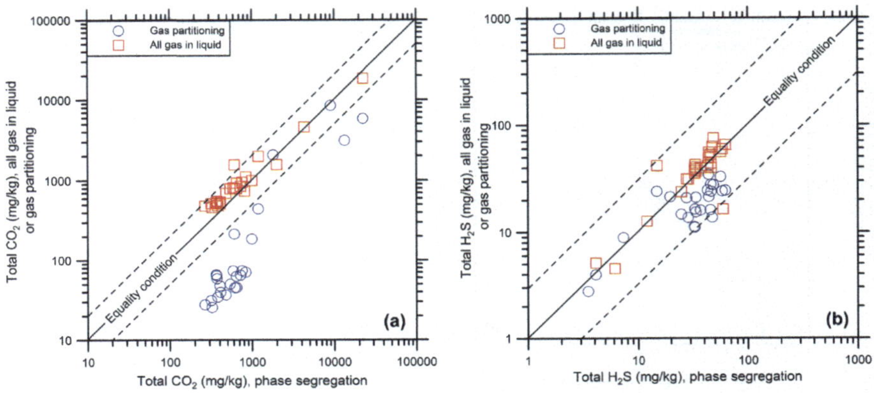

Fig. 3.4 **a** Total CO_2 and **b** total H_2S concentrations of the Olkaria reservoir liquids given by the open-system model with liquid retained in the formation (phase segregation) are compared with those given by the isolated-system model (gas partitioning) and the all-gas-in-liquid model. The dashed lines show deviations from the equality condition

In detail, the absolute average deviation of the isolated-system (gas partitioning) model from the open-system (phase segregation) model is 140 ± 45% for total CO_2 and 58 ± 29% for total H_2S, whereas the absolute average deviation of the all-gas-in-liquid model from the open-system (phase segregation) model is 32 ± 18% for total CO_2 and 21 ± 27% for total H_2S. These figures are similar to those for alkalinity (see above).

The pH of reservoir liquids is computed by WATCH starting from the low-temperature pH value and based on the assumption of constant alkalinity, apart from the increase brought about by steam separation (Arnórsson et al. 1982). In WATCH calculations, alkalinity is the sum of the concentrations (in equivalent units) of OH^-, $H_3SiO_4^-$, $H_2SiO_4^{2-}$, HCO_3^-, CO_3^{2-}, HS^- and S^{2-}, and related complexes with Na^+, Ca^{2+}, and Mg^{2+} ions, that is $CaHCO_3^+$, $MgHCO_3^+$, $CaCO_3°$, $MgCO_3°$, and $NaH_3SiO_4°$, minus H^+ molality.

Consequently, the discrepancies between the pH values computed by the three models are chiefly due to the distinct values of total alkalinity which, in turn, are essentially controlled by the differences in total CO_2 and total H_2S, whereas the differences in non-volatiles components and temperature are less important.

As shown by Fig. 3.5, the pH values given by the isolated-system model (gas partitioning) are significant higher than those of the open-system (phase segregation) model, whereas the pH values given by the all-gas-in-liquid model are similar to or somewhat lower than those of the open-system (phase segregation) model. In detail, the absolute average deviation of the isolated-system (gas partitioning) model from the open-system (phase segregation) model is 0.67 ± 0.26 pH units, whereas the absolute average deviation of the all-gas-in-liquid model from the open-system (phase segregation) model is 0.22 ± 0.15 pH units.

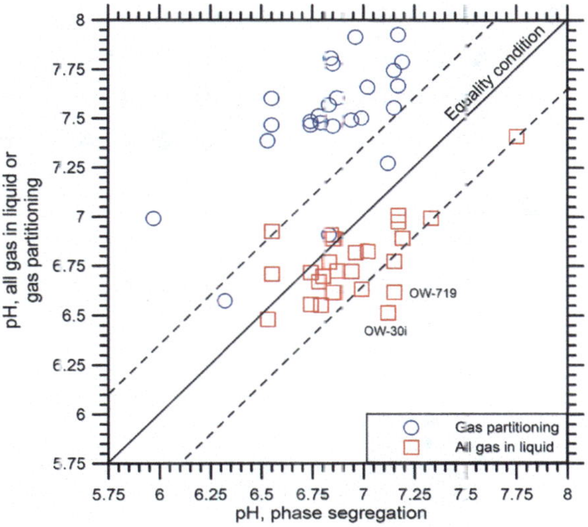

Fig. 3.5 pH values of the Olkaria reservoir liquids given by the open-system model with liquid retained in the formation (phase segregation) are compared with those given by the isolated-system model (gas partitioning) and the all-gas-in-liquid model. The dashed lines show deviations from the equality condition

Summing up, results of the all-gas-in-liquid model are not significantly different from those of the open-system model with liquid retained in the formation (phase segregation). However, the second model requires to specify the temperature (or pressure) in the intermediate zone or depressurization zone, where phase segregation occurs, whereas this information is not required by the first model. Therefore, the first model was preferred and the second model was abandoned. The next step is to choose the best model between the isolated-system model (gas partitioning) and the all-gas-in-liquid model. To this purpose, the mineral-solution equilibrium pH (definition in Sect. 3.9) was taken as reference. The isolated-system (gas partitioning) model was chosen if the difference between the pH computed using this model and the mineral-solution equilibrium pH is less than one pH unit, whereas the all-gas-in-liquid model was adopted when this pH difference resulted to be higher than one pH unit.

In this work, modeling of reservoir liquid chemistry for excess enthalpy wells was carried out using WATCH. The same calculations can be performed by other computer programs, such as SOLVEQ-XPT and CHIM-XPT (Reed 1998; Reed et al. 2014) or PHREEQC Interactive. Using PHREEQC Interactive, the initial four simulations are exactly the same as for the liquid enthalpy wells (see Fig. 3.2 and related discussion in Sect. 3.1.1). Again, the following keyword data blocks are used (Fig. 3.6):

- SOLUTION_1 to define the temperature and chemical composition of the separated liquid cooled at room temperature and to perform related speciation-saturation calculations (simulation 1);
- SOLUTION_2 to define the temperature and chemical composition of the separated vapor condensed and cooled at room temperature and to perform related speciation-saturation calculations (simulation 2);

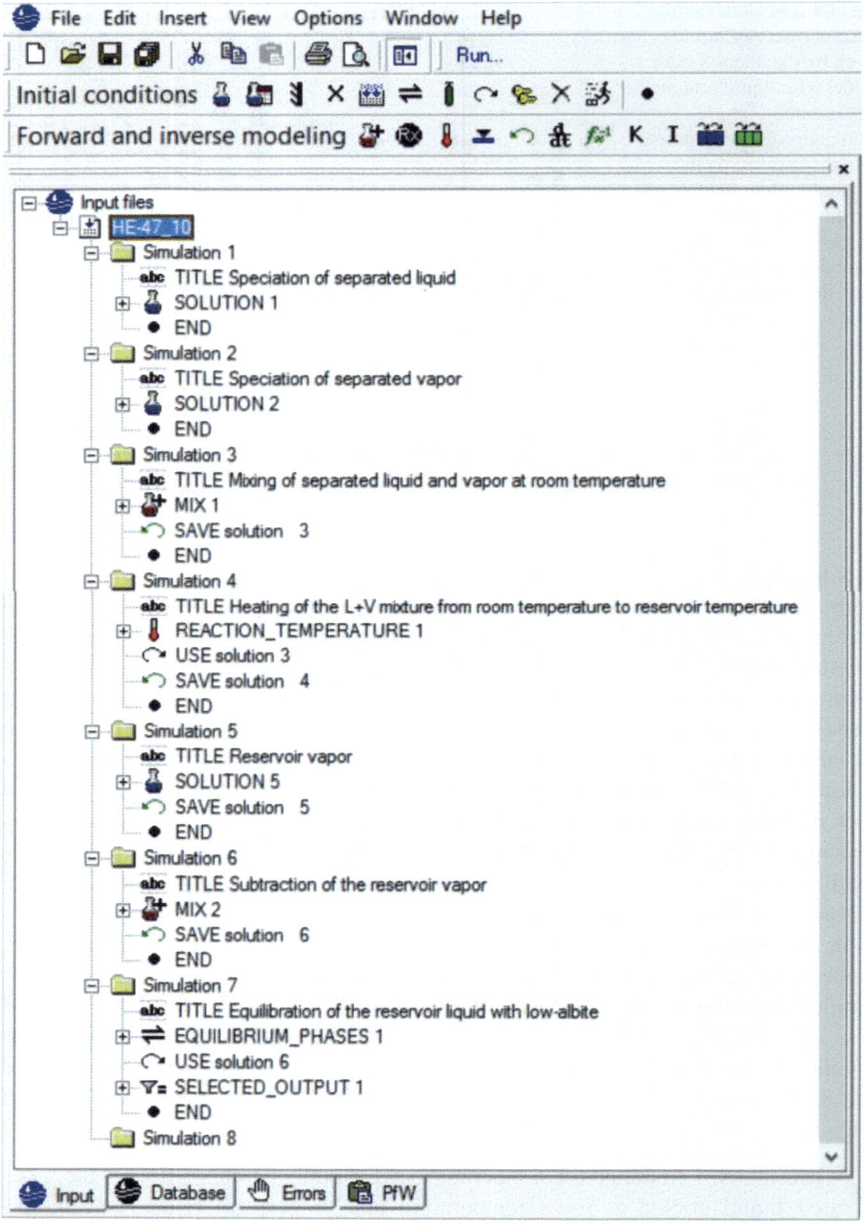

Fig. 3.6 Keyword data blocks used to compute the chemical composition of the liquid phase under reservoir condition for excess enthalpy wells by means of PHREEQC Interactive

- MIX_1 to mix together the separated liquid and condensed vapor at room temperature, considering the steam fraction at collection (simulation 3);
- REACTION_TEMPERATURE_1 to heat the liquid + condensed vapor mixture from room temperature to reservoir temperature (simulation 4).

Initially, only the 4 simulations mentioned so far are performed and the concentrations of aqueous CO_2, H_2S, and NH_3 in total discharge (obtained through simulation 4) are used to compute the concentrations of gaseous CO_2, H_2S, and NH_3 in the equilibrium vapor phase by means of the relation:

$$C_{i,V} = \frac{C_{i,O}}{y + \frac{1-y}{B_i}}. \quad (3.4)$$

Equation (3.4) is obtained by inserting the vapor-liquid gas distribution coefficient of the i-th gas species, $B_i = C_{i,V}/C_{i,L}$, into Eq. (3.3) and rearranging it. The vapor-liquid gas distribution coefficients of CO_2, H_2S, and NH_3 depend on temperature according to the following equations (T in °C; Giggenbach 1980):

$$\log B_{CO_2} = 4.7593 - 0.01092 \cdot T \quad (3.5)$$

$$\log B_{H_2S} = 4.0547 - 0.00981 \cdot T \quad (3.6)$$

$$\log B_{NH_3} = 1.4113 - 0.00292 \cdot T \quad (3.7)$$

which are valid from 100 to 340 °C. The concentrations of gaseous CO_2, H_2S, and NH_3 in the equilibrium vapor phase thus computed are then used as input data in simulation 5 to model the condensed reservoir vapor, whose pH is computed imposing the charge balance. The keyword data block SOLUTION_5 is used.

Then, in simulation 6, the keyword data block MIX_2 is utilized to subtract the condensed reservoir vapor from the total reservoir fluid, considering the aquifer steam fraction, which is obtained by plugging the enthalpies of total discharge and of vapor and liquid H_2O at reservoir temperature into Eq. (3.1).

Finally, in simulation 7, the reservoir liquid is equilibrated with low-albite, as in the last simulation for the liquid enthalpy wells, utilizing the keyword data block EQUILIBRIUM_PHASES_1.

3.1.3 Further Details on the Reconstruction of Reservoir Liquid Chemistry

The choice of the reservoir temperature depends largely on the available information. Temperatures measured in the producing sector of the well were used when available.

Enthalpy-temperatures were used for liquid enthalpy wells when measured temperatures were not available and silica temperatures appeared to be unreliable. Nevertheless, in most cases, the reservoir temperatures were considered equal to silica saturation temperatures, which were computed using the quartz/chalcedony function, i.e., Eq. (4.57), above 180 °C or the chalcedony geothermometer of Fournier (1977), i.e., Eq. (5.18), at lower temperatures. When the computed pH of the reservoir liquid resulted to be relatively high and the computed concentration of silicic acid resulted to be significantly lower than that of total dissolved silica, an iterative procedure was used, running again WATCH at the reservoir temperature given by the silica geothermometric function for the silicic acid concentration computed in the previous run. The procedure was stopped for reservoir temperatures differing by less than 0.1 °C in two consecutive runs. Usually four or five iterations are needed to attain this condition.

The reservoir temperatures of Miravalles fluids were computed averaging (1) in-hole temperature measurements performed in the same period of sampling, (2) temperatures corresponding to total discharge enthalpy, assuming the presence of a single saturated liquid in the reservoir, and (3) results of the quartz/chalcedony function, i.e., Eq. (4.57), because of the small deviations between these different temperature types.

When the liquid sample was collected at the weir box, under atmospheric pressure, and the vapor phase was obtained at higher pressure, simple preliminary calculations were carried out to compute the composition of the vapor phase at atmospheric pressure (Arnórsson et al. 2005). The alternative approach, that is the calculation of the liquid phase composition at the temperature and pressure of vapor sampling, was not adopted because it requires use of geochemical modeling for pH calculation.

Only boiled liquids were generally collected for the Icelandic systems of medium temperature (100–175 °C) whereas steam samples are not available. Therefore, the boiling spring model was adopted, selecting a suitable degassing factor to approach calcite saturation for the deep liquids, because *it is reasonable to assume that deep geothermal waters are calcite saturated* (Arnórsson et al. 1983b). Aquifer temperatures were reconstructed on the basis of in-hole temperature measurements. The chalcedony geothermometer was not used due to the usual presence of considerable amounts of silicate ion at the relatively high pH of these reservoir liquids.The citation "Arnórsson et al. (1982)" has been changed to "Arnórsson (1983a, b)" to match the author name/date in the reference list. Please check here and in subsequent occurrences, and correct if necessary. Here the correct citation is Arnórsson (1983b). I tried to correct the text but I could not. Here and in subsequent occurrences I added a comment.

Although WATCH performs speciation-saturation calculations, the chemical compositions computed by WATCH, including pH and the redox potential fixed by the HS^-/SO_4^{2-} couple, were given as input data to PHREEQC Interactive 3.1.1 to improve data processing. PHREEQC Interactive was run using a modified version the LLNL database containing the thermodynamic data of solid phases, including variably ordered adularia, from Helgeson et al. (1978). In PHREEQC Interactive runs, aluminum was assumed to be fixed by saturation with low-albite. The reasons for this choice are discussed in Sect. 4.2.3.

In most cases, the extended Debye-Hückel equation, also known as B-dot equation (see Sect. 2.3.3), was adopted to calculate activity coefficients of individual ions. However, the high-salinity geothermal liquids of Salton Sea and Asal were processed directly by EQ3, without running WATCH beforehand, and using a thermodynamic database which is consistent with the Pitzer's theory (see Sect. 2.3.4) and comprises the data of solid phases, including variably ordered adularia, from Helgeson et al. (1978).

3.1.4 *Presentation of the Main Results of Speciation-Saturation Calculations and Approach Adopted for the Chemical Classification of Reservoir Liquids*

Only the aqueous solutions with percent error on total charge lower than 10% and reservoir pH (from speciation calculations) corresponding with the mineral-solution equilibrium pH within ±1.3 pH unit were considered in data interpretation. The analytical data of reservoir liquids, main results of speciation-saturation calculations and outcomes of traditional geothermometers and f_{CO_2}-indicators as well as of the theoretical activity-based geoindicators derived in this book are reported as electronic supplementary material (file Reservoir_Liquids.XLS). The ordering parameter of hydrothermal adularia in hypothetical equilibrium with these reservoir liquids is also given. The results of speciation-saturation calculations listed in this file comprise: pH, Eh, pe, carbonate alkalinity,[2] Alk_C, the activities and molalities of H^+, Na^+, K^+, Ca^{2+}, Mg^{2+}, Al^{3+} and HCO_3^- ions, $SiO_{2(aq)}$, and $CO_{2(aq)}$, the logarithm of CO_2 fugacity, and the concentration of total dissolved Al constrained by equilibrium with low-albite.

In the following sections of this chapter, the chemical composition of reservoir liquids is investigated by means of the triangular plots of major anions (i.e., Cl, SO_4, and carbonate alkalinity) and major cations (i.e., Na, K, and Ca). Magnesium was excluded because it is generally present in very low concentrations in geothermal liquids and the relative concentration of Mg is negligible with respect to those of Na, K, and Ca. Since these triangular plots do not convey any information on total ionic salinity (Σ_{eq}), this parameter is inspected by using the binary diagrams of Cl versus $SO_4 + Alk_C$. In fact, the Σ_{eq} of the considered reservoir liquids can be appreciated in these diagrams, by comparing the position of each reservoir liquid with the lines of slope −1 which are iso-Σ_{eq} lines (see Tonani et al. 1998 for further details). All these graphs were prepared using concentrations in equivalent units, in contrast with the

[2]Carbonate alkalinity is the sum of the concentrations, in equivalent units, of HCO_3^- and CO_3^{2-} ion, including related aqueous complexes, e.g., $CaHCO_3^+$, $MgHCO_3^+$, $CaCO_3°$, and $MgCO_3°$.

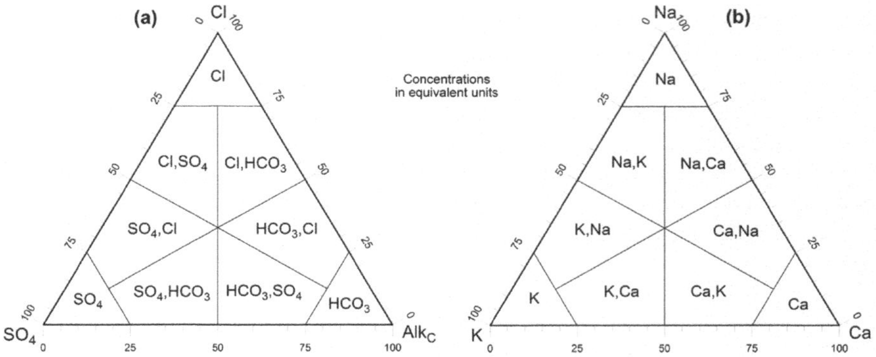

Fig. 3.7 Triangular plots of **a** major anions and **b** major cations showing the different sectors used for the chemical classification of reservoir liquids

consolidated procedure which is based on the use of concentrations in weight units. The reason for our choice is that equivalent units are proportional to the amount of electrical charges carried by each ion and, therefore, are more suitable than weight units for the chemical classification of natural waters as recognized long ago (e.g., Zaporozec 1972 and references therein). Average seawater composition (Nordstrom et al. 1979) is also represented in the relevant plots for comparison.

The terminology adopted for the chemical classification of reservoir liquids is consistent with the two triangular diagrams of Fig. 3.7. For instance a reservoir liquid positioned in the [Cl,SO_4] sector in Fig. 3.7a and in the [Na] sector in Fig. 3.7b belongs to the Na–Cl,SO_4 chemical type or chemical facies, whereas a reservoir liquid situated in the [HCO_3,SO_4] sector in Fig. 3.7a and in the [Ca] sector in Fig. 3.7b is attributed to the Ca–HCO_3,SO_4 chemical type or chemical facies, and so on.

3.2 The Reservoir Liquids from the Geothermal Systems in Iceland

Iceland is located along the Mid-Atlantic Ridge between 63°20′ N, 18°44′ W and 66°11′ N, 18°50′ W and is the largest emerged sector of a mid-oceanic ridge system all over the world. Iceland belongs to a large volcanic province which extends from Greenland to Scotland and is associated to a hotspot or a deep mantle plume (Wilson 2007).

The high-temperature (>175 °C) geothermal systems of Iceland considered here are Hellisheidi (data from Scott 2011), Hveragerdi (data from Arnórsson 1978b, Arnórsson et al. 1978, 1983b, Ping 1991; Zhanxue 1998), Krafla (data from Arnórsson et al. 1983b; Ping 1991; Gudmunsson and Arnórsson 2002; Giroud 2008; Hermanska et al. 2019), Námafjall (data from Arnórsson 1978b; Arnórsson et al. 1978, 1983b; Gudmunsson and Arnórsson 2002; Giroud 2008; Malimo 2012),

Fig. 3.8 Google Earth map showing the location of the high-temperature geothermal fields of Krafla and Námafjall in northeastern Iceland

Nesjavellir (data from Arnórsson 1978b; Arnórsson et al. 1978; Giroud 2008), Reykjanes (data from Arnórsson 1978a, b; Arnórsson et al. 1978, 1983b; Giroud 2008; Hardardóttir 2011; Berehannu 2014), and Svartsengi (data from Arnórsson 1978a, b; Arnórsson et al. 1978, 1983b; Giroud 2008). The number of accepted chemical analyses is 153.

Krafla and Námafjall are located at short distance from each other, in the northern part of Iceland (Fig. 3.8). The Krafla–Námafjall area was affected by volcanic eruptions in 1975–1984.

Hellisheidi, Nesjavellir, Hveragerdi, Svartsengi, and Reykjanes are situated in the southern part of the country (Fig. 3.9). Hellisheidi, Nesjavellir, and Hveragerdi are next to each other, in the Hengill volcano area, on the inland termination of the Reykjanes peninsula. Reykjanes and Svartsengi are positioned in the extreme southwest of Iceland, on the tip of the Reykjanes peninsula, towards the ocean, and are recharged by sea-water, at least partly.

A thorough description of the Icelandic high-temperature geothermal fields is given in the review paper of Ármansson (2016) and references therein.

For what concerns the medium-temperature (100 < T < 175 °C) geothermal systems of Iceland taken into account in this work, most data come from Bakki, with 18 entries (data from Arnórsson et al. 1983b; Arnórsson 1995b, Ping 1991, and Zhanshi 2001), whereas the other 11 data (from Arnórsson 1978b; Arnórsson et al. 1983b; Arnórsson 1995b) refer to different sites, for a total of 29 chemical analyses.

Fig. 3.9 Google Earth map showing the location of the high-temperature geothermal fields of Hellisheidi, Nesjavellir, Hveragerdi, Svartsengi, and Reykjanes in southwestern Iceland

3.2.1 Chemistry of the Reservoir Liquids from the High-Temperature Geothermal Systems in Iceland

In the triangular diagram of main cations (Fig. 3.10), all the high-temperature reservoir liquids of Iceland are positioned in the [Na] sector, indicating that sodium is by far the prevailing cation in all these aqueous solutions. The reservoir liquids of Krafla, Námafjall, Hellisheidi, Nesjavellir, and Hveragerdi are situated close to the Na vertex, whereas those of Reykjanes and Svartsengi have somewhat lower Na/Ca molar ratio, due to acquisition of Ca during high-temperature basalt-seawater interaction. In the triangular diagram of major anions for the high-temperature geothermal systems of Iceland there is a large scatter of sample points (Fig. 3.11). Based on both triangular plots the considered reservoir liquids can be attributed to eight different chemical types. In detail:

1. The Na–Cl chemical type comprises 62 of the 152 reservoir liquids of interest, that is 41% of the cases. The Na–Cl reservoir liquids come from Hellisheidi, Námafjall, Nesjavellir, Reykjanes and Svartsengi. In particular, the reservoir liquids of Reykjanes and Svartsengi are positioned exactly on the chloride vertex in Fig. 3.11.
2. The Na–Cl,HCO_3 chemical facies includes 32 reservoir liquids (21% of the total) proceeding from Hellisheidi, Hveragerdi, Krafla, Námafjall, and Nesjavellir.

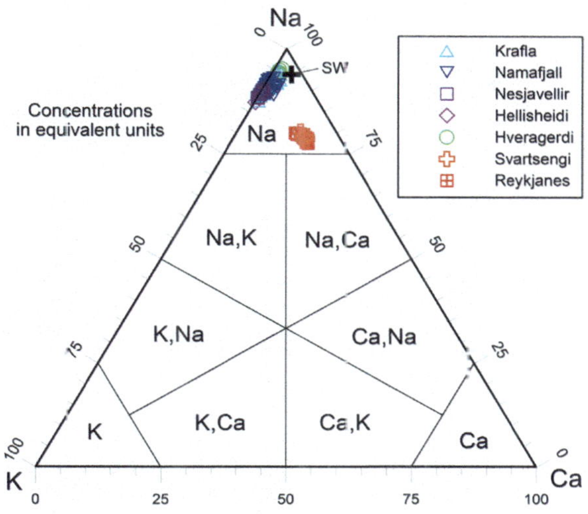

Fig. 3.10 Triangular plot of major cations for the reservoir liquids from the high-temperature Icelandic geothermal systems. Average seawater is also shown for comparison

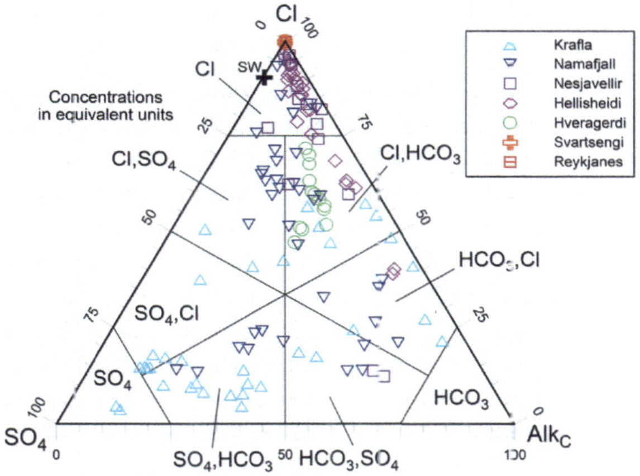

Fig. 3.11 Triangular plot of major anions for the reservoir liquids from the high-temperature Icelandic geothermal systems. Average seawater is also shown for comparison

3. The Na–HCO_3,Cl chemical facies is represented by 12 reservoir liquids (8% of the total), coming from Hellisheidi, Krafla, and Námafjall.
4. The Na–HCO_3,SO_4 chemical type groups 7 reservoir liquids (5% of the cases), proceeding from Krafla, Námafjall, and Nesjavellir.
5. The Na–SO_4,HCO_3 chemical facies includes 18 reservoir liquids (12% of the total) coming from Krafla and Námafjall.
6. The Na–SO_4 chemical facies is represented by 2 reservoir liquids (1% of the total), both from Krafla.
7. The Na–SO_4,Cl chemical type comprises 7 reservoir liquids (5% of the cases) proceeding from Krafla.

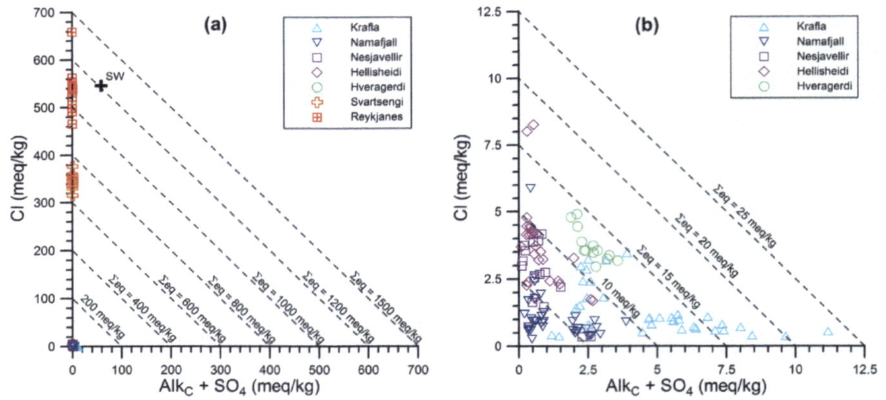

Fig. 3.12 Correlation diagrams of Cl versus SO_4 + Alk_C for the reservoir liquids from the high-temperature geothermal systems in Iceland. Plot (**b**) shows a zoomed view of the zone close to the origin of the axes in plot (**a**). Average seawater is also displayed for comparison in plot (**a**)

8. The Na–Cl,SO_4 chemical type groups 12 reservoir liquids (8% of the cases) coming from Krafla and Námafjall.

As shown by the binary diagrams of Cl versus SO_4 + Alk_C (Fig. 3.12), the reservoir liquids of Krafla, Námafjall, Hellisheidi, Nesjavellir, and Hveragerdi have low Σ_{eq}, ranging from 5.0 to 23.5 meq/kg[3]. These low values are not surprising because these geothermal circuits are recharged by meteoric waters and the extent of water-rock interaction is limited (Arnórsson 1995a). In contrast, the reservoir liquids of Reykjanes and Svartsengi have high Σ_{eq}, in the range 946–1312 meq/kg, and 643–732 meq/kg, respectively, due to the variable involvement of seawater and meteoric waters in the recharge of these geothermal circuits (Arnórsson 1978a, 1995a).

In fact, the seawater-dominated Reykjanes geothermal system is positioned along the Mid-Atlantic Ridge and is considered to be the subaerial equivalent to mid-ocean-ridge hydrothermal systems (Hardardóttir et al. 2009).

Chloride is of marine origin at Reykjanes and Svartsengi, whereas it is mainly derived from rock leaching at Hellisheidi, Nesjavellir, Hveragerdi, Námafjall and Krafla, where dissolved chloride is low because local basaltic rocks have low Cl contents (Arnórsson 1995a). Nevertheless, supply of magmatic HCl occurred at Krafla when fresh magma was intruded into the roots of Krafla volcano (Ármannsson et al. 1982).

Sulfur species are chiefly contributed by magma degassing at Námafjall and Krafla. However, the high SO_4 concentrations of the reservoir liquids of lower temperature and enthalpy from Krafla are probably due to dissolution of anhydrite, whose solubility increases with decreasing temperature (Gudmundsson and

[3]Incidentally, these low Σ_{eq} values are similar to those of bottled oligomineral waters. However, these reservoir liquids have very high concentrations of SiO_2, representing 32–84% of the TDS. Therefore, TDS values are significantly higher than those of oligomineral waters.

Arnórsson 2002). Carbon dioxide is probably supplied by magma degassing to all the high-temperature Icelandic geothermal systems (Arnórsson 1995a).

3.2.2 Chemistry of the Reservoir Liquids from the Medium-Temperature Geothermal Systems in Iceland

In the triangular plot of main cations (Fig. 3.13), all the medium-temperature reservoir liquids of Iceland, except those of Seltjarnarnes, are found in the [Na] sector, indicating that sodium is by far the prevailing cation in most cases. Of the two Seltjarnarnes reservoir liquids, one is positioned in the [Ca,Na] sector, whereas the other is situated in the [Na,Ca] sector of Fig. 3.13.

In the triangular diagram of main anions (Fig. 3.14), the reservoir liquids from the medium-temperature geothermal systems in Iceland are located in the sectors of [Cl], [Cl,SO_4], [Cl,HCO_3] and [HCO_3,SO_4]. The reservoir liquids of Bakki and Seltjarnarnes are positioned close to the seawater point in this plot. Six different chemical types are recognized based on both triangular diagrams. In detail:

1. The Na–Cl chemical type comprises 18 of the 29 reservoir liquids of interest (62% of the cases), all from Bakki.
2. The Na,Ca–Cl chemical facies includes one of the two reservoir liquids of Seltjarnarnes (3% of the cases).
3. The Ca,Na–Cl chemical type is represented by the other reservoir liquid of Seltjarnarnes (3% of the cases).

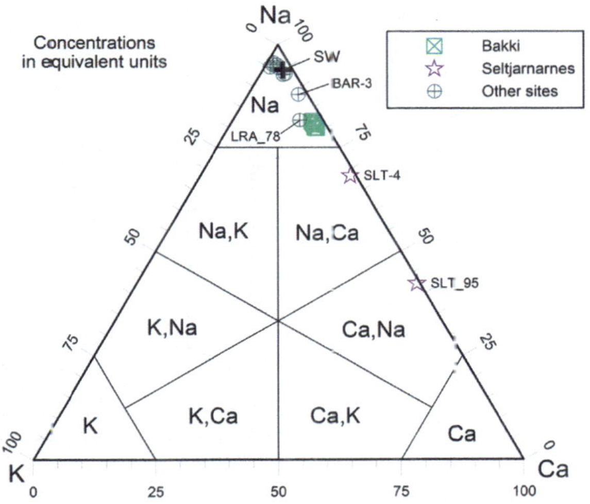

Fig. 3.13 Triangular plot of major cations for the reservoir liquids from the medium-temperature geothermal systems in Iceland. Average seawater is also shown for comparison

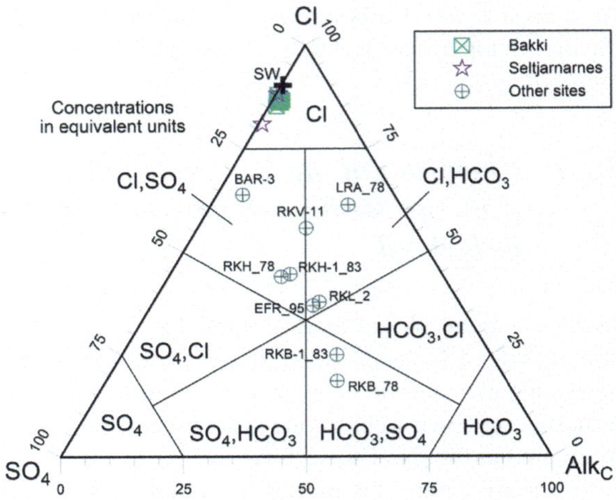

Fig. 3.14 Triangular plot of major anions for the reservoir liquids from the medium-temperature geothermal systems in Iceland. Average seawater is also shown for comparison

4. The Na–Cl,HCO_3 chemical facies groups 4 reservoir liquids (14% of the total) proceeding from Efri-Reykir, Leirà, Reykholar, and Reykjavik.
5. The Na–Cl,SO_4 chemical type comprises 3 reservoir liquids (10% of the cases) coming from Baer and Reykholt.
6. The Na–HCO_3,SO_4 chemical facies includes 2 reservoir liquids (7% of the cases), both proceeding from Reykjabol.

As shown by the correlation diagram of Cl versus SO_4 + Alk_C (Fig. 3.15), the

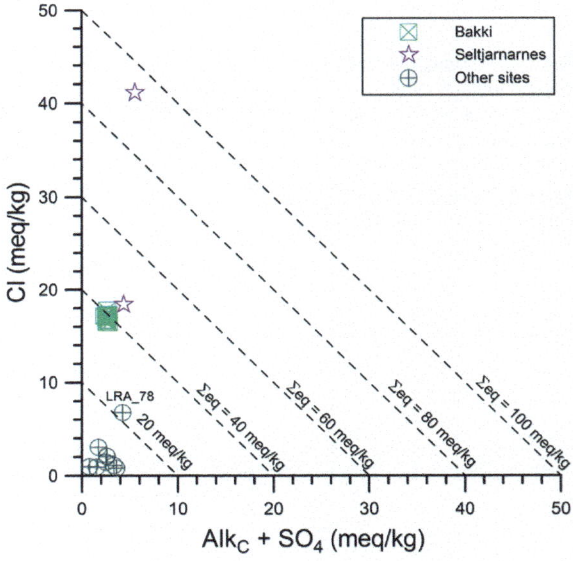

Fig. 3.15 Correlation diagram of Cl versus SO_4 + Alk_C for the reservoir liquids from the medium-temperature geothermal systems in Iceland

reservoir liquids of most geothermal sites have low Σ_{eq} values, ranging from 4 to 10 meq/kg, whereas somewhat higher Σ_{eq} values are displayed by the reservoir liquids of Leirà (code LRA_78), with Σ_{eq} of 22 meq/kg, Bakki, with Σ_{eq} of 39–41 meq/kg, and Seltjarnarnes, with Σ_{eq} values of 45 and 96 meq/kg.

The peculiar chemistry of Bakki is probably due to an appreciable seawater contribution in the recharge of this geothermal circuit or to leaching of halite contained in local marine sediments alternating with volcanic rocks (Zhanshi 2001). Similar explanations apply to Seltjarnarnes, whose high Ca concentrations are likely acquired through interaction with basalts.

3.3 The Reservoir Liquids from the Geothermal Systems in Northern and Central America

The geothermal systems in Northern and Central America taken into account in this work are: Dixie Valley (data from Reed 1989; Goff et al. 2002), Long Valley (data from White and Peterson 1991), Coso (data from Moore et al. 1989), Valles (data from Truesdell and Janik 1986; White 1986), Salton Sea (data from Helgeson 1968b; Michels 1986; Thompson and Fournier 1988; Williams and McKibben 1989) and Heber (data from Adams et al. 1989) in the USA, Cerro Prieto (data from Reed 1976; Mañon et al. 1977; Truesdell et al. 1981) and Los Azufres (data from González-Partida et al. 2000, 2005; Arellano et al. 2005) in México, Berlín in El Salvador (data from Renderos 2002), and Miravalles in Costa Rica (data from Giggenbach and Corrales Soto 1992; Yock Fung 1998; Gherardi et al. 2002; Marini et al. 2003). The accepted chemical analyses are 35 from Dixie Valley, 10 from Long Valley, 5 from Coso, 7 from Valles, 23 from Salton Sea, 16 from Heber, 19 from Cerro Prieto, 26 from Los Azufres, 55 from Berlín, and 105 from Miravalles, for a total of 301 entries.

3.3.1 Dixie Valley

The Dixie Valley in central Nevada is the site of a fault-controlled geothermal system of the Basin and Range tectonic province (Fig. 3.16). In the Dixie Valley geothermal system, hot fluids are produced from the permeable sectors of a normal fault, known as Stillwater, and associated fractured rocks at depths of 2800–3050 m (Goff et al. 2002 and references therein). The local stratigraphic sequence includes marine quartzite, siltstone, and shale, silicic and basaltic volcanic and volcaniclastic rocks as well as granodiorite of Cretaceous age (Waibel 1987). Local rocks were affected by several thermal and metasomatic events characterized by temperatures ranging from 50 to 250 °C, which compare with the interval of present-day temperatures.

Fig. 3.16 Google Earth map showing the location of the Dixie Valley geothermal field

3.3.2 *Long Valley*

The Long Valley geothermal system in east-central California is hosted within the homonymous caldera, a 17 by 32 km elliptical depression on the eastern front of the Sierra Nevada (Fig. 3.17). The Long Valley caldera formed 730 ka ago during the Bishop Tuff eruption that emitted ~600 km^3 of rhyolite magma (Bailey et al. 1976). Rhyolites and rhyodacites were erupted during the subsequent intracaldera volcanic activity. The currently active geothermal system is hosted in Paleozoic and Mesozoic metasedimentary rocks situated at depths of at least 3 km. The geothermal system is recharged by meteoric waters infiltrating in the western rim of the Long Valley caldera along the Sierra Nevada front. Thermal fluids upflow in the western part of the caldera, flow eastwards in a confined, shallow aquifer and mix progressively with dilute meteoric groundwaters before discharging through the hot springs located in the southeastern parts of the caldera (Tempel et al. 2011 and references therein).

3.3.3 *Coso*

The Coso geothermal field in east-central California is situated between the Basin and Range and Sierra Nevada tectonic provinces (Fig. 3.18), at a releasing bend stepover in a dextral strike-slip fault system (Monastero et al. 2005).

Fig. 3.17 Google Earth map showing the location of the Long Valley geothermal field

Fig. 3.18 Google Earth map showing the location of the Coso geothermal field

Permeability is probably controlled by active normal faults accommodating the regional dextral transtension. Although the reservoir does not appear to be confined to a specific lithotype, it is prevailingly hosted in granitic rocks and subordinately in mafic and metamorphic rocks associated with the Sierra Nevada composite batholith.

Basement rocks are covered by the Late Tertiary and Quaternary volcanic rocks of the Coso volcanic field, including rhyolite domes and flows younger than 300 ka (Wohletz and Heiken 1992). A silicic magma body, possibly partially molten, underlies the Coso volcanic field at a depth of at least 8 km, acting as heat source of the geothermal system (Bacon et al. 1980).

Measured temperatures in wells used as producers vary from 340 °C, at depths of 2500 m in the south, to 230 °C, in the shallow part of the system in the north, reflecting the presence of an upflowing plume of thermal water, which is also indicated by chemical and fluid inclusion data (Moore et al. 1989).

3.3.4 Valles

The Valles caldera is a Quaternary volcanic collapse structure with a diameter of 22 km (Fig. 3.19). It was formed by the eruption of the Upper Member of the Bandelier Tuff, which occurred 1.25 Ma ago and is the latest catastrophic volcanic eruption in

Fig. 3.19 Google Earth map showing the location of the Valles geothermal field

the Jemez Mountains volcanic field of north-central New Mexico (Smith and Bailey 1966).

The Valles caldera is located at the intersection of the Jemez Lineament and the western edge of the Rio Grande Rift, a major extensional structure that stretches over 1000 km. A geothermal system at temperatures of 225–330 °C is hosted within the Valles caldera, in caldera-fill ignimbrites and pre-caldera rocks, at depths of 600–2500 m (Nielson and Hulen 1984).

The Valles geothermal system comprises two drilled reservoirs: the Redondo Creek (formerly called Baca) reservoir and the Sulphur Springs reservoir. Deep reservoir fluids have neutral sodium-chloride chemical composition, TDS lower than 8000 mg/kg, and gas content of 0.4–1.5 wt%. About 10 km outside the Valles caldera, along the pre-caldera Jemez fault zone, there are two sets of neutral-chloride hot springs representing the outflows of the Valles geothermal system (Goff et al. 1988).

3.3.5 *Salton Sea, Heber, and Cerro Prieto*

Several active geothermal systems, including Salton Sea, Heber, and Cerro Prieto, are located into the extensional basins linking different sectors of strike-slip faulting within the Salton Trough, which comprises the Imperial Valley in southern California, U.S.A, and the Mexicali Valley in the northern part of Baja California, México (Fig. 3.20).

Since 4 Ma ago, the Imperial Valley has been isolated from the Gulf of California by the Colorado River sedimentation, with the consequent deposition of evaporite rocks, whose dissolution explains the high or relatively high salinity of local geothermal fluids (Williams and McKibben 1989 and references therein). The stratigraphic sequence of the region comprises Pliocene and Pleistocene fluvial, deltaic, lacustrine and evaporite sediments. These sediments were locally intruded by both mafic and felsic igneous rocks and experienced widespread metamorphism generating greenschist and even amphibolite facies hornfelses.

Hypersaline geothermal brines (20–27 wt% TDS) occupy the deep hot (260–340 °C) central sectors of the Salton Sea geothermal system, whereas lower salinity fluids (<10 wt% TDS) generally occur in the cooler (<260 °C) shallow or boundary portions of the field (Williams and McKibben 1989). The two fluid types are separated by a few hundred meter thick interface, roughly coinciding with the 260 °C isotherm.

The Heber geothermal system comprises three major units, characterized by distinct permeability, namely: (1) the low permeability capping clays from 150 to 550 m, (2) the high matrix permeability sandstone "outflow reservoir" from 550 to 1700 m, and (3) the indurated sediments, at depths of 1700 to ~3200 m, whose high permeability is controlled by both strike-slip and normal faults (James et al. 1987). A maximum bottom-hole temperature of 199 °C was measured in well GTW-6, which is probably the closest to the upflow zone in the south. The geothermal fluids traveling northeastward have higher B and Li concentrations, lower Ca and SO_4 concentrations, as well as somewhat different deuterium and oxygen-18 compared to the geothermal

Fig. 3.20 Google Earth map showing the location of the Salton Sea, Heber, and Cerro Prieto geothermal fields

fluids moving northwestward. These differences suggest that the northwestern fluids mix with small amounts of low-temperature waters and re-equilibrate with anhydrite at weakly lower temperatures than the northeastern fluids (Adams et al. 1989). Apart from these limited differences, Heber geothermal brines have close to 15,000 mg/kg TDS and low gas contents.

The wells drilled at Cerro Prieto to depths of 1250–3550 m have encountered the same temperature interval of Salton Sea and similar rocks (Reed 1976; Izquierdo et al. 2006). Moreover, the distribution of hydrothermal minerals as a function of temperature compares with that recognized at Salton Sea. In contrast, the reservoir liquids of Cerro Prieto have salinities of 17,000–36,000 mg/kg, which are significantly lower than those of the hypersaline geothermal brines of Salton Sea. The lower salinity of Cerro Prieto reservoir liquids is chiefly controlled by mixing of Colorado River water with seawater evaporated to about six times its initial salinity (Truesdell et al. 1981). During deep circulation, this mixture acquired heat as well as Li, K, Ca, B, SiO_2 and rare alkalis released from rocks, whereas Mg, SO_4, and a minor amount of Na were transferred from the aqueous solution to the rocks. After these processes, the brine was finally diluted to its present interval of temperature and salinity.

3.3.6 *Los Azufres*

The geothermal field of Los Azufres in the Michoacán state, México comprises two distinct production areas, separated by an intermediate zone without surface manifestations (Fig. 3.21). Multiple caldera collapses occurred during pre-Quaternary times in the area of the geothermal field (Ferrari et al. 1991) Based on the evidence provided by deep wells of total depth up to 3600 m, the local stratigraphic sequence includes:

1. Late Miocene to Early Pliocene volcanic rocks, primarily andesitic and basaltic lava flows and breccias, locally interstratified with pyroclastic deposits, and
2. younger rhyodacites, rhyolites, and dacites, which were emplaced between 1 and 0.15 Ma ago (Arellano et al. 2005 and references therein).

In the natural-state, the reservoir hosted a single-phase liquid below 1280 m asl, overlain by a two-phase liquid-dominated region, ranging from 1280 to 1830 m asl, and a two-phase steam-dominated region extending upward to approximately 2400 m asl (Iglesias et al. 1985). The chloride concentrations of the reservoir liquids varied from about 1750–2050 mg/kg.

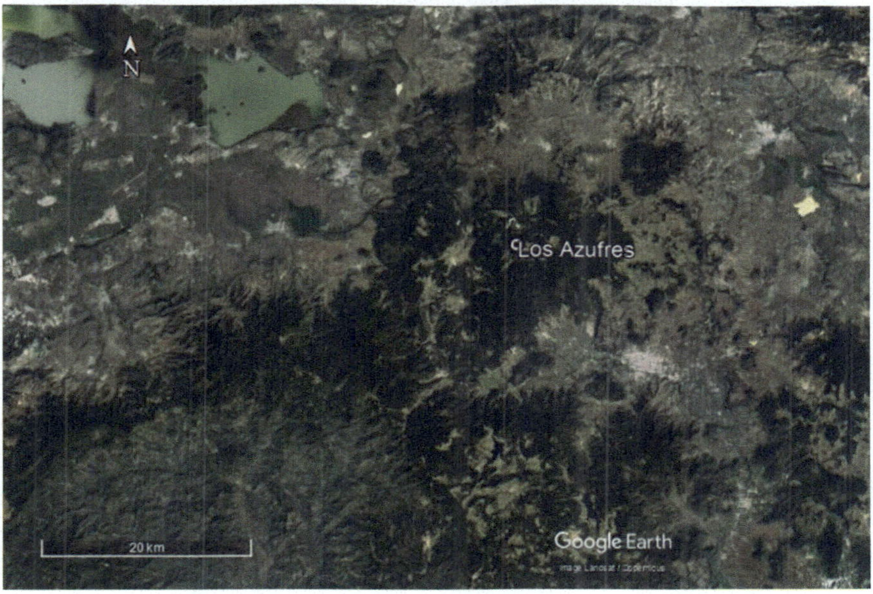

Fig. 3.21 Google Earth map showing the location of the geothermal field of Los Azufres

3.3.7 Berlin

The Berlin geothermal field is positioned on the northern slopes of the Berlin-Tecapa volcanic complex (Fig. 3.22). Reservoir rocks consist of altered lava flows and pyroclastics. An elliptical caldera with major axis close to 6–7 km and a 3–4 km wide NNW-SSE trending graben are the main geological features of the area, controlling the circulation, uprising, and migration of geothermal fluids.

The wells used as producers are clustered within an area of about 2 km^2. Reservoir temperatures generally vary from 270 to 305 °C. Deep wells meet neutral chloride waters, with pH of 5.30–6.70, Cl concentration from 3100 to 5600 mg/kg and low SO_4 concentration, 2.5–19 mg/kg, under reservoir conditions.

As observed by D'Amore and Mejia (1999), the isotopic composition of produced fluids is intermediate between that of local meteoric waters recharging the geothermal reservoir and that of injected fluids, because reinjection was carried out since the beginning of production. Consequently, available data are not representative of the natural pre-exploitation condition.

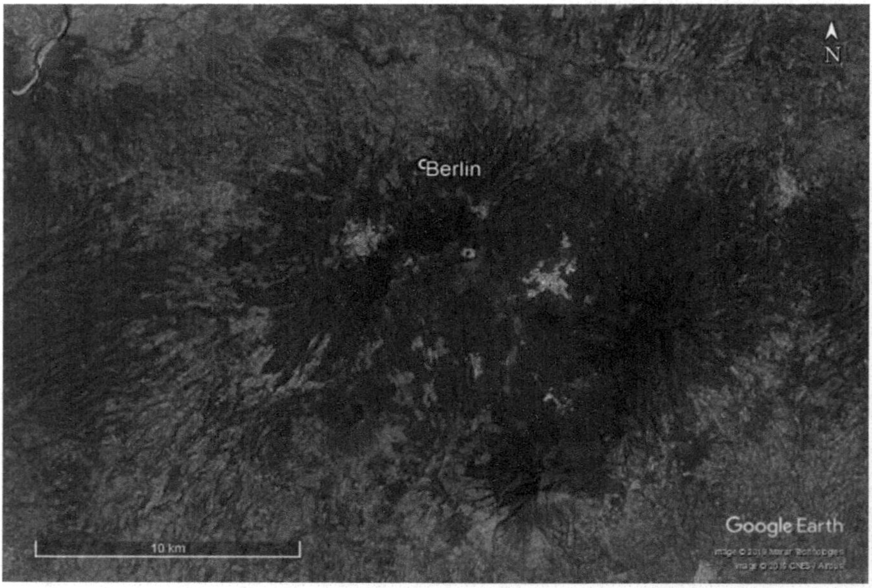

Fig. 3.22 Google Earth map showing the location of the Berlin geothermal field

3.3.8 Miravalles

The Miravalles geothermal field extends over an area greater than 21 km^2 on the southwestern flanks of the Miravalles volcano (Fig. 3.23), which belongs to the Guanacaste Cordillera in north-western Costa Rica. The geothermal reservoir is confined within the Guayabo caldera and main production zones are chiefly correlated with andesitic and dacitic lavas.

Reservoir temperatures usually vary from 230 to 240 °C, but the highest measured value is 255 °C. Most deep wells encountered neutral chloride waters, with pH of 5.27–6.31, Cl concentration from 2500 to 3300 mg/kg and low SO_4, 20–70 mg/kg, under reservoir conditions.

However, a few wells located in a small area of the field met acidic liquids, with pH of 2.80–4.05, Cl concentration of 2800–3400 mg/kg, and SO_4 concentration of 360–540 mg/kg, under reservoir conditions. These acid Cl–SO_4 liquids are not considered here because they are close to equilibrium with the hydrothermal minerals of the acid alteration suite (sensu Reyes 1990), including chalcedony, anhydrite, kaolinite (as a proxy of dickite), illite, and alunite (Marini et al. 2003).

Reservoir liquids are produced through mixing of deeply circulating meteoric waters and andesitic (arc-type) magmatic waters, accounting for 80–90% and 10–20%, respectively (Giggenbach and Corrales Soto 1992; Gherardi et al. 2002).

Fig. 3.23 Google Earth map showing the location of the Miravalles geothermal field

3.3.9 Chemistry of the Reservoir Liquids from the Geothermal Systems in Northern and Central America

Most reservoir liquids from the geothermal systems in Northern and Central America are located in the [Na] sector of the triangular diagram of main cations (Fig. 3.24). However, 20 of the 23 reservoir liquids from Salton Sea and 1 of the 16 reservoir liquids from Heber are situated in the [Na,Ca] sector. Hence, the prevailing cation in all these aqueous solutions is sodium, although most Salton Sea reservoir liquids have relative Ca concentrations varying from 30 to 39 eq% and one Heber reservoir liquid exhibits a relative Ca concentration of 22 eq%. These enrichments in calcium are expected owing to the high salinity of these aqueous solutions.

In the triangular diagram of major anions (Fig. 3.25) most reservoir liquids are positioned in the [Cl] sector, close to the Cl vertex. Only the reservoir liquids of Dixie Valley and Long Valley are distributed elsewhere, namely in the sectors of [Cl,SO_4], [Cl,HCO_3], and [HCO_3,Cl]. Five distinct chemical facies are recognized on the basis of both triangular diagrams. In detail:

1. The Na–Cl chemical type is by far the most frequent, comprising 235 of the 301 reservoir liquids of interest, that is 78% of the cases. The Na–Cl reservoir liquids come from Coso, Valles, Salton Sea, Heber, Cerro Prieto, Los Azufres, Berlin, and Miravalles.
2. The Na,Ca–Cl chemical facies comprises 21 reservoir liquids (7% of the cases), all proceeding from Salton Sea apart one from Heber.
3. The Na–Cl,SO_4 chemical type groups 34 reservoir liquids (11% of the cases), all coming from Dixie Valley.
4. The Na–Cl,HCO_3 chemical facies includes 7 reservoir liquids (2% of the total), six of which proceed from Long Valley and one from Dixie Valley.

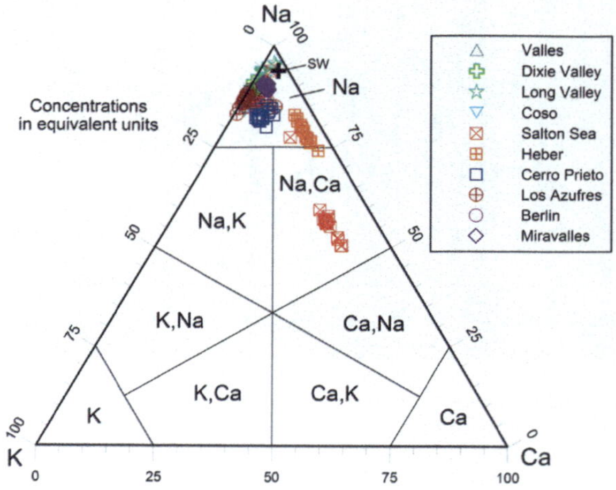

Fig. 3.24 Triangular plot of major cations for the reservoir liquids from the geothermal systems in Northern and Central America. Average seawater is also shown for comparison

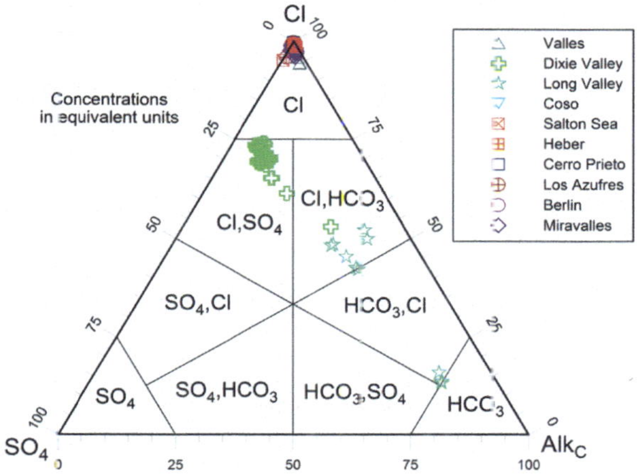

Fig. 3.25 Triangular plot of major anions for the reservoir liquids from the geothermal systems in Northern and Central America

5. The Na–HCO_3,Cl chemical type is represented by 4 reservoir liquids (1% of the total), all coming from Long Valley.

As shown by the binary diagrams of Cl versus $SO_4 + Alk_C$ (Fig. 3.26), the reservoir liquids from the geothermal systems in Northern and Central America distribute over a large range of Σ_{eq}, from the relatively low values of Na–Cl,SO_4, Na–Cl,HCO_3, and Na-HCO_3,Cl waters of Long Valley and Dixie Valley, 31–41 meq/kg (Fig. 3.26c), to the very high values of the Na,Ca–Cl hypersaline geothermal brines from Salton Sea, 6440–9080 meq/kg (Fig. 3.26a). However, most reservoir liquids, all belonging to the Na–Cl chemical type, have intermediate Σ_{eq} values, varying from 100 to 720 meq/kg (Fig. 3.26b). The distinct Σ_{eq} values of the reservoir liquids from the geothermal systems in Northern and Central America are controlled by different sources and processes.

As already recalled above, dissolution of evaporite rocks accounts for the very high Σ_{eq} values of the Na,Ca–Cl hypersaline geothermal brines from Salton Sea (Williams and McKibben 1989) whereas dilution of evaporated seawater explains the origin of the Na–Cl reservoir liquids from Cerro Prieto (Truesdell et al. 1981).

The low Σ_{eq} values of the Long Valley and Dixie Valley reservoir liquids are probably governed by meteoric recharge and rock dissolution sustained by titration of acid gases, chiefly CO_2, whose emission has experienced a considerable increase since 1989 (Farrar et al. 2003 and references therein). The reservoir liquids of Long Valley probably acquire chloride, boron and lithium through rock leaching, but only chloride has conservative behavior, whereas boron and lithium are partly incorporated in alteration minerals (White and Peterson 1991).

The intermediate Σ_{eq} values of the Na–Cl reservoir liquids from the other considered geothermal systems are probably controlled by meteoric recharge, inflow of magmatic fluids and related rock dissolution sustained by titration of acid gases. According to Gherardi et al. (2002), the reservoir liquids of Miravalles are originated through a multi-step process, consistent with the conceptual model of volcanic-hydrothermal systems situated along convergent plate boundaries proposed by

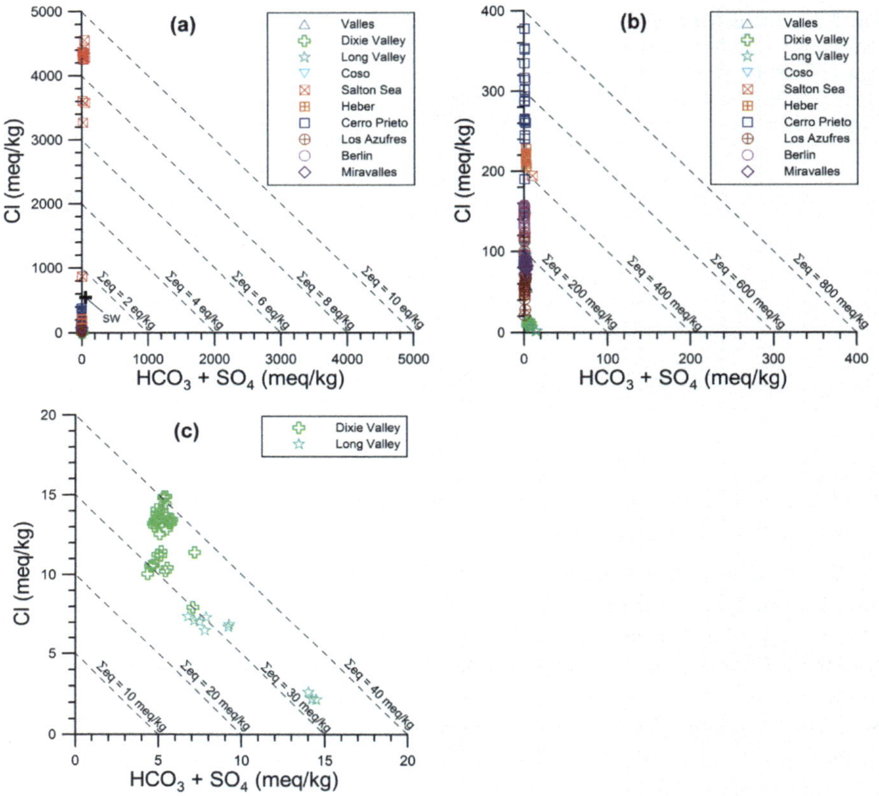

Fig. 3.26 Correlation diagrams of Cl versus SO_4 + Alk_C for the reservoir liquids from the geothermal systems of Northern and Central America. Different Σ_{eq} intervals are considered in the three diagrams **(a)**, **(b)**, and **(c)** in order to accommodate all the liquids of interest

several authors (e.g., Hedenquist 1986; Hedenquist and Lowenstein 1994; Giggenbach 1988, 1997a; Reed 1997). First, magmatic gases (mainly consisting of H_2O, CO_2, SO_2, H_2S and HCl) are absorbed into deep circulating groundwaters of meteoric origin, thus generating acid, relatively oxidized, and highly reactive fluids. Second, these fluids are partly reduced and neutralized through isochemical dissolution of wall rocks in a zone of primary neutralization. Third, further water-rock interaction converts these fluids into neutral Na–Cl waters.

3.4 The Reservoir Liquids from the Geothermal Systems in Japan

The geothermal systems in Japan considered in this work are: Mori-Nigorikawa (data from Yoshida 1991), Sumikawa (data from Ueda et al 1991; Sakai et al. 1993), Uenotai-Wasabizawa (data from Kuriyama 1985; Klein et al. 1991; Naka and Okada 1992; Takeno 2000), Onikobe (data from Truesdell and Nakanishi 2005), Oku-aizu (data from Seki 1990, 1991; Nitta et al. 1991, 1995), Takigami (data from Takenaka and Furuya 1991; Takenaka et al. 1995), Oguni (data from Sasada 1987; Yamada et al. 2000), and Fushime (data from Akaku 1990; Akaku et al. 1991; Okada et al. 2000). The accepted chemical analyses are 21 from Mori-Nigorikawa, 14 from Sumikawa, 4 from Uenotai-Wasabizawa, 7 from Onikobe, 31 from Oku-aizu, 13 from Takigami, 6 from Oguni, and 25 from Fushime, for a total of 121 entries.

3.4.1 Mori-Nigorikawa

The Nigorikawa geothermal field is located in the homonymous caldera and provides fluids to the Mori geothermal power plant (Fig. 3.27). The geochemical characteristics of Nigorikawa fluids summarized here below are from Yoshida (1991).

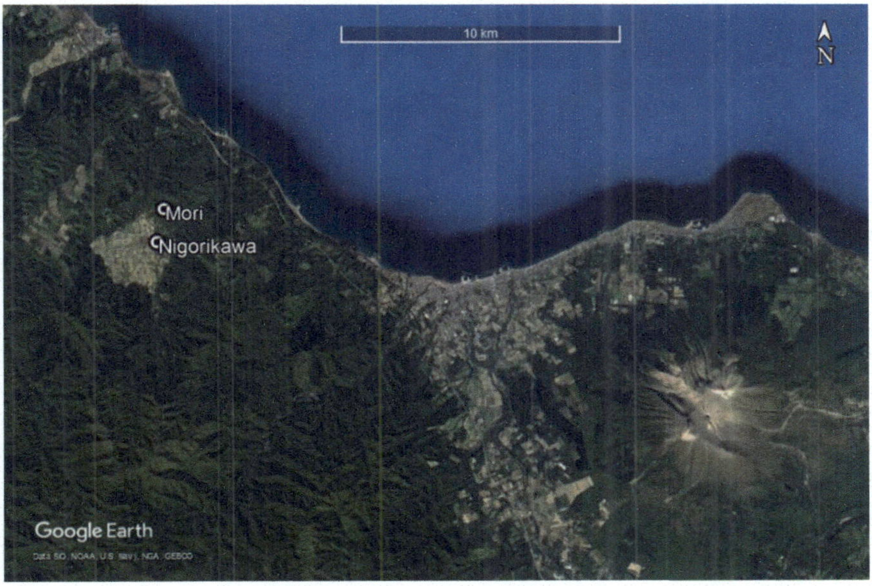

Fig. 3.27 Google Earth map showing the location of the Nigorikawa geothermal field and the Mori geothermal power plant

Deuterium and oxygen-18 data indicate that the geothermal liquids are mixtures made up of isotopically light meteoric water and isotopically heavy fluids, namely magmatic water and/or altered seawater. Relative He, Ar and N_2 contents suggest that geothermal fluids are mixtures of N_2-rich magmatic gases and air-saturated water. Enthalpy-chloride relations were initially controlled by mixing between deep reservoir liquids and Cl-free steam-heated waters, but reinjection of high-Cl brines has complicated this simple mixing pattern. Calcite scaling was experienced in the early power plant operation and was solved through injection of a scale inhibitor. Stevensite scaling was experienced in well ND-1 and related equipment, probably due to inflow of Mg-rich, low-enthalpy shallow water.

3.4.2 Sumikawa

The characteristics of the Sumikawa geothermal system reported here below are from Ueda et al. (1991) and Ariki et al. (2000). The Sumikawa geothermal field is positioned within a north–south oriented graben. To the south of the field, the graben is crossed by an east–west trending volcanic chain, comprising the Akita-Yake volcano to the west and the Hachimantai volcano to the east. At short distances from Sumikawa there are the geothermal fields of Ohnuma, Kakkonda, and Matsukawa (Fig. 3.28). The last one is the only vapor-dominated geothermal system in Japan.

Subsurface temperature increases progressively southward, that is towards the upflow zone and heat source beneath Akita-Yake, and decreases abruptly northwards, towards the outflow zone. The geothermal reservoir hosts a single liquid phase at depth, overlain by a bi-phase zone whose thickness increases southward.

The deepest Sumikawa wells met temperatures often exceeding 300 °C and neutral Na–Cl liquids. These are diluted by relatively shallow SO_4- and/or HCO_3-rich waters of enthalpy close to 850 kJ/kg heated by conductive heat transfer or steam inflow. Enthalpy-chloride relationships suggest that the Sumikawa Na–Cl parent fluid has an enthalpy of 1300 kJ/kg and chloride concentration of 280 mg/kg. Deuterium and oxygen-18 indicate that reservoir fluids are mainly of meteoric origin and have a small oxygen isotope shift, lower than 2‰ units, which might be attributed to minor contributions of magmatic fluids. This interpretation is supported by relative contents of He, Ar, and N_2 in separated vapors.

Well S-2 is unique having encountered both neutral Na–Cl liquids and acidic SO_4–Cl waters coming from two distinct productive zones at depths of 905 and 1065 m, respectively. Acidic alteration minerals, including alunite and kaolinite, are present close to the deeper feed zone. The acidic fluids are possibly of magmatic origin and are excluded from this compilation.

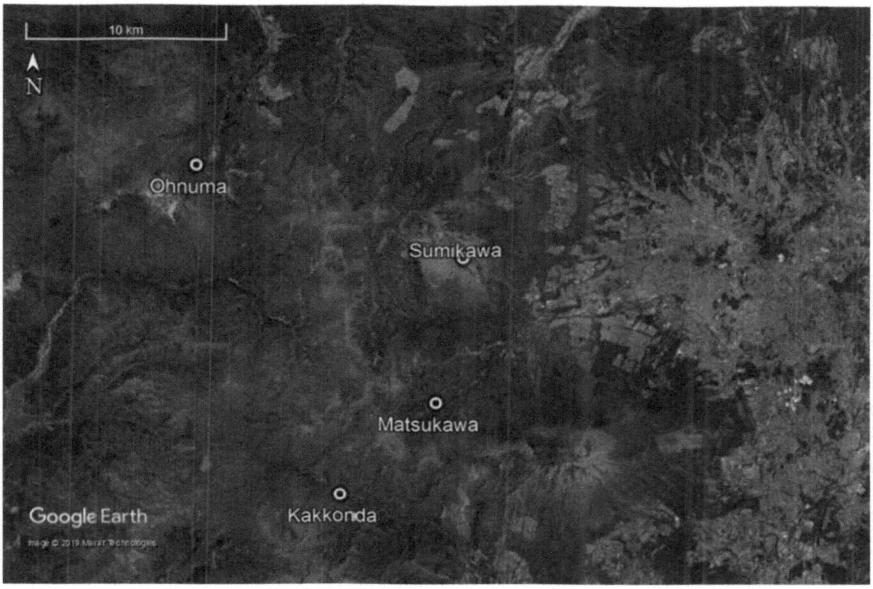

Fig. 3.28 Google Earth map showing the location of the Sumikawa geothermal field as well as the nearby geothermal fields of Ohnuma, Kakkonda, and Matsukawa

3.4.3 Uenotai-Wasabizawa

The characteristics of the Uenotai geothermal system (Fig. 3.29) presented here below are from Takeno (2000). The Uenotai geothermal system is situated 16 km west of Kurikoma volcano in an area affected by intense faulting. Uenotai was probably a single-phase liquid system in natural state, but exploitation has triggered boiling in a large, central portion of the aquifer, where several wells have excess enthalpy and some wells produce superheated steam. Measured temperatures exceed 300 °C and increase south-eastwards. Enthalpy-chloride relations suggest that reservoir liquids are derived from a deep parent liquid, with chloride concentration close to 700–750 mg/kg and temperature close to 300 °C, through vapor loss, vapor gain, and mixing with shallow vapor condensates.

Although the Wasabizawa geothermal field is situated few kilometers to the south-west of Uenotai, reservoir liquids derive from a distinct parent liquid, with chloride concentration of 1800 mg/kg and enthalpy close to 1300 kJ/kg (Suzuki et al. 2000).

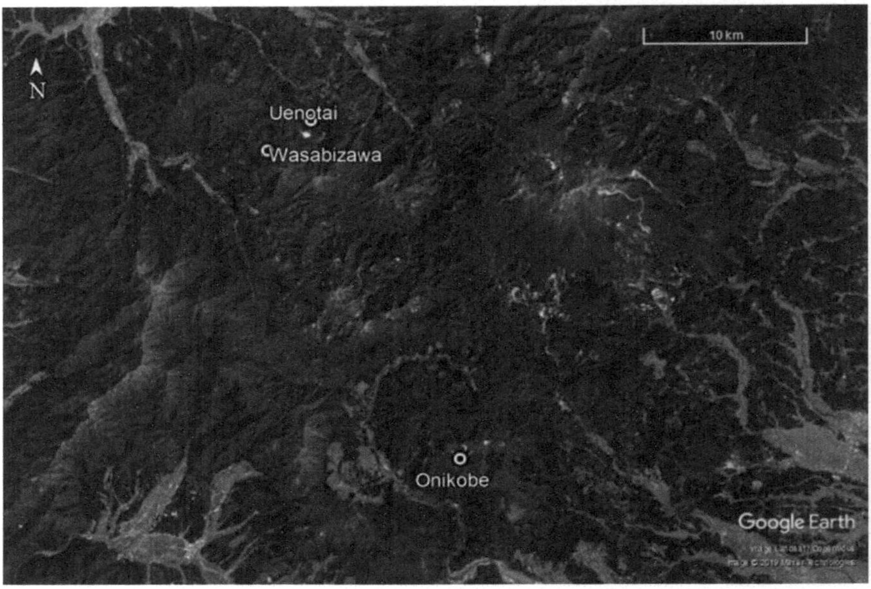

Fig. 3.29 Google Earth map showing the location of the Uenotai-Wasabizawa and Onikobe geothermal fields

3.4.4 *Onikobe*

The Onikobe geothermal field is located within the homonymous caldera, an elliptical depression with maximum axis close to 10 km, to the south of the Uenotai-Wasabizawa geothermal fields (Fig. 3.29). Seki et al. (1983) report the results obtained by the early mineralogical, petrological, and geochemical investigations performed at Onikobe on core samples collected in several boreholes drilled at depths up to 1352 m as well as temperature logs and chemical analyses of hot springs, waters and vapors discharged from the boreholes, whereas enthalpy data are not given. Unfortunately, these early data do not allow a reliable reconstruction of the chemistry of reservoir fluids.

More recently, Truesdell and Nakanishi (2005) have shown that Onikobe fluids have measured reservoir temperatures varying from 230 to 255 °C, little excess steam (with inlet vapor fraction <0.1), and chloride concentration ranging from 985 to 3920 mg/kg. The largest variations are observed for pH, which ranges from 2.8 to 8.0 for the waters collected at atmospheric pressure. The main acidity source is HCl, but H_2SO_4 could act as subordinate acidity source. Part of the acidity might have been suppressed through dissolution of reservoir rocks and well casings, leading to acquisition of Fe (0.01–371 mg/kg), Mg (0.39–58.3 mg/kg), and Ca (8.4–721 mg/kg). Large variations are observed also for volatile species with pH-dependent volatility, such as H_3BO_3 (from 3.2 to 77.2 mg/kg as B). Re-injected waters enriched in dissolved solids and depleted in gases are contributing to production fluids. In this work, only the samples not impacted by acidic fluids and related processes were selected, but some Onikobe reservoir liquids might not be representative of the mineral-solution equilibrium condition at aquifer temperatures.

Both deuterium and oxygen-18 values of separated brines discharged from Onikobe wells are inversely correlated with pH, suggesting that the enrichment in heavy isotopes is related to the proportion of unneutralized magmatic water (Pang 2006).

3.4.5 Oku-Aizu

The characteristics of the Oku-aizu geothermal system (Fig. 3.30) reported here below are chiefly from Seki (1991). The Oku-aizu field is located in the "Green Tuff region", which was affected by submarine volcanic activity during the Neogene. The nearest Quaternary volcano, Numazawa, is located 10 km to the west. There, a Plinian eruption occurred about 4600 years BP emitting the Numazawako pumice flow and causing the formation of the 1.5 km × 2 km Numazawako caldera, now largely filled by a lake.

The main feed zones of production wells are at depths of 1000–2600 m, where measured formation temperatures vary from 200 to 350 °C (Nitta et al. 1995). The

Fig. 3.30 Google Earth map showing the location of the Oku-aizu geothermal field

Oku-aizu geothermal aquifer is considered to have experienced little boiling in its natural state before the beginning of production. However, due to the pressure drop induced by exploitation, reservoir boiling occurred close to the geothermal wells. Therefore, most wells show excess enthalpy and their discharge enthalpy increases progressively with time, suggesting extension of the boiling front around the wells into the reservoir. As a consequence of steam separation in the aquifer, the total discharge composition is rarely representative of the reservoir liquid chemistry prior to boiling (Seki 1990).

The steam separated at 6.5 bar contains 2–10 vol.% of non-condensable gases, mostly CO_2 (95–99 vol.%) and H_2S (0.1–4 vol.%). The brine separated at atmospheric pressure has the second highest salinities of the geothermal systems in Japan after Fushime, with chloride concentrations of 5300 to 22,000 mg/kg, sodium concentrations of 2700–10,600 mg/kg, and lower but significant contents of potassium (180–2550 mg/kg), calcium (26–1750 mg/kg), and silica (440–1010 mg/kg). A peculiarity of the Oku-aizu reservoir liquids is the high manganese concentrations, up to 420 mg/kg, and the appreciable contents of Cu, Zn, and Pb. Consequently, sulfide scales were found on the casing wall of well 84N-2t, and in the two-phase line of well 87N-15T (Seki 1991; Nitta et al. 1991).

According to the calculations performed by Seki using the chemical speciation code PECS, reservoir liquids are expected to have pH ranging from 4.1 to 5.5, in substantial agreement with the results obtained in this work using WATCH.

Enthalpy-chloride relations indicate that the deep parent fluid (1) has temperature close to 320 °C and chloride concentration of 11,500 mg/kg and (2) mixes with a shallower water with nil to negligible chloride content and temperature less than 200 °C, heated by conductive heat transfer and/or steam inflow.

Deuterium versus oxygen-18 relationships suggest that the Oku-aizu reservoir liquids are produced by mixing of meteoric waters with either arc-type magmatic water or fossil seawater.

3.4.6 Takigami

This brief synthesis on Takigami is based on Takenaka and Furuya (1991) and Furuya et al. (2000). The Takigami geothermal field is located northeast of the Hatchobaru geothermal field at a distance of about 15 km (Fig. 3.31). The Takigami geothermal reservoir comprises two distinct portions. The south-western portion is deeper, with elevations of −1500 to −600 m asl, lower permeability and higher temperatures, from 230 to 260 °C, whereas the north-eastern portion is shallower, with elevations of −600 to 0 m asl, has high secondary permeability and temperatures of 160–210 °C. The reservoir liquids of the south-western reservoir have relatively high chloride concentrations and comparatively low sulfate concentrations. In contrast, the reservoir liquids of the north-eastern reservoir have relatively low chloride concentrations and comparatively high sulfate concentrations. These findings indicate the upflow of a hot, chloride-rich, sulfate-poor fluid in the southwest, which is gradually cooled

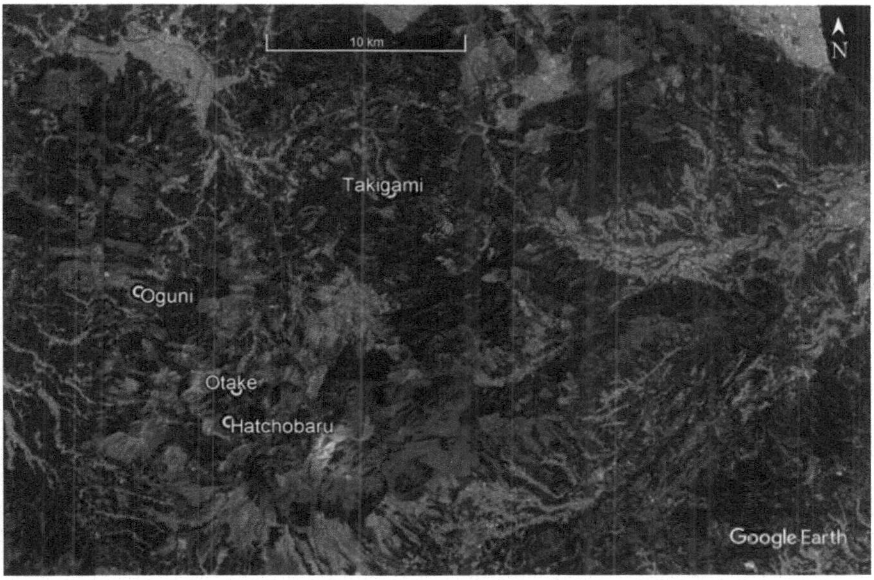

Fig. 3.31 Google Earth map showing the location of the Takigami and Oguni geothermal fields. Also shown are the nearby geothermal fields of Otake and Hatchobaru

as it flows northeast through mixing with shallow chloride-poor, sulfate-rich waters of temperature close to 100 °C. The mixing trend is clearly recognizable in the enthalpy-chloride diagram, which leads to exclude occurrence of boiling, and in the chloride plot of boron, owing to the conservative behaviour of both Cl and B. In contrast, considerable deviations from the mixing trend are observed in the chloride plots of other chemical components, which are evidently controlled by water-rock interaction and possible attainment of equilibrium between the aqueous solution and hydrothermal minerals, including anhydrite and calcite (Chiba 1991). Deuterium and oxygen-18 indicate that the Takigami geothermal fluids derive from meteoric waters.

3.4.7 *Oguni*

The main features of the Oguni geothermal system (Fig. 3.31) reported here below are from Abe et al. (1995). Oguni is located in the Hohi region, together with the geothermal fields of Takigami (see above), Hatchobaru (supplying steam to the main geothermal power plant in Japan), and Otake. The central part of the Oguni geothermal field is crossed by the northwest to southeast striking Takenoyu fault and related parallel faults forming altogether a nearly vertical high-permeability zone of several hundred meters in width. The wells drilled in this fault zone (e.g., GH-10, GH-11, GH-12, and GH-20) encountered the hottest, undiluted reservoir fluids with temperatures of 240–250 °C and chloride concentrations of 1050–1150 mg/kg. Wells drilled elsewhere (e.g., wells BS-3 and DY-1) met diluted, cooler reservoir fluids.

The Oguni geothermal reservoir is composed by two different parts, separated by a nearly vertical impermeable zone situated to the south of the Takenoyu fault. Both reservoirs hosts relatively similar neutral Na–Cl fluids with low concentrations of non-condensable gases. Carbonate scaling occurs in some production wells.

3.4.8 Fushime

The brief synthesis given here below on the Fushime geothermal field (Fig. 3.32) is based on Akaku et al. (1991). The geothermal system comprises: (1) a shallow hot zone in the upper 800 m, with temperatures up to 240 °C and (2) a deep hot zone, at depths greater than 1200 m, with temperatures between 300 and 360 °C. These two hot zones are separated by a cooler zone, marked by strong inversions in temperature profiles, at intermediate depths.

Since the temperatures in the deep production zone lie close to the boiling point curve, discharge-induced depressurization triggers boiling in the deep production zone, with development of excess enthalpy conditions in most Fushime wells.

The reservoir fluids have high salinity and maximum Cl concentration similar to the seawater value. Moreover, they are depleted in Mg and SO_4, but are enriched in

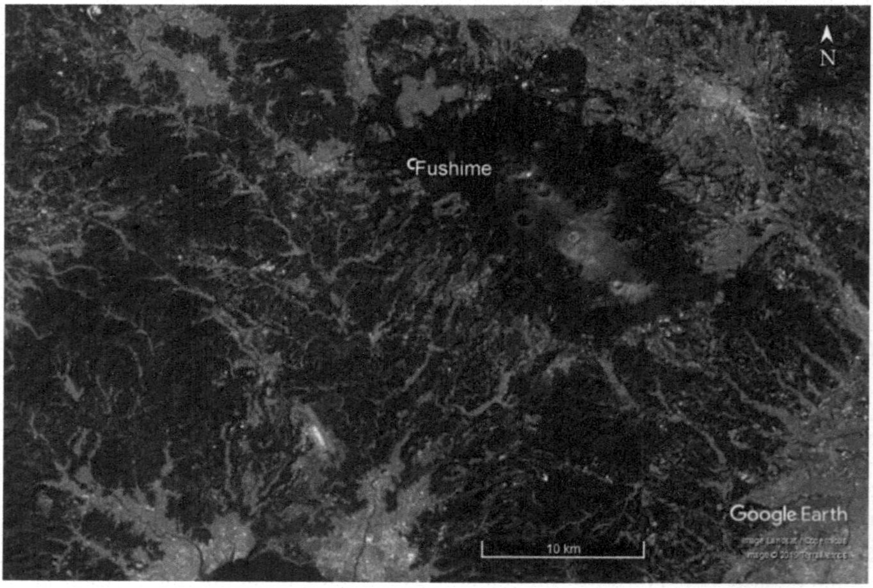

Fig. 3.32 Google Earth map showing the location of the Fushime geothermal field

several chemical components (e.g., SiO_2, K, Ca, Fe, Mn, Zn, and Pb) with respect to seawater, indicating that the reservoir fluids are generated through high-temperature seawater-rock interaction.

Chloride-enthalpy relationships suggest the occurrence of steam loss, steam gain, and dilution of reservoir liquids. The latter process probably involves the cold groundwaters overlying the geothermal system. The parent water has temperature of 330 °C and chloride concentration of 17,000 mg/kg.

The variations in K and Ca concentrations of reservoir fluids suggest occurrence of both (1) precipitation of K-bearing hydrothermal minerals, most likely K-feldspar and sericite, which are abundant around the feed zones of the Fushime wells, and (2) dissolution of at least one Ca-bearing mineral, probably anhydrite, which is also present close to the feed zones of the Fushime wells.

3.4.9 *Chemistry of the Reservoir Liquids from the Geothermal Systems in Japan*

The majority of the reservoir liquids from the geothermal systems in Japan (that is 104 of the 121 entries) are situated in the [Na] sector of the triangular diagram of major cations (Fig. 3.33). However, there are 17 exceptions, comprising 15 reservoir liquids from Oku-aizu, Onikobe, and Fushime which are located in the [Na,Ca] sector and 2 reservoir liquids from Oku-aizu which are positioned in the [Na,K] sector. Therefore, sodium is the prevailing cation in all these aqueous solutions, although calcium and/or potassium are significant in some cases.

In the triangular diagram of main anions (Fig. 3.34) most reservoir liquids are located in the [Cl] sector, and several are very close to the chloride vertex. Only some reservoir liquids of Takigami and Sumikawa are situated in the [Cl,SO_4] sector.

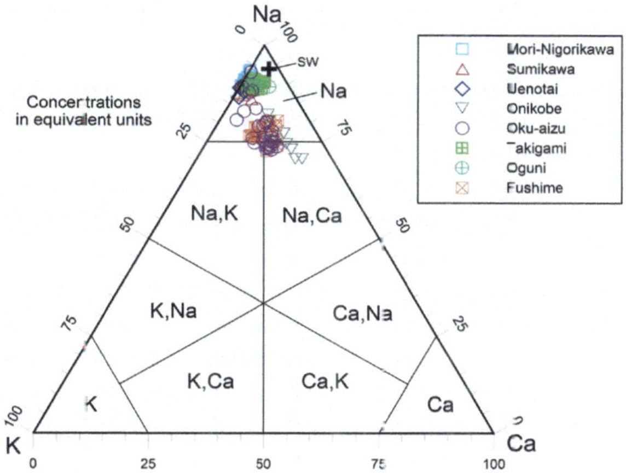

Fig. 3.33 Triangular plot of major cations for the reservoir liquids from the geothermal systems in Japan. Average seawater is also shown for comparison

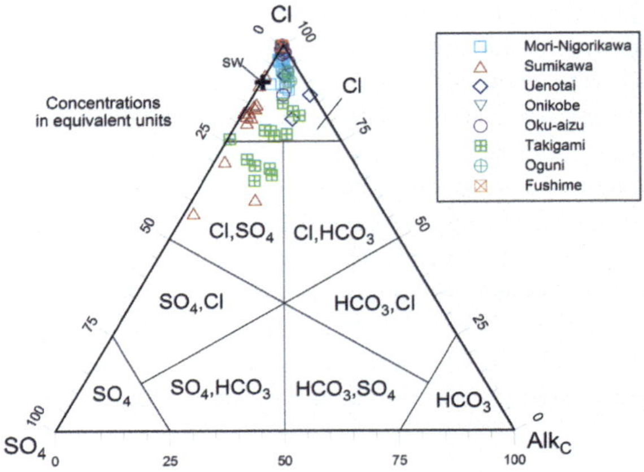

Fig. 3.34 Triangular plot of major anions for the reservoir liquids from the geothermal systems in Japan. Average seawater is also shown for comparison

Based on both triangular diagrams, it is possible to recognize four different chemical facies. In detail:

1. The Na–Cl chemical facies is by far the most common, comprising 96 of the 121 reservoir liquids of interest, that is 79% of the entries. Moreover, the Na–Cl reservoir liquids are found in all the geothermal systems in Japan.
2. The Na,Ca–Cl chemical type includes 15 reservoir liquids (12% of the cases), proceeding from Oku-aizu, Onikobe, and Fushime.
3. The Na,K–Cl chemical type is represented by 2 reservoir liquids only (2% of the total), both from Oku-aizu.
4. The Na–Cl,SO_4 chemical type groups 8 reservoir liquids (7% of the cases), coming from Takigami and Sumikawa.

The reservoir liquids from the geothermal systems in Japan span a large interval of Σ_{eq} (Fig. 3.35). Low to relatively low Σ_{eq} values are encountered at Sumikawa, 12–43 meq/kg, Takigami, 36–41 meq/kg, Oguni, 52–67 meq/kg, and Uenotai, 12–29 meq/kg, apart from the Wasabizawa well N57-T0-3 which has Σ_{eq} of 95 meq/kg. The reservoir liquids of Fushime exhibit the highest Σ_{eq} values, varying from 612 to 1190 meq/kg, whereas intermediate Σ_{eq} values are found in the geothermal systems of Oku-aizu (118–789 meq/kg), Mori-Nigorikawa (243–509 meq/kg), and Onikobe (74–166 meq/kg).

The distinct Σ_{eq} values of the reservoir liquids from the geothermal systems in Japan are governed by different processes and sources. As already noted above, the highest Σ_{eq} values of Fushime are similar to that of seawater (Fig. 3.35a), which recharges the deep reservoir of this geothermal system. In contrast, the low Σ_{eq} values of the reservoir liquids of Sumikawa, Takigami, Oguni, and Uenotai (Fig. 3.35b) are probably controlled by the prevailingly meteoric recharge and by the occurrence of limited rock dissolution processes sustained by titration of acid gases.

The intermediate Σ_{eq} values of the reservoir fluids of Oku-aizu and Mori-Nigorikawa (Fig. 3.35a) are ascribable to a mixed recharge, with significant contributions of meteoric water as well as of Cl-rich magmatic waters and/or fossil/altered

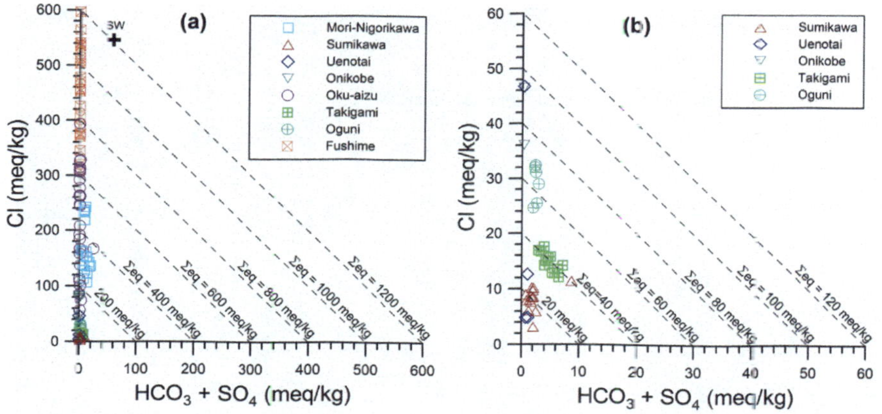

Fig. 3.35 Correlation diagrams of Cl versus SO_4 + Alk_C for the reservoir liquids from the geothermal systems of Japan. Different Σ_{eq} intervals are considered in the two diagrams **(a)** and **(b)** to accommodate all the liquids of interest

seawater, as indicated by deuterium and oxygen-18 data (see above). In both sites, a further supply of solutes is provided by rock dissolution sustained by titration of acid gases. The Onikobe geothermal reservoir is recharged by both meteoric waters and magmatic fluids. The acidity of these meteoric-magmatic fluids is partly titrated through dissolution of reservoir rocks and well casings (see above).

3.5 The Reservoir Liquids from the Geothermal Systems in the Philippines

The geothermal systems in the Philippines taken into account in this work are: Bacon-Manito (data from Ruaya et al. 1995; See 1995), Tongonan-Mahanagdong (data from Baltasar 1980; Balmes 1994; Angcoy 2010), Alto Peak (data from Reyes et al. 1993), and Palinpinon (data from Jordan 1982; Rae 2002). The accepted chemical analyses are 51 from Bacon-Manito, 39 from Tongonan-Mahanagdong, 27 from Alto Peak, and 29 from Palinpinon, for a total of 146 entries.

3.5.1 Bacon-Manito

The Bacon-Manito (Bacman) geothermal field is situated in the southwestern portion of the Luzon Island, along the volcanic chain of the Bicol Arc, between the two active volcanoes Bulusan, which is found 30 km to the SSE, and Mayon, which is located 30 km to the NW (Fig. 3.36).

Fig. 3.36 Google Earth map showing the location of the Bacon-Manito geothermal field

The deepest and hottest reservoir fluid has a Cl concentration of 8000 ± 500 mg/kg and a temperature of 325 °C (Ruaya et al. 1995). It is probably produced through mixing between Cl-poor meteoric waters and Cl-rich (15,000–18,000 mg/kg Cl) arc-type (andesitic) magmatic waters, with contributions of 50–60% and 40–50%, respectively. Steam loss, steam gain, and mixing with both steam condensates and shallow groundwaters cause large changes in Cl concentration, which spans the 1500–8700 mg/kg range. Sulfate concentration varies from 5 to 35 mg/kg in most neutral chloride liquids but attains 107 mg/kg in one sample of well Pal-2D. Sulfate concentration is higher in the acid fluid discharges of wells Pal-2D, CN-2D, and CN3RD, which are disregarded in this work, being not representative of the equilibrium condition with the typical hydrothermal minerals of the neutral alteration suite (sensu Reyes 1990).

3.5.2 Tongonan-Mahanagdong

The Greater Tongonan geothermal field is positioned on Leyte Island, along a Tertiary north-west/south-east trending volcanic arc related to the Philippine fault zone. The Greater Tongonan geothermal field comprises the Tongonan geothermal field in the north and the Mahanagdong geothermal field in the south (Fig. 3.37). The two geothermal fields are probably separated by a cold impermeable block

Fig. 3.37 Google Earth map showing the location of the Tongonan-Mahanagdong and Alto Peak geothermal fields

(Angcoy 2010). Many wells have excess enthalpy due to contribution of steam from a two-phase zone situated above the deep single-phase liquid zone.

In the Tongonan geothermal field, wells discharge neutral Na–Cl waters with reservoir chloride concentrations usually ranging from 6200 to 11,000 mg/kg, whereas the temperatures measured in the main production zones before discharge vary from 280 to 314 °C (Baltasar 1980).

The reservoir liquids of the Mahanagdong geothermal field are originated from a parent fluid with quartz temperature of ~300 °C, chloride concentration of ~4000 mg/kg, and $\delta^{18}O$ of $-1.0‰$, representing the isotopically richest value (Angcoy 2010). Close to the main outflow zone in the southeast, reservoir liquids have quartz temperature of 260–280 °C, chloride concentration of 2000–3000 mg/kg and $\delta^{18}O$ values of -2.0 to $-3.0‰$. The northern part of the reservoir hosts acid Cl–SO_4 waters with chloride concentrations and $\delta^{18}O$ values similar to the wells in the upflow zone, but with high concentrations of sulfate (>100 mg/kg), iron (>10 mg/kg) and magnesium (>5 mg/kg).

The geothermal fluids of both Tongonan and Mahanagdong are produced through mixing of local meteoric waters and arc-type magmatic water (Alvis-Isidro et al. 1993).

3.5.3 Alto Peak

The Alto Peak geothermal field is situated on Leyte Island to the south-east of Tongonan-Mahanagdong (Fig. 3.37). On the basis of the evidence provided by hydrothermal mineral petrology and the chemical characteristics of both fluid inclusions and fluids discharged from five deep wells, Reyes et al. (1993) postulated that the Alto Peak geothermal system comprises vapor-dominated and liquid-dominated sectors.

A gas-rich vapor phase, with CO_2 concentrations of 1.1–5.6 mol/kg, occupies the central core or chimney with a diameter of ~1 km and a height of ~3 km, connecting more extensive, very high temperature vapor-dominated zones at depth to a shallow zone occupied by steam-heated ground-waters. The central core is surrounded by a geothermal aquifer hosting neutral Na–Cl waters, with chloride concentration close to 7000 mg/kg and temperatures between 250 and 350 °C. Two-phase zones, increasingly wider with increasing depth and hosting fluid mixtures of highly variable compositions, are interposed between the vapor-dominated central chimney and the surrounding liquid-dominated zone.

Based on this distribution of fluids within the Alto Peak geothermal system, the changes with time in the composition of the well discharges is ascribable to the entrainment of variable amounts of fluids hosted in the liquid- and vapor-dominated zones crossed by the wells.

Most neutral Na–Cl waters of relatively low temperature were probably generated during an earlier phase in the evolution of the Alto Peak geothermal system, whereas the fluids at temperatures >300 °C and the hydrothermal, magmatic-hydrothermal, and contact-metamorphism alteration minerals encountered in wells AP-1D and AP-2D may be explained in terms of re-heating caused by recent emplacement of magma batches at shallow depths. The deuterium and oxygen-18 of well discharges suggest that they are mixtures made up of 40 to 50% of arc-type magmatic water and 60–50% of local groundwaters.

3.5.4 Palinpinon

The Palinpinon liquid-dominated, high-temperature geothermal system is located on Negros Island (Rae 2002; Rae et al. 2011 and references therein). The geothermal field is producing from two distinct sectors: one is called Puhagan, is situated in the central part of the field, and corresponds to the upflow zone; the other sector is known as Nasuji-Sogongon and is positioned in the south-west, where one of the two outflow zones is found (Fig. 3.38). The other outflow zone occurs to the north-east. The Palinpinon production wells cross multiple zones of permeability and discharge two-phase fluids. Measured down-hole temperatures vary from 220 to over 300 °C, with the highest temperature, 329 °C, measured at 1925 m below sea level in well PN20D. Most Palinpinon wells discharge typical neutral to slightly

Fig. 3.38 Google Earth map showing the location of the Palinpinon geothermal field. The place cards P, S, and N identify the Puhagan, Sogongon, and Nasuji sectors, respectively

alkaline chloride liquids with Cl < 7200 mg/kg and substantial concentrations of silica, sodium, potassium, and boron. These chloride liquids are produced from a parent reservoir liquid with Cl close to 4000 mg/kg and temperature of 328 °C through variable combinations of mixing, boiling, and/or conductive cooling. Mixed acid sulfate-chloride waters are discharged by some wells drilled to the south in the Nasuji-Sogongon production area, but are not considered in this work.

3.5.5 Chemistry of the Reservoir Liquids from the Geothermal Systems in the Philippines

Sodium is the prevailing cation in all the reservoir liquids from the geothermal systems in the Philippines, being all located in the [Na] sector of the triangular plot of major cations (Fig. 3.39).

In the triangular diagram of main anions (Fig. 3.40) all the reservoir liquids of interest except one are situated in the [Cl] sector, indicating that chloride is the dominant anion in all the samples apart from that collected in 2009 from well MG32D of Mahanagdong and identified by code MG32D_09. This sample is unique, being located in the [SO_4,Cl] sector.

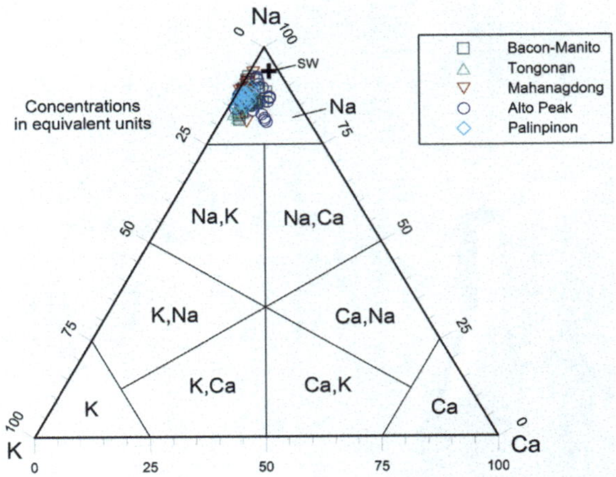

Fig. 3.39 Triangular plot of major cations for the reservoir liquids from the geothermal systems in the Philippines. Average seawater is also shown for comparison

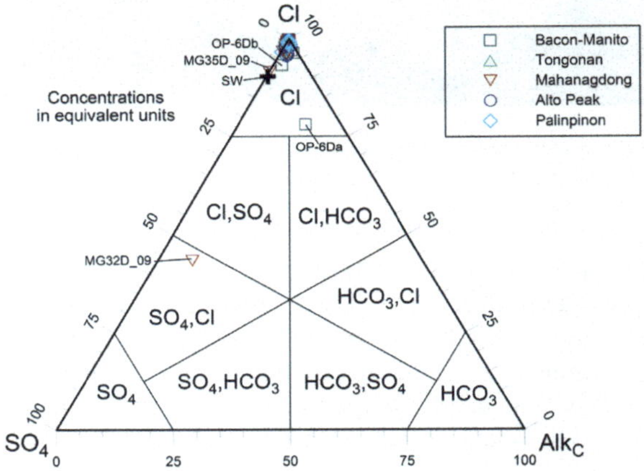

Fig. 3.40 Triangular plot of major anions for the reservoir liquids from the geothermal systems in the Philippines. Average seawater is also shown for comparison

Moreover, all the Cl-rich reservoir liquids are positioned close to the chloride vertex, with the sole exception of the sample collected on 14/02/1991 from well OP-6D of Bacon-Manito (code OP-6 Da), which has unusually high relative contents of HCO_3, 14.4 eq%, and SO_4, 7.5 eq%, as well as an abnormally low relative content of Cl, 78.1 eq%, compared to the other Cl-rich reservoir liquids.

On the basis of the evidence provided by both triangular diagrams, all the reservoir liquids of interest belong to the Na–Cl chemical facies, apart from sample MG32D_09 which has Na–SO_4,Cl composition.

The reservoir liquids from the geothermal systems in the Philippines span a relatively wide interval of Σ_{eq}, from the 25 meq/kg of sample OP-6 Da of Bacon-Manito to the 600–620 meq/kg of the two samples collected from well 407 of Tongonan

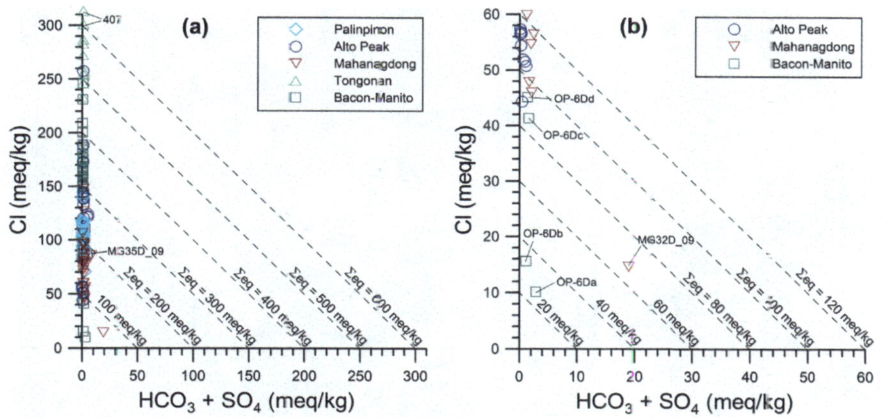

Fig. 3.41 Correlation diagrams of Cl versus $SO_4 + Alk_C$ for the reservoir liquids from the geothermal systems in the Philippines. Different Σ_{eq} intervals are considered in the two diagrams **(a)** and **(b)** in order to accommodate all the liquids of interest

(Fig. 3.41). In addition to sample OP-6 Da, also the other three samples from well OP-6D have low Σ_{eq} values, namely 33, 84, and 95 meq/kg. Considering the total discharge enthalpy of well OP6D, ~2450 kJ/kg, the low salinities of these four samples may be due to improper liquid/vapor separation during sampling, with entrainment of vapor in the liquid phase. Similar considerations apply to sample MG32D_09 which has Σ_{eq} of 69 meq/kg and discharge enthalpy of ~2400 kJ/kg. Even if these samples are potentially not representative of reservoir liquids, they are kept in the following discussion to ascertain whether the geochemical techniques of interest are applicable or not.

Apart from these five unusual aqueous solutions, the Σ_{eq} values in the range 100–600 meq/kg of most reservoir fluids from the geothermal systems in the Philippines can be explained by mixing, in variable proportions, of Cl-poor meteoric waters and Cl-rich magmatic waters, as indicated by deuterium and oxygen-18 data (see above). Moreover, a further contribution of solutes is provided by rock dissolution sustained by titration of acid gases.

3.6 The Reservoir Liquids from the Geothermal Systems in New Zealand

All the high-temperature geothermal systems in New Zealand are located in the Taupo Volcanic Zone, TVZ (Fig. 3.42), apart from Ngawha which is situated in the Northland Peninsula (Fig. 3.43).

The geothermal systems of the TVZ were divided into arc- and rift-types by Giggenbach (1995), based on both the chemical and isotopic characteristics of gases

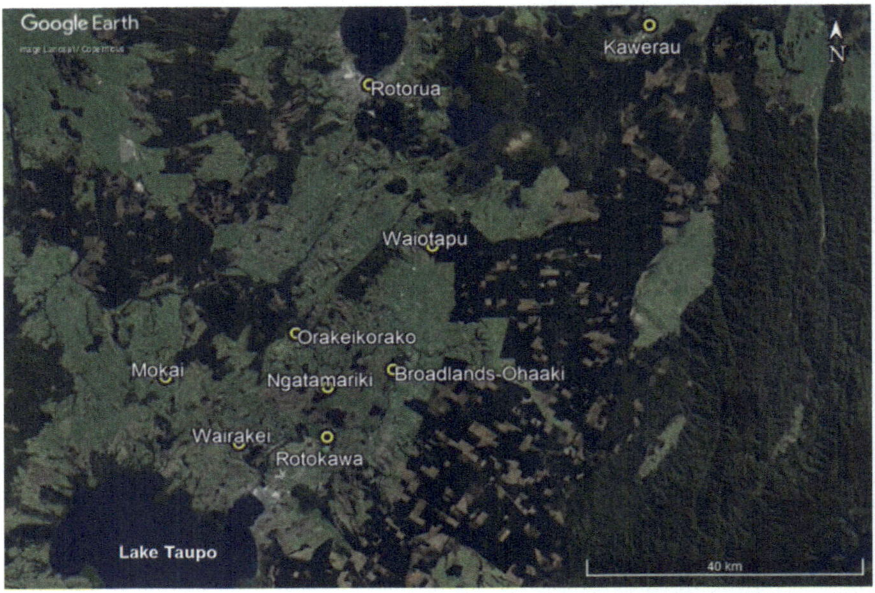

Fig. 3.42 Google Earth map showing the location of the geothermal fields of Wairakei, Rotokawa, Mokai, Ngatamariki, Broadlands-Ohaaki, Orakeikorako, Waiotapu, Rotorua, and Kawerau situated to the north-east of Lake Taupo

Fig. 3.43 Google Earth map showing the location of the Ngawha geothermal field

and waters discharged from hot springs and geothermal wells, namely the relative Cl, B, Li and Cs contents, the $CO_2/^3He$ and N_2/Ar ratios, and the δ^2H and $\delta^{18}O$ values of H_2O. The reservoir fluids of the systems along the Eastern side of the TVZ, such as Kawerau, Broadlands-Ohaaki and Rotokawa, are affected by addition of volatiles from subducted sediments and are richer in magmatic components than the fluids from the systems along the Western side of the TVZ, such as Wairakei and Mokai, which are similar to mantle-derived fluids. To corroborate his analysis, Giggenbach (1995) provided a number of data on the vapor and liquid phases discharged from the geothermal wells of Broadlands-Ohaaki, Kawerau, Mokai, Ngatamariki, Rotokawa, and Wairakei. These Giggenbach's data are considered in this work together with those produced by Mahon and Finlayson (1972), Hedenquist (1990) and Christenson et al. (2002) for Broadlands-Ohaaki, Chambefort et al. (2016) for Ngatamariki, Sheppard and Giggenbach (1980) and Sheppard (1984) for Ngawha, Sheppard and Lyon (1984) for Orakeikorako, Krupp and Seward (1987) and Reyes et al. (2003) for Rotokawa, Hedenquist and Browne (1989) for Waiotapu, and Truesdell and Singers (1974), Bruton (1995), Stefansson and Arnórsson (2000), and Clearwater et al. (2015) for Wairakei. The accepted chemical analyses are 41 from Broadlands-Ohaaki, 4 from Kawerau, 4 from Mokai, 7 from Ngatamariki, 12 from Ngawha, 1 from Orakeikorako, 9 from Rotokawa, 6 from Waiotapu, and 9 from Wairakei, for a total of 93 entries.

3.6.1 Broadlands-Ohaaki

This brief synthesis on the Broadlands-Ohaaki geothermal system (location in Fig. 3.42) is based on Hedenquist (1990) and Christenson et al. (2002). The geological framework of the Broadlands-Ohaaki field comprises a thick sequence of volcanic rocks resting unconformably onto the greywackes and argillites of the Mesozoic basement. Faulting affects the basement rocks causing local discontinuities and large changes in the elevation of the basement top but does not extend significantly into the overlying volcanic rocks which show, in general, good lateral continuity. Fluid circulation appears to be controlled by fractures and faults in the basement, whereas it occurs through inherently permeable rocks and formation contacts in the volcanic sequence. The geothermal system is characterized by the presence of two distinct upflow zones of reservoir fluids, both controlled by fault structures cutting the greywackes of the Mesozoic basement.

Both upflow zones have comparable salinity, but the fluids of the eastern upflow zone are enriched in B and F compared to those of the western upflow zone. The deep reservoir fluids originate from a parent (preboiled) fluid with a temperature of ~300 °C and a CO_2 content of ~2.6% by weight. Boiling is common in both upflow zones as indicated by the chemical characteristics of reservoir fluids and measured temperatures. CO_2-rich steam-heated waters are present above the boiling zones and on the margins of the system, where they dilute the deep chloride fluids. The main thermal manifestation is the Ohaaki pool, discharging neutral chloride water at 95 °C and contributing about 10 MWt to the total natural heat flow from the system, which is estimated at ~100 MWt. Surface manifestations comprise a few warm springs on the banks of the Waikato River and some steaming grounds, with subsurface seepages also contributing to the natural heat flow. Exploration wells at depths varying from 366 to 2587 m were drilled in 1965 -1984. Temperatures of 260 ± 20 °C were measured in the feed zones of production wells.

3.6.2 Kawerau

The geological framework of the Kawerau field (location in Fig. 3.42) comprises a >1 km thick sequence of volcanic products resting unconformably onto the Mesozoic greywacke complex (Christenson 1997 and references therein). Permeability is fault- and fracture-controlled within both the greywackes and the overlying volcanics. Lacustrine and marine sediments act as discontinuous aquicludes in the volcanic sequence. Reservoir fluids flow up through steeply dipping normal faults from the greywackes into the overlying volcanic rocks, where they spread laterally entering distinct productive aquifer levels.

The chemical characteristics of the reservoir fluids circulating in both the greywackes and the volcanic rocks are affected by both boiling and dilution. Pressure draw-down and inflow of cold fluids from the marginal recharge have been induced by production from volcanic rocks whereas production from fractured greywackes has been generally stable with time. Boiling prompted by production has caused calcite scaling in boreholes and locally in reservoir rocks. The surface manifestations include steaming grounds and neutral to acid pools, springs and seeps. Reservoir fluids are close to boiling and attains temperatures >315 °C. Shallow reservoir fluids are affected by steam heating as indicated by the high concentrations of HCO_3 and/or SO_4 and slightly to strongly acidic pH, whereas the deep reservoir fluids are of the near-neutral chloride-type and gas-rich. The base fluid has estimated Cl concentration of 925 mg/kg and gas content of 2.8% by weight.

3.6.3 Mokai

This brief account on Mokai geothermal field (location in Fig. 3.42) is based on Henley and Middendorf (1985), Henley and Plum (1985), and Hedenquist et al. (1990). Six exploration wells were drilled in the Mokai field at depths of 600–2800 m. Different volcanic units (from top to bottom: shallow ignimbrites, rhyolitic domes, deep ignimbrites and tuffs) were crossed by the wells but the Mesozoic greywacke basement was not attained. Measured maximum temperatures are in the interval 194–323 °C. The first well, MK1, was drilled to a depth of 606 m and found a maximum temperature of 194 °C above a temperature inversion. Temperatures in the range 250–278 °C were measured in the main circulation loss zone of well MK2 (total depth 1654 m) before and during discharge. High temperatures were also found in wells MK3 (total depth 1678 m) and MK5 (total depth 2592 m) up to 305 and 298 °C, respectively. Down-hole measurements indicate considerable cooling of the system close to well MK4 probably due to inflow of cooler waters. The natural heat discharge through the steam-heated thermal manifestations located within the drilled area in the Mokai field (one large fumarole, steaming grounds, acid-sulfate mud pools, and a few neutral-pH steam-heated pools) is very small, with 6 MWt only. However, a natural thermal flux of at least 80 MWt is related to the dilute chloride springs, with temperatures up to 67 °C, which are distributed in a fault-controlled gorge, 6–10 km NNE of the drilled area in the Mokai field. Hence, the upflow zone of the Mokai geothermal system is located to the south, but the major component of the surface discharge of thermal fluids is displaced to the north, largely because of the northerly hydraulic gradient of >300 m over a distance of 12 km. During this northwards travel, the deep-upflow chloride fluids are progressively diluted by addition of cold groundwater. The deep undiluted parent fluid has estimated temperature of 335 °C, chloride concentration of 2050 mg/kg, and low gas content, similar to that of Wairakei fluids.

3.6.4 Ngatamariki

The current knowledge and understanding of the Ngatamariki geothermal field (location in Fig. 3.42) has been presented by Chambefort et al. (2016). At Ngatamariki the greywackes of the Mesozoic basement have been reached only by well NM6 at a depth of ~3.4 km. They are covered by a thick sequence of volcanic and sedimentary rocks which have been attributed to different stratigraphic units. An intrusive complex, comprising a microdiorite, a quartz-diorite and a tonalite, has been encountered in wells NM4, NM8, and NM9 at depths >2.4 km approximately. The Ngatamariki geothermal system comprises a deep geothermal reservoir at temperatures of 260–285 °C covered by a clay-cap which is overlain by two separate shallow aquifers.

The fluid circulating in the Ngatamariki reservoir has Na–Cl composition, Cl concentration of 900–1000 mg/kg, near neutral pH, ~6, and low gas content, ~0.3% by weight. The reservoir fluid is chiefly made up of meteoric water with a limited contribution of magmatic water. Most natural manifestations discharge mixed chloride-bicarbonate waters with low SO_4 and slightly alkaline pH. However, 10–15 L/s of chloride waters are discharged by the most important hot spring, the South One pool, which has a surface area of ~1000 m^2. This spring is hosted into the vent of the hydrothermal eruption which occurred in 2005. Sinter is deposited by the hot spring waters in several sites. The natural background thermal flux is estimated at 40 MWt.

3.6.5 *Ngawha*

The Ngawha geothermal field is situated in the Northland Peninsula (Fig. 3.43). The geothermal reservoir is hosted in quartz-feldspathic greywackes and argillites, which have low inherent permeability and porosity close to zero (Cox and Browne 1998). Locally, faults and joints enhance rock permeability creating the conditions needed for the circulation of reservoir fluids which move within a myriad of joint channels interconnected in three dimensions (Browne 1980). The cap-rock overlying the geothermal reservoir is represented by a 500–600 m thick sequence of low-permeability marine sediments.

The Ngawha reservoir fluids are of neutral Na–Cl type but have unusually high concentrations of HCO_3, B, and NH_4 as well as high gas contents, of which 95% is CO_2. Boron isotope data suggest that high B concentrations are the result of rock leaching under low water/rock ratios in the capped hydrothermal system (Aggarwal et al. 2003).

The large oxygen isotope shift of Ngawha geothermal fluids, 11‰ units, was attributed to isotope exchange with the local rocks assuming either a low water/rock ratio or a very young age for the system, but without any involvement of magmatic waters (Aggarwal et al. 2003 and references therein).

3.6.6 *Orakeikorako*

The Orakeikorako geothermal system (location in Fig. 3.42) is situated at the south-western termination of the Paeroa Fault and might be associated to the adjacent Te Kopia geothermal system. In the mid 1960s, the four exploration wells OK-1, OK-2, OK-4 and OK6 were drilled at Orakeikorako outside the area of hot spring activity (Simpson and Bignall 2016 and references therein). These wells reached depths of 1155–1404 m, found a maximum temperature of 265 °C, but poor permeability conditions. About ¾ of the Orakeikorako natural manifestations, including many spectacular geysers, were submerged by the damming of the Waikato River in 1961 for hydroelectric power production. Today, the Orakeikorako natural manifestations consist of hot springs, geysers, minor steaming grounds, and few mud pools and cover an area of ~1.8 km^2 mainly on the eastern bank of the Waikato River. The discharged fluids have generated the Umukuri sinter, which is the most extensive deposit of this type in the TVZ, with an area close to 1 km^2, and is locally up to 18 m thick. The natural heat flow is estimated at 340 MWt.

3.6.7 Rotokawa

Similar to the other geothermal systems of the TVZ, the Rotokawa system (location in Fig. 3.42) is hosted into a sequence of Pleistocene and Recent volcanic rocks and the underlying Mesozoic greywacke basement. At least thirteen different hydrothermal explosion breccia have been recognized in surface outcrops. The largest crater is 1.5 km in diameter, was formed by a hydrothermal eruption which occurred ~6060 a ago, and is currently occupied by Lake Rotokawa, whose waters have pH close to 2. Between 1965 and 2010, a total of 31 geothermal wells have been drilled at Rotokawa reaching depths between 2500 and 3000 m. Thirteen of these wells have been drilled in 2007–2010 (Price et al. 2011). As proposed by Winick et al. (2009), the Rotokawa geothermal system comprises three main aquifers, namely the shallow aquifer, the intermediate aquifer and the deep reservoir.

Neutral-chloride waters, with Cl concentrations of 450–850 mg/kg and temperatures of 300–340 °C, circulate in the deep reservoir, which is hosted both in the volcanic units and in the underlying greywacke basement and is partially capped by a smectite-rich clay zone. In the south, reservoir fluids have higher concentrations of Cl (up to 1900 mg/kg in wells RK2 and RK3) and B (up to 62 ppm in wells RK2 and RK3) due to boiling in the natural-state conditions. Very large geochemical gradients in Cl concentration, Cl/B ratio and non-condensable gas contents occur from south to north, suggesting that the deep reservoir fluids are progressively diluted northward by cooler, marginal fluids. Based on the available dataset, it is possible to relate the deep reservoir fluids to a single parent fluid, through a combination of dilution and boiling processes, as proposed by Hedenquist et al. (1988). Nevertheless, it is not possible to exclude alternative multiple-parent models, for instance the presence of two distinct fluid sources, one in the north and the other in the south, as suggested by Giggenbach (1995). Conspicuous surface hydrothermal activity, comprising steaming grounds, fumaroles, and acid-sulfate springs, occurs in the southern part of the geothermal field. Mixed chloride-sulfate springs are found along the northern shores of Lake Rotokawa and partially contribute to the acid lake water. Springs discharging neutral chloride-bicarbonate waters are distributed along the Waikato River that dissects the geothermal field. Geysers and extensive silica terraces are absent.

3.6.8 Waiotapu

The Waiotapu geothermal field (location in Fig. 3.42) has the highest natural heat loss, up to 540 MWt, and its surface manifestations occupy the largest area in the TVZ, 17 km^2 (Simpson and Bignall 2016 and references therein). Between 1956 and 1959, seven exploratory wells (WT-1 to WT-7) were drilled at Waiotapu, to depths varying between 435 and 1100 m. Although temperatures of 200–295 °C were encountered, geothermal exploration was stopped because of the low flow rates of discharged fluids, the occurrence of calcite scaling and the choice to concentrate geothermal development at Wairakei. The Waiotapu thermal manifestations have different characteristics depending on their location, with waters affected by acid condensates in the north and mixed chloride-sulfate springs in the south. The most famous surface manifestation, Champagne Pool, occupies a hydrothermal eruption vent and discharges undiluted, boiled reservoir fluids cooled at 75 °C. The fluids discharged from Champagne Pool generated the Primrose silica sinter terrace extending over an area of ~220 m × 90 m.

3.6.9 Wairakei

This brief synthesis on Wairakei is based on Bixley et al. (2009) and references therein. The Wairakei geothermal field is mostly positioned on the west bank of the Waikato River, just north of Lake Taupo (Fig. 3.42) and is part of the larger Wairakei–Tauhara geothermal system. The hydrological link between these two fields has been demonstrated by the close relation between the pressure changes at Wairakei and those at the Tauhara geothermal field, about 10 km to the southeast. In contrast, the Rotokawa geothermal field, 10 km to the northeast, has not responded to the pressure changes at Wairakei. Wairakei was a liquid-dominated system with a base temperature close to 260 °C, before the onset of geothermal exploitation, as pointed out by both physical measurements and geothermometric techniques. Based on the data acquired through the early wells, it was recognized from that the maximum measured temperatures approximated the boiling point for depth down to 400 m depth, roughly corresponding to sea level (Banwell 1957).

Neutral chloride waters from the geothermal reservoir were mainly discharged at the surface through the hot springs and geysers in the Geyser Valley. The chemistry of these thermal manifestations indicated that deep chloride water had experienced both dilution and boiling before discharging at the surface (Glover and Mroczek 2009). The other major area of surface activity was in the Waiora Valley where acidic chloride–sulfate or acid sulfate waters were discharged due to the higher elevation of this site. Steam discharges occurred in the Karapiti Thermal Area, in the southern part of the geothermal field. The natural heat discharge from Wairakei has been evaluated by several authors with average values close to 400 MWt. From 1950 to 2009, drilling activity resulted in more than 200 wells, completed to depths ranging from 300 to 2750 m and spread over the geothermal field, with some wells outside its boundaries. Thanks to this large number of wells, it was possible to monitor and understand the variations in the reservoir that have been caused by fluid extraction and, more recently, by injection.

3.6.10 *Chemistry of the Reservoir Liquids from the New Zealand Geothermal Systems*

All the New Zealand reservoir liquids are situated in the [Na] sector of the triangular diagram of major cations. Therefore, sodium is by far the prevalent cation with relative concentrations varying from 82 to 96 eq% (Fig. 3.44). It is followed by potassium, with relative concentrations ranging from 3 to 18 eq%. The relative K concentration increases with temperature and, in fact, is somewhat higher at Rotokawa, Mokai, and Ngatamariki compared to the other fields. The relative concentrations of Ca are low, varying from 0.1 to 2.5 eq% only.

In the triangular diagram of main anions (Fig. 3.45) most reservoir liquids are

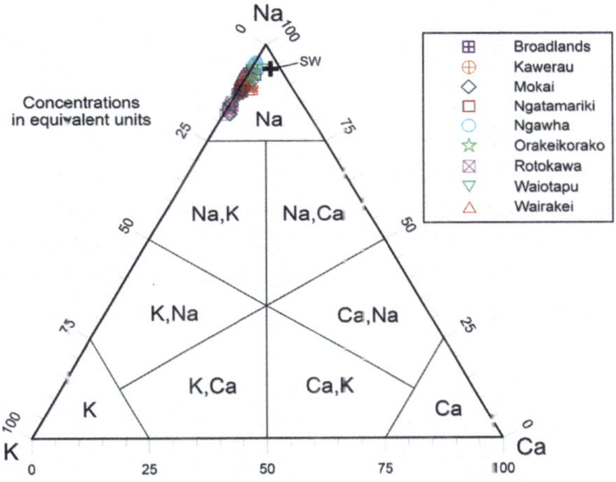

Fig. 3.44 Triangular plot of major cations for the reservoir liquids from the high-temperature geothermal fields in New Zealand

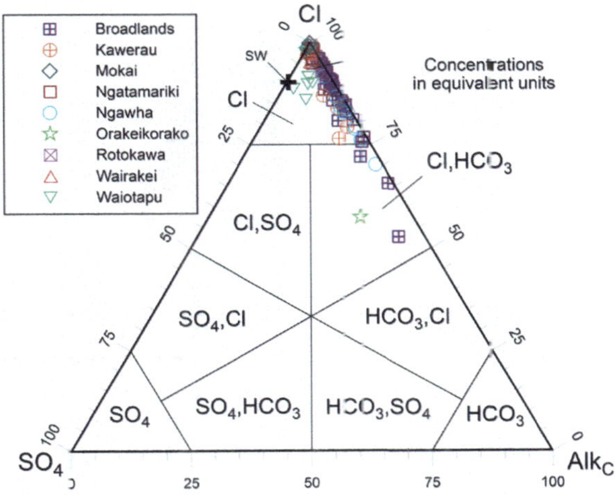

Fig. 3.45 Triangular plot of major anions for the reservoir liquids from the high-temperature geothermal fields in New Zealand

located in the [Cl] sector, and several of them, from Mokai, Ngatamariki, Rotokawa, and Wairakei, are very close to the chloride vertex. Only a few reservoir liquids from Broadlands, Orakeikorako, and Ngawha are situated in the [Cl,HCO_3] sector. Based on both triangular diagrams, it is possible to recognize the following two chemical facies:

1. The Na–Cl chemical type, which is by far the most common, comprising 88 of the 93 reservoir liquids of interest, that is 95% of the entries. Moreover, the Na–Cl reservoir liquids are found in all the considered New Zealand geothermal systems, except Orakeikorako for which one sample only is available.
2. The Na–Cl,HCO_3 chemical facies, which groups 5 reservoir liquids (5% of the cases), 3 coming from Broadlands, and 1 each from Orakeikorako and Ngawha.

The reservoir liquids from the geothermal systems in New Zealand distribute over a relatively limited interval of Σ_{eq}, from the minimum values of the Rotokawa wells, 20–48 meq/kg, to the maximum values of the Mokai wells, 88 to 133 meq/kg (Fig. 3.46).

These low to relatively low Σ_{eq} values are not surprising because the geothermal aquifers of the TVZ are mainly recharged by meteoric waters with subordinate contributions of magmatic (arc-type) waters, as indicated by isotopic data. According to Giggenbach (1995), the magmatic component contribution is $14 \pm 5\%$ for the eastern systems of the TVZ and $6 \pm 2\%$ for the western systems of the TVZ. The contribution of magmatic waters is nil at Ngawha, which is recharged by meteoric waters only as already recalled in Sect. 3.6.5.

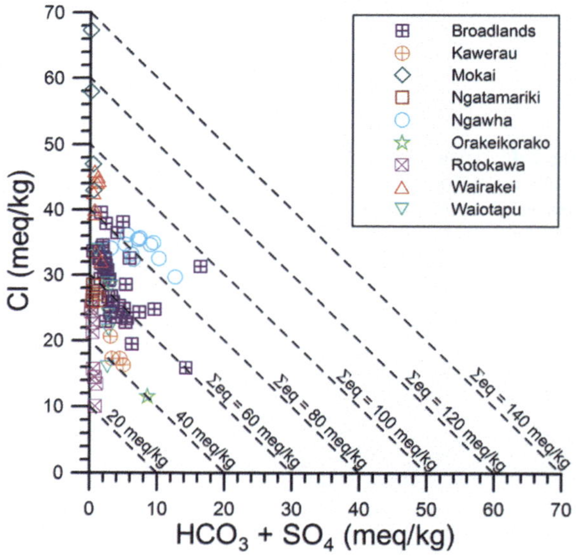

Fig. 3.46 Correlation diagram of Cl versus SO_4+ Alk_C for the reservoir liquids from the high-temperature geothermal fields in New Zealand

3.7 The Reservoir Liquids from Miscellaneous Geothermal Systems

The miscellaneous geothermal fields taken into account in this work are: Yangbajing in China (data from Ping et al. 1998a, b; Guo et al. 2014), Kizildere in Turkey (data from Guidi et al. 1988, 1990; Haizlip and Haklidir 2011; Haklidir et al. 2015; Tarcan et al. 2016), the three Italian fields of Bagnore (data from Ruggieri et al. 2004), Latera (data from Gianelli and Scandiffio 1989), and Mofete (data from Balducci and Chelini 1992), Ribeira Grande, São Miguel, Azores (data from Carvalho et al. 2006), as well as the following geothermal systems in the East African Rift System (EARS), from N to S: Asal, Djbouti (data from D'Amore et al. 1998; Sanjuan et al. 1990; Sanjuan 2010), Tendaho, Ethiopia (data from D'Amore et al. 1997; Ali 2005), Aluto-Langano, Ethiopia (data from Gizaw 1993, 1996; Teklemariam et al. 1996 and references therein), and Olkaria, Kenya (data from Muna 1982; Karingithi et al. 2010). Data on reservoir liquids from Menengai, Kenya were also compiled (from Kipng'ok 2011; Sekento 2012; Malimo 2013; Auko 2014) but were disregarded because these aqueous solutions are probably not representative of the equilibrium condition with relevant hydrothermal minerals.

The accepted chemical analyses are 32 for Yangbajing, 28 for Kizildere, 2 for Bagnore, 5 for Latera, 3 for Mofete, 15 for Ribeira Grande, 4 for Asal, 36 for Tendaho, 3 for Aluto-Langano, and 42 for Olkaria, for a total of 170 entries.

3.7.1 *Yangbajing*

The Yangbajing geothermal field is located on the Tibetan plateau at an altitude of 4300–4500 m asl. From the structural point-of-view, Yangbajing is positioned at an inflection point of the Nyainquentanglha slip-fault zone where its strike changes from N 60° E to N 30° E (Fig. 3.47). This strike change caused pervasive rock fracturing creating the permeability conditions needed for thermal water circulation (Ji and Ping 2000).

The Yangbajing geothermal system comprises a shallow reservoir hosted in altered granite and Quaternary sediments at depths of 180–280 m, and a deep reservoir hosted in fractured granite at depths of 950–1850 m. Temperature ranges from 150 to 165 °C in the shallow reservoir, whereas the deep reservoir is subdivided into two parts: the upper part with temperatures of 250–275 °C between 950 and 1350 m depth, and the lower part with temperatures >300 °C below 1500 m depth, (Xiaoping 2002).

Before the Yangbajain geothermal power plants began operating, the field was affected by hydrothermal explosions (e.g., in November 1975 and December 1977) and several manifestations were present. However, all primary thermal springs and pools disappeared in response to the intensive exploitation of the geothermal resource (Guo et al. 2014).

Fig. 3.47 Google Earth map showing the location of the Yangbajing geothermal field

Based on isotopic evidence, the Yangbajing geothermal fluids originate from meteoric waters. Both the shallow and the deep reservoirs host neutral chloride fluids, with chloride concentration of 1.5 and 2.8 g/kg, respectively. The fluid of the deep reservoir ascends and enters the shallow reservoir where it mixes with cooler meteoric water (Xiaoping 2002).

3.7.2 *Kizildere*

The Kizildere geothermal field is located in the northern sector of the Buyuk Menderes graben (Fig. 3.48), a continental rift zone whose activity begun in Early-Middle Miocene.

Quaternary alluvia crop out in the central part of the graben, where they reach a maximum thickness of some hundred meters, whereas both a Neogene series and the underlying Paleozoic basement crop out in the horsts (Guidi et al. 1988 and references therein). The basement is made up of gneisses and micaschists with lenses of marbles. The Neogene series includes prevailing clastic sediments and a sequence of limestones and marly limestones, with the local occurrence of gypsum and anhydrite. The limestones and the metamorphic basement exhibit good permeability, particularly where fractured, whereas the other Neogene deposits are impervious.

Fig. 3.48 Google Earth map showing the location of the Kizildere geothermal field and the famous Pamukkale thermal springs characterized by intense travertine deposition

Deuterium and oxygen-18 data (e.g., Özgür 1998; Tarcan et al. 2016 and references therein) indicate that the Kizildere reservoir liquids have meteoric origin, whereas the considerable oxygen isotope shift, up to ~5‰ units, is due to isotope exchange with carbonate rocks.

Temperatures are 148–198 °C in the first (shallow) reservoir, 200–212 °C in the second (middle) reservoir, and close to 240 °C in the third (deepest) reservoir, at depth of ca. 2.2 km (Tarcan et al. 2016). The Kizildere reservoir liquids have Na–HCO_3 to Na–HCO_3,SO_4 composition and high f_{CO_2}, usually in the range 35–50 bar (Guidi et al. 1988, 1990).

Today, the only thermal manifestations present in the area of the geothermal field are the fumaroles located at the northern boundary of the drilled zone, where hot springs were once found (Dominco and Samilgil 1970). This change indicates a substantial deepening of the piezometric surface caused by exploitation. Hot springs, with temperature up to the boiling point, and an old drilled well are present at Tekke Hamam, in front of Kizildere, on the opposite side of the Buyuk Menderes graben. Several hot springs are found in nearby areas. At Pamukkale there are thermal springs and extensive deposits of white travertines, visible in the Google Earth map of Fig. 3.48. These white travertines are responsible for the name of this place. In fact, pamuk is Turkish for cotton.

3.7.3 Bagnore

This brief synthesis on Bagnore is based on Calamai et al. (1970), Gianelli et al. (1988), and Bertini et al. (1995). There are two high-temperature geothermal fields in the southern sector of the Quaternary Mt. Amiata volcano (Principe et al. 2017 and references therein), one is at Bagnore while the other one is at Piancastagnaio (Fig. 3.49).

Two distinct reservoirs are present in the Bagnore geothermal field. The shallow reservoir is situated at depths of 400–1000 m, is hosted in carbonate-evaporite (anhydrite-bearing) rocks, and was exploited before the 1970s. The deep reservoir is positioned at depths of 1300–3000 m, is hosted in fractured metamorphic rocks, and has been exploited more recently.

The maximum temperature recorded in the shallow reservoir is 160 °C. The deep reservoir is water-dominated, has a hydrostatic pressure of ~200 bar at 3000 m depth, and temperatures close to 300–330 °C. The produced fluid is a two-phase mixture. The steam phase separated at wellhead at 20 bar has a high content of non-condensable gases, up to 15% by weight. The whole system is recharged by meteoric waters circulating through fractured rocks. Heat is probably contributed by a granitic intrusion which is located at an estimated depth of about 7000 m and has an expected temperature of 750–800 °C.

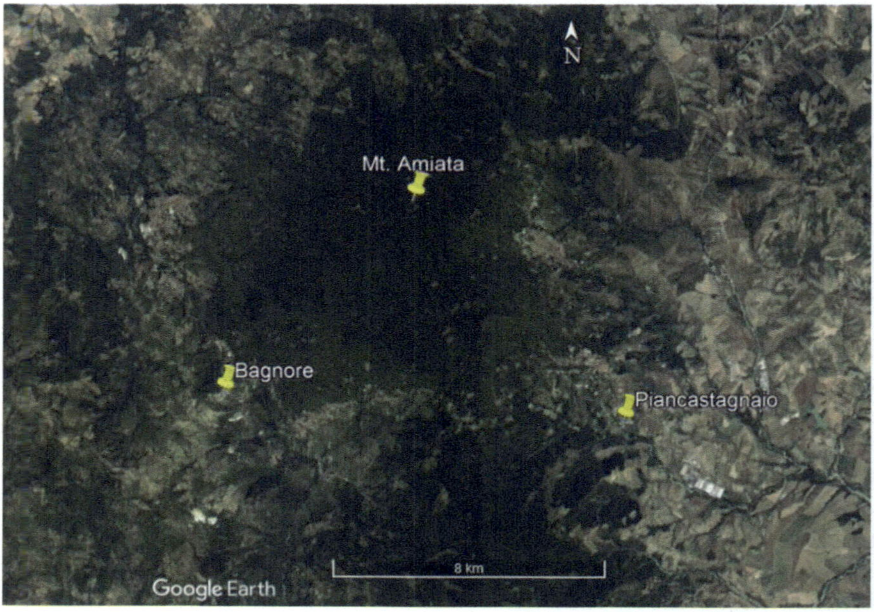

Fig. 3.49 Google Earth map showing the location of the Bagnore and Piancastagnaio geothermal fields in the southern sector of Mt. Amiata volcano

3.7.4 Latera

The Latera geothermal field is located within the homonymous caldera of 7 km × 9 km and roughly elliptical shape (Fig. 3.50). The eastern rim of the Latera caldera intersects a wider volcano-tectonic depression, which is partly filled by Lake Bolsena.

The Latera geothermal reservoir occurs at 1000–1500 m depth in a structural high made up of carbonate rocks (Cavarretta et al. 1985; Gianelli and Scandiffio 1989). The productive zones of the Latera geothermal system have hydrostatic pressures and temperatures of 200–240 °C, whereas higher temperatures were measured in some unproductive wells, such as well L10, in which the bottom-hole temperature is >400 °C at ~3300 m depth, and well L1, in which a temperature of 340 °C was measured at ~2800 m depth.

Available deuterium and oxygen-18 data (Battaglia et al. 1992) suggest that the Latera reservoir liquids have meteoric origin. Their extensive oxygen isotope shift, up to ~8‰ units, is ascribable to isotope exchange with carbonate rocks.

The conceptual model proposed by Gianelli and Scandiffio (1989) comprises a regional warm aquifer of $Ca–SO_4,HCO_3$ composition and an upflow of hot Na–Cl waters, with Cl concentration of 1500–3300 mg/kg and total CO_2 concentration of 2000–32,000 mg/kg. This model is supported by well L14, which encountered two

Fig. 3.50 Google Earth map showing the location of the Latera geothermal field within the homonymous caldera. The Latera caldera is situated to the west of the Bolsena volcano-tectonic depression, which is occupied by the homonymous lake

distinct productive zones, the first one delivering hot Na–Cl waters at 430 m depth, the second one supplying warm Ca–SO_4,HCO_3 waters at 800 m depth.

All the wells produce two-phase fluids, with discharge enthalpy ranging from 791 to 1026 kJ/kg and temperature varying from 186 to 238 °C. The water analyses of wells L2, L3D and L4 are more reliable than those of other wells because of the longer production time, in the order of several months.

The chemistry of the Latera reservoir liquids was obtained from Gianelli and Scandiffio (1989) without running WATCH because data for the separated liquid and vapor phases are not available.

3.7.5 *Mofete*

The Mofete geothermal field is situated within the Campi Flegrei caldera, extending west of Naples (Fig. 3.51). The caldera has a diameter of ~12 km and formed ~35,000 years ago after the Campanian Ignimbrite eruption (Rosi et al. 1983). The southern half of this volcano-tectonic depression is submerged, whereas its northern half occurs on land and includes several intra-caldera small volcanic apparatuses.

The geothermal reservoir is hosted into a volcanic sequence made up of alternating tuffs and lavas as well as quartz-feldspathic siltites below ~700 m depth. The Mofete

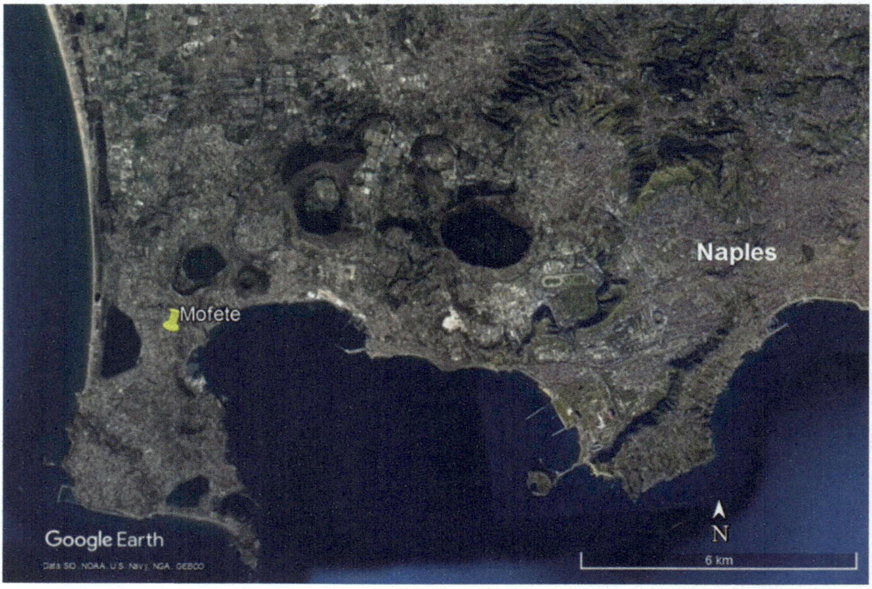

Fig. 3.51 Google Earth map showing the location of the Mofete geothermal field within the Campi Flegrei caldera, close to Naples

geothermal system includes a shallow reservoir at temperatures of 230–300 °C, which was encountered by five of the seven wells drilled in the late 1970s to early 1980s, and an intermediate-deep reservoir at temperatures of 340–350 °C which was found by two wells only (Carella and Guglielminetti 1983).

Neutral Na–Cl reservoir liquids circulate in fractured zones not cemented by hydrothermal minerals. The brines separated at atmospheric pressure exhibit considerable differences in Cl concentration, which ranges from 21,000 to 44,000 mg/kg for those coming from the shallow and intermediate reservoirs, whereas those delivered from the deep reservoir have Cl concentration of ~310,000 mg/kg (Carella and Guglielminetti 1983) and are close to saturation with halite. The system is recharged by seawater, modified by steam separation and by the irreversible mass exchanges occurring during the high-temperature interaction with reservoir rocks (Balducci and Chelini 1992).

3.7.6 *Ribeira Grande*

The water-dominated geothermal field of Ribeira Grande is situated within the Água de Pau stratovolcano in the central part of São Miguel Island, Azores (Fig. 3.52). The stratovolcano is truncated by an outer 4 km × 7 km caldera that formed ~33,600 years ago and an inner 2.5 km × 3 km caldera that originated ~15,200 years ago (Carvalho et al. 2006 and references therein). The geothermal system is recharged by meteoric waters, infiltrating in the upper parts and flanks of the Água de Pau volcano and flowing from south to north.

Fig. 3.52 Google Earth map showing the location of the Ribeira Grande geothermal field in the central part of São Miguel Island, Azores

Wells CL1, CL2, CL3, and CL4 have total depth of 2029, 1402, 1459, and 1500 m, respectively. Production from these wells is sustained by three different aquifers, all probably hosted in pillow lavas and situated at depths of 800 to 1300 m approximately, where temperature varies from 230 to 245 °C. A small temperature reversal occurs below ~1300 m depth in these four wells. Well PV1 is somewhat different, with a total depth of 811 m only and maximum temperatures of 220–236 °C at ~500 m depth, but again a temperature reversal is present below.

The deepest aquifer contains a single liquid phase, the intermediate aquifer hosts a two-phase (vapor + liquid) mixture with a steam fraction up to 0.08, whereas the shallower aquifer produces a single steam phase. All the wells have excess enthalpy.

The Ribeira Grande reservoir liquids are neutral, have Na–Cl composition, and Cl concentration of 1100–2600 mg/kg. Calcite scaling was directly observed in wells PV1 and CL1, but probably occurs in all geothermal wells upon flashing.

3.7.7 Asal

The Asal geothermal system, positioned in the Afar region between Lake Asal and the Gulf of Ghoubbet, is situated along the still emerged portion of an oceanic rift (Fig. 3.53).

Fig. 3.53 Google Earth map showing the location of the Asal geothermal field, Djibouti, between Lake Asal and the Gulf of Ghoubbet

Six deep geothermal wells were drilled in the Asal Rift area in the 70's and 80's. Well Asal 1 produced water and steam for several months, whereas well Asal 2 produced a very small amount of fluids (Fouillac et al. 1989). The Na–Cl brine discharged from well Asal 1 has Cl concentration of 115,000–120,000 mg/kg at atmospheric conditions (Sanjuan 2010).

As reported by D'Amore et al. (1998): (1) well Asal 3 has total depth of 1316 m, a maximum temperature of 265 °C and encountered high-enthalpy fluids below 1075 m depth; (2) well Asal 4 has total depth of 2011 m, a maximum temperature of 344 °C, but it met cold water at shallow depth and it is dry below 400 m; (3) well Asal 5 has total depth of 2105 m, a maximum temperature of 359 °C, but it is dry below 400 m; (4) well Asal 6 has total depth of 1761 m, a maximum temperature of 281 °C and encountered high-enthalpy fluids at 265 °C from 1100 to 1300 m depth. Wells Asal 3 and Asal 6 produced a Na–Cl brine with Cl concentration of 103,000–106,000 mg/kg and pH 4.8–4.9 at atmospheric conditions. Sphalerite and galena scaling was detected in both wells downstream of the flash level.

As recognized by previous authors, the chemistry of the Asal reservoir liquids is controlled by high-temperature basalt-seawater interaction, associated with variable evaporation. The chemistry of the Asal reservoir liquids was obtained from D'Amore et al. (1998) and Sanjuan (2010) without running WATCH because of the unavailability of data for the separated liquid and vapor phases.

3.7.8 Tendaho

The Tendaho geothermal system is positioned in the inland portion of the Afar region (Fig. 3.54), at a distance of ~150 km from the sea coast. Most information on Tendaho was derived from Battistelli et al. (2002).

Deuterium and oxygen-18 indicate that the Tendaho geothermal reservoir is recharged by meteoric waters infiltrating at elevations of 2000–3000 m asl in the Western Ethiopian Plateau. Consistently, the shallow Tendaho geothermal reservoir hosts low-salinity brines with TDS of ~2 g/kg and low gas content, ~370 mg/kg.

Available enthalpy and temperature data for wells TD-2 and TD-4 suggest the presence in the reservoir of either a single liquid phase or bi-phase conditions with small excess enthalpy as well as the provenance of produced fluids from different feed zones. Reservoir conditions and production characteristics of wells TD-5 and TD-6 are similar to those of well TD-4.

Owing to the limited information available on enthalpy and gas chemistry, a single liquid phase was hypothesized to be present in the reservoir, at the temperature indicated by the quartz/chalcedony geothermometer, and the CO_2 and H_2S contents of all the reservoir liquids of wells TD-5 and TD-6 were assumed to be 307 mg/kg and 2.14 mg/kg, respectively, based on the gas chemistry data of wells TD-2 and TD-4 from D'Amore et al. (1997).

Fig. 3.54 Google Earth map showing the location of the Tendaho geothermal system in the inland portion of the Afar region, Ethiopia

3.7.9 Aluto-Langano

The Aluto-Langano geothermal field is found in the central sector of the Main Ethiopian Rift, between Lake Ziway to the north and Lake Abiyata and Lake Langano to the south (Fig. 3.55).

The Aluto-Langano geothermal system is associated with the Quaternary Aluto volcanic complex. The stable isotope values indicate that the geothermal well discharges consist essentially of meteoric water infiltrating in the volcanic complex (Rango et al. 2010 and references therein). The upflow of geothermal fluids is controlled by an active fault zone known as Wonji Fault Belt (Gianelli and Teklemariam 1993).

According to Teklemariam et al. (1996), the two hottest wells, LA-3 and LA-6, have total discharge enthalpies of 1600–1650 kJ/kg and aquifer temperatures of 320–335 °C, respectively. These data indicate the presence of bi-phase conditions with mass steam fractions of 0.11–0.15 and 0.037–0.083 in the reservoir zones encountered by wells LA-3 and LA-6, respectively. The reservoir liquids of Aluto-Langano wells LA-3 and LA-6 were disregarded because computed pH values resulted to be higher than the mineral-solution equilibrium pH values by 1.8 pH unit. A single liquid phase seems to occur in the aquifer zones connected with wells LA-4, LA-7, and LA-8, with measured temperatures of 233, 226, and 270 °C, respectively.

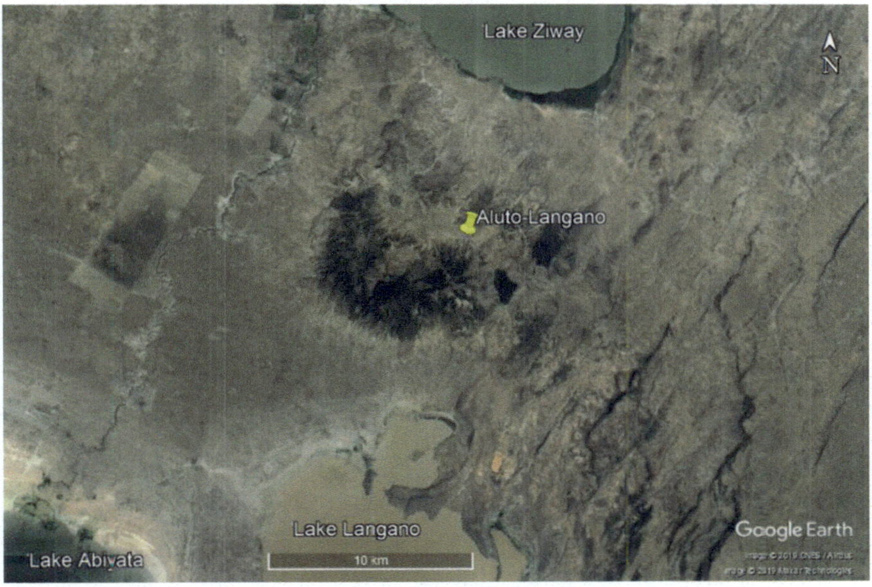

Fig. 3.55 Google Earth map showing the location of the Aluto-Langano geothermal system in the Main Ethiopian Rift

The chemistry of the Aluto-Langano reservoir liquids encountered by wells LA-4, LA-7, and LA-8 was directly taken from Teklemariam et al. (1996) without running WATCH because data for the separated liquid and vapor phases are unavailable.

3.7.10 Olkaria

The Olkaria geothermal field is situated to the south of Lake Naivasha in the central portion of the Kenyan Rift Valley (Fig. 3.56). The Olkaria geothermal system is related with the Quaternary Olkaria volcanic complex, where magmatic activity began during the late Pleistocene and continued until 180 $\pm$ 50 years ago when the Ololbutot comendite was erupted (Clarke et al. 1990). Incidentally, comendites are present only in the Olkaria area within the whole Kenyan Rift Valley.

The following brief synthesis on Olkaria is based on Karingithi et al. (2010). The over 100 wells drilled at Olkaria have variable discharge enthalpy. Most of the wells have excess enthalpy which is due to depressurization boiling in producing aquifers. However, contribution of steam from the steam cap present above the bi-phase reservoir of Olkaria East is responsible for the excess enthalpy of the wells drilled in this sector of the field.

Fig. 3.56 Google Earth map showing the location of the Olkaria geothermal system in the Kenyan Rift Valley

The fluids in Olkaria West and Domes sectors have high carbonate alkalinity and low chloride concentration in sharp contrast with those in the Olkaria East and Olkaria Northeast sectors which have lower alkalinity and higher chloride concentration. Moreover, the fluids encountered by the wells of Olkaria West and Domes sectors have high concentrations of CO_2, which are controlled by considerable gas fluxes from a magmatic heat source. In contrast, CO_2 concentration is relatively low in the other sectors of the Olkaria geothermal field, being buffered by the epidote-prehnite-calcite-quartz hydrothermal mineral assemblage.

Under-saturation with respect to the Ca-bearing solid phases, which are known to be present in the geothermal reservoir as hydrothermal minerals, is probably an artifact due to loss of Ca through calcite precipitation occurring in the depressurization zone around the wells.

3.7.11 Chemistry of the Reservoir Liquids from Miscellaneous Geothermal Systems

One hundred sixty six of the one hundred seventy reservoir liquids considered in this section are situated in the [Na] sector of the triangular diagram of major cations (Fig. 3.57). The four exceptions are the reservoir liquids from Asal, which are located

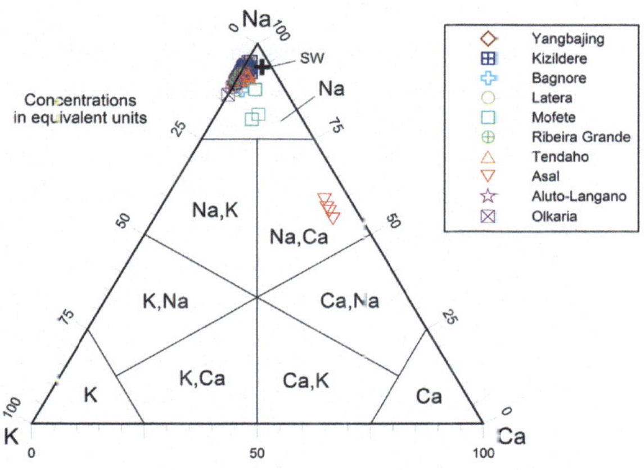

Fig. 3.57 Triangular plot of major cations for the reservoir liquids from miscellaneous geothermal systems (see legend). Average seawater is also shown for comparison

in the [Na,Ca] sector, due to gain of Ca through basalt-seawater interaction at elevated temperatures. Therefore, sodium is the prevailing cation in all the reservoir liquids of interest, with relative Na concentrations >75 eq% at Yangbajing, Kizildere, Bagnore, Latera, Mofete, Ribeira Grande, Tendaho, Aluto-Langano, and Olkaria. The Asal reservoir liquids have relative Na concentrations of 54–59 eq%, relative Ca concentrations of 35–40 eq%, relative K concentrations of 5.5–6.2 eq%, and relative Mg concentrations ≤0.1 eq%.

In the triangular diagram of main anions (Fig. 3.58) most reservoir liquids are located in the [Cl] sector and some of them are close to the chloride vertex. Nevertheless, several reservoir liquids are situated elsewhere, namely in the [Cl,HCO_3], [Cl,SO_4], [HCO_3,Cl], [HCO_3,SO_4], and [HCO_3] sectors, whereas no sample is found in the [SO_4,Cl], [SO_4,HCO_3], and [SO_4] sectors. Based on both triangular diagrams, it is possible to recognize eight different chemical facies. In detail:

1. The Na,Ca–Cl chemical type includes only the 4 reservoir liquids of Asal, but is the only chemical facies present in this field.

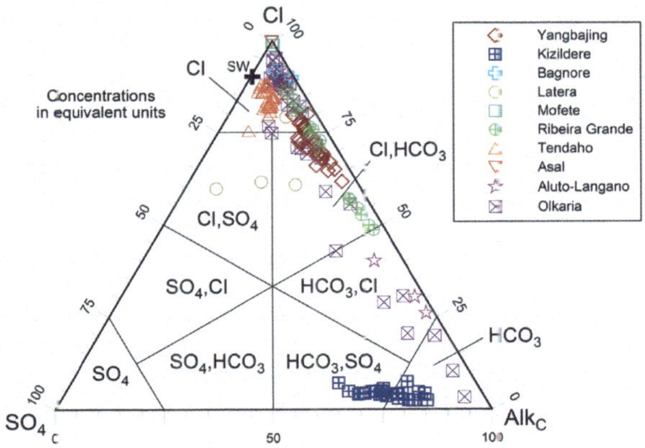

Fig. 3.58 Triangular plot of major anions for the reservoir liquids from miscellaneous geothermal systems (see legend). Average seawater is also shown for comparison

2. The Na–Cl chemical facies is by far the most common, with 88 reservoir liquids. Thirty six are from Tendaho, three are from Mofete, and two are from Bagnore, but they constitute 100% of the samples from these three geothermal fields. Twenty eighth are from Olkaria, where they represent 67% of the entries. Six are from Ribeira Grande, where they make up 40% of the items. Two are from Latera, where they account for 40% of the records. Eleven are from Yangbajing, where they represent 34% of the entries.
3. The Na–Cl,HCO_3 chemical type is rather widespread with 37 reservoir liquids. It is present at Yangbajing with 21 samples, corresponding to 66% of the entries of this site, Ribeira Grande with 8 samples, representing 53% of the items of this place, Olkaria with 7 samples, corresponding to 17% of the records of this field, and Latera with 1 sample, representing 20% of the entries of this site.
4. The Na–Cl,SO_4 chemical type groups 2 reservoir liquids only, both from Latera, where they account for 40% of the items.
5. The Na–HCO_3,Cl chemical facies comprises 8 reservoir liquids. Three are from Aluto-Langano, where they explain 100% of the items. Four are from Olkaria, where they make up 10% of the entries. One is from Ribeira Grande, where it constitutes 7% of the records.
6. The Na–HCO_3,SO_4 chemical type includes 15 reservoir liquids, all from Kizildere, where they account for 54% of the items.
7. The Na–HCO_3 chemical facies comprises 16 reservoir liquids. Thirteen are from Kizildere, where they represent 46% of the records. Three are from Olkaria, where they make up 7% of the entries.

The binary diagram of Cl versus $SO_4 + Alk_C$ (Fig. 3.59a) shows that the reservoir liquids of Asal have Σ_{eq} values varying from 3780 to 4210 meq/kg, which are 3–3.5

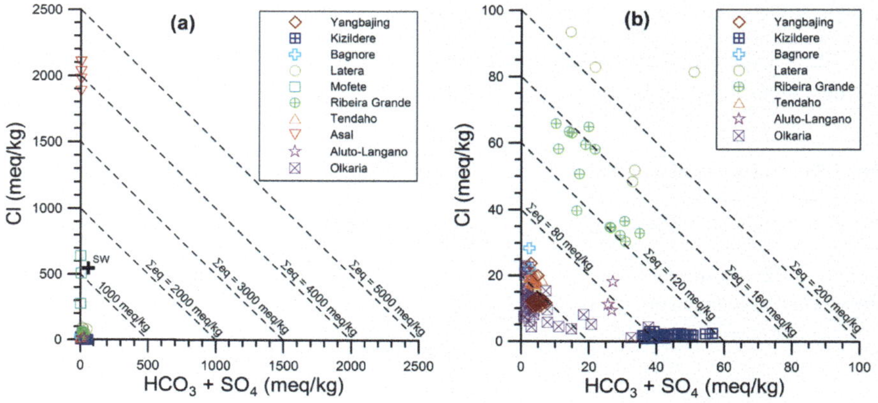

Fig. 3.59 Correlation diagrams of Cl versus $SO_4 + Alk_C$ for the reservoir liquids from miscellaneous geothermal systems (see legend). Average seawater is also shown for comparison in **(a)**. Different Σ_{eq} intervals are considered in the two diagrams **(a)** and **(b)** in order to accommodate all the liquids of interest

times higher than the average seawater value, 1210 meq/kg. These very high ionic salinities are essentially controlled by the predominant role of evaporated seawater in the recharge of this geothermal circuit, in spite of the subsequent occurrence of important chemical changes caused by water-rock interaction (D'Amore et al. 1998; Sanjuan et al. 1990; Sanjuan 2010).

Two of the three reservoir liquids of Mofete have Σ_{eq} of 1020 and 1310 meq/kg, comparing with the value of seawater, which recharges this geothermal system, as already mentioned above (Fig. 3.59a). The other reservoir liquid of Mofete has lower ionic salinity, 542 meq/kg, due to dilution with meteoric waters.

The other reservoir liquids considered in this section have lower Σ_{eq} values, namely 28 – 57 meq/kg for Yangbajing, 79–110 meq/kg for Kizildere, 50–59 meq/kg for Bagnore, 170–265 meq/kg for Latera, 116–165 meq/kg for Ribeira Grande, 41–48 meq/kg for Tendaho, 76–96 meq/kg for Aluto-Langano, and 20–86 meq/kg for Olkaria (Fig. 3.59b). These low to relatively low Σ_{eq} values are expected, because deuterium and oxygen-18 data indicate that these geothermal circuits are recharged by meteoric waters (see above).

Carbon dioxide is entirely or chiefly contributed by deep sources, either mantle/magma degassing, as suggested by the $\delta^{13}C$ values of CO_2 for Yangbajing (Zhao et al. 1998), Olkaria (Darling 1998), Aluto-Langano (Darling 1998), Tendaho (D'Amore et al. 1997), and Asal (Darling 1998; D'Amore et al. 1997), or high-temperature decomposition of carbonate rocks, as indicated by the $\delta^{13}C$ values of CO_2 for Kizildere (Özgür 1998). Where the CO_2 flux from these deep sources is high, reservoir liquids have high f_{CO_2} values and Na–HCO_3 or Na–HCO_3,Cl or Na–HCO_3,SO_4 compositions, reflecting the occurrence of water-rock interaction governed by conversion of CO_2 to HCO_3^- ion.

In contrast, reservoir liquids have Na–Cl or Na–Cl,HCO_3 composition where Cl sources are more important than CO_2 sources. Dissolution of carbonate-evaporite rocks is probably responsible of the relatively high SO_4 concentrations of the Na–Cl,SO_4 reservoir liquids of Latera, as suggested by the $\delta^{34}S$ values of the hydrothermal anhydrite samples from the deep geothermal wells L1 and L2 (Cavarretta et al. 1985) and the Na–HCO_3,SO_4 reservoir liquids of Kizildere (Guidi et al. 1988, 1990).

3.8 Main Results of Speciation Calculations for the Reservoir Liquids and Implications

3.8.1 The pH Value

As already mentioned in Sect. 3.1.2, the pH value of a reservoir liquid is computed by WATCH considering the low-temperature pH value and assuming conservation of alkalinity upon cooling from the reservoir temperature to the temperature of measurement of pH and alkalinity, apart from the increment caused by steam separation (Arnórsson et al. 1982).

Looking at the pH of reservoir liquids from a different point of view, Ellis (1970) recognized that the a_{K^+}/a_{H^+} ratio is controlled by coexistence of K-feldspar, muscovite, and a silica mineral, according to the following reaction:

$$1.5\,KAlSi_3O_8 + H^+ = 0.5\,KAl_3Si_3O_{10}(OH)_2 + 3\,SiO_2 + K^+ \tag{3.8}$$

and the a_{Na^+}/a_{H^+} ratio is buffered by coexistence albite, K-feldspar, muscovite, and a silica mineral, as described by the reaction:

$$0.5\,KAlSi_3O_8 + NaAlSi_3O_8 + H^+ = 0.5\,KAl_3Si_3O_{10}(OH)_2 + 3\,SiO_2 + Na^+. \tag{3.9}$$

Assuming that salinity is approximated by the sum of the molalities of Na^+ and K^+ ions, Ellis (1970) underscored that the pH of reservoir liquids unaffected by boiling is a function of salinity and temperature. In this way, he was able to explain the low pH values of the deep waters of high salinity and high temperature, such as those of Salton Sea in California, as well as the high pH values of the deep waters of low salinity and low temperature, such as those of Hveragerdi in Iceland.

The Ellis approach was later expanded by Chiodini et al. (1991), who used a mineral-solution equilibrium model to calculate the pH value and the concentrations of compatible solutes (Na, K, Mg, Ca, Al, SiO_2, HCO_3, SO_4, and F) for the aqueous solutions in equilibrium with the hydrothermal mineral assemblage comprising low-albite, adularia, either a Ca–Al-silicate and/or calcite (depending on CO_2 fugacity and temperature), clinochlore, muscovite, quartz, anhydrite, and fluorite, at specified temperature, CO_2 fugacity, and concentration of chloride, the only considered conservative (mobile) component (further details in Sect. 5.8). Through multiple regression analysis (MRA) of the results of the mineral-solution equilibrium model, Chiodini et al. (1991) derived the following equation:

$$pH_{mse} = 1.757 - 0.822 \cdot \log \Sigma_{eq} + 1846/T - 0.0171 \cdot \log f_{CO_2}, \tag{3.10}$$

relating the pH fixed by mineral-solution equilibrium, pH_{mse}, to total ionic salinity (Σ_{eq}, in eq/kg), temperature (in K), and CO_2 fugacity (in bar), which were treated as independent variables in MRA. MRA results also indicated that the pH fixed by mineral-solution equilibrium is controlled by total ionic salinity, temperature, and CO_2 fugacity, in order of decreasing importance. In agreement with Ellis findings, Eq. (3.10) indicates that pH_{mse} decreases with increasing Σ_{eq} and temperature and vice versa. Moreover, pH_{mse} decreases with increasing CO_2 fugacity.

As mentioned in Sect. 3.1.4, the correspondence, within ± 1.3 pH unit, between the pH computed through speciation calculations usually carried out using WATCH, pH_{sc}, and mineral-solution equilibrium pH is one of the two criteria adopted to accept the analytical data of the compiled reservoir liquids. The range spanned by

pH_{sc}, 4.17–8.13, is somewhat smaller than the interval covered by pH_{mse}, 4.03–8.63. The correlation diagrams of Fig. 3.60, in which these two pH values are contrasted, provide the following indications.

1. All or most reservoir liquids of some geothermal systems have $pH_{sc} < pH_{mse}$. This is the case of the medium-temperature geothermal systems in Iceland (Fig. 3.60a), Long Valley (Fig. 3.60b), Onikobe and Oku-aizu (Fig. 3.60c), Alto Peak (Fig. 3.60d), Ngawha (Fig. 3.60e), Kizildere and Latera (Fig. 3.60f). These differences between pH_{sc} and pH_{mse} values are probably ascribable to excess of acidic species, in most cases CO_2 from deep sources. The amount of CO_2 entering these geothermal aquifers from below is so large that relevant mineral-solution reactions are not able to consume all the CO_2, whose concentration and fugacity in the reservoir liquid become higher than the values fixed by mineral-solution equilibrium. Consequently, the pH of these reservoir liquids decreases becoming lower than the value controlled by equilibrium with the minerals of the neutral alteration suite, but it does not attain the low values causing the development of the acid alteration suite (sensu Reyes 1990). The only possible exception is Onikobe, where the main acidity source is HCl, with H_2SO_4 acting as potential subordinate acidity source, as noted in Sect. 3.4.4.
2. In contrast, all or most reservoir liquids of other geothermal systems have $pH_{sc} > pH_{mse}$. Examples are Krafla, Námafjall, Hellisheidi, and Nesjavellir, (Fig. 3.60a), Salton Sea, Cerro Prieto, Los Azufres, and Berlin (Fig. 3.60b), Sumikawa, Uenotai, and Oguni (Fig. 3.60c), Tongonan (Fig. 3.60d), Wairakei (Fig. 3.60e), Ribeira Grande, Olkaria and Tendaho (Fig. 3.60f). These divergences between pH_{sc} and pH_{mse} values might be caused by CO_2 loss through aquifer boiling either in the natural (pre-exploitation) state or in response to the depressurization induced by exploitation.
3. There are also geothermal systems whose reservoir liquids have pH_{sc} values similar to pH_{mse} values, as is the case of Hveragerdi, Svartsengi, and Reykjanes (Fig. 3.60a), Dixie Valley, Heber, and Miravalles (Fig. 3.60b), Takigami (Fig. 3.60c), Palinpinon (Fig. 3.60d), Kawerau and Ngatamariki (Fig. 3.60e) and Bagnore (Fig. 3.60f). These reservoir liquids with pH_{sc} not too different from pH_{mse} are probably close to the mineral-solution equilibrium condition, which is apparently unaffected or negligibly affected by gain/loss of CO_2. Therefore, strictly speaking, only these reservoir liquids should be considered for investigating the traditional geoindicators and elaborating theoretical, activity-based geothermometers and f_{CO_2} indicators, as we plan to do in this work, whereas the reservoir liquids with pH_{sc} significantly different from pH_{mse} should be disregarded because they are not (or not completely) representative of this equilibrium condition. However, such a choice would cause a considerable loss of information due to the large number of reservoir liquids with pH_{sc} considerably different from pH_{mse}.

However, the discrepancies between pH_{sc} and pH_{mse} values could also be caused, at least in part, by uncertainties in the reconstruction of reservoir liquid chemistry and related speciation calculations, especially for the excess enthalpy wells. Accepting

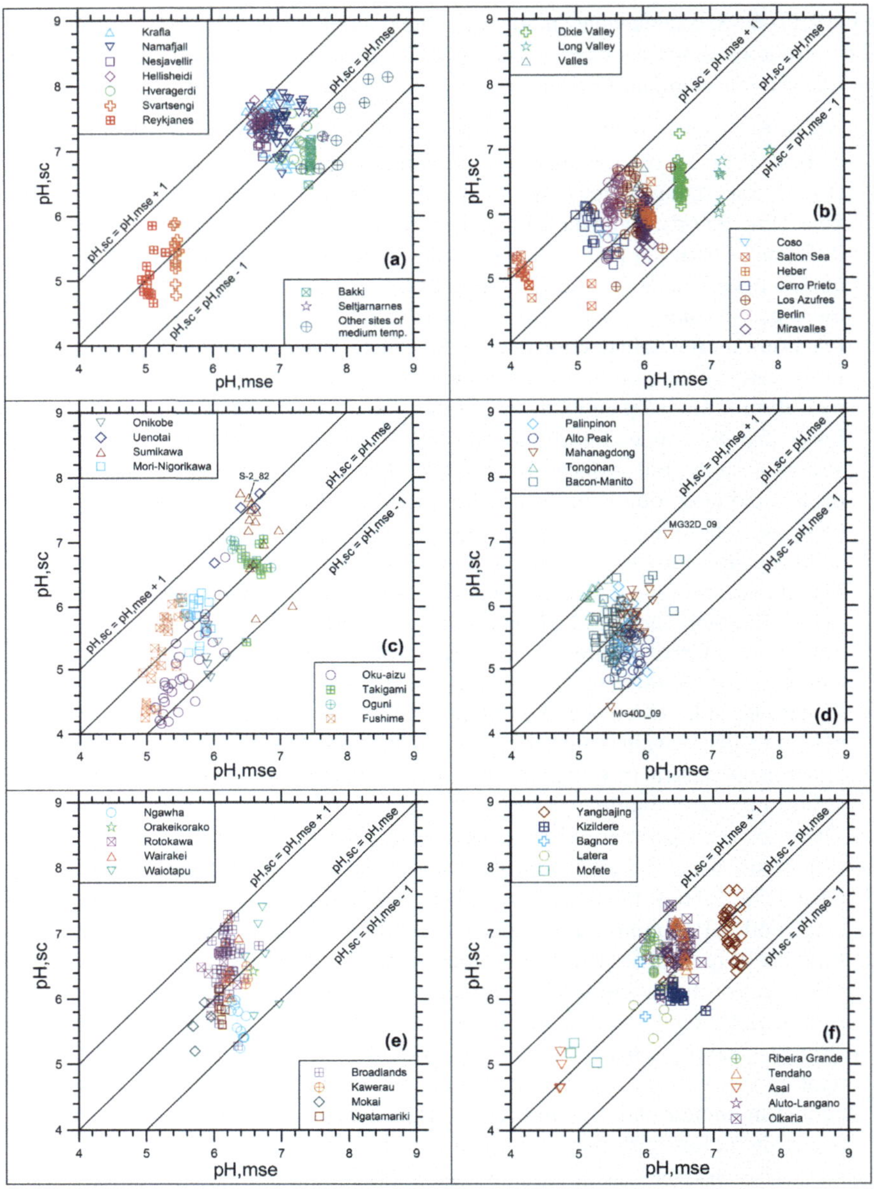

Fig. 3.60 Correlations diagrams between the pH values computed through speciation calculations usually carried out using WATCH, pH_{sc}, and the corresponding mineral-solution equilibrium pH values, pH_{mse}, for the reservoir liquids of **a** Iceland, **b** Northern and Central America, **c** Japan, **d** The Philippines, **e** New Zealand, and **f** Miscellaneous geothermal fields

this explanation, it is legitimate to take into consideration also the reservoir liquids with pH_{sc} differing from pH_{mse} by up to ±1.3 pH unit, as long as we aim at elaborating theoretical, activity-based geothermometers and f_{CO_2} indicators which do not involve pH and are based on variables which are not influenced or negligibly influenced by pH, like the concentrations and activities of non-volatile solutes. These variables, in fact, are not impacted or are marginally impacted by uncertainties in the calculation of reservoir liquid chemistry (see Sect. 3.1.2). Traditional geothermometers satisfy these criteria, but have other weaknesses, which will be discussed in Chap. 5.

3.8.2 The Fraction of Undissociated $SiO_{2(aq)}$

The fraction of undissociated $SiO_{2(aq)}$ is the ratio of the molal concentration of undissociated silicic acid, $SiO_{2(aq)}$, over the total (or analytical) molal concentration of SiO_2. The fraction of undissociated $SiO_{2(aq)}$ is strongly correlated with reservoir pH and departs significantly from unity for pH values higher than 7.5–8.0 (Fig. 3.61), as expected based on the pK of silicic acid dissociation, which varies from 8.8 to 10.1, in the 100–350 °C range (see Sect. 5.2.1 for details). Hence, for pH > 7.5–8.0, it is advisable to consider undissociated $SiO_{2(aq)}$ instead of total dissolved SiO_2 in geothermometric calculations, in spite of possible uncertainties on the pH values computed through speciation calculations. Actually, fractions of undissociated $SiO_{2(aq)}$ lower than 0.95 are rather uncommon, as they pertain to 25 reservoir liquids only, of which 7 each from Námafjall and Krafla, 2 from Yangbajing, and 1 each from Baer, Bakki, Hveragerdi, Reykholar, Reykholt, Reykjavik, Seltjarnarnes, Sumikawa, and Waiotapu.

3.8.3 The Fractions of Free Na^+, K^+, Ca^{2+}, and Mg^{2+} Ions

The fractions of free (or uncomplexed) Na^+, K^+, Ca^{2+}, and Mg^{2+} ions are the ratios of the molal concentrations of free Na^+, K^+, Ca^{2+}, and Mg^{2+} ions over the total (or analytical) molal concentrations of Na, K, Ca, and Mg, respectively.

The fractions of free Na^+, K^+, Ca^{2+}, and Mg^{2+} ions are contrasted with aquifer temperature in the correlation diagrams of Figs. 3.62, 3.63, 3.64 and 3.65, respectively. The fractions of free cations depend on the concentrations of relevant anionic ligands and the stability of pertinent ion complexes which, in turn, increases with increasing reservoir temperature. Owing to variable composition, salinity and temperature of the considered reservoir liquids there is a considerable scatter of sample points in the binary plots of Figs. 3.62, 3.63, 3.64 and 3.65.

The fractions of free Na^+ and K^+ ions and especially those of free Ca^{2+} and Mg^{2+} ions exhibit large variations (see below) because the aqueous speciation of most reservoir liquids was modeled using the B-dot equation (see Sect. 2.3.3) for

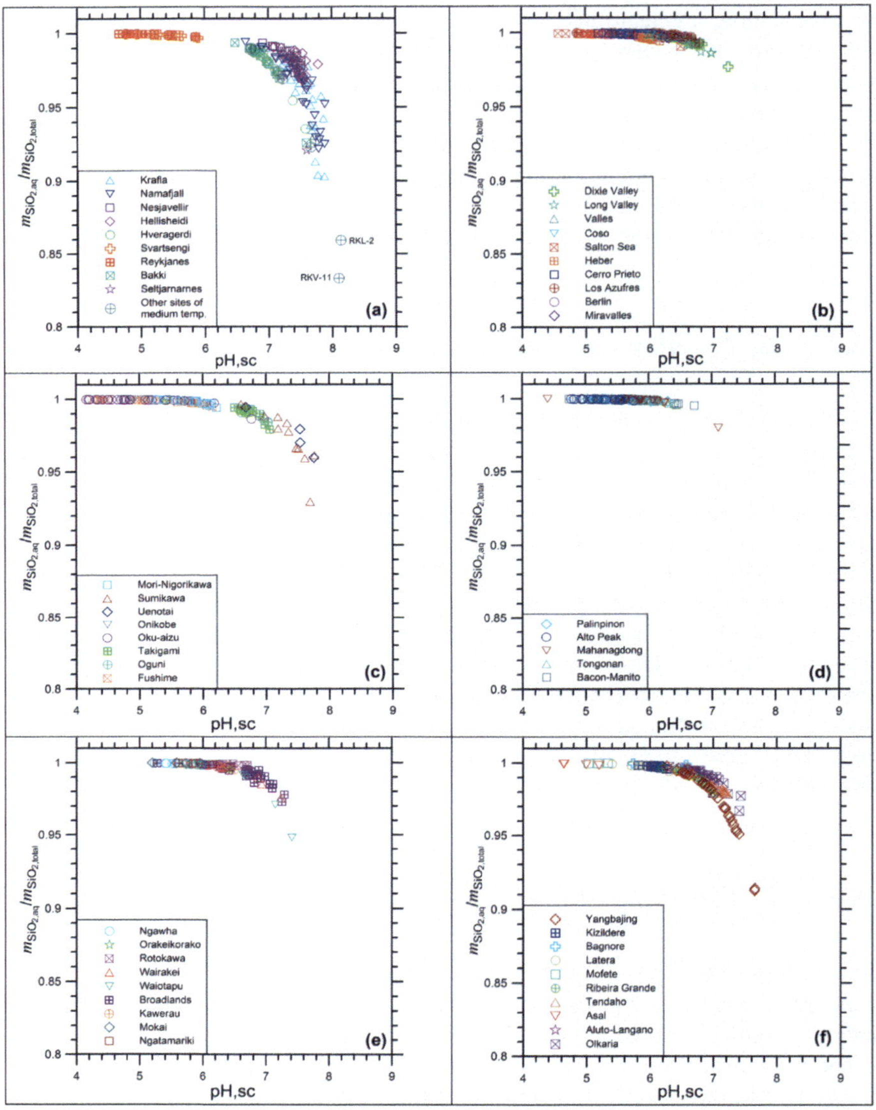

Fig. 3.61 Correlations diagrams showing the fraction of undissociated silica as a function of aquifer pH (from speciation calculations) for the reservoir liquids of **a** Iceland, **b** Northern and Central America, **c** Japan, **d** The Philippines, **e** New Zealand, and **f** Miscellaneous geothermal fields

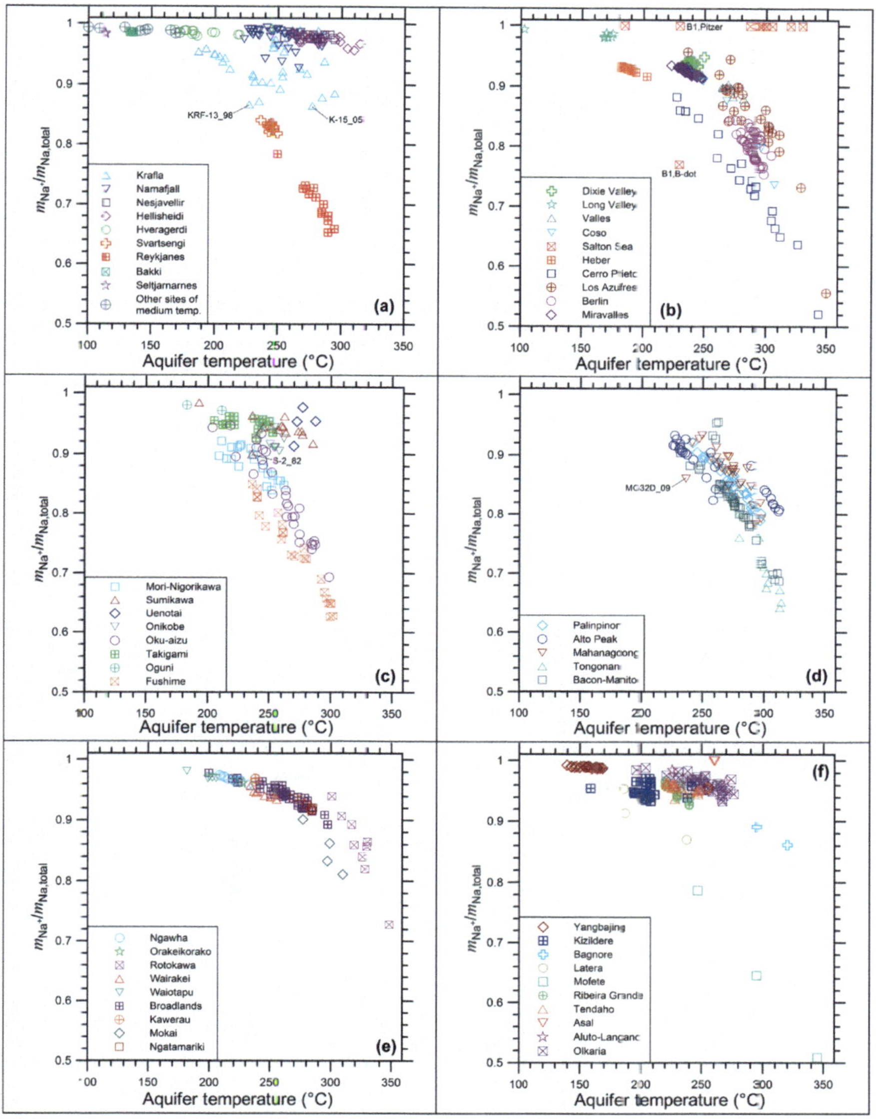

Fig. 3.62 Correlations diagrams showing the fraction of free Na^+ ion as a function of aquifer temperature for the reservoir liquids of **a** Iceland, **b** Northern and Central America, **c** Japan, **d** The Philippines, **e** New Zealand, and **f** Miscellaneous geothermal fields

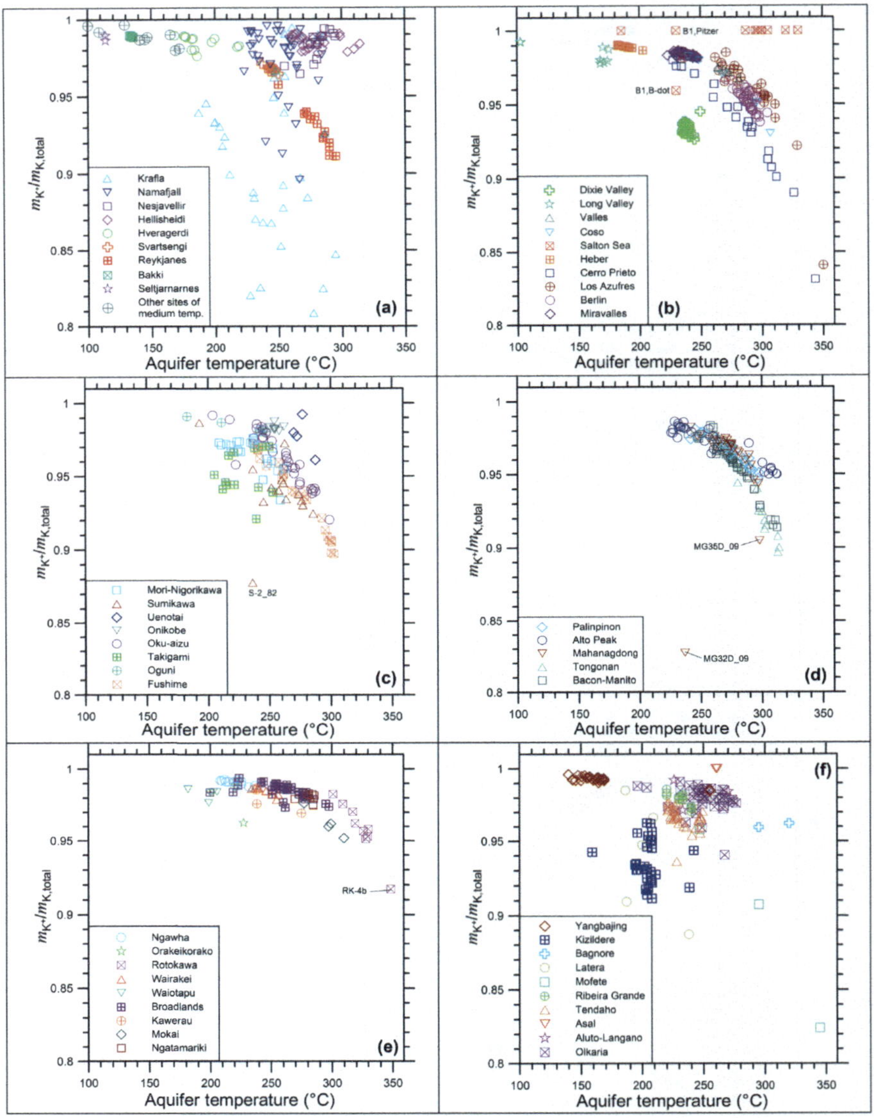

Fig. 3.63 Correlations diagrams showing the fraction of free K^+ ion as a function of aquifer temperature for the reservoir liquids of **a** Iceland, **b** Northern and Central America, **c** Japan, **d** The Philippines, **e** New Zealand, and **f** Miscellaneous geothermal fields

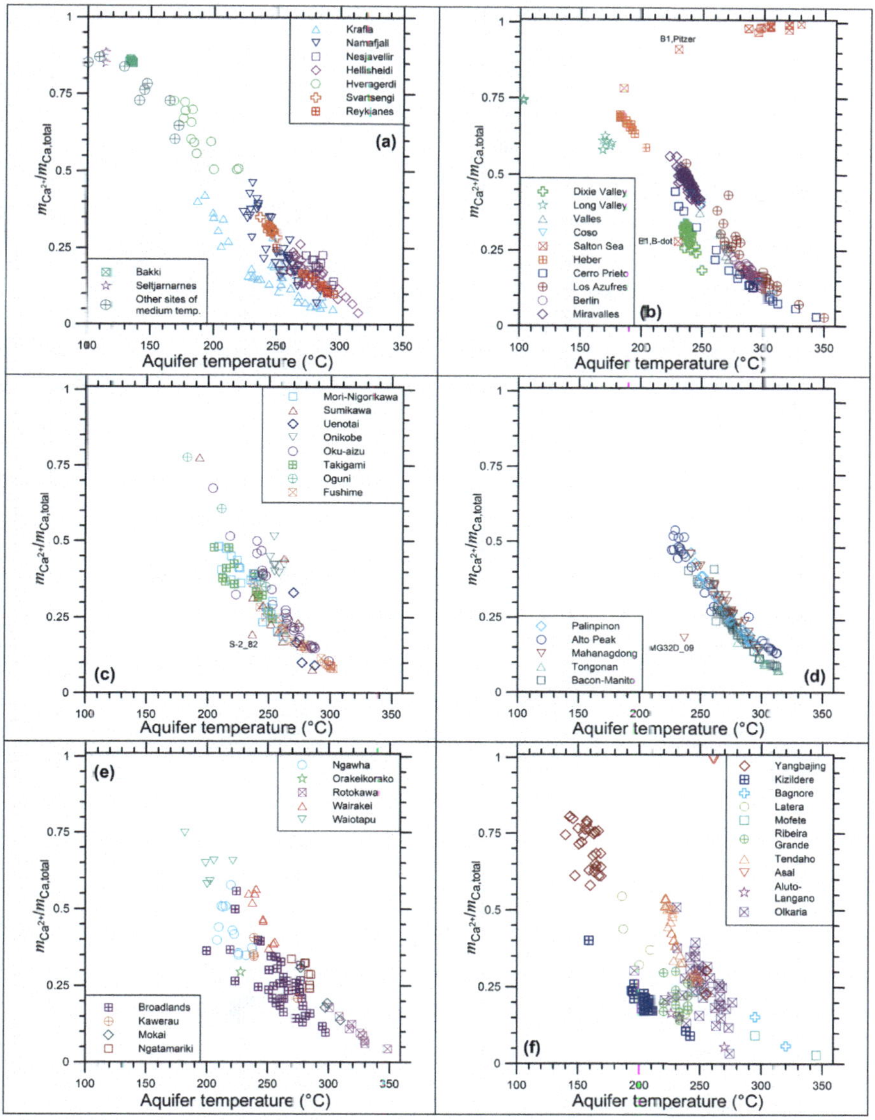

Fig. 3.64 Correlations diagrams showing the fraction of free Ca^{2+} ion as a function of aquifer temperature for the reservoir liquids of **a** Iceland, **b** Northern and Central America, **c** Japan, **d** The Philippines, **e** New Zealand, and **f** Miscellaneous geothermal fields

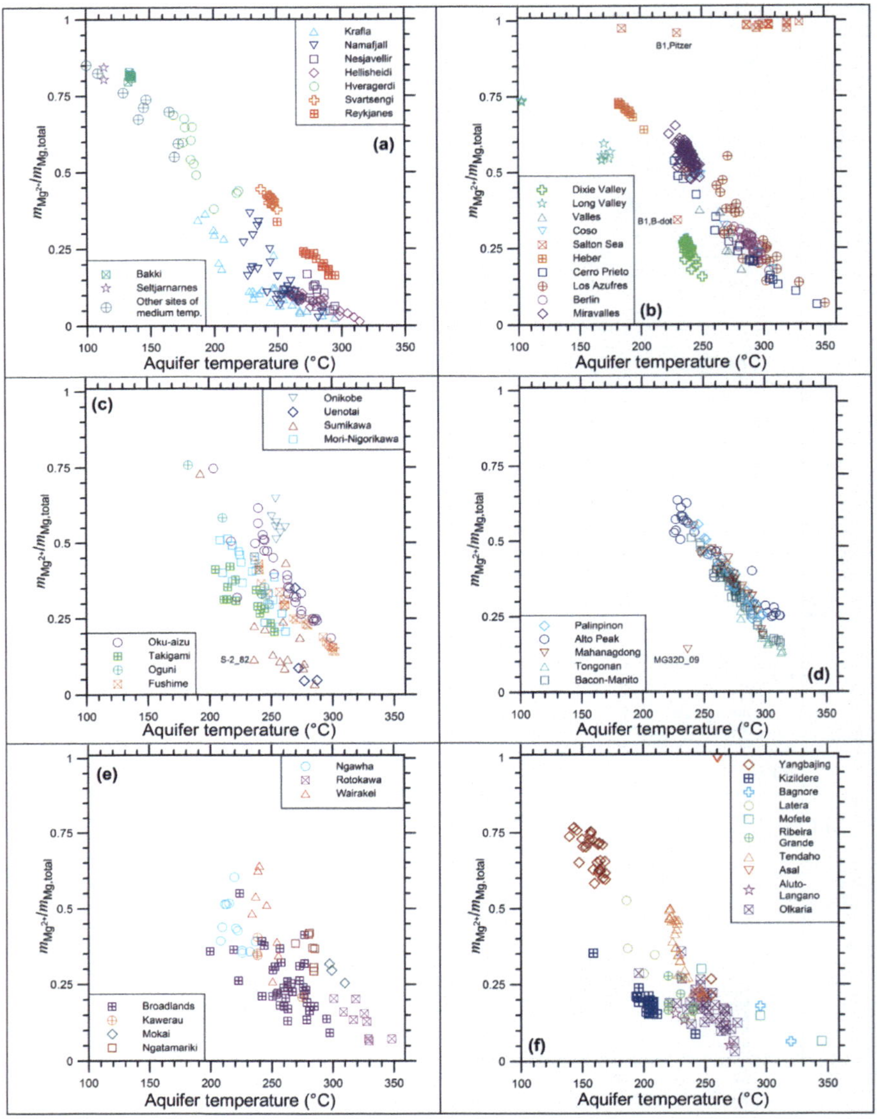

Fig. 3.65 Correlations diagrams showing the fraction of free Mg^{2+} ion as a function of aquifer temperature for the reservoir liquids of **a** Iceland, **b** Northern and Central America, **c** Japan, **d** The Philippines, **e** New Zealand, and **f** Miscellaneous geothermal fields

computing the activity coefficients of ionic species and several ion pairs were taken into account, as typically done in this approach.

In contrast, the speciation of the Salton Sea and Asal reservoir liquids were modeled using the Pitzer equations (see Sect. 2.3.4). In the Pitzer's approach, solute-solvent and solute-solute interactions are mainly explained by activity coefficients and few complex species are considered, namely NaF° for sodium, $CaCO_3{}^\circ$, $CaHCO_3{}^+$, $CaOH^+$, and $CaSO_4{}^\circ$ for calcium, and $MgCO_3{}^\circ$, $MgHCO_3{}^+$, and $MgOH^+$, for magnesium. Consequently, in the Salton Sea and Asal reservoir liquids: (1) the fraction of free K^+ is equal to 1 because no aqueous complex of K^+ ion is taken into account; (2) the fraction of free Na^+ and Mg^{2+} ions are close to 1 due to the very small importance of the considered ion pairs in these aqueous solutions; (3) the fraction of free Ca^{2+} ion departs from unity in the aqueous solution of lowest temperature, from well Commercial # 113 of Salton Sea (code Com #113 in the file Reservoir_Liquids.XLS), due to the moderate importance of the $CaSO_4{}^\circ$ aqueous complex.

To be noted that the reservoir liquid of well Commercial # B1 of Salton Sea (indicated by codes Com #B1 in the file Reservoir_Liquids.XLS and B1 in the diagrams), characterized by a Cl concentration of 31,000 mg/kg and an aquifer temperature of 230 °C (which are low values for Salton Sea), was modeled adopting both approaches to appreciate the difference in the fractions of free cations.

The Fraction of Free Na^+ Ion The fraction of free Na^+ ion varies from 0.65 to 0.994 in the reservoir liquids of Iceland, indicating that free Na^+ ion is the prevailing aqueous species (Fig. 3.62a). The main aqueous complexes are: (1) the $NaSO_4{}^-$ ion pair in 112 reservoir liquids, of which 33 from Námafjall, 32 from Krafla, 18 from Bakki, 13 from Hveragerdi, 5 from Nesjavellir, 3 from Hellisheidi, 2 each from Reykholt and Reykjabol, and 1 each from Baer, Efri-Reykir, Leirà, and Seltjarnarnes; (2) the $NaCl^\circ$ ion pair in 67 reservoir liquids, of which 25 from Svartsengi and Reykjanes (these aqueous solutions have Na–Cl composition and high salinity), 24 from Hellisheidi, 4 from Krafla, 3 from Námafjall, 10 from Nesjavellir, and 1 from Seltjarnarnes; (3) the $NaHSiO_3{}^\circ$ ion pair in three reservoir liquids from Námafjall, Reykholar, and Reykjavik.

Neglecting the hypersaline reservoir liquids of Salton Sea that were modeled using the Pitzer's equations, all the other reservoir liquids of Northern and Central America have fraction of free Na^+ ion ranging from 0.52 to 0.994, pointing out that free Na^+ ion is the dominating aqueous species (Fig. 3.62b). The main aqueous complexes are: (1) the $NaSO_4^-$ ion pair in 45 reservoir liquids, of which 35 from Dixie Valley and 10 from Long Valley, of relatively low salinity and variable composition, Na–Cl,SO_4 or Na–Cl,HCO_3 or Na–HCO_3,Cl; (2) the $NaCl^\circ$ ion pair in the remaining 233 reservoir liquids, of which 105 from Miravalles, 55 from Berlin, 26 from Los Azufres, 19 from Cerro Prieto, 16 from Heber, 7 from Valles, and 5 from Coso, all of medium-high salinity and Na–Cl composition.

The reservoir liquids of Japan have fractions of free Na^+ ion varying from 0.98 to 0.63, suggesting that free Na^+ ion is the principal aqueous species (Fig. 3.62c). The lowest fractions of free Na^+ ion are encountered in the high-salinity, Na–Cl reservoir liquids of Fushime and Oku-aizu. The main aqueous complexes are: (1) the $NaCl^{\circ}$ ion pair in 98 reservoir liquids, of which 31 from Oku-aizu, 25 from Fushime, 21 from Mori-Nigorikawa, 7 from Onikobe, 6 from Oguni, 3 each from Takigami and Uenotai, and 1 each from Sumikawa and Wasabizawa; (2) the $NaSO_4^-$ ion pair in the remaining 23 reservoir liquids, of which 13 from Sumikawa and 10 from Takigami.

In the reservoir liquids of the Philippines, the fraction of free Na^+ ion ranges from 0.96 to 0.64. The lowest values pertain to the Na–Cl reservoir liquids with the highest salinities, mostly from Tongonan and some from Bacon-Manito (Fig. 3.62d). The main aqueous complex after the free Na^+ ion is the $NaCl^{\circ}$ ion pair in all the reservoir liquids apart from one, the Na–SO_4,Cl reservoir liquid MG32D_09 from Mahanagdong, whose speciation is dominated by free Na^+ ion followed by the $NaSO_4^-$ ion pair.

The reservoir liquids of New Zealand have high fractions of free Na^+ ion, varying from 0.98 to 0.73. There are only eleven reservoir liquids with values lower than 0.9, of which 1 is from Broadlands, 3 are from Mokai, and 7 are from Rotokawa (Fig. 3.62e). The $NaCl^{\circ}$ ion pair is the main aqueous species after the free Na^+ ion in all the reservoir liquids of New Zealand, apart from OK-2 from Orakeikorako in which the main aqueous complex is the $NaSO_4^-$ ion pair.

Leaving aside the reservoir brines of Asal that were modeled adopting the Pitzer's approach, most miscellaneous reservoir liquids have fractions of free Na^+ ion in the interval 0.927–0.993. Only 7 reservoir liquids have lower fractions of free Na^+ ion, with the minimum value of 0.509 pertaining to reservoir liquid MF-2 from Mofete (Fig. 3.62f). The $NaCl^{\circ}$ ion pair is the major aqueous species after the free Na^+ ion in 111 reservoir liquids, of which 30 from Olkaria, 29 from Tendaho, 27 from Yangbajing, 15 from Ribeira Grande, 3 each from Aluto-Langano and Mofete, and 2 each from Bagnore and Latera. The $NaSO_4^-$ ion pair is the prevailing aqueous species after the free Na^+ ion in 55 reservoir liquids, of which 28 from Kizildere, 12 from Olkaria, 7 from Tendaho, 5 from Yangbajing, and 3 from Latera.

The Fraction of Free K^+ Ion The fraction of free K^+ ion ranges from 0.81 to 0.997 in the reservoir liquids of Iceland, indicating that free K^+ ion is the prevailing aqueous species (Fig. 3.63a). The main aqueous complexes are: (1) the KSO_4^- ion pair in 154 reservoir liquids, of which 26 from Hellisheidi, 13 from Hveragerdi, 36 from Krafla, 37 from Námafjall, 13 from Nesjavellir, 18 from Bakki, 2 each from Reykholt, Reykjabol, and Seltjarnarnes, 1 each from Baer, Efri-Reykir, Leirà, Reykholar, and Reykjavik; (2) the KCl° ion pair in the remaining 28 cases, comprising the 25 high-salinity, Na–Cl reservoir liquids of Svartsengi and Reykjanes, 2 reservoir liquids from Nesjavellir and 1 from Hellisheidi.

Ignoring the reservoir brines of Salton Sea that were modeled adopting the Pitzer's approach, all the other reservoir liquids of Northern and Central America have fraction of free K^+ ion fluctuating from 0.83 to 0.993, pointing out that aqueous speciation is dominated by free K^+ ion (Fig. 3.63b). The main aqueous complexes are: (1) the KCl° ion pair in 232 reservoir liquids, of which 55 from Berlin, 19 from Cerro Prieto, 5 from Coso, 16 from Heber, 25 from Los Azufres, 105 from Miravalles, and 7 from Valles, all characterized by medium-high salinity and Na–Cl composition; (2) the

KSO_4^- ion pair in the other 46 reservoir liquids, including 35 of Dixie Valley and 10 of Long Valley, all of comparatively low salinity and varying composition, Na–Cl,SO_4 or Na–Cl,HCO_3 or Na-HCO_3,Cl, as well as the Na–Cl reservoir liquid AZ-14 of Los Azufres that has a relatively low temperature, 236 °C, compared to the other fluids of this field.

Most reservoir liquids of Japan have fractions of free K^+ ion varying from 0.99 to 0.92. Somewhat lower values, in the interval 0.91–0.90, occur in a few high-salinity, Na–Cl reservoir liquids from Fushime (Fig. 3.63c). A Na–Cl,SO_4 reservoir liquid of relatively low salinity from Sumikawa, code S-2_82, with 406 mg Cl/kg and 401 mg SO_4/kg has an even lower fraction of free K^+ ion, 0.88. Therefore, free K^+ ion prevails in all the reservoir liquids of Japan and is accompanied by the following main aqueous complexes: (1) the KCl° ion pair in 78 reservoir liquids, of which 30 from Oku-aizu, 25 from Fushime, 14 from Mori-Nigorikawa, 7 from Onikobe, 1 each from Uenotai and Wasabizawa; (2) the KSO_4^- ion pair in the other 43 reservoir liquids, of which 14 from Sumikawa, 13 from Takigami, 7 from Mori-Nigorikawa, 6 from Oguni, 2 from Uenotai, and 1 from Oku-aizu.

In the reservoir liquids of the Philippines, the fraction of free K^+ ion varies from 0.99 to 0.83. Values lower than 0.93 are found in eight Na–Cl reservoir liquids from Tongonan, six Na–Cl reservoir liquids from Bacon-Manito, and two reservoir liquids from Mahanagdong (Fig. 3.63d). Of these two Mahanagdong reservoir liquids, MG32D_09 has Na–SO_4,Cl composition and low salinity, while MG35D_09 belongs to the Na–Cl facies, but has an unusually high SO_4 concentration. Therefore, free K^+ ion dominates the speciation of all the reservoir liquids of the Philippines and is complemented by: (1) the KCl° ion pair in 138 cases, of which 49 from Bacon-Manito, 28 from Palinpinon, 26 from Alto Peak, 24 from Mahanagdong, and 11 from Tongonan; (2) the KSO_4^- ion pair in the remaining 8 cases, of which 4 from Mahanagdong (including samples MG32D_09 and MG35D_09), 2 from Bacon-Manito, and 1 each from Alto Peak and Palinpinon.

The reservoir liquids of New Zealand are characterized by high fractions of free K^+ ion, ranging from 0.99 to 0.92. One reservoir liquid only, RK-4b from Rotokawa, has fraction of free K^+ ion lower than 0.95 (Fig. 3.63e). The KCl° aqueous complex is the main species after the free K^+ ion in 66 of these 93 reservoir liquids, of which 34 from Broadlands, 9 from Rotokawa, 7 each from Ngatamariki and Wairakei, 5 from Ngawha and 4 from Mokai. The KSO_4^- ion pair is the main species after the free K^+ ion in 27 reservoir liquids, of which 7 each from Broadlands and Ngawha, 6 from Waiotapu, 4 from Kawerau, 2 from Wairakei, and 1 from Orakeikorako.

Disregarding the high-salinity reservoir liquids of Asal that were modeled using the Pitzer's equations, all miscellaneous reservoir liquids except two have fractions of free K^+ ion higher than 0.90 up to a maximum of 0.996. The two exceptions have fractions of free K^+ ion of 0.887 and 0.825 and are from Latera and Mofete, respectively (Fig. 3.63f). The KSO_4^- ion pair is the dominant aqueous species after the free K^+ ion in 159 reservoir liquids, of which 41 from Olkaria, 36 from Tendaho, 31 from Yangbajing, 28 from Kizildere, 15 from Ribeira Grande, 5 from Latera, and 3 from Aluto-Langano. The KCl° ion pair is the principal aqueous species after the free K^+ ion in the remaining 7 reservoir liquids, of which 3 from Mofete, 2 from Bagnore, and 1 each from Olkaria and Yangbajing.

The Fraction of Free Ca^{2+} Ion The medium-temperature reservoir liquids of Iceland and those of Hveragerdi have fractions of free Ca^{2+} ion in the interval 0.89–0.51, whereas all the other high-temperature reservoir liquids of Iceland have lower fractions of free Ca^{2+} ion, in the range 0.46–0.04 (Fig. 3.64a). The Ca speciation varies from site to site, as a function of the chemical characteristics and temperature. The main aqueous species are: (1) the $CaOH^+$ aqueous complex in 66 reservoir liquids, of which 24 from Hellisheidi, 21 from Námafjall, 14 from Nesjavellir, and 7 from Krafla; (2) the free ion Ca^{2+} in 53 reservoir liquids, of which 18 from Bakki, 13 from Hveragerdi, 10 from Námafjall, 2 each from Reykholt, Reykjabol, and Seltjarnarnes, and 1 each from Baer, Efri-Reykir, Krafla, Leirà, Reykholar, and Reykjavik; (3) the $CaSO_4°$ ion pair in 30 reservoir liquids, of which 24 from Krafla and 6 from Námafjall; (4) the $CaCl^+$ ion pair in the 25 high-salinity, Na–Cl reservoir liquids from Reykjanes and Svartsengi; (5) the $CaHCO_3^+$ aqueous complex in 8 reservoir liquids, of which 4 from Krafla, 3 from Hellisheidi, and 1 from Nesjavellir.

Apart from the reservoir liquids of Salton Sea that were modeled using the Pitzer's approach, all the other reservoir liquids of Northern and Central America have fractions of free Ca^{2+} ion varying from 0.75 to 0.03 (Fig. 3.64b). The free ion Ca^{2+} is the prevailing aqueous species in 113 reservoir liquids, of which 85 from Miravalles, 16 from Heber, 10 from Long Valley, and 1 each from Coso and Los Azufres. The ion pair $CaCl^+$ is the major aqueous species in 130 reservoir liquids, of which 55 from Berlin, 25 from Los Azufres, 20 from Miravalles, 19 from Cerro Prieto, 7 from Valles, and 4 from Coso. The aqueous complex $CaSO_4°$ is the main aqueous species in the 34 reservoir liquids of low-salinity and Na–Cl,SO_4 composition from Dixie Valley, while the $CaHCO_3^+$ ion pair prevails in the other reservoir liquid of low-salinity and Na–Cl,HCO_3 composition from this field.

Most reservoir liquids of Japan have fractions of free Ca^{2+} ion lower than 0.5 (Fig. 3.64c), with the lowest values, from 0.12 to 0.08 in 12 samples, of which 8 from Fushime, 2 from Uenotai, and 1 each from Ozu-aizu and Sumikawa. The free Ca^{2+} ion is the main aqueous species in 32 reservoir liquids, of which 9 from Mori-Nigorikawa, 9 from Takigami, 6 from Oguni, 5 from Oku-aizu, 2 from Sumikawa, and 1 from Onikobe. The ion pair $CaCl^+$ is the major aqueous species in 69 reservoir liquids, of which 26 from Oku-aizu, 25 from Fushime, 11 from Mori-Nigorikawa, 6 from Onikobe, and 1 from Wasabizawa. The aqueous complex $CaSO_4°$ is the prevailing aqueous species in 11 reservoir liquids, of which 8 from Sumikawa and 3 from Takigami. The ion pair $CaOH^+$ is the prevalent aqueous species in 6 reservoir liquids, of which 3 from Sumikawa and 3 from Uenotai. The $CaHCO_3^+$ ion pair prevails in 2 reservoir liquids only, 1 from Mori-Nigorikawa and 1 from Takigami.

All the Na–Cl reservoir liquids of the Philippines distribute in a tight trend in the diagram of aquifer temperature versus the fraction of free Ca^{2+} ion (Fig. 3.64d), but the Na–SO_4,Cl reservoir liquid MG32D_09 from Mahanagdong deviates from this trend. The fraction of free Ca^{2+} ion ranges from 0.54 to 0.07 in these reservoir liquids. The main ion complex is the $CaCl^+$ ion pair in all these aqueous solutions apart from the reservoir liquids OP-6 Da from Bacon-Manito and MG32D_09 from Mahanagdong, in which the prevailing aqueous complexes are the $CaHCO_3^+$ and $CaSO_4^\circ$ ion pairs, respectively.

The 93 reservoir liquids of New Zealand have fractions of free Ca^{2+} ion varying from 0.75 to 0.05, 77 have fractions of free Ca^{2+} ion lower than 0.5, while the remaining 16 entries (6 from Waiotapu, 4 from Ngawha, 4 from Wairakei, and 2 from Broadlands) have fractions of free Ca^{2+} ion higher than 0.5 (Fig. 3.64e). The main aqueous species are: (1) the $CaHCO_3^+$ ion pair in 38 reservoir liquids, of which 27 from Broadlands, 5 from Ngawha, 3 from Kawerau, 2 from Rotokawa, and 1 from Orakeikorako; (2) free Ca^{2+} ion in 29 reservoir liquids, of which 9 from Wairakei, 7 from Ngawha, 6 each from Broadlands and Waiotapu, and 1 from Kawerau; (3) the $CaCl^+$ ion pair in 26 reservoir liquids, of which 8 from Broadlands, 7 each from Ngatamariki and Rotokawa, and 4 from Mokai.

Leaving aside the reservoir brines of Asal that were modeled using the Pitzer's approach, the considered miscellaneous reservoir liquids have fractions of free Ca^{2+} ion ranging from 0.81 to 0.03, with 43 entries higher than 0.50 and 124 cases lower than this threshold (Fig. 3.64f). The main aqueous species are: (1) free Ca^{2+} ion in 86 reservoir liquids, of which 35 from Tendaho, 30 from Yangbajing, 18 from Olkaria, 2 from Latera, and 1 from Kizildere; (2) the $CaHCO_3^+$ ion pair in 66 reservoir liquids, of which 27 from Kizildere, 15 each from Olkaria and Ribeira Grande, 3 each from Aluto-Langano and Latera, 2 from Yangbajing, and 1 from Bagnore; (3) the CaF^+ ion pair in 7 reservoir liquids from Olkaria; (4) the $CaCl^+$ ion pair in 4 reservoir liquids, of which 3 from Mofete and 1 from Bagnore; (5) the $CaSO_4^\circ$ ion pair in 2 reservoir liquids, one from Olkaria and one from Tendaho; (6) the $CaOH^+$ ion pair in 1 reservoir liquid from Olkaria.

The Fraction of Free Mg^{2+} Ion The fraction of free Mg^{2+} ion is higher than 0.50 in all the reservoir liquids of medium-temperature of Iceland and in most of Hveragerdi, up to a maximum of 0.85. In contrast, all the other high-temperature reservoir liquids of Iceland have fraction of free Mg^{2+} ion in the interval 0.45–0.02 (Fig. 3.65a). The Mg speciation of the 165 reservoir liquids with Mg concentration higher than detection limit is different from one field to another, depending on chemistry and temperature. The main aqueous species are: (1) the $MgOH^+$ ion pair in 82 reservoir liquids, of which 29 from Námafjall, 20 from Hellisheidi, 19 from Krafla, 13 from Nesjavellir, and 1 from Hveragerdi; (2) the free Mg^{2+} ion in 43 reservoir liquids, of which 18 from Bakki, 11 from Hveragerdi, 3 from Námafjall, 2 each from Reykholt, Reykjabol, and Seltjarnarnes,1 each from Baer, Efri-Reykir, Leirà, Reykholar, and Reykjavik; (3) the $MgCl^+$ ion pair in the 25 high-salinity, Na–Cl reservoir liquids from Reykjanes and Svartsengi; (4) the $MgSO_4^\circ$ ion pair in 15 reservoir liquids, of which 14 from Krafla and 1 from Námafjall.

Excluding the reservoir liquids of Salton Sea that were modeled adopting the Pitzer's equations, all the reservoir liquids of Northern and Central America have fractions of free Mg^{2+} ion ranging from 0.74 to 0.07 (Fig. 3.65b). The free ion Mg^{2+} is the predominant aqueous species in 145 reservoir liquids, of which 105 from Miravalles, 16 from Heber, 10 from Long Valley, 7 from Los Azufres, 3 from Valles, and 2 each from Cerro Prieto and Coso. The $MgCl^+$ ion pair is the principal aqueous species in 94 reservoir liquids, of which 55 from Berlin, 18 from Los Azufres, 17 from Cerro Prieto, 3 from Coso, and 1 from Valles. The $MgSO_4°$ ion pair is the dominant aqueous species in 34 reservoir liquids from Dixie Valley, characterized by low-salinity and Na–Cl,SO_4 composition, whereas the $MgHCO_3^+$ ion pair is the predominant species in the other reservoir liquid of low-salinity and Na–Cl,HCO_3 composition from this site as well as in 2 reservoir liquids from Valles. The aqueous speciation of the remaining reservoir liquid from Valles is dominated by the $MgOH^+$ ion pair.

Most reservoir liquids of Japan have fractions of free Mg^{2+} ion lower than 0.5, with the lowest values, from 0.13 to 0.04 in 11 reservoir liquids, of which 8 from Sumikawa and 3 from Uenotai (Fig. 3.65c). The free ion Mg^{2+} is the principal aqueous species in 43 reservoir liquids, of which 13 from Mori-Nigorikawa, 10 from Oku-aizu, 7 from Onikobe, 5 from Oguni, 4 from Takigami, 2 from Sumikawa, and 1 from Wasabizawa. The $MgCl^+$ ion pair is the prevailing aqueous species in 49 reservoir liquids, of which 25 from Fushime, 20 from Oku-aizu, and 4 from Mori-Nigorikawa. The $MgOH^+$ ion pair is the main aqueous species in 13 reservoir liquids, of which 9 from Sumikawa, 3 from Uenotai, and 1 from Oguni. The $MgSO_4°$ ion pair is the predominant aqueous species in 11 reservoir liquids, of which 8 from Takigami and 3 from Sumikawa. The $MgHCO_3^+$ ion pair is the predominant species in the remaining 6 reservoir liquids, of which 4 from Mori-Nigorikawa and 1 each from Oku-aizu and Takigami.

A tight trend comprises all the Na–Cl reservoir liquids of the Philippines in the binary plot of aquifer temperature versus the fraction of free Mg^{2+} ion (Fig. 3.65d), but the Na-SO_4,Cl reservoir liquid MG32D_09 from Mahanagdong does not belong to this trend. The fraction of free Mg^{2+} ion varies from 0.64 to 0.13 in these 145 aqueous solutions with concentration of Mg higher than detection limit. The $MgCl^+$ ion pair is the main aqueous species in 104 reservoir liquids, of which 42 from Bacon-Manito, 21 from Palinpinon, 17 from Mahanagdong, 13 from Alto Peak, and 11 from Tongonan. The free ion Mg^{2+} is the major aqueous species in 40 reservoir liquids, of which 14 from Alto Peak, 10 from Mahanagdong, and 8 each from Bacon-Manito and Palinpinon. The $MgSO_4°$ ion pair prevails only in the reservoir liquid MG32D_09 from Mahanagdong.

The 81 reservoir liquids of New Zealand with Mg concentrations higher than the detection limit have fractions of free Mg^{2+} ion varying from 0.64 to 0.07, but apart from 9 entries (4 from Ngawha, 4 from Wairakei, and 1 from Broadlands), the other 72 have fractions of free Mg^{2+} ion lower than 0.5 (Fig. 3.65e). The main aqueous species are: (1) the $MgHCO_3^+$ ion pair in 36 reservoir liquids, of which 23 from Broadlands, 7 from Ngawha, and 3 each from Kawerau and Rotokawa; (2) the free Mg^{2+} ion in 26 reservoir liquids, of which 7 each from Ngatamariki and Wairakei, 6 from Broadlands, 5 from Ngawha, and 1 from Kawerau; (3) the $MgOH^+$ ion pair in 12 reservoir liquids, of which 9 from Broadlands, 2 from Rotokawa, and 1 from Wairakei; (4) the $MgCl^+$ ion pair in 7 reservoir liquids, of which 4 from Rotokawa and 3 from Mokai.

Disregarding the high-salinity reservoir liquids of Asal that were modeled adopting the Pitzer's equations and the 9 reservoir liquids with Mg concentration lower than detection limit, the other 157 miscellaneous reservoir liquids have fractions of free Mg^{2+} ion ranging from 0.77 to 0.03, with 34 cases higher than 0.50 and 123 entries lower than 0.50 (Fig. 3.65f). The main aqueous species are: (1) free Mg^{2+} ion in 56 reservoir liquids, of which 30 from Yangbajing, 24 from Tendaho, and 1 each from Latera and Kizildere; (2) the $MgHCO_3^+$ ion pair in 51 reservoir liquids, of which 26 from Kizildere, 8 from Ribeira Grande, 7 from Olkaria, 3 each from Aluto-Langano and Latera, and 2 each from Yangbajing and Bagnore; (3) the MgF^+ ion pair in 30 reservoir liquids from Olkaria; (4) the $MgOH^+$ ion pair in 15 reservoir liquids, of which 11 from Tendaho and 4 from Olkaria; (5) the $MgCl^+$ ion pair in 3 reservoir liquids from Mofete; (6) the $MgSO_4^\circ$ ion pair in 2 reservoir liquids, one from Latera and one from Tendaho.

A Final Remark The effects of ion association becomes more and more important with increasing ionic strength and temperature and, consequently, the free ion fractions deviate more and more from unity. Nevertheless, free Na^+ and K^+ ions generally represent by far the largest fraction of total dissolved Na and K, whereas free Mg^{2+} and Ca^{2+} ions often account for a minor fraction of total dissolved Mg and Ca.

3.8.4 *The Activity Coefficient of Undissociated $SiO_{2(aq)}$*

Consistent with the Garrels and Christ (1965) recommendation, the activity coefficients of $SiO_{2(aq)}$ is set to unity for the relatively dilute solutions, that is when the activity coefficients of individual ions are calculated by means of the B-dot equation (see Sect. 2.3.3). This approach was adopted for all reservoir liquids except the brines of Asal and Salton Sea.

In contrast, the Pitzer equations (see Sect. 2.3.4) were used to model the speciation of these brines, representing the focus of the present discussion. The activity coefficients of $SiO_{2(aq)}$ and the ionic strengths of the Asal and Salton Sea brines are contrasted in the two binary plots of Fig. 3.66a, b, in which the other reservoir liquids of Northern and Central America and the other miscellaneous reservoir liquids, respectively, are also displayed as usual. These two graphs show that the Asal brines have activity coefficient of $SiO_{2(aq)}$ oscillating from 0.915 to 1.09 and ionic strengths of 0.22–2.5 mol/kg, whereas the Salton Sea brines have activity coefficient of $SiO_{2(aq)}$ varying from 0.859 to 0.670 and ionic strengths of 4.2–5.5 mol/kg. These significant deviations from unity of the activity coefficient of $SiO_{2(aq)}$ are caused by the increasing solute-solvent and solute-solute interactions with increasing ionic strength.

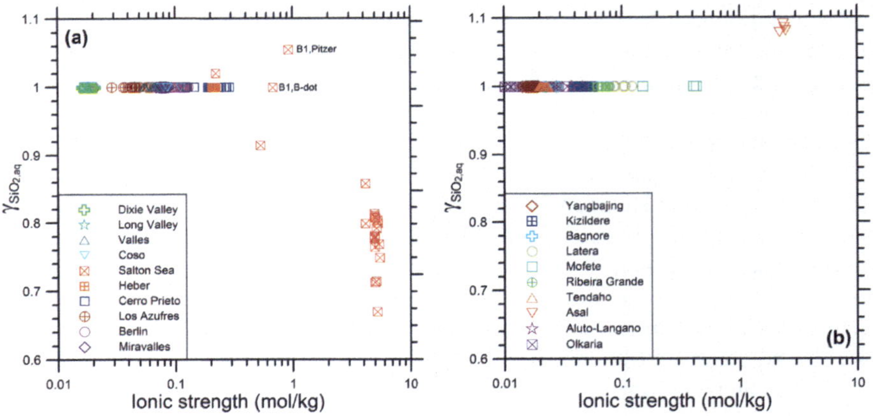

Fig. 3.66 Correlations diagrams showing the activity coefficient of undissociated $SiO_{2(aq)}$ as a function of ionic strength for the reservoir liquids of **a** Northern and Central America and **b** Miscellaneous geothermal fields

3.8.5 The Activity Coefficients of Free Na^+, K^+, Ca^{2+}, and Mg^{2+} Ions

The activity coefficients of free Na^+, K^+, Ca^{2+}, and Mg^{2+} ions are plotted versus aquifer temperature in the binary graphs of Figs. 3.67, 3.68, 3.69 and 3.70, respectively. In each diagram there are also two lines displaying the dependence of individual-ion activity coefficients on temperature at two fixed ionic strengths. These two lines were generated using the B-dot equation (see Sect. 2.3.3) and choosing two suitable ionic strengths to bracket most reservoir liquids shown in each plot. In detail:

1. For the geothermal fields in Iceland, the two selected ionic strengths are 0.0069 mol/kg, which is the average of the reservoir liquids from Krafla, Námafjall, Hellisheidi, Nesjavellir, Hveragerdi, and the medium temperature sites other than Bakki and Seltjarnarnes, and 0.34 mol/kg, which is the mean of the reservoir liquids from Reykjanes and Svartsengi.
2. For the geothermal fields in Northern and Central America, the two chosen ionic strengths are 0.018 mol/kg, which is the average of the reservoir liquids from Dixie Valley and Long Valley, and 0.22 mol/kg, which is the mean of the reservoir liquids from Cerro Prieto and Heber.
3. For the geothermal fields in Japan, the two adopted ionic strengths are 0.015 mol/kg, which is the average of the reservoir liquids from Sumikawa and Uenotai, and 0.34 mol/kg, which is the mean of the reservoir liquids from Fushime.
4. For the geothermal fields of the Philippines, the two selected ionic strengths are 0.046 mol/kg, which is the mean of the reservoir liquids encountered by well

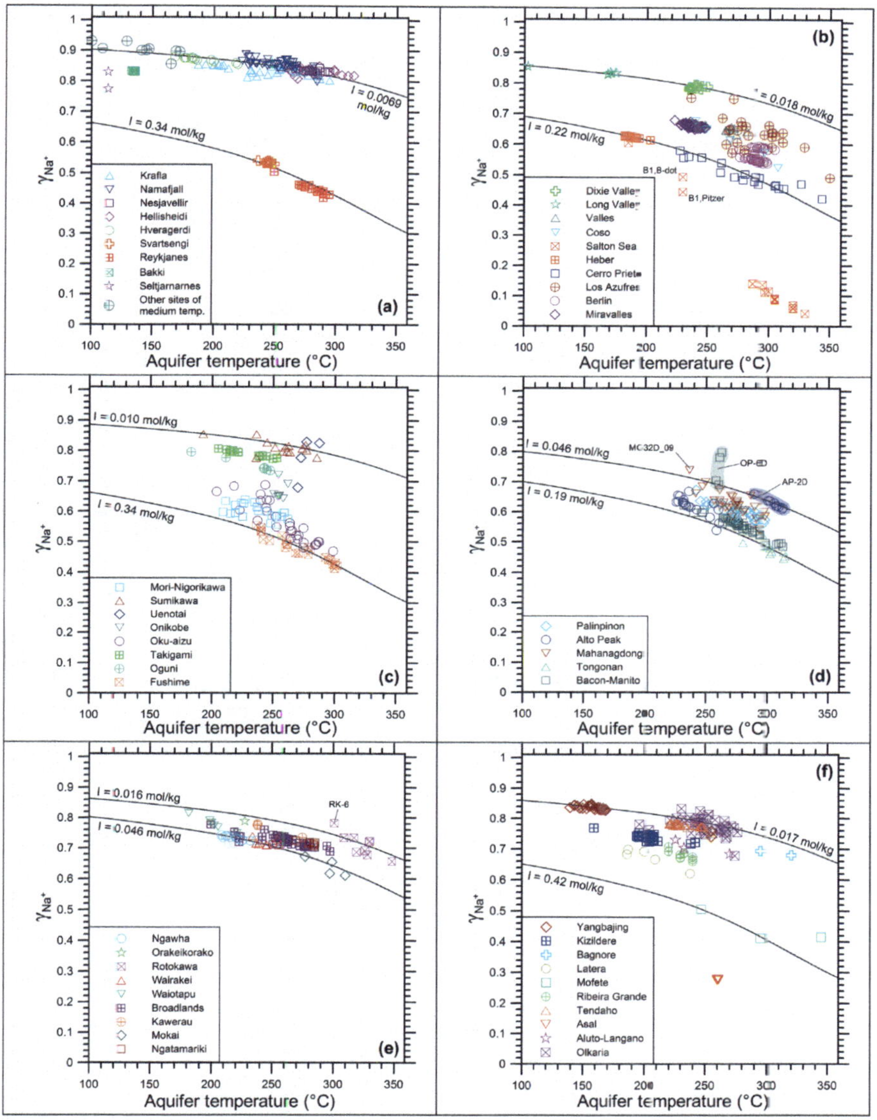

Fig. 3.67 Correlations diagrams showing the activity coefficient of free Na^+ ion as a function of aquifer temperature for the reservoir liquids of **a** Iceland, **b** Northern and Central America, **c** Japan, **d** The Philippines, **e** New Zealand, and **f** Miscellaneous geothermal fields

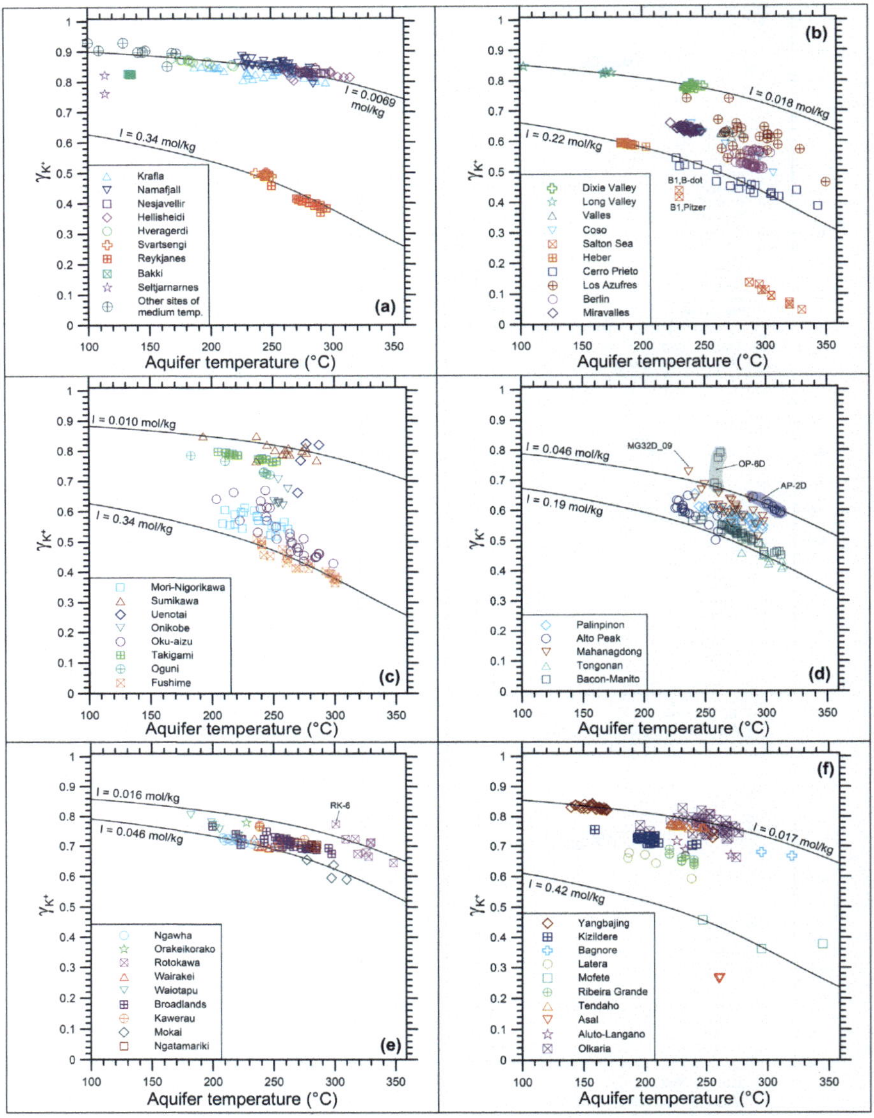

Fig. 3.68 Correlations diagrams showing the activity coefficient of free K^+ ion as a function of aquifer temperature for the reservoir liquids of **a** Iceland, **b** Northern and Central America, **c** Japan, **d** The Philippines, **e** New Zealand, and **f** Miscellaneous geothermal fields

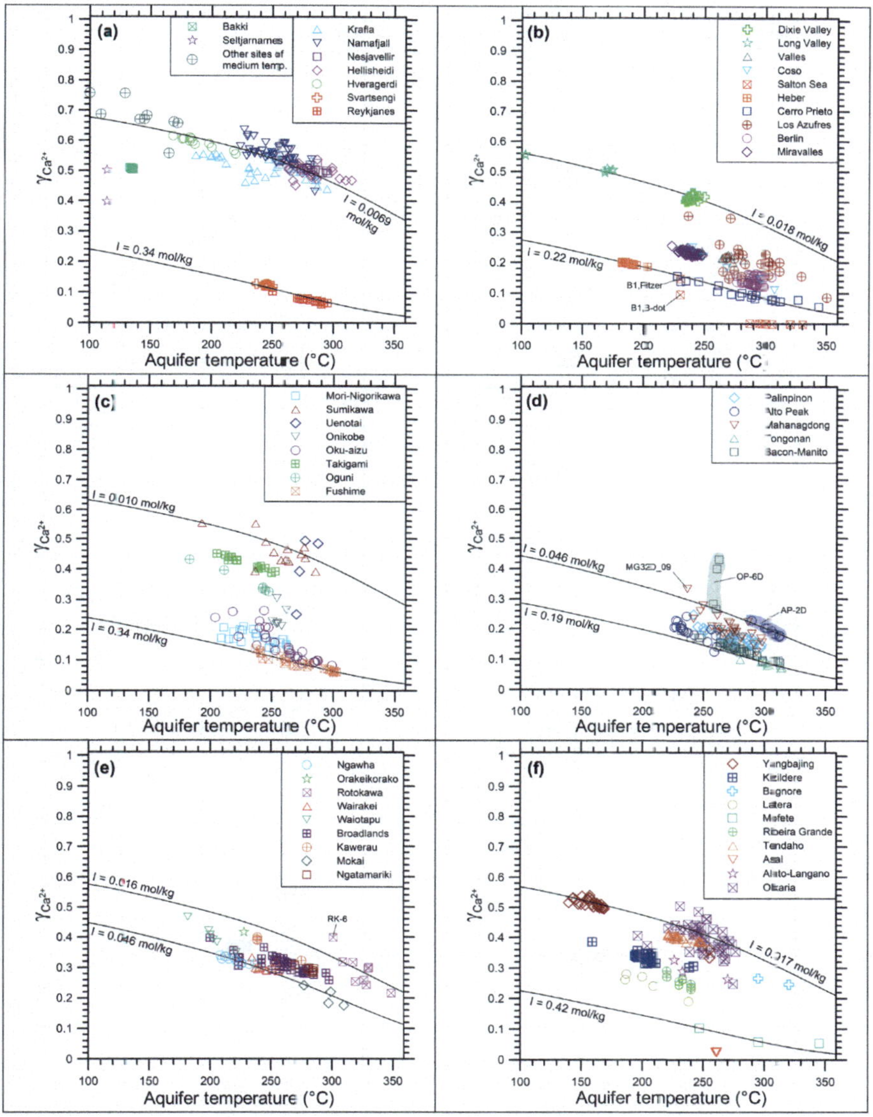

Fig. 3.69 Correlations diagrams showing the activity coefficient of free Ca^{2+} ion as a function of aquifer temperature for the reservoir liquids of **a** Iceland, **b** Northern and Central America, **c** Japan, **d** The Philippines, **e** New Zealand, and **f** Miscellaneous geothermal fields

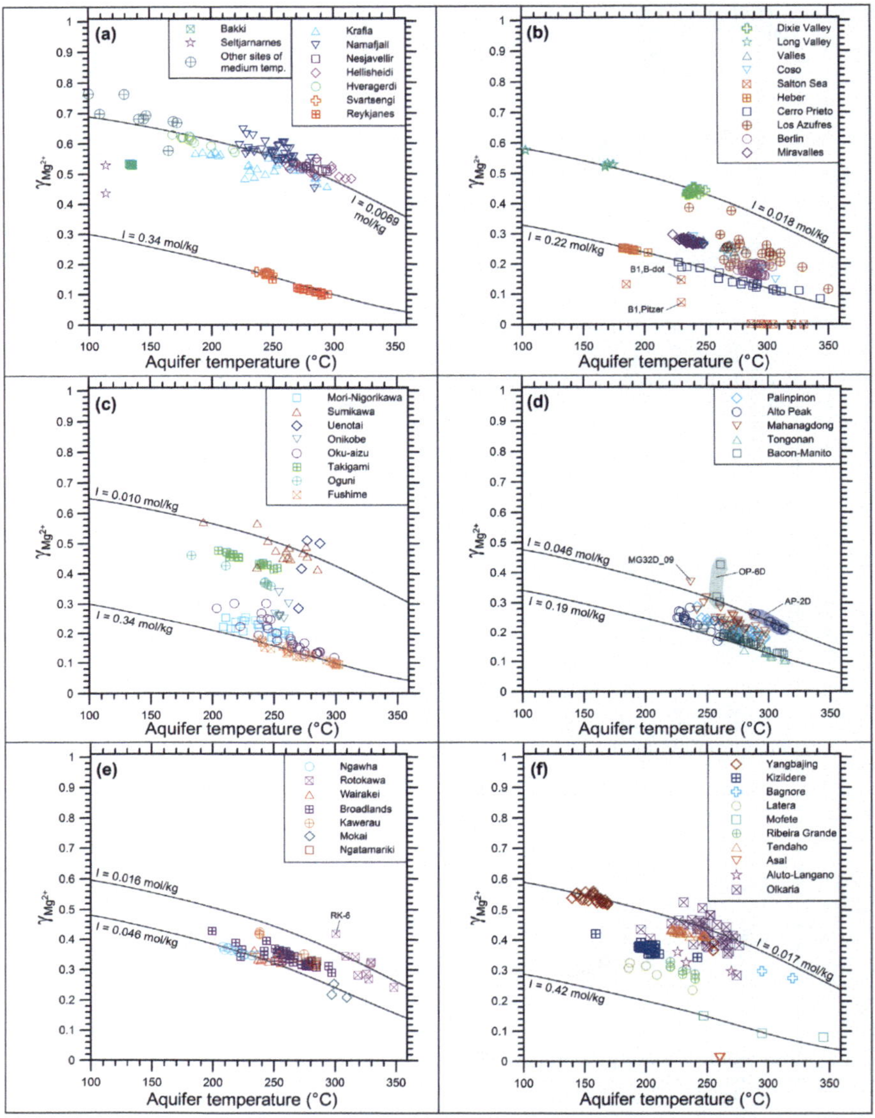

Fig. 3.70 Correlations diagrams showing the activity coefficient of free Mg^{2+} ion as a function of aquifer temperature for the reservoir liquids of **a** Iceland, **b** Northern and Central America, **c** Japan, **d** The Philippines, **e** New Zealand, and **f** Miscellaneous geothermal fields

AP2D of Alto Peak, and 0.19 mol/kg, which is the average of the reservoir liquids from Tongonan.

5. For the New Zealand geothermal fields, the two chosen ionic strengths are 0.016 mol/kg, which is the mean of the reservoir liquids from Rotokawa, and 0.046 mol/kg, which is the average of the reservoir liquids from Mokai and Ngawha.
6. For the miscellaneous geothermal fields, the two adopted ionic strengths are 0.017 mol/kg, which is the mean of the reservoir liquids from Olkaria and Yangbajing, and 0.42 mol/kg, which is the average of the reservoir liquids met by well MF-1 of Mofete.

The Activity Coefficient of Free Na^+ Ion The activity coefficient of free Na^+ ion varies from 0.420 to 0.931 for the reservoir liquids of Iceland, from 0.423 to 0.856 for the reservoir liquids of Northern and Central America apart from Salton Sea, from 0.413 to 0.855 for the reservoir liquids from the geothermal fields in Japan, from 0.447 to 0.798 for the reservoir liquids from the geothermal fields in the Philippines, from 0.611 to 0.815 for the reservoir liquids from the New Zealand geothermal fields, and from 0.412 to 0.849 for the reservoir liquids from the miscellaneous geothermal fields disregarding Asal (Fig. 3.67).

The Activity Coefficient of Free K^+ Ion The activity coefficient of free K^+ ion ranges from 0.370 to 0.930 for the reservoir liquids of Iceland, from 0.386 to 0.850 for the reservoir liquids of Northern and Central America except Salton Sea, from 0.365 to 0.852 for the reservoir liquids from the geothermal fields in Japan, from 0.407 to 0.791 for the reservoir liquids from the geothermal fields in the Philippines, from 0.591 to 0.807 for the reservoir liquids from the New Zealand geothermal fields, and from 0.363 to 0.844 for the reservoir liquids from the miscellaneous geothermal fields ignoring Asal (Fig. 3.68).

The Activity Coefficient of Free Ca^{2+} Ion The activity coefficient of free Ca^{2+} ion varies from 0.0619 to 0.758 for the reservoir liquids of Iceland, from 0.0571 to 0.558 for the reservoir liquids of Northern and Central America disregarding Salton Sea, from 0.0589 to 0.554 for the reservoir liquids from the geothermal fields in Japan, from 0.0704 to 0.430 for the reservoir liquids from the geothermal fields in the Philippines, from 0.176 to 0.467 for the reservoir liquids from the New Zealand geothermal fields, and from 0.0543 to 0.540 for the reservoir liquids from the miscellaneous geothermal fields apart from Asal (Fig. 3.69).

The Activity coefficient of free Mg^{2+} Ion The activity coefficient of free Mg^{2+} ion ranges from 0.0989 to 0.765 for the reservoir liquids of Iceland, from 0.0849 to 0.580 for the reservoir liquids of Northern and Central America neglecting Salton Sea, from 0.0941 to 0.572 for the reservoir liquids from the geothermal fields in Japan, from 0.104 to 0.426 for the reservoir liquids from the geothermal fields in the Philippines, from 0.210 to 0.431 for the reservoir liquids from the New Zealand geothermal fields, and from 0.0816 to 0.562 for the reservoir liquids from the miscellaneous geothermal fields ignoring Asal (Fig. 3.70).

Individual-Ion Activity Coefficients for the Asal and Salton Sea Reservoir Brines The individual-ion activity coefficients for the reservoir brines from Asal, of ionic strength of 2.2–2.5 mol/kg, and Salton Sea, of ionic strength of 4.2–5.5 mol/kg, whose speciation was modeled using the Pitzer's approach (see Sect. 2.3.4), are lower or even much lower than those calculated by the B-dot equation for the other reservoir liquids. This is not surprising because solute-solvent and solute-solute interactions are chiefly described by activity coefficients and considering few ion pairs, in the Pitzer's approach, as already recalled in previous section. In detail:

1. The activity coefficient of free Na^+ ion varies from 0.276 to 0.281 for the Asal reservoir brines and from 0.0436 to 0.143 for the Salton Sea reservoir brines.
2. The activity coefficient of free K^+ ion ranges from 0.262 to 0.268 for the Asal reservoir brines and from 0.0478 to 0.137 for the Salton Sea reservoir brines.
3. The activity coefficient of free Ca^{2+} ion varies from 0.0252 to 0.0276 for the Asal reservoir brines and from 5.66×10^{-5} to 3.04×10^{-3} for the Salton Sea reservoir brines.
4. The activity coefficient of free Mg^{2+} ion varies from 0.0140 to 0.0152 for the Asal reservoir brines and from 1.09×10^{-4} to 2.87×10^{-3} for the Salton Sea reservoir brines.

A Final Remark All in all, the activity coefficients of free Na^+ and K^+ ions and especially those of free Mg^{2+} and Ca^{2+} ions exhibit large deviations from unity, which become more and more important with increasing ionic strength and temperature.

3.9 Final Considerations on the Reservoir Liquids

All the 1013 reservoir liquids of high and medium temperature coming from different geothermal systems worldwide are probably mature waters, irrespective of the sources and processes contributing the main solutes and controlling the different chemical compositions, salinities, and pH values of these reservoir liquids. In other terms, all these reservoir liquids have probably attained or closely approached the condition of chemical equilibrium with the hydrothermal (secondary) minerals occurring in the geothermal reservoirs of provenance, possibly excluding a few cases, such as some reservoir liquids of Onikobe, whose chemistry is impacted by magmatic-derived HCl and H_2SO_4.

If so, it is necessary to reconsider the axiom that neutral sodium-chloride reservoir liquids are the only mature waters in equilibrium with hydrothermal minerals. This axiom applies to the geothermal systems situated along convergent plate boundaries, such as those of Central America (e.g., Miravalles and Berlin in Sect. 3.3), Japan (Sect. 3.4), the Philippines (Sect. 3.5), and New Zealand (Sect. 3.6), as well as to other geothermal systems, as shown by Ellis and Mahon (1977) and Giggenbach (1988) among the others. However, this axiom does not apply to some geothermal systems positioned along the oceanic rift zone in Iceland, such as Krafla, Námafjall, Hveragerdi and, to some extent, Hellisheidi and Nesjavellir (Sect. 3.2), some geothermal systems located along the continental rift zone in East Africa, such as Aluto-Langano and, to some extent, Olkaria (Sect. 3.7), some geothermal systems of the Basin and Range Province in Northern America, such as Dixie Valley and Long Valley (Sect. 3.3), the fault-controlled geothermal systems of Turkey, such as Kizildere (Sect. 3.7) and, at least in part, the Ribeira Grande geothermal system in the Azores islands, occupying the triple junction between the North American, African and Eurasian Plates (Sect. 3.7).

Therefore, we suggest to adopt a comprehensive approach to water classification, including not only the triangular diagram of major anions but also the triangular diagram of main cations and suitable salinity plots, as shown in previous sections. These graphs should be prepared from the concentrations in equivalent units and a chemical terminology should be adopted. Chloride-solute diagrams and enthalpy-chloride plots (not shown here) are also essential for identifying mixing and boiling processes, as well as to reconstruct the behavior of the solutes of interest. Based on a sufficiently complete analysis of the existing geochemical data, it should be possible to distinguish mature waters suitable for geothermometry from immature waters unsuitable for geothermometry, also in surface exploration projects, in which fluids delivered from deep wells are not available.

Total concentrations of cations and SiO_2 are normally inserted into traditional geothermometers and f_{CO_2} indicators, but mineral-solution equilibrium fixes the activities of free ions and undissociated $SiO_{2(aq)}$. The activities of Na^+, K^+, Mg^{2+}, and Ca^{2+} ions and undissociated $SiO_{2(aq)}$ deviate from the total concentrations of Na, K, Mg, Ca, and SiO_2, respectively, due to the formation of ion pairs and dissociation of silicic acid, as indicated by the fractions of free Na^+, K^+, Mg^{2+}, and Ca^{2+} ions and $SiO_{2(aq)}$ (Sects. 3.8.2 and 3.8.3), and because of solute-solvent and solute-solute interactions, as described by activity coefficients (Sects. 3.8.4 and 3.8.5). Above, it was shown that the effects of these processes can cause significant differences between activities and total concentrations, a fact that must be taken in due account, as we intend to do in this work, through the elaboration of theoretical, activity-based geothermometers and f_{CO_2} indicators.

Above, it was also shown that pH is strongly impacted by uncertainties in the computation of reservoir liquids chemistry (see Sect. 3.1.2). Therefore, the new theoretical geothermometers and f_{CO_2} indicators should involve variables which are not influenced or negligibly influenced by pH, like the concentrations and activities of non-volatile solutes.

References

Abe M, Yamada M, Kawano Y, Todaka N, Tezuka S (1995) Development of the Oguni geothermal field, Japan. In: Proceedings world geothermal congress, Florence, Italy, pp 1319–1322

Adams MC, Lemieux M, Moore JN, Johnson SD (1989) Fluid chemistry and hydrology of the Heber geothermal system, California. In: Proceedings, fourteenth workshop on geothermal reservoir engineering. Stanford University, Stanford, California, 24–26 Jan 1989. SGP-TR-122, pp 81–86

Aggarwal JK, Sheppard D, Mezger K, Pernicka E (2003) Precise and accurate determination of boron isotope ratios by multiple collector ICP-MS: origin of boron in the Ngawha geothermal system, New Zealand. Chem Geol 199:331–342

Akaku K (1990) Geochemical study on mineral precipitation from geothermal waters at the Fushime field, Kyushu, Japan. Geothermics 19:455–467

Akaku K, Reed MH, Yagi M, Kai K, Yasuda Y (1991) Chemical and physical processes occurring in the Fushime geothermal system, Kyushu, Japan. Geochem J 25:315–333

Ali S (2005) Geochemical studies of the Tendaho geothermal field. In: Proceedings world geothermal congress. Antalya, Turkey

Alvis-Isidro RR, Solana RR, D'Amore F, Nuti S, Gonfiantini R (1993) Hydrology of the Greater Tongonan geothermal system, Philippines, as deduced from geochemical and isotopic data. Geothermics 22:435–449

Angcoy EC (2010) Geochemical modelling of the high-temperature Mahanagdong geothermal field, Leyte, Philippines. UNU-GTP report 2010-1, 71p

Arellano VM, Torres MA, Barragán RM (2005) Thermodynamic evolution of the Los Azufres, Mexico, geothermal reservoir from 1982 to 2002. Geothermics 34:592–616

Ariki K, Kato H, Ueda A, Bamba M (2000) Characteristics and management of the Sumikawa geothermal reservoir, northeastern Japan. Geothermics 29:171–189

Ármannsson H, Gíslason G, Hauksson T (1982) Magmatic gases in well fluids aid the mapping of the flow pattern in a geothermal system. Geochim Cosmochim Ac 46:167-177

Ármansson H (2016) The fluid geochemistry of Icelandic high temperature geothermal areas. Appl Geochem 66:14–64

Arnórsson S (1978a) Major element chemistry of the geothermal sea-water at Reykjanes and Svartsengi, Iceland. Mineral Mag 42:209–220

Arnórsson S (1978b) Precipitation of calcite from flashed geothermal waters in Iceland. Contrib Mineral Petr 66:21–28

Arnórsson S (1995a) Geothermal systems in Iceland: structure and conceptual models—I. High-temperature areas. Geothermics 24:561–602

Arnórsson S (1995b) Geothermal systems in Iceland: Structure and conceptual models—II. Low-temperature areas. Geothermics 24:603–629

Arnórsson S, Stefánsson A (2005b). Wet-steam well discharges. II. Assessment of aquifer fluid compositions. In: Proceedings world geothermal congress. Antalya, Turkey

Arnórsson S, Grönvold K, Sigurdsson S (1978) Aquifer chemistry of four high-temperature geothermal systems in Iceland. Geochim Cosmochim Acta 42:523–536

Arnórsson S, Sigurdsson S, Svavarsson H (1982) The chemistry of geothermal waters in Iceland. I. Calculation of aqueous speciation from 0 to 370 °C. Geochim Cosmochim Acta 46:1513–1532

Arnórsson S, Gunnlaugsson E, Svavarsson H (1983a) The chemistry of geothermal waters in Iceland. III. Chemical geothermometry in geothermal investigations. Geochim Cosmochim Acta 47:567–577

Arnórsson S, Gunnlaugsson E, Svavarsson H (1983b) The chemistry of geothermal waters in Iceland. II. Mineral equilibria and independent variables controlling water compositions. Geochim Cosmochim Acta 47:547–566

Arnórsson S, Björnsson S, Muna ZW, Bwire-Ojiambo S (1990) The use of gas chemistry to evaluate boiling processes and initial steam fractions in geothermal reservoirs with an example from the Olkaria field, Kenya. Geothermics 19:497–514

Arnórsson S, Stefánsson A, Bjarnason JÖ (2005) Wet-steam well discharges. I. Sampling and calculation of total discharge compositions. Proceedings World Geothermal Congress, Antalya, Turkey

Arnórsson S, Stefánsson A, Bjarnason JO (2007) Fluid-fluid interactions in geothermal systems. Rev Mineral Geochem 65:259–312

Arnórsson S, Angcoy E, Bjarnason JÖ, Giroud N, Gunnarsson I, Kaasalainen H, Karingithi C, Stefánsson A (2010) Gas chemistry of volcanic geothermal systems. In: World geothermal congress, Bali, Indonesia

Auko LO (2014) Evaluation of fluid-mineral interaction in the Menengai geothermal system, Central Rift, Kenya. UNU-GTP report 2014-8, pp 39–64

Bacon CR, Duffield WA, Nakamura K (1980) Distribution of quaternary rhyolite domes of the Coso Range, California: implications for extent of the geothermal anomaly. J Geophys Res-Sol Earth 85(B5):2425–2433

Bailey RA, Dalrymple GB, Lanphere MA (1976) Volcanism, structure, and geochronology of Long Valley caldera, Mono County, California. J Geophys Res-Solid 81(5):725–744

Balducci S, Chelini W (1992) Hydrothermal equilibria in the active Mofete geothermal system, Phlegraean Fields, Naples, Italy. Acta Vulcanol Marinelli 2:17–34

Balmes CP (1994) The geochemistry of the Mahanagdong sector, Tongonan geothermal field, Philippines. UNU-GTP report 1994-2, pp 31–52

Baltasar AJ (1980) Interpretation of the water and gas chemistry from three geothermal areas in the Philippines—Manito in Albay, Biliran Island and Tongonan in Leyte. UNU-GTP report 1980-3, 55p

Banwell CJ (1957) Borehole measurements. In: Physics of the New Zealand thermal area, bulletin 123. DSIR, Wellington, New Zealand, pp 39–72

Battaglia A, Ceccarelli A, Ridolfi A, Frohlich K, Panichi C (1992) Radium isotopes in geothermal fluids in central Italy. Proceedings of international symposium on isotope techniques in water resources development, I.A.E.A., 11–15 Mar 1991, Vienna, pp 363–383

Battistelli A, Yiheyis A, Calore C, Ferragina C, Abatneh W (2002) Reservoir engineering assessment of Dubti geothermal field, northern Tendaho rift, Ethiopia. Geothermics 31:381–406

Berehannu MM (2014) Geochemical interpretation of discharge from Reykjanes well 29 and well 32. UNU-GTP report 2014-19, pp 351–368

Bertini G, Cappetti G, Dini I, Lovari F (1995) Deep drilling results and updating of geothermal knowledge of the Monte Amiata area. In: Proceedings of the world geothermal congress, Florence, Italy, vol 2, pp 1283–1286

Bixley PF, Clotworthy AW, Mannington WI (2009) Evolution of the Wairakei geothermal reservoir during 50 years of production. Geothermics 38:145–154

Bjarnason JÖ (2010) The chemical speciation program WATCH, version 2.4. ISOR–Iceland GeoSurvey, Reykjavik, Iceland

Browne PRL (1980) Joint channels in reservoir rocks of the Ngawha geothermal field, Northland, New Zealand. In: N.Z. geothermal workshop, proceedings. University of Auckland, pp 81–84

Bruton CJ (1995) Testing EQ3/6 and GEMBOCHS using fluid-mineral equilibria in the Wairakei geothermal system. Lawrence Livermore National Laboratory report UCRL-ID-129280, 21p

Calamai A, Cataldi R, Squarci P, Taffi L (1970) Geology, geophysics and hydrogeology of the Mt. Amiata geothermal field. Geothermics 1(Special Issue):1–9

Carella R, Guglielminetti M (1983) Multiple reservoirs in the Mofete field, Naples, Italy. In: Proceedings ninth workshop geothermal reservoir engineering. Stanford University, Stanford, California, Dec 1983, SGP-TU-74, pp 53–64

Carvalho MR, Forjaz VH, Almeida C (2006) Chemical composition of deep hydrothermal fluids in the Ribeira Grande geothermal field (São Miguel, Azores). J Volcanol Geother Res 156:116–134

Cavarretta G, Gianelli G, Scandiffio G, Tecce F (1985) Evolution of the Latera geothermal system II: metamorphic, hydrothermal mineral assemblages and fluid chemistry. J Volcanol Geother Res 26:337–364

Chambefort I, Buscarlet E, Wallis IC, Sewell S, Wilmarth M (2016) Ngatamariki geothermal field, New Zealand: Geology, geophysics, chemistry and conceptual model. Geothermics 59:266–280
Charlton SR, Parkhurst DL (2002) PhreeqcI-A graphical user interface to the geochemical model PHREEQC. U.S. Geological Survey Fact Sheet FS-031-02, 2p
Chiba H (1991) Attainment of solution and gas equilibrium in Japanese geothermal systems. Geochem J 25:335–355
Chiodini G, Cioni R, Guidi M, Marini L (1991) Chemical geothermometry and geobarometry in hydrothermal aqueous solutions: A theoretical investigation based on a mineral-solution equilibrium model. Geochim Cosmochim Acta 55:2709–2727
Christenson BW (1997) Kawerau geothermal field: geochemical structure of the reservoir and its response to exploitation. Geother Res T 21:17–24
Christenson BW, Mroczek EK, Kennedy BM, van Soest MC, Stewart MK, Lyon G (2002) Ohaaki reservoir chemistry: characteristics of an arc-type hydrothermal system in the Taupo Volcanic Zone, New Zealand. J Volcanol Geother Res 115:53–82
Clarke GC, Woodhall DG, Allen D, Darling G (1990) Geological, volcanological and hydrogeological controls on the occurrence of geothermal activity in the area surrounding Lake Naivasha. Kenya, Ministry of Energy report, Kenya, 245p
Clearwater E, Seastres J Jr, Newson J, Mulusa G (2015) Modelling of scaling in a Tauhara production well. In: Proceedings world geothermal congress 2015, Melbourne, Australia, 19–25 Apr 2015, 9p
Cox ME, Browne P (1998) Hydrothermal alteration mineralogy as an indicator of hydrology at the Ngawha geothermal field, New Zealand. Geothermics 27:259–270
D'Amore F, Mejia JT (1999) Chemical and physical reservoir parameters at initial conditions in Berlin geothermal field, El Salvador: a first assessment. Geothermics 28:45–73
D'Amore F, Giusti D, Gizaw B (1997) Geochemical assessment of the Northern Tendaho Rift, Ethiopia. In: Proceedings of twenty-second workshop on geothermal reservoir engineering. Stanford University. SGP-TR-155, pp 435–445
D'Amore F, Giusti D, Abdallah A (1998) Geochemistry of the high-salinity geothermal field of Asal, Republic of Djibouti, Africa. Geothermics 27:197–210
Darling WG (1998) Hydrothermal hydrocarbon gases: 2. Application in the East African Rift System. Appl Geochem 13:825–840
Dominco E, Samilgil E (1970) The geochemistry of the Kizildere geothermal field, in the framework of the Saraykoy-Denizli geothermal area. Geothermics 2:553–560
Ellis AJ (1970) Quantitative interpretation of chemical characteristics of hydrothermal systems. Geothermics 2:516–528
Ellis AJ, Mahon WAJ (1977) Chemistry and geothermal systems. Academic Press, 392p
Farrar CD, Sorey ML, Roeloffs E, Galloway DL, Howle JF, Jacobson R (2003) Inferences on the hydrothermal system beneath the resurgent dome in Long Valley Caldera, east-central California, USA, from recent pumping tests and geochemical sampling. J Volcanol Geoth Res 127:305–328
Ferrari L, Garduño VH, Pasquarè G, Tibaldi A (1991) Geology of Los Azufres Caldera, Mexico, and its relationships with regional tectonics. J Volcanol Geoth Res 47:129–148
Fouillac AM, Fouillac C, Cesbron F, Pillard F, Legendre O (1989) Water-rock interaction between basalt and high-salinity fluids in the Asal Rift, Republic of Djibouti. Chem Geol 76:271–289
Fournier RO (1977) Chemical geothermometers and mixing models for geothermal systems. Geothermics 5:41–50
Furuya S, Aoki M, Gotoh H, Takenaka T (2000) Takigami geothermal system, northeastern Kyushu, Japan. Geothermics 29:191–211
Garrels RM, Christ CL (1965) Solutions, minerals, and equilibria. Harper & Row, New York, p 450
Gherardi F, Panichi C, Yock A, Gerardo-Abaya J (2002) Geochemistry of the surface and deep fluids of the Miravalles volcano geothermal system (Costa Rica). Geothermics 31:91–128
Gianelli G, Scandiffio G (1989) The Latera geothermal system (Italy): chemical composition of the geothermal fluid and hypotheses on its origin. Geothermics 18:447–463

Gianelli G, Teklemariam M (1993) Water-rock interaction processes in the Aluto-Langano geothermal field (Ethiopia). J Volcanol Geother Res 56:429–445
Gianelli G, Puxeddu M, Batini F, Bertini G, Dini I, Pandeli E, Nicolich R (1988) Geological model of a young volcano-plutonic system: the geothermal region of Monte Amiata (Tuscany, Italy). Geothermics 17:719–734
Giggenbach WF (1980) Geothermal gas equilibria. Geochim Cosmochim Acta 44:2021–2032
Giggenbach WF (1988) Geothermal solute equilibria. derivation of Na–K–Mg–Ca geoindicators. Geochim Cosmochim Acta 52:2749–2765
Giggenbach WF (1995) Variations in the chemical and isotopic composition of fluids discharged from the Taupo Volcanic Zone, New Zealand. J Volcanol Geoth Res 68:89–116
Giggenbach WF (1997a) The origin and evolution of fluids in magmatic-hydrothermal systems. In: Geochemistry of hydrothermal ore deposits, 3rd edn. In: Barnes HL (ed) Wiley, pp 737–796
Giggenbach WF, Corrales Soto R (1992) Isotopic and chemical composition of water and steam discharges from volcanic-magmatic-hydrothermal systems of the Guanacaste Geothermal Province, Costa Rica. Appl Geochem 7:309–332
Giroud N (2008) A chemical study of arsenic, boron and gases in high-temperature geothermal fluids in Iceland. Ph.D. thesis, Faculty of Science, University of Iceland, 110p
Gizaw B (1993) Aluto-Langano geothermal field, Ethiopian Rift Valley: physical characteristics and the effects of gas on well performance. Geothermics 22:101–116
Gizaw B (1996) The origin of high bicarbonate and fluoride concentrations in waters of the Main Ethiopian Rift Valley, East African Rift system. J Afr Earth Sci 22(4):391–402
Glover RB, Mroczek EK (2009) Chemical changes in natural features and well discharges in response to production at Wairakei. New Zealand. Geothermics 38:117–133
Glover RB, Lovelock B, Ruaya JR (1981) A novel way of using gas and enthalpy data. In: Proceedings of 3rd N.Z. geothermal workshop. University of Auckland, pp 163–169
Goff F, Shevenell L, Gardner JN, Vuataz F-D, Grigsby CO (1988) The hydrothermal outflow plume of Valles Caldera, New Mexico, and a comparison with other outflow plumes. J Geophys Res-Sol Earth 93(B6):6041–6058
Goff F, Bergfeld D, Janik CJ Counce D, Murrell M (2002) Geochemical data on waters, gases, scales, and rocks from the Dixie Valley Region, Nevada (1996–1999). Los Alamos report LA-13972-MS, 71p
González-Partida EG, Tello-Hinojosa ET, Pal-Verma M (2000) Análisis geoquímico e isotópico de aguas geotérmicas y manantiales para definir el estado de equilibrio agua-roca del reservorio de Los Azufres, Michoacán, México. Ing Hidraul Mex 15:88–99
González-Partida E, Carrillo-Chávez A, Levresse G, Tello-Hinojosa E, Venegas-Salgado S, Ramirez-Silva G, Pal-Verma M, Tritlla J, Camprubi A (2005) Hydro-geochemical and isotopic fluid evolution of the Los Azufres geothermal field, Central Mexico. Appl Geochem 20:23–39
Gudmundsson BT, Arnórsson S (2002) Geochemical monitoring of the Krafla and Námafjall geothermal areas, N-Iceland. Geothermics 31:195–243
Guidi M, Marini L, Cioppi D (1988) Geotermometria chimica su acque geotermiche provenienti da serbatoi carbonatici. CNR-PFE report, 38p
Guidi M, Marini L, Principe C (1990) Hydrogeochemistry of Kizildere geothermal system and nearby region. Geother Res T 14:901–908
Guo Q, Nordstrom DK, McCleskey RB (2014) Towards understanding the puzzling lack of acid geothermal springs in Tibet (China): insight from a comparison with Yellowstone (USA) and some active volcanic hydrothermal systems. J Volcanol Geother Res 288:94–104
Haizlip JR, Haklidir FST (2011) High noncondensible gas liquid dominated geothermal reservoir Kizildere, Turkey. Geother Res T 35:615–618
Haklidir FT, Sengun R, Haizlip JR (2015) The geochemistry of the deep reservoir wells in Kizildere (Denizli City) geothermal field (Turkey). In: Proceedings world geothermal congress 2015—Melbourne, Australia, 19–25 Apr 2015, 4p

Hardardóttir V (2011) Metal-rich scales in the Reykjanes geothermal system, SW Iceland. Sulfide minerals in a seawater-dominated hydrothermal environment. Ph.D. thesis, Faculty of Science, University of Ottawa, 288p
Hardardóttir V, Brown KL, Fridriksson T, Hedenquist JW, Hannington MD, Thorhallsson S (2009) Metals in deep liquid of the Reykjanes geothermal system, southwest Iceland: implications for the composition of seafloor black smoker fluids. Geology 37:1103–1106
Hedenquist JW (1986) Geothermal systems in the Taupo volcanic zone; their characteristics and relation to volcanism and mineralisation. In: Smith IEM (ed) Late Cenozoic volcanism in New Zealand, vol 23. Bulletin—Royal Society of New Zealand, pp 134–168
Hedenquist JW (1990) The thermal and geochemical structure of the Broadlands-Ohaaki geothermal system, New Zealand. Geothermics 19:151–185
Hedenquist JW, Browne PR (1989) The evolution of the Waiotapu geothermal system, New Zealand, based on the chemical and isotopic composition of its fluids, minerals and rocks. Geochim Cosmochim Acta 53:2235–2257
Hedenquist JW, Lowenstern JB (1994) The role of magmas in the formation of hydrothermal ore deposits. Nature 370:519–526
Hedenquist JW, Mroczek EK, Giggenbach WF (1988) Geochemistry of the Rotokawa geothermal system: summary of data, interpretation and appraisal for energy development. Chemistry Division DSIR Technical Note 88/6, 63p
Hedenquist JW, Goff F, Phillips FM, Elmore D, Stewart MK (1990) Groundwater dilution and residence times, and constraints on chloride source, in the Mokai geothermal system, New Zealand, from chemical, stable isotope, tritium, and ^{36}Cl data. J Geophys Res-Sol Earth 95(B12):19365–19375
Helgeson HC (1968) Geologic and thermodynamic characteristics of the Salton Sea geothermal system. Am J Sci 266:129–166
Helgeson HC, Delany JM, Nesbitt HW, Bird DK (1978) Summary and critique of the thermodynamic properties of rock-forming minerals. Am J Sci 278A:229p
Henley RW, Middendorf KI (1985) Geothermometry in the recent exploration of Mokai and Rotokawa geothermal fields, New Zealand. Geoth Res T 9:317–324
Henley RW, Plum H (1985) Chemistry of geothermal fluids discharged from exploration wells at Mokai/New Zealand. Z Dtsch Geol Ges 136:235–251
Henley RW, Truesdell AH, Barton JP (1984) Fluid-mineral equilibra in hydrothermal systems. Rev Econ Geol 1:268p
Heřmanská M, Stefánsson A, Scott S (2019) Supercritical fluids around magmatic intrusions: IDDP-1 at Krafla, Iceland. Geothermics 78:101–110
Iglesias ER, Arellano VM, Garfias A, Miranda C, Aragon A (1985) A one-dimensional vertical model of the Los Azufres, Mexico, geothermal reservoir in its natural state. Geother Res T 9:331–336
Izquierdo G, Aragón A, Portugal E, Arellano VM, de León J, Álvarez J (2006) Mineralogía de la zona mineralizada de sílice-epidota (ZMSE) del yacimiento geotérmico de Cerro Prieto, BC, México. Geotermia 19:2–12
James ED, Hoang VT, Epperson IJ (1987) Structure, permeability and production characteristics of the Heber, California geothermal field. Ln: Proceedings, twelfth workshop on geothermal reservoir engineering. Stanford University, Stanford, California, 20–22 Jan 1987. SGP-TR-109, pp 267–271
Ji D, Ping Z (2000) Characteristics and genesis of the Yangbajing geothermal field, Tibet. In: Proceedings of the world geothermal congress, Kyushu-Tohoku, Japan, vol 28, pp 1083–1088
Jordan OT (1982) Implications of solution-mineral equilibria on the exploitation of the S-Negros geothermal field, Philippines. UNU-GTP report 1982-7, 67p
Karingithi CW, Arnórsson S, Grönvold K (2010) Processes controlling aquifer fluid compositions in the Olkaria geothermal system, Kenya. J Volcanol Geoth Res 196:57–76
Kipng'ok J (2011) Fluid chemistry, feed zones and boiling in the first geothermal exploration well at Menengai, Kenya. UNU-GTP report 2011-15, pp 281–302

Klein CW, Iwata S, Takeuchi R, Naka T (1991) Prediction and prevention of silica scaling at low levels of oversaturation: case studies, and calculations for Uenotai geothermal field, Akita Prefecture, Japan. In: Proceedings, sixteenth workshop on geothermal reservoir engineering. Stanford University, Stanford, California, 23–25 Jan 1991. SGP-TR-134, pp 165–176

Krupp RE, Seward TM (1987) The Rotokawa geothermal system, New Zealand: an active epithermal gold-depositing environment. Econ Geol 82:1109–1129

Kuriyama T (1985) Geothermal system in the Yuzawa-Ogachi area, Northern Honshu. J Geotherm Res Soc Jpn 7:311–328 (in Japanese with English abstract)

Lemmon EW, McLinden MO, Friend DG (2017) Thermophysical properties of fluid systems. In: Linstrom PJ, Mallard WG (eds) NIST chemistry Webbook, NIST standard reference database number 69. National Institute of Standards and Technology, Gaithersburg MD. http://webbook.nist.gov/

Mahon WAJ, Finlayson JB (1972) The chemistry of the Broadlands geothermal area, New Zealand. Am J Sci 272:48–68

Malimo SJ (2012) Aquifer fluid modeling and assessment of mineral-gas-liquid equilibria in the Namafjall geothermal system. NE-Iceland. UNU-GTP report 2012-3, 57p

Malimo SJ (2013) Fluid chemistry of Menengai geothermal wells, Kenya. Geother Res T 37:425–430

Mañon A, Mazor E, Jimenez M, Sanchez A, Fausto J, Zenizo C (1977) Extensive geochemical studies in the geothermal field of Cerro Prieto, Mexico. US-DOE report LBL-7019, contract W-7405-ENG-48, 113p

Marini L, Yock Fung A, Sanchez E (2003) Use of reaction path modeling to identify the processes governing the generation of neutral Na–Cl and acidic Na–Cl–SO_4 deep geothermal liquids at Miravalles geothermal system, Costa Rica. J Volcanol Geoth Res 128:363–387

Michels DE (1986) SSSDP fluid compositions at first flow test of State 2-14. Geother Res T 10:461–465

Monastero FC, Katzenstein AM, Miller JS, Unruh JR, Adams MC, Richards-Dinger K (2005) The Coso geothermal field: a nascent metamorphic core complex. Geol Soc Am Bull 117:1534–1553

Moore JN, Adams MC, Bishop BP, Hirtz P (1989) A fluid flow model of the Coso geothermal system: data from production fluids and fluid inclusions. In: Proceedings, fourteenth workshop on geothermal reservoir engineering. Stanford University, Stanford, California, 24–26 Jan 1989, SGP-TR-122, pp 139–144

Muna ZW (1982) Chemistry of well discharges in the Olkaria geothermal field, Kenya. UNU-GTP report 1982–8, 38p

Naka T, Okada H (1992) Exploration and development of Uenotai geothermal field, Akita Prefecture, northeastern Japan. Resour Geol 42:223–240 (in Japanese with English abstract)

Nielson DL, Hulen JB (1984) Internal geology and evolution of the Redondo Dome, Valles Caldera, New Mexico. J Geophys Res Sol Earth 89(B10):8695–8711

Nitta T, Adachi M, Takahashi M, Inoue K, Abe Y (1991) Heavy metal precipitation from geothermal fluid of 87N-15T production well in the Okuaizu geothermal field, Tohoku district, Japan. Resour Geol 41:231–242 (in Japanese with English abstract)

Nitta T, Tsukagoshi S, Adachi M, Seo K (1995) Exploration and development in the Okuaizu geothermal field, Japan. Resour Geol 45:201–212 (in Japanese with English abstract)

Nordstrom DK, Plummer LN, Wigley TML, Wolery TJ, Ball JW, Jenne EA, Bassett RL, Crerar DA, Florence TM, Fritz B, Hoffman M, Holdren GR, Lafon GM, Mattigod SV, McDuff RE, Morel F, Reddy MM, Sposito G, Thrailkill J (1979) A comparison of computerized chemical models for equilibrium calculations in aqueous systems. In: Jenne EA (ed) Chemical modeling in aqueous systems, ACS symposium series, vol 93, pp 857–892

Okada H, Yasuda Y, Yagi M, Kai K (2000) Geology and fluid chemistry of the Fushime geothermal field, Kyushu, Japan. Geothermics 29:279–311

Özgür N (1998) Geochemical signature of the Kizildere geothermal field, western Anatolia, Turkey. Int Geol Rev 44:153–163

Pang Z-H (2006) pH dependant isotope variations in arc-type geothermal waters: new insights into their origins. J Geochem Explor 89:306–308
Panichi C (2004) Geochemical impact of re-injecting geothermal waste waters: example, Larderello, Italy. In: Gieré R, Stille P (eds) Energy, waste, and the environment: a geochemical perspective, vol 236. Geological Society, London, Special Publications, pp 337–354
Parkhurst DL, Appelo CAJ (2013) Description of input and examples for PHREEQC version 3: a computer program for speciation, batch-reaction, one-dimensional transport, and inverse geochemical calculations. Techniques and Methods 6-A43. U.S. Geological Survey
Ping Z (1991) Gas geothermometry and chemical equilibria of fluids from selected geothermal fields. UNU-GTP report 1991-14, 46p
Ping Z, Jian J, Haizheng Z, Ji D, Tingli L (1998a) Chemical composition of thermal water in the Yangbajing geothermal field, Tibet. Sci Geol Sin 33:61–72 (in Chinese with English abstract)
Ping Z, Tingli L, Jian J, Haizheng Z (1998b) Characteristics of gas geochemistry in Yangbajing geothermal field, Tibet. Chin Sci Bull 43:1770–1777
Price L, Powell TS, Atkinson L (2011) Geothermal fluid evolution at Rotokawa: hydrothermal alteration indicators. Geother Res T 35:977–982
Principe C, Lavorini G, Vezzoli LM (eds) (2017) Il Vulcano di Monte Amiata. Edizioni Scientifiche e Artistiche, 399p
Rae AJ (2002) Alteration systematics and mineralising potential of the Palinpinon geothermal field, Negros Island, Philippines. Ph.D. thesis, University of Tasmania, Australia, 243p
Rae AJ, Cooke DR, Brown KL (2011) The trace metal chemistry of deep geothermal water, Palinpinon geothermal field, Negros Island, Philippines: implications for precious metal deposition in epithermal gold deposits. Econ Geol 106:1425–1446
Rango T, Petrini R, Stenni B, Bianchini G, Slejko F, Beccaluva L, Ayenew T (2010) The dynamics of central Main Ethiopian Rift waters: Evidence from δD, $\delta^{18}O$ and $^{87}Sr/^{86}Sr$ ratios. Appl Geochem 25:1860–1871
Reed MJ (1976) Geology and hydrothermal metamorphism in the Cerro Prieto geothermal field, Mexico. In: Proceedings 2nd U.N. symposium on development and use of geothermal resources, vol 1, San Francisco, pp 539–547
Reed MJ (1989) Thermodynamic calculations of calcium carbonate scaling in geothermal wells, Dixie Valley geothermal field, U.S.A. Geothermics 18:269–277
Reed MH (1997) Hydrothermal alteration and its relationship to ore fluid composition. In: Barnes HL (ed) Geochemistry of hydrothermal ore deposits, 3d edn. Wiley, pp 517–611
Reed MH (1998) Calculation of simultaneous chemical equilibria in aqueous-mineral-gas systems and its application to modeling hydrothermal processes. In: Richards J, Larson P (eds) Techniques in hydrothermal ore deposits geology. Reviews in economic geology, vol 10, pp 109–124
Reed MH, Spycher NF (1984) Calculation of pH and mineral equilibria in hydrothermal waters with application to geothermometry and studies of boiling and dilution. Geochim Cosmochim Acta 48:1479–1492
Reed MH, Palandri J, Clemente V, Cabahug R (2014) Computation of reservoir geochemical conditions from excess-enthalpy wellhead samples. Geother Res T 38:461–464
Remoroza AI (2010) Calcite mineral scaling potentials of high-temperature geothermal wells. Master's thesis, Faculty of Science, University of Iceland, 99p
Renderos RE (2002) Chemical characterization of the thermal fluid discharge from well production tests in the Berlín geothermal field, El Salvador. UNU-GTP report 2002-12, pp 205–231
Reyes AG (1990) Petrology of Philippine geothermal systems and the application of alteration mineralogy to their assessment. J Volcanol Geoth Res 43:279–309
Reyes AG, Giggenbach WF, Saleras JRM, Salonga ND, Vergara MC (1993) Petrology and geochemistry of Alto Peak, a vapor-cored hydrothermal system, Leyte Province, Philippines. Geothermics 22:479–519
Reyes AG, Trompetter WJ, Britten K, Searle J (2003) Mineral deposits in the Rotokawa geothermal pipelines, New Zealand. J Volcanol Geoth Res 119:215-239

Rosi M, Sbrana A, Principe C (1983) The Phlegraean fields: structural evolution, volcanic history and eruptive mechanisms. J Volcanol Geoth Res 17:273–288
Ruaya JR, Buenviaje MM, Solis RP, Gonfiantini R (1995) Chemical and isotopic studies of fluids in the Bacon-Manito geothermal field, Philippines. In: Isotope and geochemical techniques applied to geothermal investigations. IAEA-TECDOC-788, pp 185–208
Ruggieri G, Giolito C, Gianelli G, Manzella A, Boiron MC (2004) Application of fluid inclusions to the study of Bagnore geothermal field (Tuscany, Italy). Geothermics 33: 675–692
Sakai Y, Matsunaga E, Kubota Y (1993) Geothermal energy development in the Sumikawa field, Northeast Japan. Resour Geol 43:409–425 (in Japanese with English abstract)
Sanjuan B (2010) Use of a new sodium/lithium (Na/Li) geothermometric relationship for High-Temperature (HT) geothermal fluids derived from seawater/basalt interaction processes: application to the Djibouti case. In: Third East African Rift Geothermal Conference ARGEO-C3, 21p
Sanjuan B, Michard G, Michard A (1990) Origine des substances dissoutes dans les eaux des sources thermales et des forages de la région Asal-Ghoubbet (République de Djibouti). J Volcanol Geother Res 43:333–352
Sasada M (1987) Fluid inclusions in calcite scale from DY-1 drill hole, Hchi geothermal area, Japan. J Geotherm Res Soc Jpn 9:197–205
Scott SW (2011) Gas chemistry of the Hellisheidi geothermal field. Master's thesis. REYST/Faculty of Science, University of Iceland, 81p
Scott S, Gunnarsson I, Arnórsson S, Stefánsson A (2014) Gas chemistry, boiling and phase segregation in a geothermal system, Hellisheidi, Iceland. Geochim Cosmochim Acta 124:170–189
See FS (1995) Anhydrite deposition in Cawayan wells, Bacman geothermal field Philippines: prediction and possible remedies. UNU-GTP report 13, pp 321–348
Sekento LR (2012) Geochemical and isotopic study of the Menengai geothermal field, Kenya. UNU-GTP report 2012-31, pp 769–792
Seki Y (1990) Gas concentration in aquifer fluid prior to boiling in the Oku-aizu geothermal system, Fukushima, Japan. Geochem J 24:105–121
Seki Y (1991) The physical and chemical structure of the Oku-aizu geothermal system, Japan. Geochem J 25:245–265
Seki Y, Liou JG, Guillemette R, Sakai H, Oki Y, Hirano T, Onuki H (1983) Investigation of geothermal systems in Japan I. Onikobe geothermal area. Hydroscience and Geotechnology Laboratory, Saitama University, Memoir No. 3, 206p
Sheppard DS (1984) Fluid chemistry of Ngawha reservoir. In: Proceedings 6th New Zealand geothermal workshop, pp 151–154
Sheppard DS, Giggenbach WF (1980) Chemistry of the well discharges at Ngawha. In: Proceedings of New Zealand geothermal workshop, University of Auckland, pp 91–95
Sheppard DS, Lyon GL (1984) Geothermal fluid chemistry of the Orakeikorako field, New Zealand. J Volcanol Geother Res 22:329–349
Simpson MP, Bignall G (2016) Undeveloped high-enthalpy geothermal fields of the Taupo Volcanic Zone, New Zealand. Geothermics 59:325–346
Smith RL, Bailey RA (1966) The Bandelier Tuff: a study in ash-flow eruption cycles from zoned magma chambers. Bull Volcanol 29:83–104
Stefánsson A, Arnórsson S (2000) Feldspar saturation state in natural waters. Geochim Cosmochim Acta 64:2567–2584
Suzuki M, Futagoishi M, Inoue T, Yamada K, Obara K, Fujino T (2000) Conceptual hydrogeological model of the Wasabizawa geothermal field, Akita Prefecture, Japan. In: Proceedings of the world geothermal congress 2000, Kyushu-Tohoku, Japan, pp 2241–2245
Takenaka T, Furuya S (1991) Geochemical model of the Takigami geothermal system, northeast Kyushu, Japan. Geochem J 25:267–281
Takenaka T, Gotoh H, Yamamoto Y, Furuya S (1995) Exploration and development of the Takigami geothermal system, Kyshu, Japan. J Soc Resour Geol 45:361–376 (in Japanese with English abstract)

Takeno N (2000) Thermal and geochemical structure of the Uenotai geothermal system, Japan. Geothermics 29:257–277
Tarcan G, Özen T, Gemici Ü, Çolak M, Karamanderesi İH (2016) Geochemical assessment of mineral scaling in Kızıldere geothermal field, Turkey. Environ Earth Sci 75:1317, 19p
Teklemariam M, Battaglia S, Gianelli G, Ruggieri G (1996) Hydrothermal alteration in the Aluto-Langano geothermal field, Ethiopia. Geothermics 25:679–702
Tempel RN, Sturmer DM, Schilling J (2011) Geochemical modeling of the near-surface hydrothermal system beneath the southern moat of Long Valley Caldera, California. Geothermics 40:91–101
Thompson JM, Fournier RO (1988) Chemistry and geothermometry of brine produced from the Salton Sea scientific drill hole, Imperial Valley, California. J Geophys Res-Sol Earth 93(B11):13165–13173
Tonani FB, Nagao K, Moore J, Natale G, Sperry T (1998) Water and gas geochemistry of the Cove-Fort Sulphurdale geothermal system. In: Proceedings, twenty-third workshop on geothermal reservoir engineering. Stanford University, Stanford, California, 26–28 Jan 1998
Truesdell AH (1979) Aquifer boiling may be normal in exploited high-temperature geothermal systems. In: Proceedings, fifth workshop on geothermal reservoir engineering. Stanford University, Stanford, California, 12–14 Dec 1979
Truesdell AH, Janik CJ (1986) Reservoir processes and fluid origins in the Baca geothermal system, Valles Caldera, New Mexico. J Geophys Res-Sol Earth 91(B2):1817–1833
Truesdell AH, Nakanishi S (2005) Chemistry of neutral and acid production fluids from the Onikobe geothermal field, Miyagi Prefecture, Honshu, Japan. In: Use of isotope techniques to trace the origin of acidic fluids in geothermal systems. IAEA-TECDOC-1448, pp 169–193
Truesdell AH, Singers W (1974) The calculation of aquifer chemistry in hot-water geothermal systems. J Res US Geol Surv 2:271–278
Truesdell AH, Thompson JM, Coplen TB, Nehring NL, Janik CJ (1981) The origin of the Cerro Prieto geothermal brine. Geothermics 10:225–238
Truesdell AH, Lippmann MJ, Quijano JL, D'Amore F (1995) Chemical and physical indicators of reservoir processes in exploited high-temperature, liquid-dominated geothermal fields. In: Proceedings of the world geothermal congress, Florence, Italy, vol 3, pp 1933–1938
Ueda A, Kubota Y, Katoh H, Hatakeyama K, Matsubaya O (1991) Geochemical characteristics of the Sumikawa geothermal system, northeast Japan. Geochem J 25:223–244
Waibel AF (1987) An overview of the geology and secondary mineralogy of the high temperature geothermal system in Dixie Valley, Nevada. Geot Res Council Bull 11:479–486
White AF (1986) Chemical and isotopic characteristics of fluids within the Baca geothermal reservoir, Valles Caldera, New Mexico. J Geophys Res-Sol Earth 91(B2):1855–1866
White AF, Peterson ML (1991) Chemical equilibrium and mass balance relationships associated with the Long Valley hydrothermal system, California, USA. J Volcanol Geoth Res 48:283–302
Williams AE, McKibben MA (1989) A brine interface in the Salton Sea geothermal system, California: fluid geochemical and isotopic characteristics. Geochim Cosmochim Acta 53:1905–1920
Wilson BM (2007) Igneous petrogenesis. A global tectonic approach. Springer, 466p
Winick J, Powell T, Mroczek E (2009) The natural-state geochemistry of the Rotokawa reservoir. In: New Zealand geothermal workshop 2009, Proceedings, 16–18 Nov 2009, Rotorua, New Zealand, 8p
Wohletz K, Heiken G (1992) Volcanology and geothermal energy. University of California Press, 432p
Wolery TJ (1992) EQ3NR, a computer program for geochemical aqueous speciation-solubility calculations: Theoretical manual, user's guide and related documentation (version 7.0). Report UCRL-MA-110662 PT III. Lawrence Livermore National Laboratory, Livermore
Wolery TJ, Daveler SA (1992) EQ6, A computer program for reaction path modeling of aqueous geochemical systems: Theoretical manual, user's guide, and related documentation (version 7.0). Report UCRL-MA-110662 PT IV. Lawrence Livermore National Laboratory, Livermore

Xiaoping F (2002) Conceptual model and assessment of the Yangbajing geothermal field, Tibet, China. UNU-GTP report 2002-5, pp 27–52

Yamada M, Iguchi K, Nakanishi S, Todaka N (2000) Reservoir characteristics and development plan of the Oguni geothermal field, Kyushu, Japan. Geothermics 29:151–169

Yock Fung A (1998) Chemical and isotopic studies in the Miravalles geothermal field, Costa Rica. UNU-GTP report 1998-17, pp 461–499

Yoshida Y (1991) Geochemistry of the Nigorikawa geothermal system, southwest Hokkaido, Japan. Geochem J 25:203–222

Zaporozec A (1972) Graphical interpretation of water-quality data. Ground Water 10:32–43

Zhanshi Z (2001) Water-rock interaction in the Bakki low-temperature geothermal field, SW-Iceland. UNU-GTP report 2001-17, pp 405–434

Zhanxue S (1998) Geothermometry and chemical equilibria of geothermal fluids from Hveragerdi, SW-Iceland, and selected hot springs, Jiangxi province, SE-China. UNU-GTP report 1998-14, pp 373–402

Zhao P, Dor J, Liang T, Jin J, Zhang H (1998) Characteristics of gas geochemistry in Yangbajing geothermal field, Tibet. Chin Sci Bull 43:1770–1777

Chapter 4
The Hydrothermal Minerals

Abstract Over 2200 chemical analyses of hydrothermal minerals collected at depth in different active geothermal systems were compiled and processed in this work. In agreement with the outcomes of previous studies, it turned out that hydrothermal quartz, calcite, adularia, albite, anorthite, and laumontite are virtually pure solid phases. Therefore, their activities can be assumed to be equal to one. In contrast, the other hydrothermal minerals of interest, i.e., white mica, chlorite, epidote, prehnite, wairakite, and garnet, are solid solutions of variable composition. Therefore, the activities of pertinent endmembers (i.e., muscovite, 7Å-clinochlore, clinozoisite, prehnite, wairakite, and grossular) were computed, under the assumption of random mixing of atoms on energetically equivalent sites. Then, following the approach delineated in Sect. 2.2.1, the Gibbs free energy and the thermodynamic equilibrium constant of the dissolution reaction of the solid solutions with average activity of these pertinent endmembers were computed. These log K values are points of reference more representative than the log K of the dissolution reactions of the corresponding pure minerals.

4.1 The Hydrothermal Alteration Suites and Their Zones

A general relationship between the occurrence of hydrothermal alteration minerals and temperature was recognized long ago, although the formation of these secondary solid phases is controlled also by fluid chemistry, permeability, and duration of hydrothermal activity (e.g., Browne and Ellis 1970; Kristmannsdóttir 1979; Browne 1978; Bird et al. 1984). As pointed out by Reyes (1990), two distinct hydrothermal alteration suites occur in the high-temperature geothermal systems of the Philippines, namely (a) the neutral-pH suite, associated with the neutral-pH, alkali chloride liquids and (b) the acidic alteration suite, related with sulfur-rich aqueous solutions of low pH.

Based on the key phyllosilicate minerals, the neutral-pH suite is divided in the following four zones:

R. Cioni and L. Marini, *A Thermodynamic Approach to Water Geothermometry*, Springer Geochemistry, https://doi.org/10.1007/978-3-030-54318-1_4

(1) Smectite zone, extending up to temperatures of 180 °C. Smectite is accompanied by chlorite, above 120 °C, and low-temperature zeolites, such as heulandite, stilbite, and ptilolite, although laumontite occurs above 120 °C.
(2) Transition zone, corresponding to temperatures of 180–220 °C. It is characterized by the increase in chlorite, the appearance of illite-smectite, and the persistence of laumontite, which is the typical Ca–Al-silicate of this zone. Albite, adularia, and sphene appear in this zone.
(3) Illite zone, corresponding to temperatures of 220 to 270–280 °C, although illite is stable at higher temperatures as well. Wairakite, epidote, and prehnite are the representative Ca–Al-silicates of this zone. They are accompanied by chlorite, adularia, albite, sphene, and sulfide minerals.
(4) Biotite zone, extending at temperatures higher than 270–280 °C. It is characterized by the appearance of high-temperature minerals, such as garnet (e.g., andradite and grossular), amphibole (e.g., actinolite and tremolite), and pyroxene (e.g., diopside), as well as by the persistence of illite, chlorite, albite, adularia, wairakite, epidote, sphene, and sulfide minerals.

Some minerals, such as quartz, calcite, and pyrite, occur at any temperature.

Based on the key hydroxy silicate of aluminum, the acidic alteration suite is divided in the following four zones: (i) The kaolinite zone, extending up to temperatures of 120 °C. (ii) The dickite ± kaolinite zone, corresponding to temperatures of 120–200 °C. (iii) The dickite ± phyrophyllite zone, corresponding to temperatures of 200–260 °C. (iv) The phyrophyllite ± illite zone, corresponding to temperatures of 230–320 °C. These hydroxy silicates of aluminum are typically accompanied by sulfur, alunite, anhydrite, diaspore, and pyrite as well as by other less common solid phases.

The same hydrothermal alteration zones present in active geothermal systems are also recognizable in hydrothermal ore deposits, which are fossil geothermal systems (Pirajino 2009). This homogeneity in the sequence of alteration minerals is not surprising considering that rocks act as suppliers of chemical elements during water-rock interaction and there are relatively limited differences in the concentrations of major (rock-forming) elements in different lithotypes (e.g., Turekian and Wedepohl 1961), apart from the extreme case of monomineralic rocks such as limestones, dolomites, and quartzites.

As pointed out by Giggenbach (1980, 1984, 1988), the non-conservative elements transferred to the aqueous solution by CO_2-driven rock dissolution remain in part in the fluid and are partly incorporated in hydrothermal alteration minerals. Under favorable conditions (e.g., constant temperature, low fluid/rock ratio), the aqueous solution is expected to approach and possibly attain the condition of chemical equilibrium with the hydrothermal alteration minerals, after a lapse of time largely depending on the kinetics of relevant dissolution-precipitation reactions.

Focusing on the felsic rocks prevailing in the continental crust, the thermodynamically stable hydrothermal paragenesis is expected to comprise quartz, muscovite (or illite), albite, adularia, clinochlore (or chlorite), and either a Ca–Al-silicate, if CO_2 fugacity is relatively low, or calcite, if CO_2 fugacity is comparatively high, in accordance with mineralogical evidence (e.g., Ellis 1970; Ellis and Mahon 1977; Giggenbach 1984, 1988; Arnórsson et al. 1983a, b; Arnórsson and Gunnlaugsson 1985). If so, the activities of undissociated SiO_2 and Al^{3+}, Na^+, K^+, Mg^{2+}, and Ca^{2+} ions are fixed. In addition to the solid phases listed above, anhydrite, fluorite, and calcite may also be part of the hydrothermal mineral assemblage, thus fixing the activities of $SO_4{}^{2-}$, F^-, and $HCO_3{}^-$ ions, respectively (Guidi et al. 1990; Chiodini et al. 1991 and references therein).

Therefore, the activity of all these compatible[1] aqueous species can be involved in geothermometers and f_{CO_2}-indicators, referring to suitable mineral-solution reactions. For performing correctly this task, it is necessary to know the main characteristics of these minerals and if they occur in active geothermal systems either as pure (or nearly pure) solid phases or as components of solid solutions. In fact, in the first case their activity can be assumed to be equal to one (based on the standard state convention for solids, see Sect. 2.1.2), whereas in the second case their activity must be properly calculated and involved in some way in geothermometers and f_{CO_2}-indicators (see Sect. 2.2.1). Therefore, the next sections of this chapter are devoted to summarize the main characteristics of relevant hydrothermal alteration minerals, especially their chemical composition, which is used to compute the activity of solid solution components of interest.

4.2 Feldspars

4.2.1 Main Characteristics of Feldspars

There is a number of books completely dedicated to feldspar minerals, such as Ribbe (1983), Smith and Brown (1988), and Deer et al. (2001).

Feldspars have either monoclinic or triclinic symmetry. The crystal structure of feldspars consists of a three dimensional framework of (Si, Al)–O or T–O tetrahedra, with each oxygen atom constituting the vertex of two adjacent tetrahedra. Each Si atom balances half charges of the four oxygen atoms in its Si–O tetrahedron, but one negative charge is unbalanced in each Al–O tetrahedron. Therefore, large monovalent and bivalent cations (collectively denoted M) occupy the interstices to maintain the charge balance. Because there is one interstice for every four tetrahedra and the interstices are completely filled up, the general crystallochemical formula of feldspars is MT_4O_8, where M is commonly Na^+, K^+ or Ca^{2+} and T is usually Al^{3+} or Si^{4+}, with $1 \leq Al \leq 2$ and $2 \leq Si \leq 3$.

Most feldspars are ternary solid solutions of albite [Ab, $NaAlSi_3O_8$], orthoclase [Or, $KAlSi_3O_8$], and anorthite [An, $CaAl_2Si_2O_8$]. The solid solutions predominantly

[1]Following Arnórsson et al. (1983b), the term "compatible" is utilized to indicate the chemical components whose activity is limited by incorporation in the lattice of hydrothermal alteration minerals.

composed of albite and orthoclase form the alkali feldspar series, whereas the solid solutions primarily made up of albite and anorthite form the plagioclase series.

At high temperatures (>1000 °C), feldspars exhibit remarkable solid solution. Accordingly, the plagioclase series comprises high-anorthite, with An > 90%; high-bytownite, with 70 < An < 90%; high-labradorite, with 50 < An < 70%; high-andesine with 30 < An < 50%; high-oligoclase, with 10 < An < 30%; and high-albite, with An < 10%. According to Parsons (2010), the alkali feldspar series comprises K-sanidine, with Or >80%; sanidine, with 50 < Or < 80%; Na-sanidine with 35 < Or < 50%; and anorthoclase, with Or < 35%, although different nomenclatures are adopted by other authors. High-albite and anorthoclase are triclinic at room temperature but they assume monoclinic symmetry when observed at elevated temperatures due to shearing transformation. Sanidine has monoclinic symmetry at all temperatures.

At low temperatures (<300 °C) feldspars exhibit very limited solid solution and the stable phases are triclinic low-microcline, low-albite and anorthite, all with nearly stoichiometric composition. In fact, low-temperature natural feldspars are frequently made up of intergrowths of two or more phases. Ideally, low-microcline, low-albite and anorthite are expected to exhibit Al–Si order. Nevertheless, disordered structures are frequent because of metastable crystallization.

There are two mechanisms for the transition from monoclinic to triclinic symmetry. In alkali feldspars with Ab > 60% it occurs by a quick shearing phase transformation brought about by twisting of (Si, Al)–O tetrahedra. In contrast, a slow 'diffusive' order-disorder transformation occurs in the Or-rich alkali feldspars, in which the Al ions, distributed randomly between four equivalent sites at high temperatures, become ordered on one of these sites during cooling.

For what concerns the substitutions, the M-site can be occupied by Rb^+, Tl^+, and NH_4^+ and in part by Li^+ and Cs^+ in alkali feldspars and by Sr^{2+}, Ba^{2+}, Pb^{2+}, and Eu^{2+}, as well as by (La^{3+}, Na^+) in alkali-earth feldspars. The ammonium alkali feldspar is buddingtonite, $NH_4AlSi_3O_8 \cdot ½ H_2O$, whereas the barium plagioclase is celsian, Cls, $BaAl_2Si_2O_8$. The main substitutions in T-sites of natural feldspars are B and Fe. The boron analog of albite is reedmergnerite, $NaBSi_3O_8$.

4.2.2 The Chemistry of Hydrothermal Feldspars

Both adularia (i.e., hydrothermal orthoclase) and hydrothermal albite are widespread in active geothermal systems where they usually occur at temperatures higher than 120–150 °C, in spite of some notable exceptions (e.g., Browne 1970, 1978; Reyes 1990). For instance, only adularia occurs in the Cerro Prieto geothermal field, whereas hydrothermal albite is absent although primary plagioclase usually represents up to 10% of the sandstone mineral constituents at all depths (Bird et al. 1984; Schiffman et al. 1985). Conversely, hydrothermal albite replaces primary plagioclase at temperatures higher than 120 °C in the geothermal system of Onikobe, whereas adularia is absent (Seki et al. 1983). Hydrothermal anorthite and plagioclases of intermediate composition are uncommon.

In this work, several chemical analyses of hydrothermal feldspars obtained from deep boreholes drilled in active geothermal systems were compiled. 480 chemical analyses were considered of acceptable quality, having sum of oxides from 98 to 102 wt%, and processed to compute the percentages of albite, orthoclase, and anorthite.

The Ab–Or–An triangular plot of Fig. 4.1 displays the composition of 119 hydrothermal feldspars from the geothermal fields of Roosevelt Hot Springs, Utah (Ballantyne 1978), Salton Sea, California (Shearer et al. 1988), Heber, California (Browne 1977), Cerro Prieto, Mexico (Schiffman et al. 1985), Miravalles, Costa Rica (Milodowski et al. 1989), Berlin, El Salvador (Ruggieri et al. 2006), Kilauea, Hawaii (Bargar et al. 1996), Onikobe, Japan (Seki et al. 1983), Broadlands-Ohaaki, New Zealand (Lonker et al. 1990), Waiotapu, New Zealand (Hedenquist and Browne 1989) Milos, Greece (Liakopoulos 1987), as well as Latera (Cavarretta et al. 1985), Larderello (Cavarretta et al. 1982), and Pantelleria (Fulignati et al. 1997), Italy. Among these 119 hydrothermal feldspars, there are 60 adularias with Ab < 15 mol%,

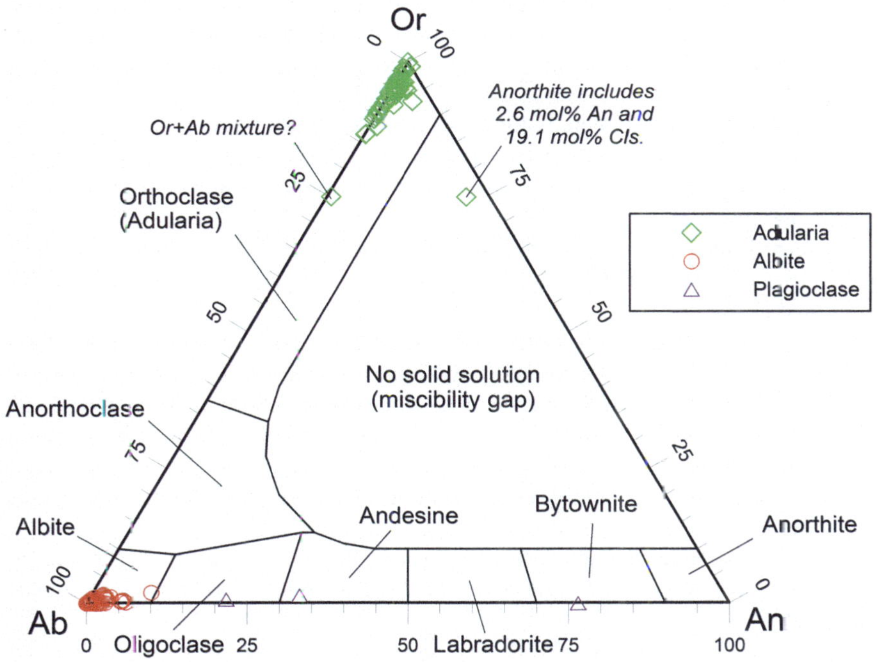

Fig. 4.1 Ab-Or-An triangular plot showing the composition of 119 hydrothermal feldspars from different active geothermal fields, including Roosevelt Hot Springs (Ballantyne 1978), Salton Sea (Shearer et al. 1988), Heber (Browne 1977), Cerro Prieto (Schiffman et al. 1985), Miravalles (Milodowski et al. 1989), Berlin (Ruggieri et al. 2006), Kilauea (Bargar et al. 1996), Onikobe (Seki et al. 1983), Broadlands-Ohaaki (Lonker et al. 1990), Waiotapu (Hedenquist and Browne 1989), Milos (Liakopoulos 1987), Latera (Cavarretta et al. 1985), Larderello (Cavarretta et al. 1982), and Pantelleria (Fulignati et al. 1997)

54 albites, and 5 anomalous samples, namely 1 probable mixture (intergrowth) of adularia and albite (with ~25 mol% Ab), 3 plagioclases of variable composition (1 oligoclase, 1 andesine, and 1 bytownite), and 1 celsian-rich adularia from Berlin, formally constituted by 74.7 mol% Or, 19.1 mol% Cls, 3.6 mol% Ab, and 2.6 mol% An.

Neglecting the 5 anomalous samples, the composition of adularias and hydrothermal albites from these different geothermal systems is described by the following statistical parameters:

(i) X_{Or} in adularia (N = 60): Min = 0.864, Max = 1.000, Mean = 0.946, Median = 0.951, Std. Dev. = 0.030.
(ii) X_{Ab} in albite (N = 54): Min = 0.889, Max = 1.000, Mean = 0.981, Median = 0.989, Std. Dev. = 0.022.

The composition of adularia shows limited variations in well State 2-14 at Salton Sea, with X_{Or} from 0.90 to 0.98 (Shearer et al. 1988). The composition of hydrothermal plagioclase is close to end-member albite at depths <1828.8 m, whereas the hydrothermal plagioclase is richer in An at depths >1828.8 m. Three hydrothermal feldspars coexists in samples below 3017.5 m depth, K-feldspar with X_{Or} ~ 0.91, albite-oligoclase with X_{Ab} 0.95–0.86, and oligoclase-andesine with X_{Ab} 0.75–0.63.

Both hydrothermal albite and oligoclase, with X_{Ab} from 0.883 to 0.850, occur in an altered diabase dike or sill at Heber (Browne 1977). Textural relations suggest that oligoclase is an early mineral, probably not in equilibrium with present-day geothermal fluids.

At Cerro Prieto, adularia has X_{Or} varying from 0.91 to 0.97, with a mean close to 0.95, whereas hydrothermal albite is absent (Schiffman et al. 1985), as already mentioned above.

In most of the Miravalles reservoir, hydrothermal plagioclase is almost pure albite (Milodowski et al. 1989). Uncommon hydrothermal oligoclase (with X_{Ab} close to 0.85) might be explained by the presence of primary plagioclase relicts in albite. Rare hydrothermal bytownite (with X_{An} of 0.77) occurs at 1453 m depth in association with grandite garnet (60–65 mol% andradite) and magnetite. Adularia shows small concentrations of Na and variable concentrations of Ba, which are higher in the cores of crystals.

At Berlin, hydrothermal albite has X_{Ab} from 0.90 to 0.98, whereas a Ba-bearing feldspar with X_{Cls} up to 0.191 occurs in well TR-2 at about 1500 m depth (Ruggieri et al. 2006).

At Onikobe (Seki et al. 1983), the replacement of primary plagioclase by hydrothermal albite (with X_{Ab} > 0.95) begins at 120–180 °C and the process comes to completion at temperatures higher than 230–240 °C. Adularia is absent as already mentioned above.

Adularia and hydrothermal albite span small compositional intervals at Broadlands-Ohaaki, with X_{Or} in adularia from 0.94 to 0.995 and X_{Ab} in hydrothermal albite from 0.96 to 0.99, irrespective of primary plagioclase composition (Lonker et al. 1990). The intermediate compositions of adularia in well BR8 at 774–781 m depth cannot be explained involving primary sanidine because it is absent at Broadlands-Ohaaki (Browne and Ellis 1970). They are likely due to fine intergrowths of hydrothermal albite with adularia (Lonker et al. 1990).

At Waiotapu, hydrothermal albite and adularia are common as alteration products of primary andesine at temperatures higher than ~180 °C (Hedenquist and Browne 1989). Adularia is later than albite as indicated by textural evidence. Most hydrothermal plagioclase has X_{Ab} from 0.90 to 0.99, but some samples range from albite to andesine. The X_{Or} in adularia varies from 0.94 to 0.995.

At Milos, albite is practically pure, with X_{Ab} ranging from from 0.999 to 1.000, whereas X_{Or} in adularia varies from 0.926 to 0.974 (Liakopoulos 1987).

At Larderello, adularia has X_{Or} ranging from 0.950 to 0.959 in the Sasso-22 well at depths of 1600–2263 m based on microprobe analyses, whereas plagioclase compositions (as X_{Ab}) determined by optical microscopy are: (i) from 0.97 to 0.95 in the San Ippolito 122 well at 783 m depth; (ii) from 0.87 to 0.86 in the Sasso 22 well at 2770 m depth; (iii) from 0.66 to 0.60 in the Anqua well at 2635 m depth (Cavarretta et al. 1982).

At Pantelleria, hydrothermal albite has X_{Ab} varying from 0.982 to 0.992 and adularia has X_{Or} ranging from 0.946 to 1.000 (Fulignati et al. 1997).

As shown in the Ab–Or–An triangular plot of Fig. 4.2, the 361 hydrothermal feldspars from Reykjanes, Iceland (mostly from Libbey and Williams-Jones 2016 and subordinately from Marks et al. 2010; Fowler et al. 2015) comprise 150 adularias with Or >85 mol%, 182 albite-oligoclase samples, and 18 anorthites as well as 14 anomalous samples which are probable mixtures (i.e., intergrowths of two or more phases, possibly including relicts of primary phases), 8 rich in adularia and 6 rich in oligoclase.

Neglecting these 14 anomalous samples, the composition of the hydrothermal feldspars from Reykjanes is defined by the following statistical parameters:

(i) X_{Or} in adularia (N = 150): Min = 0.870, Max = 0.992, Mean = 0.954, Median = 0.963, Std. Dev. = 0.025.
(ii) X_{Ab} in albite-oligoclase (N = 179): Min = 0.697, Max = 0.977, Mean = 0.889, Median = 0.910, Std. Dev. = 0.060.
(iii) X_{An} in anorthite (N = 18): Min = 0.900, Max = 0.991, Mean = 0.955, Median = 0.963, Std. Dev. = 0.030.

Reykjanes is unique among the considered geothermal systems for the presence of hydrothermal anorthite, which occurs only in the deepest parts of the geothermal system, where temperature is >300 °C. Hydrothermal anorthite is distinguishable from its igneous counterpart being more An-rich and having much lower concentrations of K and Mg (Libbey and Williams-Jones 2016).

In contrast, hydrothermal albite and oligoclase are abundant throughout the Reykjanes geothermal system, where they mainly occur as replacement of primary plagioclase, although they are also present as subhedral crystals in veins and vugs (Libbey and Williams-Jones 2016). Therefore, most oligoclase might actually represent the product of incomplete replacement of a more calcic primary plagioclase, which could

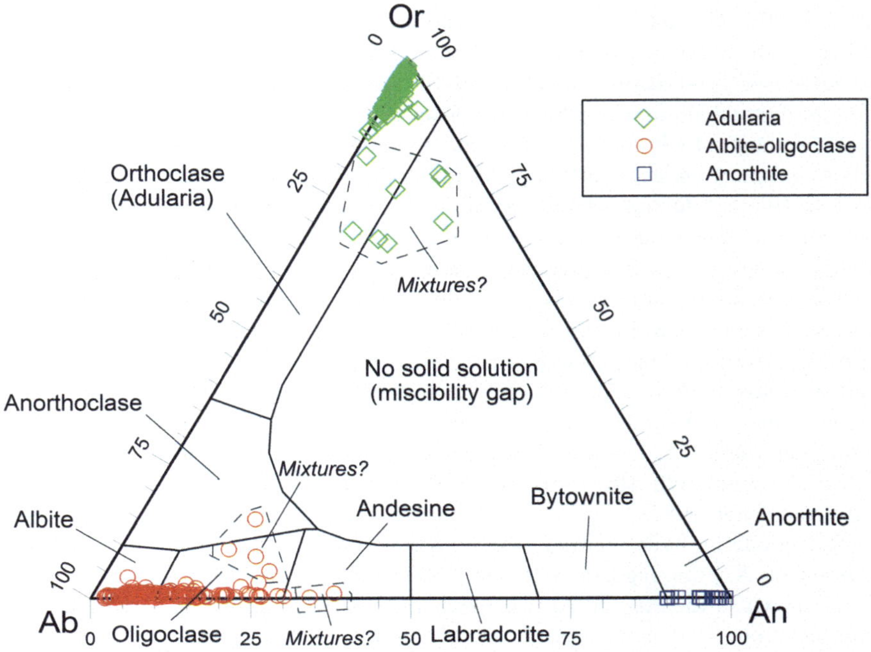

Fig. 4.2 Ab-Or-An triangular plot showing the composition of 361 hydrothermal feldspars from the Reykjanes geothermal field (data mainly from Libbey and Williams-Jones 2016 and subordinately from Marks et al. 2010 and Fowler et al. 2015)

be still present as relict. If so, oligoclase would be less frequent then evaluated above for Reykjanes, consistent with what is generally observed in active geothermal fields (Browne 1977).

Summing up, accepting that oligoclase compositions are due to the possible presence of relicts of primary plagioclase in hydrothermal albite and that mixed compositions are due to intergrowths of different endmember feldspars, it can be concluded that both adularia and albite are the most abundant hydrothermal feldspars in active geothermal systems, where they occur as pure or nearly pure solid phases. Based on this evidence, the activities of Or in adularia and Ab in albite can be reasonably assumed equal to one.

4.2.3 The Structural State and Degree of Ordering of Hydrothermal Alkali Feldspars

Only few investigations were performed so far to characterize the structural state and Al–Si order-disorder on the tetrahedral sites of authigenic K- and Na-feldspars from active geothermal systems.

The occurrence of almost pure adularia as hydrothermal mineral at temperature ≤265 °C in the Wairakei geothermal system was shown long ago by Steiner and coworkers (Steiner 1970 and references therein). Optical data indicate that hydrothermal K-feldspar is partly monoclinic and partly triclinic although X-ray diffractograms indicate only monoclinic lattice. According to Steiner (1970) either monoclinic and triclinic adularia are precipitated almost concomitantly, or triclinic K-feldspar alone is formed and the associated monoclinic structure is only apparent, being simulated by very fine twinning that is optically not detectable. If so, hydrothermal, pure potassium feldspar may have any degree of Si–Al ordering on crystallization.

McDowell (1986) investigated the structural state and Al–Si order on the tetrahedral sites of the authigenic and re-equilibrated detrital K- and Na-feldspars coexisting at temperatures of 250–360 °C in the Salton Sea geothermal reservoir and recovered from the Elmore 1 borehole. The ordering parameter Z (as defined by Thompson 1969) is 0.79 ± 0.09 at ~335 °C and 0.70 ± 0.07 at ~360 °C for albite (N = 6), whereas Z is 0.81 at ~335 °C for K-feldspar (N = 1).

Shearer et al. (1988) noted a systematic increase with depth of Al–Si disorder in K-feldspar from Salton Sea, but this information refers to allogenic-authigenic mixtures of this mineral because the XRD analyses do not distinguish allogenic and authigenic solid phases. Therefore, Shearer et al. (1988) concluded that the systematic increase in Al–Si disorder with depth probably suggests the presence of higher fractions of disordered authigenic (sanidine-like) K-feldspar in the hydrothermal assemblage.

Hedenquist and Browne (1989) studied the hydrothermal minerals at Waiotapu, concluding that (i) the albite structure is partially ordered, based on the position of the d_{204} reflection whereas (ii) adularia is almost completely ordered (maximum microcline) as indicated by the position of d_{060} and d_{204} reflections.

Adularias from low-temperature hydrothermal veins are remarkable for the extensive differences in structural state, from totally disordered monoclinic sanidine (or the highly disordered triclinic form) to highly ordered microcline, chiefly depending on nucleation and crystallization rates and temperature (Černý and Chapman 1986). To be noted that hydrothermal alkali feldspars probably formed under supersaturation conditions rather than at equilibrium. Consequently, their degree of ordering is expected to be a function of the extent of supersaturation, with the Al–Si distribution on the tetrahedral sites becoming more and more disordered with increasing supersaturation. In fact, at conditions of sufficient supersaturation, the disordered phase is kinetically favored over its ordered equivalent, irrespective of their relative thermodynamic stabilities (Carpenter and Putnis 1985).

The post-crystallization thermal regime can also affect the Al–Si distribution on the tetrahedral sites of alkali feldspar. For instance, initially disordered phases kept at high temperature could re-organize their lattice acquiring a more ordered Al–Si distribution. Owing to these potential thermal effects on the hydrothermal feldspars, it is advisable to take into consideration the authigenic alkali feldspars present in sedimentary rocks, which formed at low temperatures and were certainly not affected by annealing and, therefore, are expected to give reliable indications on the degree of ordering of adularia and albite.

According to Morad (1978), diagenetic K-feldspar crystals are monoclinic and, less commonly, triclinic with various degrees of Al–Si ordering, whereas diagenetic albite is well-ordered, triclinic. Kastner and Siever (1979) recognized that authigenic feldspars may have almost any state of Al–Si ordering, from completely ordered, triclinic microclines and albites to highly disordered monoclinic sanidines. Albites tend to be ordered (e.g., Kastner and Waldbaum 1968; Fishman et al. 1995) whereas the structural state of K-feldspar depends on rock type, probably due to kinetic factors. K-feldspars are fully ordered in carbonate rocks, are highly disordered in ash beds, whereas a wide range of ordering degree is found in sandstones and shales.

Further indications are given by the experiments performed by Flehmig (1977), who synthesized highly pure K-feldspar and albite crystals by means of hydroxide gels, mainly at 20 °C apart from a few runs at 3 and 60 °C. Strictly speaking, the factors controlling crystal growth in gels are the free solute concentrations and pH in the gel pores. However, since these parameters cannot be measured, the corresponding parameters in the supernatant solution were reported and discussed by Flehmig (1977). He found that the structural state and the degree of ordering of the synthesized feldspars depends on pH and Na and K concentrations. The growth of K-feldspar crystals took place in the pH range 5–10, with more ordered triclinic K-feldspar (microcline) at higher pH and alkali concentrations and less ordered monoclinic K-feldspar at lower pH and alkali concentrations. Albite crystals were obtained only at pH $\geq$8 and high Na concentrations and showed high degrees of ordering. Similar influences of pH and alkali concentration on the formation of low albite had previously been recognized by Martin (1969).

On the basis of the evidence summarized above, the hydrothermal alkali feldspars occurring in active geothermal systems are expected to be triclinic, fully ordered, albite and variably ordered adularia, from fully ordered microcline to completely disordered sanidine, with either triclinic or monoclinic symmetry.

These expectations are at variance with the findings of Stefánsson and Arnórsson (2000), who underscored that primary disordered plagioclases and alkali-feldspars of compositions typically found in volcanic rocks tend to dissolve in geothermal waters with simultaneous precipitation of ordered alkali-feldspars with near pure endmember compositions. However, Stefánsson and Arnórsson (2000) used the solubility constants of feldspars of Arnórsson and Stefánsson (1999) whose reliability is questionable, at least as far as the alkali feldspars are concerned, as shown in the next section.

4.2.4 *The Thermodynamic Properties of Endmember Alkali Feldspars*

As already underscored in Sect. 2.2, in the geochemical literature there is a lack of consensus on the thermodynamic data of several minerals, including the alkali feldspars. The reproduction of the monoclinic/triclinic transition temperature of alkali feldspars, which is close to 500 °C in K-feldspar and close to 660 °C in Na-feldspar at low pressures (Deer et al. 2001), represents a good way to prove the reliability of the thermodynamic data of alkali feldspars, because of the simplicity of the solid-solid reactions of interest (see below).

First, the thermodynamic data of high-albite, low-albite, sanidine, and microcline from Holland and Powell (1998) and Arnórsson and Stefánsson (1999), see Table 4.1, were inserted into SUPCRT92.

The heat capacity coefficients of Holland and Powell (1998) refer to the polynomial equation:

$$C^{\circ}_{P} = a + b \cdot T + c \cdot T^{-2} + d \cdot T^{-1/2}, \tag{4.1}$$

whereas the heat capacity coefficients of Arnórsson and Stefánsson (1999) are consistent with the polynomial relation:

$$C^{\circ}_{P} = a + 2b \cdot T + c \cdot T^{-2} + f \cdot T^{2} + g \cdot T^{-1/2}. \tag{4.2}$$

Therefore, heat capacity coefficients were recalculated to make them consistent with the Maier-Kelley relation, Eq. (2.17). The thermodynamic data of Holland and Powell (2011) were not considered because these authors give the same entropy (214.3 J K^{-1} mol^{-1}) for sanidine and microcline probably due to a typo. In fact, Holland and Powell (1998) give different entropies for sanidine (230 J mol^{-1} K^{-1}) and microcline (216 J mol^{-1} K^{-1}), as expected from configurational entropy considerations (Openshaw et al. 1976).

Second, the standard Gibbs free energy, ΔG°_{r}, of the solid-solid reaction;

$$\text{Microcline} = \text{Sanidine} \tag{4.3}$$

was calculated by means of SUPCRT92 as a function of temperature, at both 1 bar and 1 kbar, considering alternatively the thermodynamic data of Helgeson et al. (1978), Holland and Powell (1998), and Arnórsson and Stefánsson (1999).

Results are displayed in the diagram of ΔG°_{r} versus temperature (Fig. 4.3) which shows that:

(a) the thermodynamic data of Helgeson et al. (1978) are consistent with a sanidine/microcline transition temperature close to 500 °C, in agreement with the available experimental evidence (see above),

Table 4.1 Standard state molal thermodynamic properties, at 25 °C, 1 bar, of maximum-microcline, high-sanidine, low-albite, and high-albite from H78 = Helgeson et al. (1978), HP98 = Holland and Powell (1998), and AS99 = Arnórsson and Stefánsson (1999)[a]

Mineral	ΔG_f^o	ΔH_f^o	S°	V°	a	b (×10^3)	c (×10^{-5})	References
	cal mol^{-1}	cal mol^{-1}	cal K^{-1} mol^{-1}	cm^3 mol^{-1}	cal K^{-1} mol^{-1}	cal K^{-2} mol^{-1}	cal K mol^{-1}	
Max-microcline	−895,374	−949,188	51.13	108.74	63.83	12.9	−17.05	H78
Microcline	−896,317	−950,060	51.63	108.92	56.29	22.76	−12.79	HP98
Microcline	−896,163	−950,022	51.20	108.74	49.41	32.66	−9.49	AS99
High-sanidine	−893,738	−946,538	54.53	109.01	63.83	12.9	−17.05	H78
Sanidine	−894,888	−947,634	54.97	109.00	56.29	22.76	−12.79	HP98
Sanidine	−895,305	−947,832	55.66	109.01	51.09	30.86	−10.04	AS99
Low-albite	−886,308	−939,680	49.51	100.07	61.7	13.9	−15.01	H78
Low-Albite	−887,168	−940,392	50.22	100.06	57.30	21.85	−13.38	HP98
Low-albite	−887,437	−940,771	49.76	100.07	52.09	29.41	−10.42	AS99
High-albite	−884,509	−937,050	52.30	100.43	61.7	13.9	−15.01	H78
High-albite	−885,784	−938,059	53.39	101.09	57.30	21.85	−13.38	HP98
High-albite	−885,531	−937,710	53.63	100.43	51.30	31.07	−10.23	AS99

Following Helgeson et al. (1978), the names high-sanidine and high-albite connote complete disorder, whereas maximum-microcline and low-albite refer to the completely ordered phases

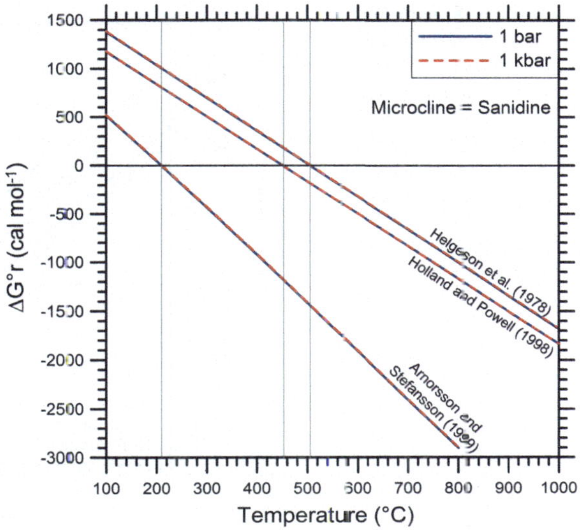

Fig. 4.3 Diagram of the standard Gibbs free energy of sanidine/microcline transition as a function of temperature, at a pressure of 1 bar and 1 kbar, based on the thermodynamic data of Helgeson et al. (1978), Holland and Powell (1998), and Arnórsson and Stefánsson (1999)

(b) as explicitly stated in the paper of Holland and Powell (1998), their thermodynamic data are consistent with a sanidine/microcline transition temperature of 452°C, which is ~50 °C lower than the experimental datum;
(c) the thermodynamic data of Arnórsson and Stefánsson (1999) are consistent with a sanidine/microcline transition temperature of 210 °C, which is ~290 °C lower than the experimental value.

Third, the $\Delta G°_r$ of the solid-solid reaction;

$$\text{High-albite} = \text{Low-albite} \tag{4.4}$$

was computed as a function of temperature, at both 1 bar and 1 kbar, taking into account alternatively the thermodynamic data of Helgeson et al. (1978), Holland and Powell (1998), and Arnórsson and Stefánsson (1999).

Results are presented in the plot of $\Delta G°_r$ versus temperature (Fig. 4.4) which shows that:

(a) the thermodynamic data of Helgeson et al. (1978) are consistent with a high-albite/low-albite transition temperature close to 660 °C, in accordance with the available experimental evidence (see above),
(b) the thermodynamic data of Holland and Powell (1998) are consistent with a high-albite/low-albite transition temperature of 460–465 °C, which is ~200 °C lower than the experimental datum;
(c) the thermodynamic data of Arnórsson and Stefánsson (1999) are consistent with a high-albite/low-albite transition temperature of 515 °C, which is ~145 °C lower than the experimental value.

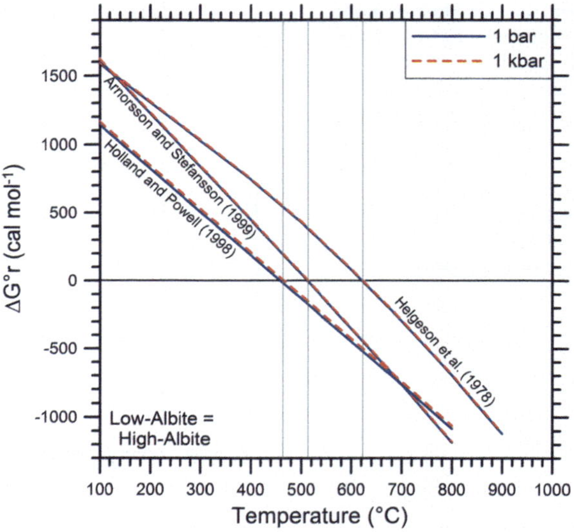

Fig. 4.4 Diagram of the standard Gibbs free energy of high-albite/low-albite transition as a function of temperature, at a pressure of 1 bar and 1 kbar, based on the thermodynamic data of Helgeson et al. (1978), Holland and Powell (1998), and Arnórsson and Stefánsson (1999)

Summing up, only the thermodynamic data of Helgeson et al. (1978) are consistent with the monoclinic/triclinic transition temperatures of alkali feldspars whereas the thermodynamic data of Holland and Powell (1998) and Arnórsson and Stefánsson (1999) are not. Consequently, the thermodynamic data of Helgeson et al. (1978) can be used to compute the log K values of the exchange reactions involving alkali feldspars and other Al-bearing minerals of interest for geothermometry and f_{CO_2} evaluation, as discussed in Chaps. 6–9, whereas the thermodynamic data of Holland and Powell (1998) and Arnórsson and Stefánsson (1999) cannot be used to this aim. In addition to these general implications, the monoclinic/triclinic transition temperatures of alkali feldspars have other specific implications on Na–K geothermometry, whose discussion is postponed to Sect. 6.1.

4.2.5 The Thermodynamic Properties of Variably Ordered Alkali Feldspars

Since the conversion between the monoclinic and triclinic crystal structures in K-feldspar has little or no effect on its thermodynamic behavior, the different thermodynamic properties of ordered microcline and disordered sanidine are entirely due to substitutional Al–Si order-disorder or nearly so (Helgeson et al. 1978). In contrast, the displacive transformation in albite from monoclinic to triclinic symmetry is accompanied by a relatively large heat of transition and, therefore, albite experiences two superimposed lambda transitions with increasing temperature, one caused by exchange of Al and Si atoms on its tetrahedral sites, and the other produced by displacive changes in its symmetry (Helgeson et al. 1978).

Accepting that variably ordered adularia occurs in active geothermal systems (see Sect. 4.2.3), the thermodynamic properties of adularia with ordering parameter Z varying from 0.9 to 0.1, at steps of 0.1 units, were calculated in this work and added to the SUPCRT92 thermodynamic database. In these calculations, maximum-microcline and high-sanidine were assumed to be representative of adularia with ordering parameter Z of 1 and 0, respectively. Following Helgeson et al. (1978), the standard molal enthalpy and the standard molal entropy of substitutional disorder for variably ordered adularia, $\Delta H^{\circ}_{ds,Adl}$ (in cal mol^{-1}) and $\Delta S^{\circ}_{ds,Adl}$ (in cal K^{-1} mol^{-1}), respectively, at 25 °C, 1 bar, were computed using the simple relations:

$$\Delta H^{\circ}_{ds,Adl} = 2650 \cdot (1 - Z) \tag{4.5}$$

$$\Delta S^{\circ}_{ds,Adl} = 3.4 \cdot (1 - Z). \tag{4.6}$$

The standard molal Gibbs free energy of substitutional disorder for variably ordered adularia, $\Delta G^{\circ}_{ds,Adl}$ (in cal mol^{-1}), at 25 °C, 1 bar, was then obtained by means of the fundamental Eq. (2.1), rewritten as:

$$\Delta G^{\circ}_{ds,Adl} = \Delta H^{\circ}_{ds,Adl} - T \cdot \Delta S^{\circ}_{ds,Adl}. \tag{4.7}$$

Helgeson et al. (1978) assumed the same heat capacity power function coefficients for high-sanidine and maximum-microcline, because the difference in the heat capacity of the two K-feldspars, at 25 °C and 1 bar, is only 0.07 cal K^{-1} mol^{-1} (Openshaw et al. 1976). Here, this hypothesis was adopted for adularia irrespective of its ordering parameter Z.

Accepting that hydrothermal albite occurring in active geothermal systems is fully ordered (see Sect. 4.2.3), there is no need to calculate the thermodynamic properties of variably ordered albite. Nevertheless, the thermodynamic properties of albite with ordering parameter Z varying from 0.9 to 0.1, at steps of 0.1 units, were also computed in this work and added to the SUPCRT92 thermodynamic database. In these calculations, low-albite and high-albite were assumed to be representative of albite with ordering parameter Z of 1 and 0, respectively. Following Helgeson et al. (1978), the standard molal enthalpy and the standard molal entropy of substitutional disorder for variably ordered albite, $\Delta H^{\circ}_{ds,Ab}$ (in cal mol^{-1}) and $\Delta S^{\circ}_{ds,Ab}$ (in cal K^{-1} mol^{-1}), respectively, at 25 °C, 1 bar, were calculated utilizing the equations:

$$\Delta H^{\circ}_{ds,Ab} = 2630 \cdot (1 - Z). \tag{4.8}$$

$$\Delta S^{\circ}_{ds,Ab} = 2.79 \cdot (1 - Z). \tag{4.9}$$

The standard molal Gibbs free energy of substitutional disorder for variably ordered albite, $\Delta G^{\circ}_{ds,Ab}$ (in cal mol^{-1}), at 25 °C, 1 bar, was then be computed by means of the fundamental relation (2.1) rewritten as:

$$\Delta G^{\circ}_{ds,Ab} = \Delta H^{\circ}_{ds,Ab} - T \cdot \Delta S^{\circ}_{ds,Ab}. \tag{4.10}$$

The same heat capacity power function coefficients were adopted for any variably ordered albite because those of high-albite are equal to those of low-albite, at least below 350 °C (Helgeson et al. 1978).

The molar volumes of variably ordered adularia were computed through linear interpolation of the values of maximum-microcline and high-sanidine reported by Helgeson et al. (1978). Similarly, the molar volumes of variably ordered albite were calculated through linear interpolation of the values of low-albite and high-albite given by Helgeson et al. (1978). The standard state molal thermodynamic properties, at 25 °C, 1 bar, of variably ordered adularia and albite are listed in Table 4.2 together with those of maximum-microcline, high-sanidine, low-albite, and high-albite reported by Helgeson et al. (1978).

4.3 White Micas

4.3.1 Main Characteristics of White Micas

The three rock-forming white micas are muscovite $[KAl_2^{VI}(AlSi_3)^{IV}O_{10}(OH)_2]$, paragonite $[NaAl_2^{VI}(AlSi_3)^{IV}O_{10}(OH)_2]$, and margarite $[CaAl_2^{VI}(AlSi_2)^{IV}O_{10}(OH)_2]$. As pointed out by Guidotti and Sassi (2002): (i) muscovite may contain up to 38 mol% of paragonite in solution, (ii) paragonite may contain up to 15 mol% of muscovite in solution, and (iii) practically no solution occurs between muscovite and margarite.

In spite of these possible mixing relations, muscovite prevails by far over paragonite and margarite in white micas (or sericites) from active geothermal systems, including phengites and illites. Phengites may be described as binary solid solutions of muscovite and celadonites $[K(M^{3+})^{VI}(M^{2+})^{VI}Si_4^{IV}O_{10}(OH)_2]$ (Ernst 1963; Velde 1977). In the chemical formula of celadonite, M^{3+} generally represents Al^{3+} or Fe^{3+}, whereas M^{2+} usually stands for Fe^{2+} or Mg^{2+}, thus leading to four limiting compositions called celadonite $[K(Fe^{3+})^{VI}(Mg^{2+})^{VI}Si_4^{IV}O_{10}(OH)_2]$, Fe-celadonite $[K(Fe^{3+})^{VI}(Fe^{2+})^{VI}Si_4^{IV}O_{10}(OH)_2]$, Fe–Al-celadonite $[KAl^{VI}(Fe^{2+})^{VI}Si_4^{IV}O_{10}(OH)_2]$ and Mg–Al-celadonite $[KAl^{VI}(Mg^{2+})^{VI}Si_4^{IV}O_{10}(OH)_2]$ (Li et al. 1997). Illites are solid solutions of muscovite, celadonites and pyrophyllite $[Al_2Si_4O_{10}(OH)_2]$ (Weaver and Pollard 1973).

Following Brigatti and Guggenheim (2002) and references therein, we recall that the 2:1 mica structure is constituted by two opposing tetrahedral (T) sheets with an octahedral (M) sheet between them to produce a TMT layer. The general chemical formula of micas is:

$$A^{XII}M^{VI}_{2-3} \square^{VI}_{1-0} T^{IV}_4 O_{10}(X)_2,$$

Table 4.2 Standard state molal thermodynamic properties, at 25 °C, 1 bar, of maximum-microcline, high-sanidine, low-albite, and high-albite (from Helgeson et al. 1978) as well as of variably ordered adularia and albite (this work)

Mineral	ΔG°_{f}	ΔH°_{f}	S°	V°	a	b (×10³)	c (×10⁻⁵)
	cal mol^{-1}	cal mol^{-1}	cal K^{-1} mol^{-1}	cm^{3} mol^{-1}	cal K^{-1} mol^{-1}	cal K^{-2} mol^{-1}	cal K mol^{-1}
Max-microcline, Z = 1	−895,374	−949,188	51.13	108.74	63.83	12.9	−17.05
Adularia, Z = 0.9	−895,210	−948,923	51.47	108.88	63.83	12.9	−17.05
Adularia, Z = 0.8	−895,047	−948,658	51.81	108.90	63.83	12.9	−17.05
Adularia, Z = 0.7	−894,883	−948,393	52.15	108.91	63.83	12.9	−17.05
Adularia, Z = 0.6	−894,720	−948,128	52.49	108.93	63.83	12.9	−17.05
Adularia, Z = 0.5	−894,556	−947,863	52.83	108.94	63.83	12.9	−17.05
Adularia, Z = 0.4	−894,392	−947,598	53.17	108.95	63.83	12.9	−17.05
Adularia, Z = 0.3	−894,229	−947,333	53.51	108.97	63.83	12.9	−17.05
Adularia, Z = 0.2	−894,065	−947,068	53.85	108.98	63.83	12.9	−17.05
Adularia, Z = 0.1	−893,902	−946,803	54.19	108.99	63.83	12.9	−17.05
High-sanidine, Z = 0	−893,738	−946,538	54.53	109.01	63.83	12.9	−17.05
Low-albite, Z = 1	−886,308	−939,680	49.51	100.07	61.7	13.9	−15.01
Albite, Z = 0.9	−886,128	−939,417	49.79	100.11	61.7	13.9	−15.01
Albite, Z = 0.8	−885,948	−939,154	50.07	100.14	61.7	13.9	−15.01
Albite, Z = 0.7	−885,768	−938,891	50.35	100.18	61.7	13.9	−15.01

(continued)

Table 4.2 (continued)

Mineral	ΔG°_{f}	ΔH°_{f}	S°	V°	a	b ($\times 10^{3}$)	c ($\times 10^{-5}$)
	cal mol^{-1}	cal mol^{-1}	cal K^{-1} mol^{-1}	cm^{3} mol^{-1}	cal K^{-1} mol^{-1}	cal K^{-2} mol^{-1}	cal K mol^{-1}
Albite, Z = 0.6	−885,588	−938,628	50.63	100.21	61.7	13.9	−15.01
Albite, Z = 0.5	−885,408	−938,365	50.91	100.25	61.7	13.9	−15.01
Albite, Z = 0.4	−885,229	−938,102	51.18	100.29	61.7	13.9	−15.01
Albite, Z = 0.3	−885,049	−937,839	51.46	100.32	61.7	13.9	−15.01
Albite, Z = 0.2	−884,869	−937,576	51.74	100.36	61.7	13.9	−15.01
Albite, Z = 0.1	−884,689	−937,313	52.02	100.39	61.7	13.9	−15.01
High-albite, Z = 0	−884,509	−937,050	52.30	100.43	61.7	13.9	−15.01

in which:

- A^{XII} refers to the twelve-coordinated interlayer sites, which are generally occupied by K, Na, Ca, Ba, and rarely by Rb, Cs, Sr, NH_4, and H_3O;
- M^{VI} indicates the six-coordinated octahedral sites, which are usually occupied by Mg, Fe^{2+}, Al, and Fe^{3+}, although Li, Ti, V, Cr, Mn, Co, Ni, Cu, and Zn may also be present in peculiar micas;
- $\square^{VI}$ indicates a vacant six-coordinated octahedral site;
- T^{IV} denotes the four-coordinated tetrahedral sites, which are generally occupied by Si, Al and Fe^{3+} and infrequently by B and Be;
- X refers to (OH), F, Cl, O, S.

The smallest structural unit of micas contains three octahedral sites. All the three octahedral sites are occupied in the so-called trioctahedral micas, such as biotite and phlogopite, whereas only two octahedral sites [usually M(2)] are occupied and one is vacant [generally M(1)], in the micas known as dioctahedral, such as muscovite, paragonite, margarite, celadonite, and illite, which are those of interest to us.

4.3.2 The Activities of Muscovite, Celadonites, and Pyrophyllite in Hydrothermal White Micas

Following Helgeson et al. (1978), Aagaard and Helgeson (1983), Helgeson and Aagaard (1985), and Aagaard and Jahren (1992), the activities of muscovite, a_{Ms}, pyrophyllite, a_{Prl}, celadonite, a_{Cel}, Fe-celadonite, $a_{Fe\text{-}Cel}$, Fe–Al-celadonite, $a_{Fe\text{-}Al\text{-}Cel}$, and Mg–Al-celadonite, $a_{Mg\text{-}Al\text{-}Cel}$, in white micas are computed using the following equations:

$$a_{Ms} = k_{Ms} \cdot X_{K,A} \cdot X^2_{Al,M(2)} \cdot X_{Al,T} \cdot X^3_{Si,T} \tag{4.11}$$

$$a_{Prl} = k_{Prl} \cdot X_{\square,A} \cdot X^2_{Al,M(2)} \cdot X^4_{Si,T} \tag{4.12}$$

$$a_{Cel} = k_{Cel} \cdot X_{K,A} \cdot X_{Mg,M(2)} \cdot X_{Fe^{3+},M(2)} \cdot X^4_{Si,T} \tag{4.13}$$

$$a_{Fe-Cel} = k_{Fe-Cel} \cdot X_{K,A} \cdot X_{Fe^{2+},M(2)} \cdot X_{Fe^{3+},M(2)} \cdot X^4_{Si,T} \tag{4.14}$$

$$a_{Fe-Al-Cel} = k_{Fe-Al-Cel} \cdot X_{K,A} \cdot X_{Fe^{2+},M(2)} \cdot X_{Al,M(2)} \cdot X^4_{Si,T} \tag{4.15}$$

$$a_{Mg\text{ - }Al\text{ - }Cel} = k_{Mg\text{ - }Al\text{ - }Cel} \cdot X_{K,A} \cdot X_{Mg,M(2)} \cdot X_{Al,M(2)} \cdot X^4_{Si,T} \tag{4.16}$$

which are based on the assumption of random mixing of atoms on energetically equivalent sites. In Eqs. (4.11)–(4.16):

$k_{Ms} = 9.4815$, $k_{Prl} = 1$, $k_{Cel} = k_{Fe\text{–}Cel} = k_{Fe\text{–}Al\text{–}Cel} = k_{Mg\text{–}Al\text{–}Cel} = 4$ are constants relating the inter- and intra-crystalline standard states;

$X_{K,A}$ and $X_{\square,A}$ are the mole fractions of potassium and vacancies on the interlayer A site;

$X_{Al,M(2)}$, $X_{Fe^{3+},M(2)}$, $X_{Mg,M(2)}$, $X_{Fe^{2+},M(2)}$ are the mole fractions of aluminum, trivalent iron, magnesium, and divalent iron on the M(2) octahedral sites;

$X_{Al,T}$ and $X_{Si,T}$ are the mole fractions of aluminum and silicon on the T tetrahedral sites.

Chemical analyses of white micas are available for several active geothermal systems, including Salton Sea, California, for the temperature range 190–322 °C McDowell and Elders (1980, 1983); Coso, California, for the temperature interval 190–250 °C (Bishop and Bird 1987); Roosevelt Hot Springs, Utah, for temperatures varying from 212 to 255 °C (Ballantyne 1978; Capuano and Cole 1982); Los Azufres, Mexico, for temperatures ranging from 190 to 270 °C (Cathelineau 1988; Cathelineau and Izquierdo 1988); Bouillante, Guadeloupe, for the temperature range 204–260 °C (Mas et al. 2006); Larderello, Italy, for the temperature interval 248–315 °C (Boyce et al. 2003); Pantelleria, Italy, at temperatures of 170 and 275 °C (Fulignati et al. 1997); Aluto-Langano, Ethiopia, for the temperature range 215–300 °C (Teklemariam et al. 1996); Waiotapu, New Zealand, for the temperature interval 170–282 °C (Hedenquist and Browne 1989); Broadlands-Ohaaki, New Zealand, for temperatures varying from 222 to 300 °C (Lonker and Gerald 1990; Lonker et al. 1990); Kakkonda, Japan, at temperatures of 205 and 300 °C (Inoue et al. 2004).

These 94 chemical analyses of white mica samples were compiled. Since the FeO and Fe_2O_3 contents are not reported and only the total Fe content is available, it is necessary to adopt some hypotheses concerning the redox state of iron to compute, first, site occupancies and, second, the activities of muscovite, pyrophyllite, and celadonite components using Eqs. (4.11)–(4.16). The adopted hypotheses are that 86.5% of total Fe is present as Fe^{3+}, which is the average value for illites according to Weaver and Pollard (1973), and that Fe^{3+} is present only in the octahedral sites. The exclusive occurrence of Fe^{3+} in the octahedral sites is a reasonable assumption considering both that Fe^{3+} occupies the octahedral sites in the ferric endmembers of celadonite (e.g., Li et al. 1997), pyrophyllite (e.g., Badaut et al. 1992 and references therein), and muscovite (e.g., Guidotti and Sassi 2002), and that Fe^{3+} tetrahedral substitution does not take place in dioctahedral micas, whereas it occurs in trioctahedral micas, such as tetra-ferriphlogopite, tetra-ferri-annite and anandite (Brigatti and Guggenheim 2002). To be noted that the alternative assumption of Fe completely present as Fe^{2+} (which was adopted by Cathelineau and Izquierdo (1988) determines changes (i) in the mole fractions of Fe^{2+} and Fe^{3+} on the M(2) octahedral sites, whereas other mole fractions are unaffected by the hypotheses on the Fe redox state and (ii) in the activities of celadonite, Fe-celadonite, and Fe–Al-celadonite, whereas the activities of Mg–Al-celadonite, muscovite, and pyrophyllite are not influenced by the assumptions on the Fe redox state.

The correlation plot of the activity of muscovite in the white mica samples versus the mole fraction of Al on the tetrahedral site (Fig. 4.5), is a good starting point to analyze the data of geothermal white micas. Since illites and white micas in general

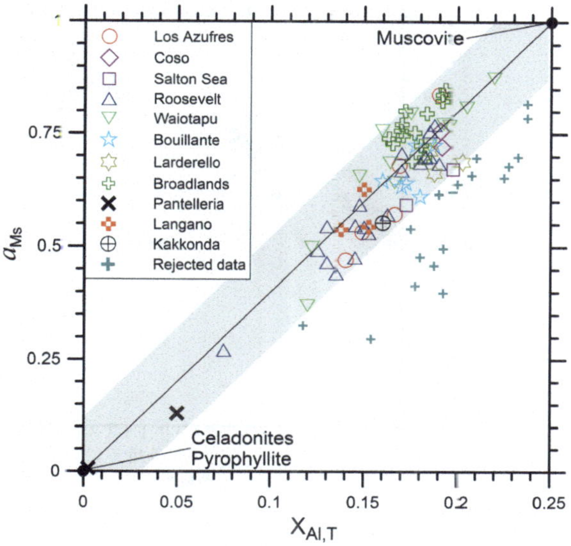

Fig. 4.5 Diagram of the activity of muscovite versus the mole fraction of Al on the T tetrahedral sites for the 94 white mica samples from different geothermal fields (see legend), also showing the idealized stoichiometry of the components of white micas, that is muscovite, pyrophyllite and celadonites

are solid solutions of muscovite, celadonites and pyrophyllite (see above), they are expected to be found along the tie line joining the muscovite point and the point representative of both celadonites and pyrophyllite or at least close to this Ms – (Prl + Cel) tie line. Allowing for a maximum deviation of ±0.12 a_{Ms} units, 75 chemical analyses were considered representative of white micas, whereas the remaining 19 were rejected. All these rejected samples show negative deviations from the Ms – (Prl + Cel) tie line, possibly because they are white micas mixed with other clay minerals or other phyllosilicates. Alternatively, assuming that all Fe is present as Fe^{3+} and occupies the tetrahedral sites, 11 samples would deviate from the Ms – (Prl + Cel) tie line by more than 0.12 a_{Ms} units, although the other 8 would exhibit smaller deviations.

Aagaard and Helgeson (1983) underscored that the distribution of Al^{3+} ion among the tetrahedral and octahedral sites of clay minerals allows one to distinguish illites from montmorillonites and mixed-layer clays. In fact, these minerals occupy different fields in the plot of Fig. 4.6, which was slightly modified adding the idealized stoichiometry of the components of white micas, that is muscovite, pyrophyllite and celadonites.

As expected: (i) the 75 accepted samples of geothermal white micas are situated inside the compositional space delimited by these three endmembers, and most of these samples are found in the field of illites or close to it, whereas (ii) the 19 rejected samples are positioned to the right of the Ms–Prl–Cel triangle. However, assuming that all Fe is present as Fe^{3+} and occupies the tetrahedral sites, also the 19 rejected samples would be found inside the Ms–Prl–Cel triangle or close to it. Since the ambiguity concerning these 19 rejected samples cannot be resolved, it is advisable to rely only on the other 75 samples.

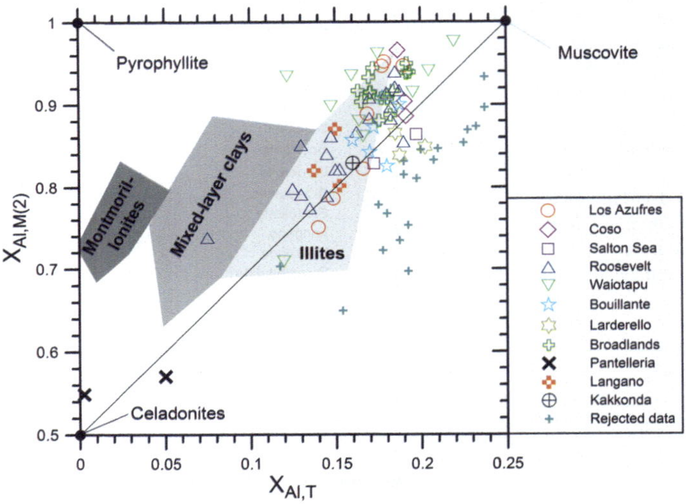

Fig. 4.6 Diagram of the mole fraction of Al on the M(2) octahedral sites versus the mole fraction of Al on the T tetrahedral sites for the 94 white mica samples from different geothermal fields (see legend). The grey fields of illites, mixed-layer clays, and montmorillonites (smectites) are from Aagaard and Helgeson (1983). The idealized stoichiometry of the components of white micas, that is muscovite, pyrophyllite and celadonites, are also shown

The activity of the muscovite endmember in the considered 75 white mica samples varies between 0.007 and 0.875, with an average of 0.654, a median of 0.688 and a standard deviation of 0.156. The sum of the activities of the celadonite components ranges from 0.033 to 0.677, with an average of 0.228, a median of 0.202 and a standard deviation of 0.130. The activity of the pyrophyllite component spans the range 0.000 to 0.156, with an average of 0.060, a median of 0.056 and a standard deviation of 0.035.

The muscovite endmember prevails in 70 of the 75 white mica samples, the celadonite components dominates the remaining 5 samples, whereas the pyrophyllite endmember is always subordinate, as shown by the triangular diagram of Fig. 4.7.

The two diagrams of Fig. 4.8 show that:

(i) the activities of the muscovite and pyrophyllite endmembers in the white mica samples of interest are not correlated, as also indicated by the very low value of the squared linear regression coefficient, $R^2 = 0.027$, whereas
(ii) there is a significant negative correlation between the activity of the muscovite component and the sum of the activities of the celadonite components, as also pointed out by the squared linear regression coefficient, $R^2 = 0.748$.

The activity of the muscovite endmember, the sum of the activities of the celadonite components and the activity of the pyrophyllite endmember have no significant relation with temperature for the 75 accepted white mica samples as pointed out by the very small R^2 values of 0.0004, 0.0191, and 0.174, respectively.

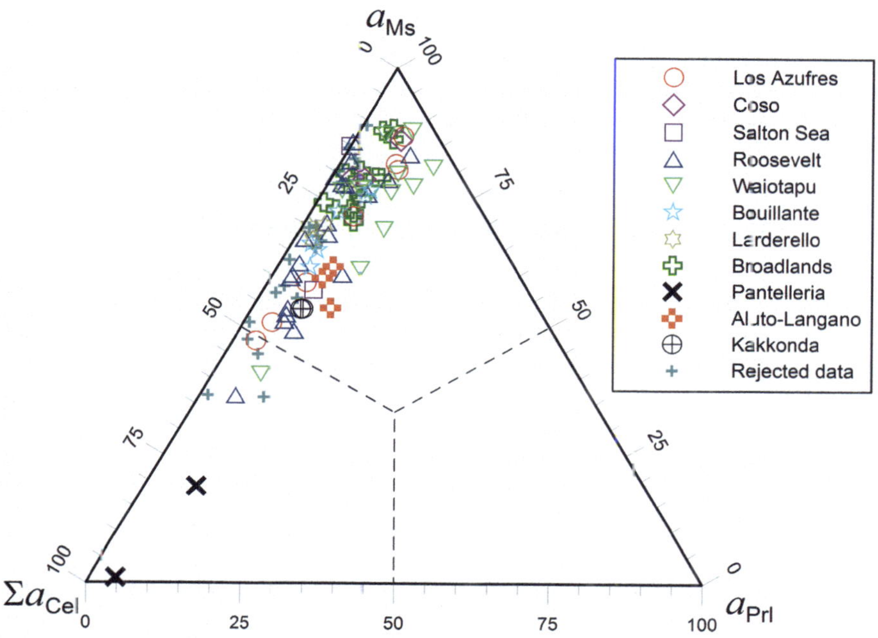

Fig. 4.7 Triangular diagram comparing the activities of muscovite, celadonites and pyrophyllite for the 75 accepted white mica samples from different geothermal fields (see legend)

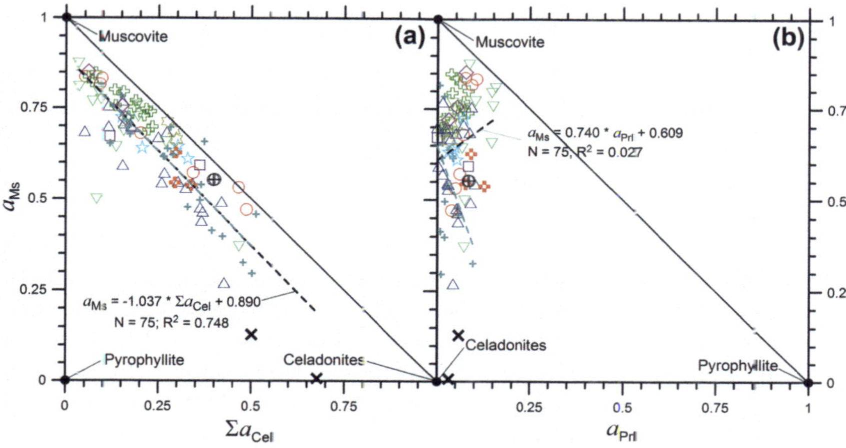

Fig. 4.8 Diagrams of the activity of the muscovite component versus **a** the sum of the activities of the celadonite components and **b** the activity of the pyrophyllite component for the 75 accepted white mica samples from different geothermal fields (symbols as in Figs. 4.5, 4.6 and 4.7). The idealized stoichiometry of muscovite, pyrophyllite and celadonites are also shown

In contrast, a_{Ms}, $a_{\Sigma Cel}$, and a_{Prl} exhibit a meaningful temperature dependence for the 19 rejected samples, with R^2 values of 0.686, 0.705, and 0.266, respectively. Again, this evidence suggests that white micas are mixed with other clay minerals or other phyllosilicates in these 19 rejected samples.

4.3.3 The Thermodynamic Properties of Muscovite

The values of the $\Delta G°_f$ and $\Delta H°_f$ of muscovite at 25 °C, 1 bar given in the first row of Table 4.3 were calculated by Helgeson et al. (1978) from the equilibrium constants derived from high pressure, high temperature experimental data for the following equilibria:

$$2 \text{ muscovite} + 6 \text{ quartz} + 2 \text{ H}^+ = 3 \text{ pyrophyllite} + 2 \text{ K}^+ \quad (4.17)$$

$$\text{muscovite} + \text{quartz} = \text{K-feldspar} + \text{andalusite} + \text{H}_2\text{O}. \quad (4.18)$$

Sverjensky et al. (1991) adopted a similar approach and considered the following equilibria:

$$1.5 \text{ K-feldspar} + \text{H}^+ = 0.5 \text{ muscovite} + 3 \text{ quartz} + \text{K}^+ \quad (4.19)$$

$$\text{muscovite} + \text{H}^+ + 1.5 \text{ H}_2\text{O} = 1.5 \text{ kaolinite} + \text{K}^+ \quad (4.20)$$

$$\text{muscovite} + \text{H}^+ + 3 \text{ quartz} = 1.5 \text{ pyrophyllite} + \text{K}^+ \quad (4.21)$$

$$\text{muscovite} + \text{H}^+ = 1.5 \text{ andalusite} + 1.5 \text{ quartz} + 1.5 \text{ H}_2\text{O} + \text{K}^+. \quad (4.22)$$

However, Sverjensky et al. (1991) used more recent thermodynamic data for the relevant aqueous species, obtaining the $\Delta G°_f$, $\Delta H°_f$, and S° values at 25 °C, 1 bar listed in the second row of Table 4.3. These $\Delta G°_f$, $\Delta H°_f$, and S° values differ from those of Helgeson et al. (1978) by 634 cal mol^{-1}, 992 cal mol^{-1}, and 1.20 cal K^{-1} mol^{-1}, respectively. Sverjensky et al. (1991) proposed also to adjust the free energies and enthalpies at 25 °C, 1 bar of all the K-bearing silicates by adding −236 cal mol^{-1} to preserve the consistency with phase equilibria at high pressures and temperatures. In principle, the thermodynamic data of muscovite of Sverjensky et al. (1991) are probably more representative than those of Helgeson et al. (1978), but their adoption and the addition of −236 cal mol^{-1} to the free energies and enthalpies at 25 °C, 1 bar of all the K-bearing silicates are not sufficient for maintaining the internal consistency of the thermodynamic database of Helgeson and coworkers unless a thorough revision is performed.

Table 4.3 Standard state molal thermodynamic properties, at 25 °C, 1 bar, of muscovite [$KAl_2(AlSi_3)O_{10}(OH)_2$] from H78 = Helgeson et al. (1978), S91 = Sverjensky et al. (1991), HP98 = Holland and Powell (1998), and HP11 = Holland and Powell (2011) as well as of pyrophyllite [$Al_2Si_4O_{10}(OH)_2$] from H78 = Helgeson et al. (1978)

Mineral	ΔG°$_f$	ΔH°$_f$	S°	V°	a	b (×10^3)	c (×10^{-5})	d	References
	cal mol^{-1}	cal mol^{-1}	cal K^{-1} mol^{-1}	cm^3 mol^{-1}	cal K^{-1} mol^{-1}	cal K^{-2} mol^{-1}	cal K mol^{-1}	cal K$^{-0.5}$ mol^{-1}	
Muscovite	−1,336,301	−1,427,408	68.8	140.71	97.56	26.38	-25.44	–	H78
Muscovite	−1,335,667	−1,426,416	70.0	–	–	–	–	–	S91
Muscovite	−1,339,319	−1,430,239	69.8	140.83	180.8	−4.7419	−5.186	−1668.1	HP98
Muscovite	−1,337,512	−1,428,432	69.8	140.83	180.8	−4.7419	−5.186	−1668.1	HP11
Pyrophyllite	−1,255,997	−1,345,313	57.2	126.6	79.432	39.214	−17.282	–	H78

The suggestion of Sverjensky et al. (1991) was questioned by Holland and Powell (1998), who proposed the ΔG°$_f$, ΔH°$_f$, and S° values at 25 °C, 1 bar given in the third row of Table 4.3 which differ from those of Helgeson et al. (1978) by 3018 cal mol^{-1}, 2813 cal mol^{-1}, and 1 cal K^{-1} mol^{-1}, respectively. Holland and Powell (2011) suggested the ΔG°$_f$ and ΔH°$_f$ at 25 °C, 1 bar listed in the fourth row of Table 4.3. These ΔG°$_f$ and ΔH°$_f$ are more similar to those of Helgeson et al. (1978) than the free energy and enthalpy proposed Holland and Powell (1998), with deviations of 1211 cal mol^{-1} and 1024 cal mol^{-1}, respectively. Because of the remarkable differences in the thermodynamic data stored for muscovite in these different thermodynamic databases, it is meaningless to pick up the data from distinct databases because it would be like mixing apples and oranges.

As shown in Sect. 4.3.2, muscovite is the main component of white mica solid solutions, whereas pyrophyllite and celadonites play a subordinate role. Among the different components of white mica solid solutions, Helgeson et al. (1978) report the thermodynamic properties of muscovite and pyrophyllite (seventh row of Table 4.3) but not of celadonites.

Following the approach delineated in Sect. 2.2.1, it is advantageous to consider the white mica solid solution with average activity of muscovite, $a_{Ms} = 0.654$, and to compute the Gibbs free energy and the thermodynamic equilibrium constant of its dissolution reaction. The results obtained taking into account the muscovite thermodynamic data of Helgeson et al. (1978) are listed in Table 4.4 as a function of temperature together with the standard molal Gibbs free energies and the natural and decimal logarithms of the thermodynamic equilibrium constant of the dissolution reaction of pure muscovite:

$$KAl_3Si_3O_{10}(OH)_2 + 10\ H^+ = K^+ + 3\ Al^{3+} + 3\ SiO_{2(aq)} + 6\ H_2O. \qquad (4.23)$$

Table 4.4 Standard molal Gibbs free energies and natural and decimal logarithms of the thermodynamic equilibrium constant of the dissolution reactions of both pure muscovite and the white mica solid solution with average activity of muscovite as a function of temperature

T	$\Delta_r G^\circ_{Ms}$	$\Delta_r G^\circ_{Ms,aa}$	ln K_{Ms}	ln $K_{Ms,aa}$	log K_{Ms}	log $K_{Ms,aa}$
°C	cal mol^{-1}	cal mol^{-1}				
0	−18,378	−18,147	33.8574	33.4328	14.7041	14.5197
25	−15,732	−15,480	26.5525	26.1279	11.5316	11.3472
60	−11,004	−10,723	16.6214	16.1968	7.2186	7.0342
100	−5084	−4769	6.8561	6.4315	2.9776	2.7932
150	2711	3068	−3.2240	−3.6486	−1.4002	−1.5846
200	10,976	11,375	−11.6735	−12.0982	−5.0698	−5.2542
250	19,962	20,403	−19.2015	−19.6261	−8.3391	−8.5235
300	30,064	30,548	−26.3958	−26.8205	−11.4636	−11.6480

4.4 Chlorites

4.4.1 *Main Characteristics of Chlorites*

The structure of chlorites consists in the regularly alternating (i) octahedral, brucite-like layers, also known as the hydroxide or O sheets, and (ii) tetrahedral-octahedral-tetrahedral, talc-like layers, also known as T–O–T or 2:1 sheets (e.g., Bailey 1988). Chlorites are, therefore, defined as T–O–T + O or 2:1 + 1 phyllosilicates.

Wiewiora and Weiss (1990) proposed a crystallochemical classification involving six octahedral positions and four tetrahedral positions, corresponding to half unit cell. The six octahedral sites are occupied by both (i) trivalent (and tetravalent) cations, such as Al^{3+}, Fe^{3+}, Cr^{3+}, (and Ti^{4+}), whose sum is indicated by R^{3+} and (ii) divalent cations, such as Mg^{2+}, Fe^{2+}, Mn^{2+}, and Ni^{2+}, whose sum is represented by R^{2+}. Vacancies are indicated by the symbol □. The four tetrahedral sites are occupied by Si and Al. Therefore, the chemistry of chlorites is described by the general crystallochemical formula (from Wiewiora and Weiss 1990, modified by De Caritat et al. 1993):

$$(R_u^{2+}R_y^{3+}\square_z)^{VI}(Si_{(4-x)}Al_x)^{IV}O_{10+w}(OH)_{8-w}$$

where $u + y + z = 6$, $z = (y - x - w)/2$, and w is usually zero or a small number. Octahedral occupancy is the sum of all cations in the octahedral sites and is given by $\Sigma^{VI} = u + y$. Based on the general crystallochemical formula of chlorites, the charge balance is expressed by the following relation:

$$2u + 3y - x - w - 12 = 0. \tag{4.24}$$

Foster (1962) compiled 150 chemical analyses of chlorites and found that: (i) Si^{IV} values range between 2.34 and 3.45 and 99% of these values are in the 2.40–3.40 interval, corresponding to $0.60 < x < 1.60$; (ii) octahedral vacancies range from −0.05 to +0.55 but are usually in the 0.0 to +0.3 interval; (iii) all the data have charge balance within the range −0.05 to +0.05, with 94% of those data falling within the smaller interval −0.03 to +0.03. The last point implies that the charge balance assumption is reasonable for the chlorite structure.

4.4.2 *Crystal Chemistry of Hydrothermal and Diagenetic Chlorites*

Referring to the general crystallochemical formula of chlorites, the chemical composition of chlorites with Mg or Fe^{2+}, Al, and □ in the octahedral positions is limited by the following four nodes:

(1) $u = 6, y = 0, z = 0, x = 0$, corresponding to the stoichiometry of either serpentine, $Mg_6Si_4O_{10}(OH)_8$, or greenalite, $Fe_6^{2+}Si_4O_{10}(OH)_8$;
(2) $u = 4, y = 2, z = 0, x = 2$, which is consistent with the formulae of either amesite, $(Mg_4Al_2)(Si_2Al_2)O_{10}(OH)_8$, or Fe^{2+}-amesite, $\left(Fe_4^{2+}Al_2\right)(Si_2Al_2)O_{10}(OH)_8$;
(3) $u = 0, y = 4.67, z = 1.33, x = 2$ or $(Al_{4.67}\square_{1.33})(Si_2Al_2)O_{10}(OH)_8$; this composition exceptionally rich in Al has no equivalent among the chlorites found in nature.
(4) $u = 0, y = 4, z = 2, x = 0$ or $(Al_4\square_2)Si_4O_{10}(OH)_8$; this chemical composition is relatively similar to that of some natural dioctahedral chlorites.

These four nodes and some intermediate compositions are shown in the chlorite classification diagram of Wiewiora and Weiss (1990) of Fig. 4.9, in which octahedral R^{2+} and tetrahedral Si are reported on the abscissa and ordinate axes, respectively, whereas octahedral Al and vacancies are represented in the grid. Both 181 chemical analyses of hydrothermal chlorites from different active geothermal systems and 54 chemical analyses of diagenetic chlorites are shown in this plot.

Chlorites, especially those from low-temperature environments, are often interstratified with smectite, vermiculite, saponite, and/or illite. Since the resolution of the scanning electron microscope is insufficient to establish the possible presence of

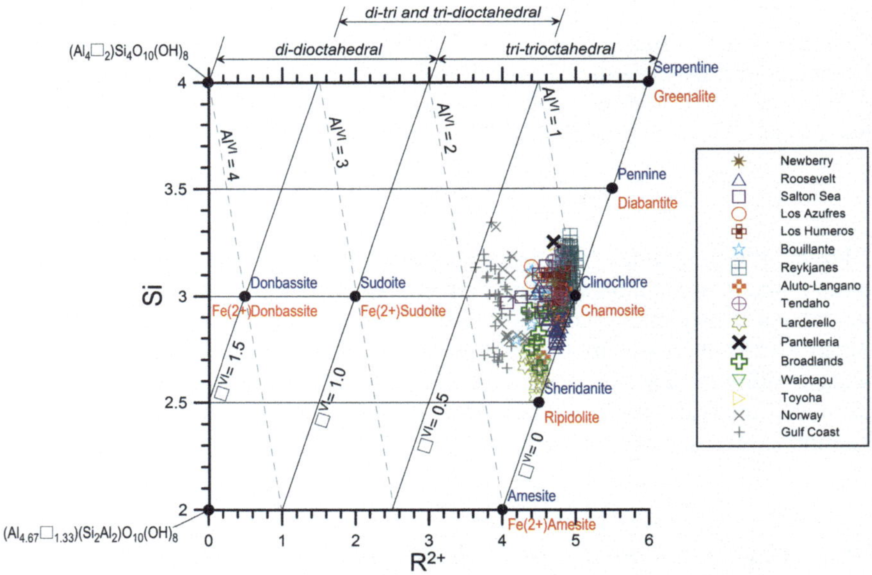

Fig. 4.9 Projection of chlorite structural formulae in the plot of tetrahedral Si versus the sum of octahedral divalent cations, R^{2+}, both expressed in atoms per half unit cell (from Wiewióra and Weiss 1990). Names of Mg-bearing chlorites are in blue whereas names of Fe^{2+}-bearing chlorites are in red. Structural formulae refer to 181 hydrothermal chlorites from different geothermal systems (see legend) and 54 diagenetic chlorites occurring in deeply buried sandstones from the Gulf Coast and the Norwegian Continental Shelf (see legend)

these minerals in the analyzed chlorites, the chemical analyses with ($CaO + Na_2O + K_2O$) > 0.5 wt% were rejected. This procedure was adopted by previous authors (e.g., Foster 1962), based on the assumption that these elements are not hosted in the chlorite lattice.

The considered diagenetic chlorites were obtained from deeply buried sandstones of both (i) the Gulf Coast area, Texas, where pressure is in the 300–1200 bar interval and temperature is in the 102–232 °C range (Bourdelle et al. 2013) and (ii) the Norwegian Continental Shelf at depths of 2393–4967 m, where temperature varies from 90 to 180 °C (Jahren and Aagaard 1992).

The considered hydrothermal chlorites were sampled from the following active geothermal systems: Newberry Caldera, Oregon, at 287 °C (Keith and Bargar 1988); Roosevelt Hot Springs, Utah, in the temperature range 228–253 °C (Ballantyne 1978; Capuano and Cole 1982); Salton Sea, California, in the temperature interval 190–322 °C (McDowell and Elders 1980, 1983); Los Azufres, Mexico, for temperatures ranging from 210 to 310 °C (Cathelineau and Nieva 1985; Cathelineau 1988); Los Humeros, Mexico, for temperatures varying from 145 to 315 °C (Martinez-Serrano and Dubois 1998; Martinez-Serrano 2002); Bouillante, Guadeloupe, in the temperature range 182–260 °C (Mas et al. 2006); Reykjanes, Iceland, in the temperature interval 170–331 °C (Libbey and Williams-Jones 2016); Aluto-Langano, Ethiopia, for temperatures ranging from 180 to 300 °C (Teklemariam et al. 1996); Tendaho, Ethiopia, for temperatures varying from 165 to 270 °C (Gianelli et al. 1998); Larderello, Italy, in the temperature range 248–315 °C (Boyce et al. 2003); Pantelleria, Italy, at 256 °C (Fulignati et al. 1997); Broadlands-Ohaaki, New Zealand, in the temperature interval 267–300 °C (Lonker et al. 1990); Waiotapu, New Zealand, at temperatures of 285–290 °C (Hedenquist and Browne 1989); Toyoha, Japan, for temperatures varying from 209 to 298 °C (Inoue et al. 2010).

All these literature data were compiled and their structural formulae were calculated on the basis of 10 oxygen atoms and 8 hydroxyl groups for a total of 28 negative charge units. Iron was distributed between Fe^{2+} and Fe^{3+} assuming that negative charges are balanced by the same number of positive charges, which is a reasonable hypothesis (see above). In this way, Fe^{2+} resulted to range between 1.08 and 4.33 atoms per half unit cell, whereas Fe^{3+} turned out to be strongly subordinated, with a maximum value of 0.39.

The diagram of Fig. 4.9 shows that:

(a) 90% of the hydrothermal chlorites have tetrahedral Si values ranging from 2.7 and 3.2, R^{2+} varying between 4.4 and 4.9, octahedral Al oscillating from 1.0 to 1.6, and octahedral occupancy of 5.8–6.0;
(b) the diagenetic chlorites have Si^{IV} values generally varying from 2.7 and 3.2, which compares with the range of the hydrothermal chlorites, but other parameters are different, with R^{2+} usually oscillating between 3.8 and 4.4, octahedral Al generally ranging from 1.5 to 1.8 and octahedral occupancy of 5.5–5.8.

These differences between the diagenetic and hydrothermal chlorites considered here (apart from the similarity in tetrahedral Si values) were already underscored by previous authors (e.g., De Caritat et al. 1993 and references therein).

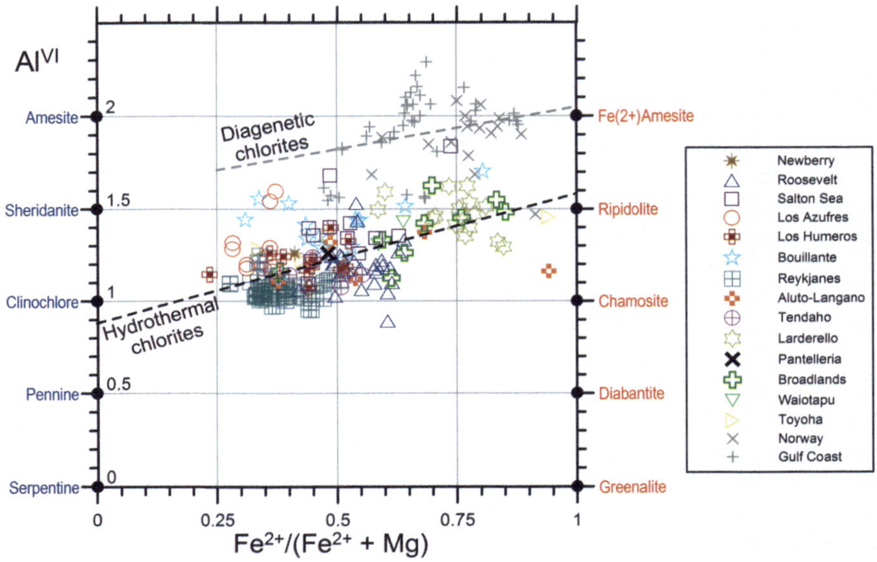

Fig. 4.10 Diagram of the $Fe^{2+}/(Fe^{2+} + Mg)$ atomic ratio versus the octahedral Al value, in atoms per half unit cell for 181 hydrothermal chlorites from different geothermal systems (see legend) and 54 diagenetic chlorites occurring in deeply buried sandstones from the Gulf Coast and the Norwegian Continental Shelf (see legend). Names of Mg-bearing chlorites are in blue whereas names of Fe^{2+}-bearing chlorites are in red

In the diagram of Fig. 4.9, the Mg-bearing chlorites (whose names are in blue) cannot be distinguished from the Fe^{2+}-bearing chlorites (whose names are in red) having the same R^{2+} value. This distinction is conveniently performed by means of the plot of Fig. 4.10, in which the $Fe^{2+}/(Fe^{2+} + Mg)$ ratio is contrasted with the octahedral Al value.

This diagram highlights that:

(i) The diagenetic chlorites are richer in ferrous iron, with the $Fe^{2+}/(Fe^{2+} + Mg)$ ratio ranging from 0.47 to 0.91 and can be considered mixtures of Fe^{2+}-amesite, amesite, and sheridanite, with the ferrous endmember generally prevailing over the magnesian components.

(ii) The hydrothermal chlorites have highly variable $Fe^{2+}/(Fe^{2+} + Mg)$ ratio, from 0.23 to 0.94, that is from Fe^{2+}-rich compositions, close to ripidolite, to Mg-rich compositions, relatively similar to clinochlore or sheridanite.

Considering separately each geothermal system which is represented by at least 4 chlorite samples ($N \geq 4$), usually the $Fe^{2+}/(Fe^{2+} + Mg)$ ratio distributes in a relatively small range, in the order of 0.21–0.30 units, as is the case of Tendaho (range = 0.21, N = 6), Roosevelt (range = 0.22, N = 29), Reykjanes (range = 0.23, N = 61), Larderello (range = 0.26, N = 15), Los Humeros (range = 0.29, N = 11), and Salton Sea (range = 0.30, N = 12). The 9 chlorite samples from Los Azufres span an even smaller $Fe^{2+}/(Fe^{2+} - Mg)$ range, of 0.09 units only, whereas a relatively large $Fe^{2+}/(Fe^{2+} + Mg)$ range characterizes the chlorites from Broadlands (range = 0.47, N = 9), Bouillante (range = 0.49, N = 15), Aluto-Langano (range = 0.56, N = 6), and Toyoha (range = 0.60, N = 4).

The relatively small $Fe^{2+}/(Fe^{2+} + Mg)$ range of most geothermal systems suggests that they are hosted in rocks of relatively uniform mineralogical and chemical composition. In particular, the hydrothermal chlorites from geothermal systems hosted in silica-rich rocks, such as granite at Larderello (Boyce et al. 2003), have Fe^{2+}-rich compositions, whereas the hydrothermal chlorites from geothermal systems associated to silica-poor rocks, such as basalt at Reykjanes (Libbey and Williams-Jones 2016) have Mg-rich compositions.

In contrast, the relatively large $Fe^{2+}/(Fe^{2+} + Mg)$ range of Broadlands, Bouillante, Aluto-Langano, and Toyoha calls for either (i) the occurrence of rocks of different mineralogical and chemical composition, for instance basalts and silicic ignimbrites at Aluto-Langano (Teklemariam et al. 1996) or (ii) the local presence of mineralized zones bearing sulfide minerals, as is the case of Broadlands (Weissberg et al. 1979) and Toyoha (Inoue et al. 2010), or (iii) the occurrence of successive alteration stages during the thermal history of the geothermal system, as is the case of Bouillante (Mas et al. 2006).

These findings agree with what was recognized long ago by Albee (1962), who pointed out that chlorite chemistry depends upon the composition of both the rock and the minerals associated with it. This matter was thoroughly discussed by De Caritat et al. (1993).

Also fluids can control to some extents the composition of chlorites. For instance, the Mg-rich chlorites of Reykjanes and Bouillante might be due to the inflow of Mg-rich seawater into these geothermal systems, as suggested by Arnórsson for Reykjanes and Sanjuan et al. (2001) for Bouillante.

4.4.3 *The Activities of Clinochlore and Chamosite in Hydrothermal and Diagenetic Chlorites*

Following Helgeson et al. (1978), Aagaard and Helgeson (1983), Helgeson and Aagaard (1985), and Aagaard and Jahren (1992), the activity of clinochlore, a_{Clc}, and the activity of chamosite, a_{Chm}, in the chlorite solid solutions of interest were calculated by means of the following relations:

$$a_{Clc} = k_{Clc} \cdot X_{Mg,O}^{5} \cdot X_{Al,O} \cdot X_{Al,T} \cdot X_{Si,T}^{3} \tag{4.25}$$

$$a_{Chm} = k_{Chm} \cdot X_{Fe^{2+},O}^{5} \cdot X_{Al,O} \cdot X_{Al,T} \cdot X_{Si,T}^{3} \tag{4.26}$$

which are based on the hypothesis of random mixing of atoms on energetically equivalent sites. In Eqs. (4.25) and (4.26):

$k_{Clc} = k_{Chm} = 141.55776$ are constants relating the inter- and intra-crystalline standard states;

$X_{Mg,O}$, $X_{Fe^{2+},O}$, and $X_{Al,O}$ are the mole fractions of magnesium, divalent iron, and aluminum, on the octahedral sites;

$X_{Al,T}$ and $X_{Si,T}$ are the mole fractions of aluminum and silicon on the tetrahedral sites.

The selected 181 hydrothermal chlorites have:

(i) activity of the clinochlore endmember varying from 4.39×10^{-7} to 0.250, with an average of 0.0553, a median of 0.0451 and a standard deviation of 0.0484, and
(ii) activity of the chamosite endmember ranging from 6.77×10^{-4} to 0.589, with an average of 0.0480, a median of 0.0173 and a standard deviation of 0.0794.

The 54 diagenetic chlorites considered in this work have:

(i) activity of the clinochlore endmember ranging from 4.04×10^{-6} to 0.0333, with an average of 0.00513, a median of 0.00216 and a standard deviation of 0.00832, and
(ii) activity of the chamosite endmember varying from 0.0194 to 0.539, with an average of 0.135, a median of 0.0875 and a standard deviation of 0.110.

As shown in the diagram of Fig. 4.11a, the activities of both endmembers in the considered chlorite solid solutions are linked by a hyperbolic relationship, which reflects the tight inverse correlation between $X_{Mg,O}$ and $X_{Fe^{2+},O}$ (Fig. 4.11b; R-squared = 0.891, N = 235).

The activities of both clinochlore and chamosite are poorly correlated with the temperature measured at the depth of provenance of the chlorite samples, as shown

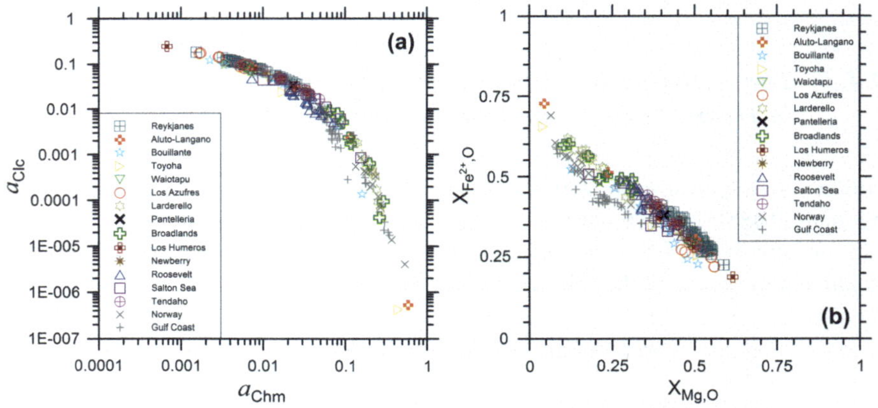

Fig. 4.11 Plots of **a** the activity of clinochlore, a_{Clc}, versus the activity of chamosite, a_{Chm}, and **b** the mole fraction of octahedral Fe^{2+} versus the mole fraction of octahedral Mg for 181 hydrothermal chlorites from different geothermal systems (see legend) and 54 diagenetic chlorites occurring in deeply buried sandstones from the Gulf Coast and the Norwegian Continental Shelf (see legend)

Table 4.5 Squared linear regression coefficients of the pairs a_{Clc}–temperature and a_{Chm}–temperature for the entire chlorite dataset as well as for each geothermal system which is represented by at least 4 chlorite samples

System	N	R^2 [a_{Clc}-T]	R^2 [a_{Chm}-T]
Reykjanes	59	0.321	0.162
Gulf Coast	38	0.046	0.003
Roosevelt	29	0.00007	0.003
Norway	16	0.157	0.488
Larderello	15	0.147	0.121
Bouillante	15	0.112	0.037
Salton Sea	12	0.203	0.491
Los Humeros	11	0.073	0.494
Broadlands	9	0.024	0.354
Los Azufres	9	0.006	0.021
Tendaho	6	0.0002	0.042
Aluto-Langano	6	0.003	0.353
Toyoha	4	0.831	0.854
All dataset	237	0.133	0.112

by the squared linear regression coefficients of the pairs a_{Clc}–temperature and a_{Chm}–temperature for the entire dataset as well as for each geothermal system which is represented by at least 4 chlorite samples (Table 4.5). The only exception is the Toyoha geothermal field, but the relatively high R^2 values are probably fortuitous, also taking into account the low number of available samples.

The lack of correlation between temperature and the activities of both clinochlore and chamosite in the chlorite solid solution is not surprising considering that, according to De Caritat et al. (1993) and other authors, there is a lack of confidence in applying the chlorite geothermometers which have been designed by several workers to obtain indications of palaeo-temperature from chlorite composition and structure.

4.4.4 *The Thermodynamic Properties of Chlorites*

The only minerals of the chlorite group considered by Helgeson et al. (1978) are 7Å-clinochlore and 14Å-clinochlore, both of stoichiometry $Mg_5Al_2Si_3O_{10}(OH)_8$. The clinochlore with 7Å basal spacing is metastable with respect to its 14Å counterpart at all temperatures and pressure. According to Helgeson et al. (1978), the reason for this is probably the relatively high activation energy required for the conversion of 7Å-clinochlore to the 14Å variety. This hypothesis could explain the persistence of 7Å-clinochlore in weathering profiles, sedimentary environments, hydrothermal alteration zones, and laboratory experiments at temperatures lower than 400–500 °C. Accepting this explanation, 7Å-clinochlore is the clinochlore variety of interest for this work.

As shown by Table 4.6, the thermodynamic data of seven chlorite minerals are

Table 4.6 Standard state molal thermodynamic properties, at 25 °C, 1 bar, of different chlorite minerals from H78 = Helgeson et al. (1978), HP98 = Holland and Powell (1998), and HP11 = Holland and Powell (2011)

Name	$\Delta G°_f$	$\Delta H°_f$	S°	V°	a	b (×10^3)	c (×10^{-5})	d	References
	cal mol^{-1}	cal mol^{-1}	cal K^{-1} mol^{-1}	cm^3 mol^{-1}	cal K^{-1} mol^{-1}	cal K^{-2} mol^{-1}	cal K mol^{-1}	cal $K^{-0.5}$ mol^{-1}	
7Å-Clinochlore	−1,957,101	−2,113,197	106.5	211.5	162.82	50.62	−40.88	–	H78
14Å-Clinochlore	−1,961,703	−2,116,964	111.2	207.11	166.50	42.10	−37.47	–	H78
Ord-Clinochlore	−1,974,988	−2,134,288	98.1	210.9	277.68	2.4218	−18.301	−2316.2	HP98
Ord-Clinochlore	−1,971,946	−2,129,357	104	211.4	279.83	−0.3604	−9.144	−2465.3	HP11
Al-free chlorite	−1,930,801	−2,089,914	97.5	216.6	274.04	2.6828	−18.776	−2229.6	HP98
Al-free chlorite	−1,929,293	−2,086,197	105	215.7	276.05	−0.0997	−9.619	−2378.8	HP11
14Å-Amesite	−2,002,481	−2,163,607	93.2	205.2	281.31	2.1609	−17.827	−2402.7	HP98
14Å-Amesite	−2,001,077	−2,160,564	98.7	207.1	283.46	−0.6212	−8.669	−2551.9	HP11
Daphnite (chamosite)	−1,562,036	−1,709,845	130	213.4	295.75	3.2490	−8.946	−2688.8	HP98
Daphnite (chamosite)	−1,555,905	−1,700,935	140	216.2	284.89	−1.4197	−11.535	−2334.7	HP11
Mn-chlorite	−1,688,695	−1,835,681	139	225.9	293.45	−6.4508	−15.057	−2502.2	HP98
Mn-chlorite	−1,694,993	−1,840,911	142	225.9	271.63	−1.2531	−13.260	−2129.9	HP11
Sudoite	−1,906,494	−2,063,662	95.4	203.0	343.24	−11.6513	−6.569	−3289.7	HP98
Sudoite	−1,904,428	−2,061,881	94.4	203.0	343.24	−11.6513	−6.569	−3289.7	HP11
Fe-Sudoite	−1,738,924	−1,891,197	109	204.0	350.45	−11.3205	−2.827	−3438.8	HP98
Fe-Sudoite	−1,735,827	−1,888,172	109	204.0	350.45	−11.3205	−2.827	−3438.8	HP11

listed by both Holland and Powell (1998) and Holland and Powell (2011), namely Ordered Clinochlore [$Mg_5Al_2Si_3O_{10}(OH)_8$], Al-free chlorite [$Mg_6Si_4O_{10}(OH)_8$], 14Å-Amesite [$Mg_4Al_4Si_2O_{10}(OH)_8$], Daphnite [$Fe_5Al_2Si_3O_{10}(OH)_8$], Mn-chlorite [$Mn_5Al_2Si_3O_{10}(OH)_8$], Sudoite [$Mg_2Al_4Si_3O_{10}(OH)_8$], and Fe-Sudoite [$Fe_2Al_4Si_3O_{10}(OH)_8$]. Incidentally, the term chamosite should be used instead of daphnite, for consistency with the chlorite classification of Wiewiora and Weiss (1990).

Since the 14Å basal spacing is specified only for amesite, we assume that all the other compositions have 7Å basal spacing. If so, the $\Delta G°_f$ and $\Delta H°_f$ values reported by Holland and Powell (1998) for clinochlore differ from those of Helgeson et al. (1978) by 17887 cal mol^{-1} and 21091 cal mol^{-1}, respectively, whereas the $\Delta G°_f$ and $\Delta H°_f$ values given by Holland and Powell (2011) deviate from those of Helgeson et al. (1978) by 14845 cal mol^{-1} and 16160 cal mol^{-1}, respectively. Moreover, for the same mineral, there are differences from 1404 to 6298 cal mol^{-1} in $\Delta G°_f$ and from 1781 to 8910 cal mol^{-1} in $\Delta H°_f$ between the database of Holland and Powell (1998) and that of Holland and Powell (2011). Again, due to these considerable differences in the thermodynamic data stored for clinochlore in the different thermodynamic databases, it is not possible to put together the data from distinct databases because it would be equivalent to mix apples and oranges.

Following the approach outlined in Sect. 2.2.1, it is advisable to consider the chlorite solid solution with average activity of clinochlore, $a_{Clc} = 0.0553$, and to compute the Gibbs free energy and the thermodynamic equilibrium constant of its dissolution reaction. The results obtained considering the 7Å-clinochlore thermodynamic data of Helgeson et al. (1978) are listed in Table 4.7 as a function of temperature together with the standard molal Gibbs free energies and the natural and decimal logarithms of the thermodynamic equilibrium constant of the dissolution reaction of pure 7Å-clinochlore:

Table 4.7 Standard molal Gibbs free energies and natural and decimal logarithms of the thermodynamic equilibrium constant of the dissolution reactions of both pure 7Å-clinochlore and the chlorite solid solution with average activity of clinochlore as a function of temperature

T	$\Delta_r G^\circ_{Clc}$	$\Delta_r G^\circ_{Clc,aa}$	ln K_{Clc}	ln $K_{Clc,aa}$	log K_{Clc}	log $K_{Clc,aa}$
°C	cal mol^{-1}	cal mol^{-1}				
0	−98,377	−96,806	181.2380	178.3431	78.7107	77.4534
25	−94,465	−92,750	159.4385	156.5435	69.2432	67.9860
60	−88,148	−86,231	133.1465	130.2515	57.8248	56.5675
100	−80,703	−78,556	108.8337	105.9387	47.2659	46.0086
150	−71,323	−68,889	84.8188	81.9238	36.8363	35.5791
200	−61,652	−58,930	65.5700	62.6750	28.4767	27.2194
250	−51,199	−48,189	49.2484	46.3534	21.3883	20.1310
300	−39,258	−35,961	34.4681	31.5731	14.9693	13.7120

$$Mg_5Al_2Si_3O_{10}(OH)_8 + 16\ H^+ = 5\ Mg^{2+} + 2\ Al^{3+} + 3\ SiO_{2(aq)} + 12\ H_2O \tag{4.27}$$

4.5 Epidotes

4.5.1 *Main Characteristics and Nomenclature of the Minerals of the Epidote Group*

The following notes on the nomenclature and main features of the minerals of the epidote group are largely based on the review paper of Franz and Liebscher (2004). The minerals of the epidote group are monoclinic sorosilicates with the general formula $A_2M_3[Si_2O_7][SiO_4]O(OH)$ indicating the presence of both SiO_4 tetrahedra and Si_2O_7 groups, as well as the M cations in six-fold octahedral coordination and the A cations in eight-fold coordination (further details in the next section). Orthorhombic zoisite is a polymorph of clinozoisite and is structurally very similar to the monoclinic minerals of the epidote group.

According to the Commission on New Minerals, Nomenclature and Classification (CNMNC) of the International Mineralogical Association (IMA) on the nomenclature of epidote minerals:

(i) The term epidote refers to the mineral, the whole mineral group and the composition $Ca_2(Al_2Fe^{3+})[Si_2O_7][SiO_4]O(OH)$. To avoid confusion, the terms epidote and epidote component (or endmember) are used in this book to indicate the mineral/mineral group and the solid solution component (endmember), respectively.
(ii) The name clinozoisite identifies the stoichiometry $Ca_2Al_3[Si_2O_7][SiO_4]O(OH)$.
(iii) The term piemontite refers to the composition $Ca_2(Al_2Mn^{3+})[Si_2O_7][SiO_4]O(OH)$.
(iv) The name mukhinite identifies the stoichiometry $Ca_2(Al_2V^{3+})[Si_2O_7][SiO_4]O(OH)$.
(v) The terms allanite-Ce, allanite-La, allanite-Nd, and allanite-Y refer to the composition $[Ca(REE)](Al_2Fe^{2+})[Si_2O_7][SiO_4]O(OH)$, in which REE = Ce, La, Nd, and Y, respectively.

The name tawmawite was used for the Cr^{3+} endmember, $Ca_2(Al_2Cr^{3+})[Si_2O_7][SiO_4]O(OH)$, but is not approved by IMA. Also the term pistacite, which was utilized to describe either Fe^{3+}-rich epidotes of stoichiometry close to $Ca_2(Al_2Fe^{3+})[Si_2O_7][SiO_4]O(OH)$ or the hypothetical endmember $Ca_2Fe_3^{3+}[Si_2O_7][SiO_4]O(OH)$ should generally be avoided.

Most monoclinic epidote minerals occurring in geological systems belong to the clinozoisite-epidote solid solution series with epidote mole fractions, X_{Ep}, usually lower than 0.4, whereas compositions with X_{Ep} as high as 0.4–0.5 were rarely found. Natural binary solid solutions along the clinozoisite-piemontite join are uncommon because most samples also contain Fe^{3+}. In addition, the octahedral M sites can be rarely occupied by Cr^{3+} and V^{3+}, whereas the concentration of the divalent cations Fe^{2+}, Mn^{2+}, and Mg^{2+} on these sites is usually very low. The A sites generally house Ca^{2+}, which can be substituted by Sr^{2+}, Ba^{2+}, Pb^{2+}, and REE^{3+}. Allanite minerals exhibit larger and more complex compositional variations, according to the formula $[(Ca,Ce,La,Y)_2(Al,Fe^{3+},Fe^{2+},Mn^{2+})_3[Si_2O_7][SiO_4]O(OH)]$.

Orthorhombic zoisite exhibits composition closer to the endmember stoichiometry $Ca_2Al_3Si_3O_{12}(OH)$ than the monoclinic minerals of the epidote group. Similar to what occurs in monoclinic epidotes, Fe^{3+}, Mn^{3+}, Cr^{3+}, and V^{3-} substitute for Al^{3+} in the octahedral M sites (although the mole fraction of Fe^{3+} is usually ≤ 0.15), whereas Sr^{2+} and Pb^{2+} substitute for Ca^{2+} in the A sites of zoisite.

Usually, hydrothermal epidote from active geothermal systems shows a wide interval in octahedral substitution of Al^{3+} for Fe^{3+}, with important oscillatory or irregular compositional zoning. These variable characteristics of hydrothermal epidote are due to the variations, in time and space, of temperature, permeability, and fluid parameters such as flux, pH, CO_2 partial pressure, redox potential, and the aqueous speciation of Al and Fe (Bird and Spieler 2004).

4.5.2 *The Crystal Chemistry of Epidote Solid Solutions*

This section is mainly focused on the equilibrium mixing model of Bird and Helgeson (1980) and, therefore, it is largely derived from their article. Epidote solid solutions can be represented by the following crystallochemical formula:

$$A_2^{VIII}M(1)^{VI}\ M(2)^{VI}\ M(3)^{VI}T_3^{IV}O_{12}(OH)$$

in which:

- A^{VIII} refers to the two large eight-fold coordinated sites, which are predominantly occupied by Ca^{2+} ions;
- $M(1)^{VI}$, $M(2)^{VI}$, and $M(3)^{VI}$ indicate the three six-fold coordinated octahedral sites, which are energetically different and are occupied by Al^{3+} and Fe^{3+} ions;
- T stands for the three tetrahedral sites, which are almost completely occupied by Si^{4+} ions.

Crystal structure site refinements and Mössbauer spectral data indicate that the Fe^{3+} ions are mainly housed by the largest and most distorted $M(3)^{VI}$ octahedral sites, although minor quantities of Fe^{3+} occur also in the $M(1)^{VI}$ sites. In contrast, the $M(2)^{VI}$ sites contain only Al^{3+} ions. According to Bird and Helgeson (1980), Fe^{3+} ions are only present in the M(3) octahedral sites if epidote is completely ordered, whereas octahedral disordering of epidote can be represented by the following reaction:

$$Ca_2Al_{M(1)}Al_{M(2)}Fe_{M(3)}Si_3O_{12}(OH) \rightarrow Ca_2(Al_{1-x}Fe_x)_{M(1)}Al_{M(2)}(Al_xFe_{1-x})_{M(3)}Si_3O_{12}(OH)$$

where x represents the mole fraction of Fe^{3+} ion on the M(1) sites. Bird and Helgeson (1980) defined an ordering parameter for epidote, $\sigma = 1 - 2x$, which assumes the value 1 if the mineral is fully ordered and becomes equal to 0 if it is entirely disordered. The intracrystalline exchange reaction of Al^{3+} and Fe^{3+} ions among the M(1) and M(3) sites:

$$Fe^{3+}_{M(3)} + Al^{3+}_{M(1)} \leftrightarrow Fe^{3+}_{M(1)} + Al^{3+}_{M(3)} \tag{4.28}$$

is regulated by the intracrystalline standard state equilibrium constant:

$$K = \frac{a_{Fe^{3+}_{M(1)}} \cdot a_{Al^{3+}_{M(3)}}}{a_{Fe^{3+}_{M(3)}} \cdot a_{Al^{3+}_{M(1)}}}, \tag{4.29}$$

which is a temperature function as expressed by the relation:

$$\log K = -1523.4 \cdot \left(\frac{1}{T} - \frac{1}{Tr}\right) - 5.0, \tag{4.30}$$

with Tr = 298.15. Assuming that the ratios between the activity coefficients of Al^{3+} and Fe^{3+} ions on the M(1) and M(3) sites are equal to 1, it follows that the ordering parameter σ is related to K by the relation:

$$\text{arctanh}\ \sigma = -(\ln K)/4. \tag{4.31}$$

Therefore, it is possible to compute σ, knowing or assuming a suitable temperature value. Following Bird and Helgeson (1980), the next step is the calculation of the mole fraction of Fe^{3+} ions on the M(3) sites using the relation:

$$X_{Fe^{3+}_{M(3)}} = -\left[\frac{X_{Ep}}{K-1} + \left(\frac{K \cdot (1 - X_{Ep}) + X_{Ep} + 1}{2 \cdot (K-1)}\right)^2\right]^{0.5} - \frac{K \cdot (1 - X_{Ep}) + X_{Ep} + 1}{2 \cdot (K-1)} \tag{4.32}$$

which is written in a simplified form, that is assuming absence of ions other than Al^{3+} and Fe^{3+} on the M(3) sites. Then, the mole fractions of Fe^{3+} and Al^{3+} ions on the M(1) sites and the mole fraction of Al^{3+} ions on the M(3) sites are computed by means of the equations:

$$X_{Fe^{3+}_{M(1)}} = X_{Ep} - X_{Fe^{3+}_{M(3)}} \tag{4.33}$$

$$X_{Al^{3+}_{M(1)}} = 1 - X_{Fe^{3+}_{M(1)}} \tag{4.34}$$

$$X_{Al^{3+}_{M(3)}} = 2 - X_{Ep} - X_{Al^{3+}_{M(1)}}. \tag{4.35}$$

The mole fractions computed by means of Eqs. (4.32) through (4.35) are then used to calculate the activities of the clinozoisite and epidote endmembers based on the hypothesis of random mixing of atoms on energetically equivalent sites (Helgeson et al. 1978; Bird and Helgeson 1980; Bird and Norton 1981; Helgeson and Aagaard 1985):

$$a_{Czo} = X^2_{Ca^{2+}_A} \cdot X_{Al^{3+}_{M(1)}} \cdot X_{Al^{3+}_{M(3)}} \cdot X^3_{Si^{4+}_T} \tag{4.36}$$

$$a_{Ep} = k_{Ep} \cdot X^2_{Ca^{2+}_A} \cdot \left(X_{Al^{3+}_{M(1)}} \cdot X_{Fe^{3+}_{M(3)}}\right)^{\frac{1+\sigma}{2}} \cdot \left(X_{Al^{3+}_{M(3)}} \cdot X_{Fe^{3+}_{M(1)}}\right)^{\frac{1-\sigma}{2}} \cdot X^3_{Si^{4+}_T} \tag{4.37}$$

where k_{Ep} is the constant relating the inter- and intra-crystalline standard states, which is given by:

$$k_{Ep} = \left(\frac{1+\sigma}{2}\right)^{-(1+\sigma)} \cdot \left(\frac{1-\sigma}{2}\right)^{-(1-\sigma)}. \tag{4.38}$$

The equilibrium model of Bird and Helgeson (1980) for partitioning of Al^{3+} and Fe^{3+} ions among the M(1) and M(3) sites in epidote was based on the data available at that time. In subsequent researches, the order-disorder degree was found either higher than theoretically predicted by the equilibrium model or in good agreement with model's predictions or lower than theoretically predicted by the model (Franz and Liebscher 2004). In spite of possible limitations, the model of Bird and Helgeson (1980) is used in the next section, first, to calculate the equilibrium distribution of Al^{3+} and Fe^{3+} ions among the M(1) and M(3) sites in epidote solid solutions coming from active geothermal systems, starting from the analytical data, and second, to compute the activities of the epidote and clinozoisite components of the epidote solid solutions of interest.

4.5.3 *The Activities of Clinozoisite and Epidote Endmembers in Hydrothermal Epidote Solid Solutions*

A total of 436 chemical analyses of epidote solid solutions collected in deep geothermal wells at variable depths and temperatures were compiled. Most samples, for a total of 259, come from the geothermal system of Reykjanes, Iceland at temperatures of 176–345 °C (mainly from Libbey and William-Jones 2016 and subordinately from Fowler et al. 2015 and Ottolini et al. 2012). Of the remaining 177 chemical analyses, 57 are from Larderello, Italy, at temperatures of 250–368 °C (Cavarretta et al. 1980, 1982), 40 are from Milos, Greece, at temperatures of 305–323 °C (Liakopoulos 1987), 23 are from Salton Sea, California, at temperatures of 250–353 °C (Keith et al. 1968; Bird et al. 1988; Cho et al. 1988; Shearer et al. 1988), 14 are from Los Humeros, Mexico, at temperatures of 235–350 °C (Martinez-Serrano 2002), 9 are from Waiotapu, New Zealand, at temperatures of 220–285 °C (Hedenquist and Browne 1989), 8 are from Cerro Prieto, Mexico, at temperatures of 325–350 °C (Schiffman et al. 1985), 8 are from Kakkonda, Japan, at temperatures of 225–285 °C (Muramatsu and Doi 2000; Sawaki et al. 2001), 8 are from Onikobe, Japan, at temperatures of 125–173 °C (Seki et al. 1983), 4 are from Broadlands, New Zealand, at temperatures of 260–300 °C (Lonker et al. 1990), 2 are from Hachimantai, Japan, at temperatures of 200–230 °C (Shimazu and Yajima 1973), 2 are from Newberry Caldera, Oregon, at temperature of 260 °C (Keith and Bargar 1988), and 2 are from Miravalles, Costa Rica, at temperature of 240 °C (Milodowski et al. 1989).

Clinozoisite is the main component of these hydrothermal epidote solid solutions, with mole fraction varying between 0.527 and 0.893, both average and median equal to 0.728, and standard deviation of 0.061. Being the complement to 1 of the clinozoisite mole fraction, the epidote component mole fraction ranges from 0.107 to 0.463, both average and median are equal to 0.272, and standard deviation is the same of clinozoisite, 0.061. Consequently, available data are positioned along a line of slope-1 in the correlation plot of the clinozoisite mole fraction versus the epidote component mole fraction (Fig. 4.12a). The composition of the considered hydrothermal epidote solid solutions compares with that of most monoclinic epidote minerals occurring in different geological systems.

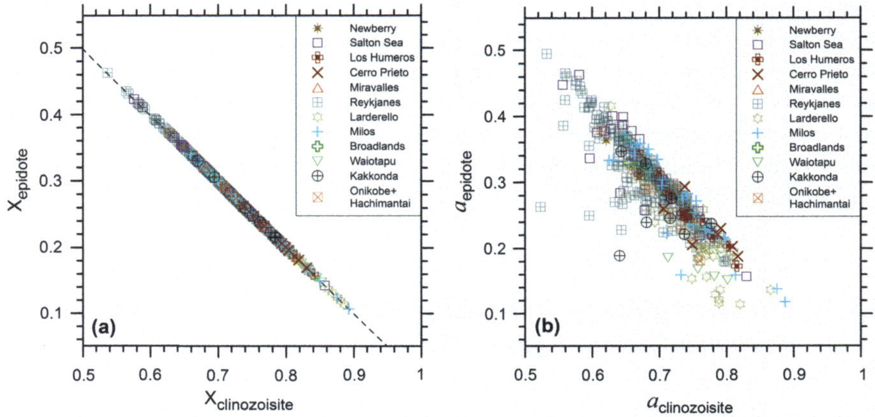

Fig. 4.12 Plots of **a** the clinozoisite mole fraction versus the epidote component mole fraction and **b** the activity of clinozoisite versus the activity of the epidote component for 436 hydrothermal epidote solid solutions from different geothermal systems (see legend)

The activity of clinozoisite varies between 0.523 and 0.888, with a mean of 0.703, a median of 0.706, and a standard deviation of 0.057, whereas the activity of the epidote component ranges from 0.115 to 0.495, with an average of 0.288, a median of 0.287 and a standard deviation of 0.066. Also in the correlation diagram of the activity of clinozoisite versus the activity of the epidote component (Fig. 4.12b), sample points are distributed along a line of slope-1, because activities of both clinozoisite and epidote endmembers are not very different from the corresponding mole fractions. Nevertheless, in Fig. 4.12b there is a considerable scatter, which is evidently due to varying deviations from ideality.

Such deviations are described by the activity coefficients of both clinozoisite and epidote endmembers which were computed by means of Eq. (2.3) solved with respect to λ. Interestingly, the activity coefficient of clinozoisite is usually <1, with minimum of 0.763 and maximum of 1.001, whereas the activity coefficient of epidote is generally >1, up to a maximum of 1.129, even though its minimum value is 0.839 (Fig. 4.13a). However, average and median are close to unity, being 0.966 and 0.973, respectively, for clinozoisite and 1.060 and 1.066, respectively, for the epidote component. The standard deviation is small in both case, with 0.028 for clinozoisite and 0.035 for the epidote component.

These deviations from ideality are ascribable to the varying degree of substitutional order/disorder which is a temperature function, as defined by Eq. (4.30) and shown in Fig. 4.13b, according to the equilibrium model of Bird and Helgeson (1980).

The temperature measured at the depth of collection of hydrothermal epidote was considered in calculations, although observed mineral alteration may have formed under thermal conditions different from contemporary temperatures. In addition, Arnórsson (1995) noted that the number of hydrothermal minerals reported in a

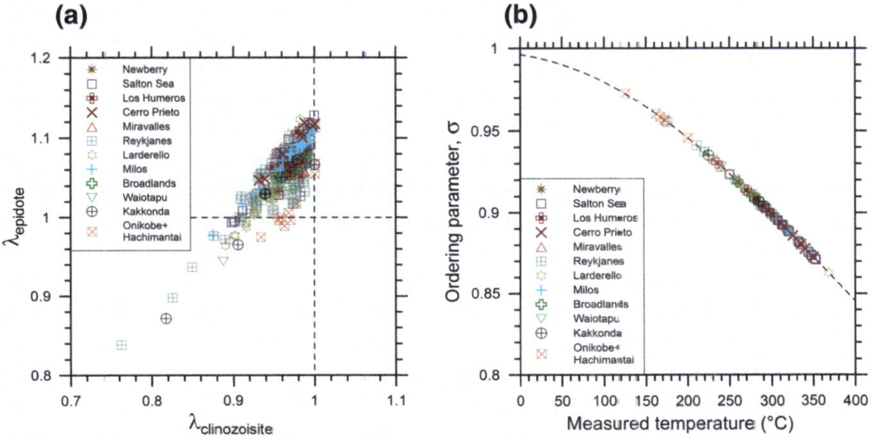

Fig. 4.13 Plots of **a** the activity coefficient of clinozoisite versus the activity coefficient of the epidote component and **b** the ordering parameter σ versus the measured temperature, for 436 hydrothermal epidote solid solutions from different geothermal systems (see legend)

single drill hole sample is commonly greater than permitted by consideration of the phase rule, indicating that secondary mineral assemblages may form over a range of temperatures and fluid compositions, and that metastable phase relations are not uncommon.

The hydrothermal epidote samples appear to be relatively ordered, as indicated by the range of the ordering parameter σ, from 0.863 to 0.973, with a mean of 0.905, a median of 0.904, and a standard deviation of 0.018. However, deviations from the equilibrium distribution predicted by the model of Bird and Helgeson (1980) are possible, due to occurrence of metastable states of substitutional order/disorder on the octahedral sites. For instance, this condition was ascertained by means of ^{57}Fe Mössbauer spectroscopy for four epidote samples from the State 2–14 drill hole, Salton Sea (Bird et al. 1988) and for the epidotes from the fossil geothermal system of Saint Martin (Patrier et al. 1991), see also Liebscher (2004).

4.5.4 The Thermodynamic Properties of Epidotes

The thermodynamic properties of four minerals of the epidote group are reported by Helgeson et al. (1978), namely clinozoisite and zoisite, both of chemical formula $Ca_2Al_3Si_3O_{12}(OH)$, as well as epidote (which refers to the stable state of substitutional order/disorder at any pressure and temperature) and ordered epidote, both of stoichiometry $Ca_2FeAl_2Si_3O_{12}(OH)$. Holland and Powell (1998) estimated the thermodynamic properties of four minerals of the epidote group, again clinozoisite and zoisite, both of chemical formula $Ca_2Al_3Si_3O_{12}(OH)$, and ordered epidote of stoichiometry $Ca_2FeAl_2Si_3O_{12}(OH)$, as well as a solid phase richer in Fe^{3+}, called Fe-epidote, of chemical composition $Ca_2Fe_2AlSi_3O_{12}(OH)$. Monoclinic epidote minerals were treated as regular solid solutions of clinozoisite, ordered epidote, and Fe-epidote extending the model of Bird and Helgeson (1980) through incorporation of non-ideal mixing. Holland and Powell (2011) revised the thermodynamic properties of the four minerals of the epidote group considered by Holland and Powell (1998) and estimated also the thermodynamic properties of ordered piemontite, of chemical formula $Ca_2MnAl_2Si_3O_{12}(OH)$. All these data are reported in Table 4.8.

The $\Delta G°_f$ and $\Delta H°_f$ reported by Holland and Powell (1998) differ from those of Helgeson et al. (1978) by 4570 and 4467 cal mol^{-1} for clinozoisite, 951 and 286 cal mol^{-1} for ordered epidote, and 4456 and 4667 cal mol^{-1} for zoisite. The $\Delta G°_f$ and $\Delta H°_f$ listed by Holland and Powell (2011) differ from those of Helgeson et al. (1978) by 3927 and 3824 cal mol^{-1} for clinozoisite, 2582 and 2843 cal mol^{-1} for ordered epidote, and 3963 and 4103 cal mol^{-1} for zoisite. Besides, for the same mineral, there are differences from 493 to 4071 cal mol^{-1} in $\Delta G°_f$ and from 564 to 6066 cal mol^{-1} in $\Delta H°_f$ between the database of Holland and Powell (1998) and that of Holland and Powell (2011).

A very detailed analysis of the thermodynamic properties of zoisite, orthoepidote, clinozoisite, and epidote was performed by Gottschalk (2004), who examined *inter alias* the experimental measurements carried out by several authors, as well as

Table 4.8 Standard state molal thermodynamic properties, at 25 °C, 1 bar, of different epidote minerals from H78 = Helgeson et al. (1978), HP98 = Holland and Powell (1998), and HP11 = Holland and Powell (2011)

Name	$\Delta G°_f$	$\Delta H°_f$	S°	V°	a	b (×10³)	c (×10⁻⁵)	d	References
	cal mol^{-1}	cal mol^{-1}	cal K^{-1} mol^{-1}	cm^3 mol^{-1}	cal K^{-1} mol^{-1}	cal K^{-2} mol^{-1}	cal K mol^{-1}	cal K$^{-0.5}$ mol^{-1}	
Clinozoisite	−1,549,680	−1,644,221	70.64	136.2	106.118	25.214	−27.145	–	H78
Clinozoisite	−1,554,250	−1,648,688	71.9	136.30	135.5	4.3172	−16.812	−622.1	HP98
Clinozoisite	−1,553,607	−1,648,045	71.9	136.30	150.8	3.2727	−15.884	−891.8	HP11
Epidote	−1,451,346	−1,544,432	75.28	139.2	117.622	12.816	−31.864	–	H78
Epidote (ord)	−1,451,346	−1,544,456	75.2	139.2	113.78	14.69	−28.92	–	H78
Epidote (ord)	−1,452,297	−1,544,742	78.4	139.10	130.2	5.9228	−26.840	−284.9	HP98
Epidote (ord)	−1,453,928	−1,547,299	75.3	139.20	146.6	5.2749	−17.113	−714.1	HP11
Fe-epidote	−1,344,250	−1,434,558	85.3	141.90	124.3	7.5284	−36.869	52.3	HP98
Fe-epidote	−1,348,320	−1,440,624	78.6	142.10	139.7	7.2770	−18.342	−536.4	HP11
Piemontite (ord)	−1,470,118	−1,563,824	81.3	138.20	136.2	6.6420	−13.009	−672.2	HP11
Zoisite	−1,549,619	−1,644,131	70.74	135.90	106.118	25.214	−27.145	–	H78
Zoisite	−1,554,075	−1,648,798	71.0	135.75	142.4	14.8893	−14.152	−811.4	HP98
Zoisite	−1,553,582	−1,648,234	71.2	135.75	158.2	2.4895	−14.356	−1018.3	HP11

Table 4.9 Standard molal Gibbs free energies and natural and decimal logarithms of the thermodynamic equilibrium constant of the dissolution reactions of both pure clinozoisite and the epidote solid solution with average activity of clinozoisite as a function of temperature

T	$\Delta_r G^\circ_{Czo}$	$\Delta_r G^\circ_{Czo,aa}$	ln K_{Czo}	ln $K_{Czo,aa}$	log K_{Czo}	log $K_{Czo,aa}$
°C	cal mol^{-1}	cal mol^{-1}				
0	−59,755	−59,564	110.0855	109.7331	47.8095	47.6565
25	−56,211	−56,002	94.8732	94.5208	41.2029	41.0499
60	−50,204	−49,971	75.8325	75.4801	32.9336	32.7806
100	−42,797	−42,536	57.7148	57.3624	25.0652	24.9122
150	−33,067	−32,771	39.3240	38.9716	17.0782	16.9251
200	−22,723	−22,392	24.1671	23.8147	10.4956	10.3426
250	−11,420	−11,054	10.9849	10.6325	4.7707	4.6176
300	1379	1780	−1.2107	−1.5631	−0.5258	−0.6789

the values reported in internally consistent datasets, underscoring the discrepancies between different sources and the missing pieces of information.

Finally, the recommended values of the thermodynamic parameters are listed by Gottschalk (2004), but no supporting motivation is given. The concluding statement of Gottschalk (2004) "*However, the specific choice of a value might be somewhat subjective and reflects the opinion of the author*" might seem disappointing at the end of such a thorough review. However, it is a further example of the impossibility to select reliable values of thermodynamic properties by focusing the attention on a single group of minerals only. A much wider analysis is needed to establish the internal consistency of the whole thermodynamic database.

Adopting the approach described in Sect. 2.2.1, it is advisable to consider the epidote solid solution with average activity of clinozoisite, $a_{Czo} = 0.703$, and to compute the Gibbs free energy and the thermodynamic equilibrium constant of its dissolution reaction. The results obtained taking into account the clinozoisite thermodynamic data of Helgeson et al. (1978) are listed in Table 4.9 as a function of temperature together with the standard molal Gibbs free energies and the natural and decimal logarithms of the thermodynamic equilibrium constant of the dissolution reaction of pure clinozoisite:

$$Ca_2Al_3Si_3O_{12}(OH) + 13\ H^+ = 2\ Ca^{2+} + 3\ Al^{3+} + 3\ SiO_{2(aq)} + 7\ H_2O. \quad (4.39)$$

4.6 Prehnite

As indicated by several authors (e.g., Papike and Zoltai 1967; Baur et al. 1990), there are two prehnite polymorphs, namely monoclinic clinoprehnite and orthorhombic orthoprehnite, which in turn have two distinct structures according to Akizuki

(1987). The two prehnite polymorphs have very similar unit cell volumes, 469.5(1) and 468.7(2) Å^3, respectively, and supposedly a tiny energetic difference, as also suggested by their close association in nature (Baur et al. 1990). Therefore, the two polymorphs of prehnite are not distinguished in the following discussion.

4.6.1 The Crystal Chemistry of Prehnite/Ferri-Prehnite Solid Solutions

Single crystal structure analyses (e.g., Preisinger 1965; Papike and Zoltai 1967; Zunić et al. 1990; Baur et al. 1990; Artioli et al. 1995; Akasaka et al. 2003) indicate that prehnite solid solutions have seven-coordinated sites A filled by Ca^{2+}, octahedral sites M(1) containing both Al^{3+} and Fe^{3+}, and two different types of tetrahedral sites T1 and T2. The T1 sites are entirely occupied by Si^{4+} or nearly so, whereas the T2 sites house Al^{3+} and Si^{4+} in variable degrees of order-disorder. The structural formula of prehnite solid solutions can thus be written as (Bird and Helgeson 1980, modified):

$$Ca_2\left[Al_{(1-y)}Fe_y\right]^{M(1)}(Si)_2^{T1}\left(Al_{(1-x)}Si_x\right)^{T2}O_{10}(OH)_2$$

where y is the mole fraction of Fe^{3+} on the octahedral sites M(1) and x is the mole fraction of Si^{4+} on the tetrahedral sites T2.

Prehnite solid solutions from natural systems show considerable octahedral substitution of Fe^{3+} for Al^{3+}, which may be as high as 60 mol percent based on the chemical analyses of this mineral from different metamorphic environments (Bird and Helgeson 1980, and references therein), whereas other compositional variations are very limited. In fact, sodium, potassium, magnesium, ferrous iron and manganese substitution for calcium on the seven-coordinated sites, as well as the substitution of Fe^{3+} or Al^{3+} for Si^{4+} on the tetrahedral sites are usually lower than 0.1 atoms per formula unit.

Under the hypotheses of random intrasite mixing and equal interaction of Al^{3+} and Fe^{3+} on the octahedral sites M(1) and Al^{3+} and Si^{4+} on the tetrahedral sites T2, the activity of the prehnite endmember $[Ca_2Al(AlSi_3)O_{10}(OH)_2]$ and the ferri-prehnite component $[Ca_2Fe(AlSi_3)O_{10}(OH)_2]$ in prehnite solid solutions are calculated by means of the relations (Helgeson et al. 1978; Helgeson and Aagaard 1985):

$$a_{Prh} = k_{Prh} \cdot X^2_{Ca^{2+}_A} \cdot X_{Al^{3+}_{M(1)}} \cdot (X_{Al^{3+}_{T2}} \cdot X_{Si^{4+}_{T2}}) \quad (4.40)$$

$$a_{Fe-Prh} = k_{Fe-Prh} \cdot X^2_{Ca^{2+}_A} \cdot X_{Fe^{3+}_{M(1)}} \cdot (X_{Al^{3+}_{T2}} \cdot X_{Si^{4+}_{T2}}) \quad (4.41)$$

where $k_{Prh} = k_{Fe\text{-}Prh} = 4$ are the constants relating the inter- and intra-crystalline standard states.

4.6.2 The Endmember Activities in Hydrothermal Prehnite/Ferri-Prehnite Solid Solutions

In this work, a total of 127 chemical analyses of prehnite solid solutions obtained from deep geothermal wells at different temperatures and depths were compiled. Most of these chemical analyses, namely 96, comes from the two geothermal systems of Darajat, Indonesia (N = 61, temperatures 225–350 °C; Herdianita 2012) and Reykjanes, Iceland (N = 35; temperatures 170–342 °C; Libbey and Williams-Jones 2016). Of the other 31 chemical analyses, 14 are from Onikobe, Japan (temperatures 112–173 °C; Seki et al. 1983), 7 are from Cerro Prieto, Mexico (temperatures 325–350 °C; Schiffman et al. 1985), 4 are from Heber (temperature 175 °C; Wheeler et al. 2001), 3 are from Kakkonda, Japan (temperature 225 °C; Muramatsu and Doi 2000), 1 is from Larderello, Italy (temperature 290 °C; Cavarretta et al. 1982), 1 is from Miravalles (temperature 240 °C; Milodowski et al. 1989), and 1 is from Krafla (temperature not specified; Kristmannsdóttir 1979).

The plots of Fig. 4.14a, b are very similar because the activities of the prehnite and ferri-prehnite endmembers computed by means of Eqs. (4.40) and (4.41) are nearly equal to the mole fractions of Al^{3+} and Fe^{3+} on the octahedral sites M(1), respectively. Since $X_{Al^{3+}_{M(1)}} + X_{Fe^{3+}_{M(1)}} \cong 1$ and $a_{Prh} + a_{Fe-Prh} \cong 1$, most prehnite solid solutions are situated close to the line of equation $X_{Fe^{3+}_{M(1)}} = 1 - X_{Al^{3+}_{M(1)}}$ in Fig. 4.14a and to the line of equation $a_{Fe\text{-}Prh} = 1 - a_{Prh}$ in Fig. 4.14b.

Prehnite is the main component of most hydrothermal prehnite solid solutions, with activities ranging between 0.373 and 0.998, average of 0.763, median of 0.779,

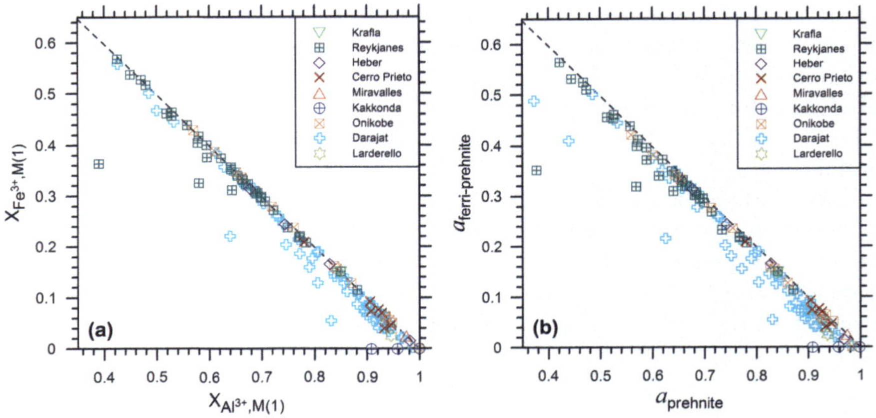

Fig. 4.14 Plots of **a** the mole fraction of Al^{3+} on the octahedral sites M(1) versus the mole fraction of Fe^{3+} on the octahedral sites M(1) and **b** the activity of prehnite versus the activity of ferri-prehnite for 127 prehnite solid solutions from different active geothermal systems (see legend). The dashed line corresponds to the equations $X_{Fe^{3+}_{M(1)}} = 1 - X_{Al^{3+}_{M(1)}}$ in **a** an $a_{Fe\text{-}Prh} = 1 - a_{Prh}$ in **b**

and standard deviation of 0.154. The activity of ferri-prehnite varies from 0.000 to 0.564, with average of 0.210, median of 0.190, and standard deviation of 0.148.

Interestingly, the prehnite solid solutions from the geothermal systems hosted in basaltic rocks, such as Reykjanes and Onikobe, are generally richer in ferri-prehnite than the prehnite samples from the geothermal systems housed in granites or clastic sedimentary rocks, such as Larderello, Cerro Prieto, and especially Kakkonda, where prehnite is Fe-free. The hydrothermal prehnite samples of Darajat distribute over the whole compositional range, although the solid solutions rich in prehnite are more abundant than those rich in ferri-prehnite.

4.6.3 The Thermodynamic Properties of Prehnite

Helgeson et al. (1978), Holland and Powell (1998), and Holland and Powell (2011) report the thermodynamic properties of prehnite of stoichiometry $Ca_2Al_2Si_3O_{10}(OH)_2$, but Holland and Powell (2011) give also the thermodynamic parameters of ferri-prehnite of composition $Ca_2FeAlSi_3O_{10}(OH)_2$ and describe deviation from ideality of prehnite solid solutions in terms of a regular solution model. According to Helgeson et al. (1978), the standard state molal $\Delta G°_f$, $\Delta H°_f$, and S° at 25 °C, 1 bar of prehnite should be regarded as provisional approximations. All these thermodynamic data of prehnite and ferri-prehnite are reported in Table 4.10, showing that the $\Delta G°_f$ and $\Delta H°_f$ of Holland and Powell (1998) differ from those of Helgeson et al. (1978) by 1702 and 502 cal mol^{-1}, respectively, whereas the $\Delta G°_f$ and $\Delta H°_f$ of Holland and Powell (2011) differ from those of Helgeson et al. (1978) by 1449 and 248 cal mol^{-1}, respectively.

Employing the approach described in Sect. 2.2.1, it is advisable to consider the prehnite/ferri-prehnite solid solution with average activity of prehnite, $a_{Prh} = 0.763$, and to compute the Gibbs free energy and the thermodynamic equilibrium constant of its dissolution reaction. The results obtained taking into account the prehnite thermodynamic data of Helgeson et al. (1978) are listed in Table 4.11 as a function of temperature together with the standard molal Gibbs free energies and the natural and decimal logarithms of the thermodynamic equilibrium constant of the dissolution reaction of pure prehnite:

$$Ca_2Al_2Si_3O_{10}(OH)_2 + 10\ H^+ = 2\ Al^{3+} + 2\ Ca^{2+} + 3\ SiO_{2(aq)} + 6\ H_2O. \quad (4.42)$$

Table 4.10 Standard state molal thermodynamic properties, at 25 °C, 1 bar of prehnite and ferri-prehnite from H78 = Helgeson et al. (1978), HP98 = Holland and Powell (1998), and HP11 = Holland and Powell (2011)

Name	ΔG°_{f}	ΔH°_{f}	S°	V°	a	b ($\times 10^{3}$)	c ($\times 10^{-5}$)	d	References
	cal mol^{-1}	cal mol^{-1}	cal K^{-1} mol^{-1}	cm^{3} mol^{-1}	cal K^{-1} mol^{-1}	cal K^{-2} mol^{-1}	cal K mol^{-1}	cal $K^{-0.5}$ mol^{-1}	
Prehnite	−1,390,537	−1,482,089	65.0	140.33	91.60	37.82	−19.60	–	H78
Prehnite	−1,392,239	−1,482,591	69.98	140.26	173.3	−3.3138	−4.9211	−1511.4	HP98
Prehnite	−1,391,986	−1,482,337	69.98	140.26	173.3	−3.3138	−4.9211	−1511.4	HP11
Ferri-prehnite	−1,288,005	−1,378,286	76.48	148.00	176.2	−4.0177	−4.6781	−1519.6	HP11

Table 4.11 Standard molal Gibbs free energies and natural and decimal logarithms of the thermodynamic equilibrium constant of the dissolution reactions of both pure prehnite and the prehnite/ferri-prehnite solid solution with average activity of prehnite as a function of temperature

T(°C)	$\Delta_r G^o_{Prh}$	$\Delta_r G^o_{Prh,aa}$	ln K_{Prh}	ln $K_{Prh,aa}$	log K_{Prh}	log $K_{Prh,aa}$
	cal mol^{-1}	cal mol^{-1}				
0	−44,958	−44,811	82.8253	82.5548	35.9706	35.8531
25	−43,057	−42,897	72.6718	72.4013	31.5610	31.4435
60	−39,407	−39,228	59.5238	59.2533	25.8509	25.7334
100	−34,749	−34,548	46.8615	46.5910	20.3517	20.2342
150	−28,569	−28,342	33.9749	33.7044	14.7551	14.6376
200	−21,965	−21,711	23.3609	23.0904	10.1455	10.0280
250	−14,682	−14,401	14.1226	13.8521	6.1334	6.0159
300	−6292	−5984	5.5243	5.2538	2.3992	2.2817

4.7 Laumontite and Wairakite

4.7.1 Main Characteristics of Ca-Zeolites

Laumontite and wairakite are the two most common zeolites in active geothermal systems (e.g., Browne 1978; Kristmannsdóttir and Tomasson 1976), together with heulandite [$(Na,K,Ca_{0.5})_{10}(Al_{10}Si_{26}O_{72}) \cdot 24H_2O$] – clinoptilolite [$(Na,K,Ca_{0.5})_{5.4}$ $(Al_{5.4}Si_{30.6}O_{72}) \cdot 20\ H_2O$], forming a continuous compositional series (Bish and Boak 2001). Less frequent zeolites comprise stilbite [$NaCa_4(Al_9Si_{27}O_{72}) \cdot 30\ H_2O$], chabazite [$(Ca_{0.5},Na,K)_4(Al_4Si_8O_{24}) \cdot 12\ H_2O$], thomsonite [$Ca_7Na_5(Al_{19}Si_{21}O_{80}) \cdot 24\ H_2O$], scolecite [$Ca_8(Al_{16}Si_{24}O_{80}) \cdot 24\ H_2O$], mordenite [$Na_3Ca_2K(Al_8Si_{40}O_{96}) \cdot 28\ H_2O$], yugawaralite [$Ca_2(Al_4Si_{12}O_{32}) \cdot 8\ H_2O$], levyne [$(Ca_{0.5},Na)_6(Al_6Si_{12}O_{36}) \cdot 18\ H_2O$], gismondine [$Ca_4(Al_8Si_8O_{32}) \cdot 16\ H_2O$], and gmelinite [$(Na,K,Ca_{0.5})_8(Al_8Si_{16}O_{48}) \cdot 22\ H_2O$].

Only laumontite and wairakite are considered in this work, because all their thermodynamic properties (and those of analcime) were retrieved by Helgeson et al. (1978). In contrast, Helgeson et al. (1978) estimated only the standard molal entropy and heat capacity power function coefficients for and other zeolites, including chabazite [$Ca(Al_2Si_4O_{12}) \cdot 6\ H_2O$], epilstilbite [$Ca(Al_2Si_6O_{16}) \cdot 5\ H_2O$], heulandite [$Ca(Al_2Si_7O_{18}) \cdot 6\ H_2O$], natrolite [$Na_2(Al_2Si_3O_{10}) \cdot 2\ H_2O$], Na-phillipsite [$Na_2(Al_2Si_5O_{14}) \cdot 5\ H_2O$], K-phillipsite [$K_2(Al_2Si_5O_{14}) \cdot 5\ H_2O$], Ca-phillipsite [$Ca(Al_2Si_5O_{14}) \cdot 5\ H_2O$], and stilbite [$NaCa_2(Al_5Si_{13}O_{36}) \cdot 14\ H_2O$].

The activity of aqueous silica and relevant cations, such as H^+, Na^+, K^+, Ca^{2+}, and Al^{3+} are essential fluid parameters determining which zeolites will form or whether zeolites will form at all. Temperature plays an important role in controlling zeolite stability, partly because of the high water content of zeolites, whereas pressure is less important (Chipera and Apps 2001).

Zeolites usually occur in the shallower parts of active geothermal systems, at temperatures <200 °C, although wairakite forms at temperatures >200 °C and is often associated with epidote (Bird et al. 1984). A peculiarity of laumontite is its occurrence also as precipitate from hot spring waters, at atmospheric pressure and temperatures from 43 to 89 °C (McCulloh et al. 1981). In contrast, wairakite occurs almost exclusively in geothermal systems, both active and fossil. Indeed, the name wairakite is after Wairakei, New Zealand, where this zeolite was first found during the study of hydrothermal alteration minerals in cores recovered from local boreholes (Steiner 1955; Coombs 1955).

4.7.2 Crystal Chemistry of Wairakite and Analcime

The identification of several samples with chemistry and physical properties (e.g., refractive indices and density) intermediate between those of wairakite [$CaAl_2Si_4O_{12} \cdot 2H_2O$] and those of analcime [$NaAlSi_2O_6 \cdot H_2O$] suggests the existence of a continuous isomorphous series between these two endmembers (Passaglia and Sheppard 2001 and references therein). These samples with intermediate properties are classified either as analcime if Na/(Na + Ca) > 0.5 or as wairakite if Na/(Na + Ca) < 0.5. The Si/(Si + Al + Be) ratio has average value of 0.67 in both analcime and wairakite and range 0.60 to 0.74 in analcime and 0.65 to 0.70 in wairakite. Subordinate extra-framework cations are Ca, K, Mg, and Cs in analcime and Na and Cs in wairakite.

Pure wairakite is monoclinic and has six tetrahedral sites, of which four are occupied by Si^{4+} and two by Al^{3+}, as well as an octahedral site housing one Ca^{2+} ion (Armbruster and Gunter 2001). Ca is coordinated by two H_2O molecules and four oxygen atoms, belonging to two AlO_4 tetrahedra. Allowing for the substitutions of Na^+ and K^+ for Ca^{2+} on the octahedral M2 sites, of Fe^{3+} for Al^{3+} on the tetrahedral T2 sites, and of Ti^{4+} for Si^{4+} on the tetrahedral T1 sites, the structural formula of wairakite/analcime solid solutions can be written as:

$$\left[Ca_{(1-x-y)}Na_xK_y\right]\left[Al_{(1-m)}Fe^{3+}_m\right]_2\left[Si_{(1-p)}Ti_p\right]_4O_{12} \cdot 2\,H_2O$$

where x and y are the mole fractions of Na^+ and K^+, respectively, on the octahedral M2 sites, m is the mole fraction of Fe^{3+} on the tetrahedral T2 sites, and p is the mole fraction of Ti^{4+} on the tetrahedral T1 sites. Under the hypothesis of random mixing of atoms on energetically equivalent sites, the activity of the wairakite endmember [$Ca(Al_2Si_4)O_{12} \cdot 2H_2O$] and the analcime component [$Na(AlSi_2)O_6 \cdot H_2O$] in the wairakite/analcime solid solutions can be calculated by means of the relations (Helgeson et al. 1978; Helgeson and Aagaard 1985):

$$a_{Wrk} = k_{Wrk} \cdot X_{Ca^{2+}_{M2}} \cdot X^2_{Al^{3+}_{T2}} \cdot X^4_{Si^{4+}_{T1}} \tag{4.43}$$

$$a_{Anl} = k_{Anl} \cdot X_{Na^{+}_{M2}} \cdot X_{Al^{3+}_{T2}} \cdot X^{2}_{Si^{4+}_{T1}} \quad (4.44)$$

where $k_{Wrk} = k_{Anl} = 1$ are the constants relating the inter- and intra-crystalline standard states.

4.7.3 *Crystal Chemistry of Laumontite*

According to Armbruster and Gunter (2001) fully hydrated laumontite has stoichiometry $CaAl_2Si_4O_{12} \cdot 4.5\ H_2O$, but previous studies, including Helgeson et al. (1978), assumed that it contains $4H_2O$ molecules. Upon exposure to low humidity at room temperature, laumontite experiences partial dehydration with production of leonhardite, of stoichiometry $CaAl_2Si_4O_{12} \cdot 3.5\ H_2O$. The reaction is reversed by soaking leonhardite in H_2O at room temperature (Coombs 1952). The Si/(Si + Al + Be) ratio of laumontite has average value of 0.67 and range 0.65–0.69, similar to analcime and wairakite (see above). Subordinate extra-framework cations are Na and K (Passaglia and Sheppard 2001).

Similar to wairakite, the laumontite structure is monoclinic and has six tetrahedral sites, of which four house Si^{4+} and two house Al^{3+} ions, as well as an octahedral site occupied by a Ca^{2+} ion (Armbruster and Gunter 2001). Ca is coordinated by three H_2O molecules and four oxygen atoms, belonging to AlO_4 tetrahedra.

Again, as for wairakite/analcime solid solutions, permitting the vicariance of Na^+ and K^+ for Ca^{2+} on the octahedral M2 sites, of Fe^{3+} for Al^{3+} on the tetrahedral T2 sites, and of Ti^{4+} for Si^{4+} on the tetrahedral T1 sites, the structural formula of laumontite solid solutions can be written as:

$$\left[Ca_{(1-x-y)}Na_xK_y\right]\left[Al_{(1-m)}Fe^{3+}_m\right]_2\left[Si_{(1-p)}Ti_p\right]_4O_{12} \cdot 4.5\ H_2O$$

where x and y are the mole fractions of Na^+ and K^+, respectively, on the octahedral M2 sites, m is the mole fraction of Fe^{3+} on the tetrahedral T2 sites, and p is the mole fraction of Ti^{4+} on the tetrahedral T1 sites. Hence, the structural formula of laumontite solid solutions is equal to that of wairakite/analcime solid solutions (see above), apart from the different number of water molecules.

Assuming random mixing of atoms on energetically equivalent sites, the activity of the laumontite endmember $[Ca(Al_2Si_4)O_{12} \cdot 4.5\ H_2O]$ and the related hypothetical alkali component $[(Na,K)(Al_2Si_4)O_{12} \cdot 4.5\ H_2O]$ in the laumontite solid solutions can be computed using the relations (Helgeson et al. 1978; Helgeson and Aagaard 1985):

$$a_{Lmt} = k_{Lmt} \cdot X_{Ca^{2+}_{M2}} \cdot X^{2}_{Al^{3+}_{T2}} \cdot X^{4}_{Si^{4+}_{T1}} \quad (4.45)$$

$$a_{Na,K-Lmt} = k_{Na,K-Lmt} \cdot (X_{Na^{+}_{M2}} + X_{K^{+}_{M2}}) \cdot X^{2}_{Al^{3+}_{T2}} \cdot X^{4}_{Si^{4+}_{T1}} \quad (4.46)$$

where $k_{Lmt} = k_{Na,K\text{-}Lmt} = 1$ are the constants relating the inter- and intra-crystalline standard states.

4.7.4 The Activity of Wairakite in Hydrothermal Wairakite/Analcime Solid Solutions

A total of 136 chemical analyses of wairakite/analcime solid solutions sampled in geothermal wells at different depths and temperatures were compiled in this work. Most of these chemical analyses, namely 119, come from the two geothermal systems of Onikobe, Japan (N = 60; temperature 110–240 °C; Seki et al. 1983) and Darajat, Indonesia (N = 59; temperature 220–267 °C; Herdianita 2012). The remaining 17 chemical analyses are from Hachimantai, Japan (N = 6; temperature 160–195 °C; Shimazu and Yajima 1973), Waiotapu, New Zealand (N = 3; temperature 220–285 °C; Hedenquist and Browne 1989), Cerro Prieto, Mexico (N = 2; temperature 320 °C; Schiffman et al. 1985), Krafla (N = 1, temperature not given; Kristmannsdottir, 1979), Reydarfjordur (N = 1; temperature not reported; Viereck et al. 1982), Yellowstone, Wyoming (N = 1; temperature 138 °C; Bargar and Beeson 1981), Kakkonda, Japan (N = 1; temperature not specified; Muramatsu and Doi 2000), Wairakei, New Zealand (N = 1; temperature not given; Steiner 1955), and Larderello, Italy (temperature 120 °C; Cavarretta et al. 1982).

The compiled wairakite/analcime solid solutions have wairakite activity ranging from 0.573 to 0.999, with average of 0.891, median of 0.920, and standard deviation of 0.090 and analcime activity varying from 0.000 to 0.427, with mean of 0.096, median of 0.068, and standard deviation of 0.091. As expected, in the diagram of Fig. 4.15a, most wairakite/analcime solid solutions are located close to the line of equation $a_{Anl} = 1 - a_{Wrk}$.

The wairakite/analcime samples coming from the two most studied geothermal systems show different characteristics, with those from Onikobe spanning a compositional range much larger than those from Darajat. In fact, at Onikobe, wairakite activity varies from 0.648 to 0.977, with average of 0.844, median of 0.846, and standard deviation of 0.083, whereas, at Darajat, wairakite activity ranges from 0.857 to 0.999, with average of 0.942, median of 0.946, and standard deviation of 0.037. The different activities of wairakite and analcime at Onikobe and Darajat might be due to differences in temperatures and in the lithological framework between the two geothermal fields.

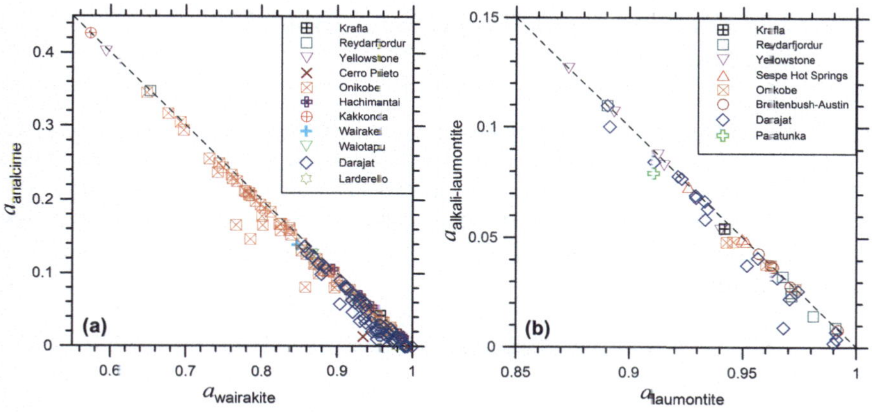

Fig. 4.15 Plots of **a** the activity of wairakite versus the activity of analcime for 136 wairakite/analcime solid solutions from different active geothermal systems (see legend) and **b** the activity of laumontite versus the activity of alkali-laumontite for 45 laumontite/alkali-laumontite solid solutions from various active geothermal systems (see legend). The dashed line corresponds to the equations $a_{\mathrm{Anl}} = 1 - a_{\mathrm{Wrk}}$ in **a** an $a_{\mathrm{Na,K\text{-}Lmt}} = 1 - a_{\mathrm{Lmt}}$ in **b**

4.7.5 The Activity of Laumontite in Hydrothermal Laumontite/Alkali-Laumontite Solid Solutions

In this work, it was possible to compile only 45 chemical analyses of laumontite/alkali-laumontite solid solutions, all obtained from geothermal wells at varying depths and temperatures, apart from those of Sespe Hot Springs which were collected at the surface.

Most of these chemical analyses, namely 19, come from the geothermal system of Darajat, Indonesia (temperature 220–231 °C; Herdianita 2012). The other 26 laumontite data are from Reydarfjordur (N = 6; temperature not reported; Mehegan et al. 1982; Viereck et al. 1982; Exley 1982), Yellowstone, Wyoming (N = 5; temperature 120–140 °C; Bargar and Beeson 1985), Breitenbush-Austin, Oregon (N = 5; temperature 110–140 °C; Oscarson and Bargar 1996), Onikobe, Japan (N = 5; temperature 118–175 °C; Seki et al. 1969, 1983), Sespe Hot Springs (N = 3, temperature 89 °C; McCulloh et al 1981), Paratunka, Kamchatka (N = 1, temperature not given, Petrova 1970), and Krafla (N = 1, temperature not given; Kristmannsdóttir 1979).

The considered laumontite/alkali-laumontite solid solutions exhibit laumontite activity varying from 0.873 to 0.992, with mean of 0.946, median of 0.951, and standard deviation of 0.031 whereas the activity of the alkali-component ranges from 0.002 to 0.127, with average of 0.051, median of 0.048, and standard deviation of 0.032. Not surprisingly, in the diagram of Fig. 4.15b, most laumontite/alkali-laumontite solid solutions are situated near the line of equation $a_{\mathrm{Na,K\text{-}Lmt}} = 1 - a_{\mathrm{Lmt}}$.

4.7.6 The Thermodynamic Properties of Wairakite and Laumontite

The thermodynamic properties of wairakite of stoichiometry $CaAl_2Si_4O_{12} \cdot 2\ H_2O$ and laumontite of composition $CaAl_2Si_4O_{12} \cdot 4\ H_2O$ given by Helgeson et al. (1978), Holland and Powell (1998), and Holland and Powell (2011) are shown in Table 4.12.

Helgeson et al. (1978) derived the thermodynamic properties of wairakite and laumontite from the experimental data of Liou (1970, 1971a), considering the following equilibria:

$$\text{wairakite} = \text{anorthite} + 2\ \text{quartz} + 2\ H_2O \tag{4.47}$$

$$\text{laumontite} = \text{wairakite} + 2\ H_2O. \tag{4.48}$$

However, the thermodynamic properties of laumontite are inconsistent with the experimental data of Thompson (1970) for the equilibrium reaction:

$$\text{laumontite} = \text{anorthite} + 2\ \text{quartz} + 4\ H_2O, \tag{4.49}$$

as pointed out by Helgeson et al. (1978).

Holland and Powell (1998) derived the enthalpy of both zeolites from the experimental results of Thompson (1970), Liou (1970, 1971a, b), and Cho et al. (1987) and estimated the entropy of laumontite and wairakite by means of the method of Holland (1989), considering an entropy contribution of zeolitic H_2O of 6.9 cal K^{-1} mol^{-1}. The differences between the thermodynamic properties reported by Helgeson et al. (1978) and those of Holland and Powell (1998) and Holland and Powell (2011) are probably due to the use of different experimental data to retrieve these thermodynamic data. To be noted also that:

(i) The $\Delta H°_f$ values of Helgeson et al. (1978), Holland and Powell (1998), and Holland and Powell (2011) differ from the calorimetric results of Kiseleva et al. (1996) by +4147, −4209, and −2782 cal mol^{-1}, respectively, for laumontite and by +9046, −4742, and −3752 cal mol^{-1}, respectively, for wairakite.
(ii) The S° values of Helgeson et al. (1978), Holland and Powell (1998), and Holland and Powell (2011) deviate from the calorimetric outcomes of Kiseleva et al. (1996) by +0.1, −6.8, and −4.9 cal K^{-1} mol^{-1}, respectively, for laumontite and by +9.3, −6.1, and −4.9 cal K^{-1} mol^{-1}, respectively, for wairakite.

A thorough discussion on the thermodynamic properties of several zeolites, including laumontite and wairakite, is provided by Chipera and Apps (2001).

Adopting the approach delineated in Sect. 2.2.1, it is advisable to take into consideration the wairakite/analcime solid solution with average activity of wairakite, $a_{Wrk} = 0.891$, and the laumontite/alkali-laumontite solid solution with mean activity of laumontite, $a_{Lmt} = 0.946$, and to calculate the Gibbs free energy and the thermodynamic equilibrium constant of their dissolution reactions. The results calculated

Table 4.12 Standard state molal thermodynamic properties, at 25 °C, 1 bar of wairakite and laumontite from H78 = Helgeson et al. (1978), HP98 = Holland and Powell (1998), HP11 = Holland and Powell (2011), and Ki96 = Kiseleva et al. (1996)

Name	ΔG°_f	ΔH°_f	S°	V°	a	b ($\times 10^3$)	c ($\times 10^{-5}$)	d	References
	cal mol^{-1}	cal mol^{-1}	cal K^{-1} mol^{-1}	cm^3 mol^{-1}	cal K^{-1} mol^{-1}	cal K^{-2} mol^{-1}	cal K mol^{-1}	cal $K^{-0.5}$ mol^{-1}	
Laumontite	−1,597,043	−1,728,884	116.1	207.55	123.2	44.47	−16.43	–	H78
Laumontite	−1,603,119	−1,737,239	109.2	203.70	242.2	−5.1178	−5.344	−2104.9	HP98
Laumontite	−1,602,262	−1,735,813	111.1	203.70	242.2	−5.1178	−5.344	−2104.9	HP11
Laumontite	–	−1,733,031 ± 2032	116.0 ± 5.0	–	–	–	–	–	Ki96
Wairakite	−1,477,652	−1,579,553	105.1	186.87	100.4	44.47	−16.43	0.0	H78
Wairakite	−1,486,628	−1,593,341	89.6	190.40	200.4	−5.1291	−5.430	−1742.9	HP98
Wairakite	−1,485,994	−1,592,352	90.8	190.40	200.4	−5.1291	−5.430	−1742.9	HP11
Wairakite	–	−1,588,599 ± 1506	95.8	–	–	–	–	–	Ki96

using the wairakite and laumontite thermodynamic data of Helgeson et al. (1978) are listed in Tables 4.13 and 4.14, respectively, as a function of temperature together with the standard molal Gibbs free energies and the natural and decimal logarithms of the thermodynamic equilibrium constant of the dissolution reactions of pure wairakite:

$$CaAl_2Si_4O_{12} \cdot 2\ H_2O + 8\ H^+ = Ca^{2+} + 2\ Al^{3+} + 4\ SiO_{2(aq)} + 6\ H_2O \quad (4.50)$$

and pure laumontite:

$$CaAl_2Si_4O_{12} \cdot 4\ H_2O + 8\ H^+ = Ca^{2+} + 2\ Al^{3+} + 4\ SiO_{2(aq)} + 8\ H_2O. \quad (4.51)$$

Table 4.13 Standard molal Gibbs free energies and natural and decimal logarithms of the thermodynamic equilibrium constant of the dissolution reactions of both pure wairakite and the wairakite/analcime solid solution with average activity of wairakite as a function of temperature

T(°C)	$\Delta_r G^\circ_{Wrk}$ cal mol^{-1}	$\Delta_r G^\circ_{Wrk,aa}$ cal mol^{-1}	ln K_{Wrk}	ln $K_{Wrk,aa}$	log K_{Wrk}	log $K_{Wrk,aa}$
0	−24,784	−24,721	45.6591	45.5437	19.8295	19.7794
25	−22,792	−22,724	38.4684	38.3530	16.7066	16.6565
60	−18,725	−18,649	28.2839	28.1685	12.2835	12.2334
100	−13,465	−13,379	18.1585	18.0431	7.8861	7.8360
150	−6529	−6432	7.7644	7.6490	3.3720	3.3219
200	736	845	−0.7828	−0.8982	−0.3400	−0.3901
250	8512	8632	−8.1877	−8.3031	−3.5559	−3.6060
300	17,160	17,291	−15.0663	−15.1817	−6.5432	−6.5933

Table 4.14 Standard molal Gibbs free energies and natural and decimal logarithms of the thermodynamic equilibrium constant of the dissolution reactions of both pure laumontite and the laumontite/alkali−laumontite solid solution with average activity of laumontite as a function of temperature

T(°C)	$\Delta_r G^\circ_{Lmt}$ cal mol^{-1}	$\Delta_r G^\circ_{Lmt,aa}$ cal mol^{-1}	ln K_{Lmt}	ln $K_{Lmt,aa}$	log K_{Lmt}	log $K_{Lmt,aa}$
0	−18,222	−18,192	33.5700	33.5145	14.5793	14.5552
25	−16,777	−16,744	28.3163	28.2608	12.2976	12.2735
60	−13,520	−13,483	20.4218	20.3663	8.8691	8.8450
100	−9247	−9206	12.4702	12.4147	5.4158	5.3916
150	−3623	−3576	4.3085	4.2530	1.8712	1.8471
200	2248	2300	−2.3909	−2.4464	−1.0383	−1.0624
250	8555	8613	−8.2291	−8.2846	−3.5738	−3.5979
300	15,662	15,725	−13.7511	−13.8066	−5.9720	−5.9961

For laumontite there is a difference of 0.024 units between the log K_{Lmt} and log $K_{Lmt,aa}$ values, corresponding to differences of 30–63 cal mol^{-1} in the Gibbs free energies of the dissolution reactions. Therefore, there is not much difference in using the thermodynamic data of pure laumontite or those of the laumontite/alkali-laumontite solid solution of average laumontite activity occurring in active geothermal systems.

4.8 Garnets

Winchell (1933) divided the silicate garnets in the following two series:

- the ugrandites, comprising the endmember garnets uvarovite [$Ca_3Cr_2Si_3O_{12}$], grossular [$Ca_3Al_2Si_3O_{12}$], and andradite [$Ca_3(Fe^{3+},Ti^{4+})_2Si_3O_{12}$], as well as intermediate compounds;
- the pyralspites, including the endmember garnets pyrope [$Mg_3Al_2Si_3O_{12}$], almandine $\left[Fe_3^{2+}Al_2Si_3O_{12}\right]$, and spessartine [$Mn_3Al_2Si_3O_{12}$], and also the intermediate compositions.

According to Deer et al. (1982), there is an almost complete and continuous variation in composition within each of these two series whereas the variation between ugrandites and pyralspites is less continuous.

4.8.1 *Crystal Chemistry of Garnets*

Garnet solid solutions are represented by the general crystallochemical formula (Novak and Gibbs 1971):

$$X_3^{VIII}Y_2^{VI}Z_3^{IV}O_{12},$$

in which, consistent with the endmembers recognized by Winchell (1933):

- X^{VIII} refers to the eight-fold dodecahedral coordinated sites, which are occupied by the divalent cations Ca^{2+}, Mg^{2+}, Mn^{2+}, and Fe^{2+}.
- Y^{VI} indicates the six-fold octahedral coordinated sites, housing the trivalent cations Al^{3+}, Fe^{3+}, and Cr^{3+} as well as Ti^{4+}.
- Z^{IV} designates the four-fold tetrahedral coordinated sites that are mainly occupied by Si^{4+} ions and subordinately by Al^{3+}.

According to Rickwood (1968), possible endmembers garnets could be of the following three types:

- Type 1: $\left\{R_3^{2+}\right\}$ $[R_2^{3+}](R_3^{4+})O_{12}$
- Type 2: $\left\{R_3^{2+}\right\}$ $[R^{3+}R^{4+}](R^{3+}R_2^{4+})O_{12}$
- Type 3: $\left\{R_3^{2+}\right\}$ $[R_2^{4+}](R_2^{3+}R^{4+})O_{12}$.

Accepting the presence of Fe^{3+} and Ti^{4+} in the tetrahedral sites in addition to Si^{4+} and Al^{3+}, but disregarding the endmembers with more Ti^{4+} in tetrahedral sites than in octahedral sites and with more than one element of any valence state, it is possible to obtain 12 endmember garnets of type (1), none of type (2), and 8 of type (3). Rickwood (1968) proposed also a procedure to recast garnet analyses into the mole fractions of 19 endmember garnets, which were suitably chosen to accommodate also Y on the dodecahedral sites, V and Zr on the octahedral sites, and H on the tetrahedral sites.

More recently, Locock (2008) suggested an alternative method, designed to recalculate garnet analyses into the mole fractions of 29 endmember garnets, including 15 natural minerals species and 14 hypothetical components. The Locock's method is consistent with the general crystallochemical formula given above, including 8 cations and 12 anions (O or F) with no vacancies in the dodecahedral and octahedral sites.

If ferric and ferrous iron are not measured separately, their proportions are computed from the total iron concentration given as FeO, using the relation of Droop (1987), which is based on the hypotheses that iron is the only multivalent element and that there are no vacant cation sites. Both assumptions are justified for most Al-silicate garnets. Based on their general crystallochemical formula (see above), the relation of Droop (1987) can be written as follows for garnets:

$$N_{Fe^{3+}} = 24 \cdot (1 - 8/S). \tag{4.52}$$

In Eq. (4.52), S is the total number of cations initially computed assuming all iron to be Fe^{2+}. If $S < 8$, then all iron is actually present as Fe^{2+}. Alternatively, if $S > 8$, Eq. (4.52) is used to calculate $N_{Fe^{3+}}$. Then, the number of Fe^{2+} ions per formula unit, $N_{Fe^{2+}}$, is computed by subtracting $N_{Fe^{3+}}$ from the number of Fe^{2+} ions per formula unit previously computed assuming all iron to be Fe^{2+}, $N_{Fe^{2+},in}$:

$$N_{Fe^{2+}} = N_{Fe^{2+},in} - N_{Fe^{3+}}. \tag{4.53}$$

If Eq. (4.53) gives a negative result, $N_{Fe^{2+}}$ is set equal to zero.

Mn^{3+} is calculated only for compositions that cannot charge balance with Fe^{3+} alone. Only fluorine and the following oxides (expressed in weight percent) are considered: SiO_2, TiO_2, ZrO_2, SnO_2, Y_2O_3, Al_2O_3, Sc_2O_3, Cr_2O_3, V_2O_3, FeO, MnO, MgO, CaO, Na_2O, H_2O. The following oxidation states are assumed: Ti^{4+}, Sn^{4+}, Cr^{3+}, V^{3+}. To be underscored that titanium occurs exclusively at the tetravalent oxidation state in garnet, as suggested by most experimental evidence (Locock 2008).

In the Locock's method cation and anion assignments are as follows: Mn^{3+} is assigned to the octahedral site. H is assumed to occur as $(H_4O_4)_3$ in substitution for $(SiO_4)_3$. H_4 is treated as one cation. Si is assigned to the tetrahedral site, but Si in excess of 3 *apfu* is assigned to the octahedral site. The tetrahedral site is filled, first with Si and H_4 and afterwards with Al and Fe^{3+}, if necessary and possible. F substitutes for O in a coupled substitution with vacancies on the tetrahedral site. Zr, Sn and Ti are assigned exclusively to the octahedral site. Any remaining Al and Fe^{3+}

as well as Sc, V and Cr are assigned to the octahedral site. The remainder of the octahedral site is filled with Fe^{2+} and Mg, if necessary and possible. Ca, Mn^{2+}, and Na are assigned to the dodecahedral site with the remaining Fe^{2+} and Mg, if any. Proportions are based on 8 cations, including F-related tetrahedral vacancies and H_4.

The proportions of the 29 endmembers are computed by the Locock's procedure in the order given in Table 4.15.

In other terms, the proportions of the Mn^{3+} endmembers henritermierite and blythite are computed first and are followed by katoite, F-garnets, and yttrogarnet. The proportions of the remaining endmembers are calculated based on the octahedral cations, in the order: Zr, Sn, Ti, Si, Sc, V, Cr, Al, Fe^{3+}. For a given octahedral cation, endmembers are usually considered in order of increasing rarity. The rigorous assignment of cations to particular sites may lead to non-ideal subtotals for that site, e.g., if (Ca + Na) > 3, or if (Ti + Zr) > 2, or if (Si + H_4) > 3. To be noted that the chemical formula of andradite reported in Table 4.15 does not comprise Ti, in contrast with the stoichiometry of andradite adopted by other authors (e.g., Deer et al. 1982 and references therein).

The Microsoft Excel spreadsheet provided by Locock (2008) was used in this work to compute the mole fractions of endmember garnets and the mole fractions of cations on the dodecahedral, octahedral, and tetrahedral sites. Cations mole fractions were then used to calculate the activities of andradite and grossular in garnet solid solutions by means of the following relations (Bird and Norton 1981):

$$a_{\mathrm{Grs}} = \mathrm{k_{Grs}} \cdot \mathrm{X}^3_{\mathrm{Ca}^{2+}_{\mathrm{X}}} \cdot \mathrm{X}^2_{\mathrm{Al}^{3+}_{\mathrm{Y}}} \cdot \mathrm{X}^3_{\mathrm{Si}^{4+}_{\mathrm{Z}}} \tag{4.54}$$

$$a_{\mathrm{Adr}} = \mathrm{k_{Adr}} \cdot \mathrm{X}^3_{\mathrm{Ca}^{2+}_{\mathrm{X}}} \cdot \mathrm{X}^2_{\mathrm{Fe}^{3+}_{\mathrm{Y}}} \cdot \mathrm{X}^3_{\mathrm{Si}^{4+}_{\mathrm{Z}}} \tag{4.55}$$

where $\mathrm{k_{Grs}} = \mathrm{k_{Adr}} = 1$ are the constants relating the inter- and intra-crystalline standard states. Equations (4.54) and (4.55) are based on the hypothesis of random mixing of atoms on energetically equivalent sites (Helgeson et al. 1978; Helgeson and Aagaard 1985).

4.8.2 The Composition of Hydrothermal Garnet Solid Solutions

A total of 201 chemical analyses of garnet solid solutions collected in deep geothermal wells at different depths and temperatures were compiled. The majority of these samples, for a total of 137, come from the geothermal systems of Darajat (N = 83; measured temperature 137–235 °C; Herdianita 2012) and Reykjanes, Iceland (N = 54; measured temperature 176–321 °C; Libbey and William-Jones 2016). Of the remaining 64 chemical analyses, 21 are from Latera, Italy (measured temperatures 220–343 °C; Cavarretta et al. 1985), 9 are from Larderello, Italy (present day temperatures 175–375 °C; Cavarretta et al. 1982; Gianelli and Ruggieri 2002), 9 are from

Table 4.15 Endmembers considered in the procedure proposed by Locock (2008) to recalculate garnet analyses

Order	Endmembers	Cation site			
		Dodecahedral	Octahedral	Tetrahedral	Anion
1	Henritermierite	Ca_3	Mn_2^{3+}	$Si_2(H_4)$	O_{12}
2	*Blythite*	Mn_3^{2+}	Mn_2^{3+}	Si_3	O_{12}
3	Katoite	Ca_3	Al_2	$(H_4)_3$	O_{12}
4	*F-Ca garnet*	Ca_3	Al_2	$(\square)_3$	F_{12}
5	*F-Mn garnet*	Mn_3	Al_2	$(\square)_3$	F_{12}
6	*Yttrogarnet (YAG)*	Y_3	Al_2	Al_3	O_{12}
7	Kimzeyite	Ca_3	Zr_2	$SiAl_2$	O_{12}
8	*Kimzeyite-Fe*	Ca_3	Zr_2	$SiFe_2^{3+}$	O_{12}
9	*Tin garnet*	Ca_3	$SnFe^{2+}$	Si_3	O_{12}
10	Schorlomite	Ca_3	Ti_2	$SiFe_2^{3+}$	O_{12}
11	*Schorlomite-Al*	Ca_3	Ti_2	$SiAl_2$	O_{12}
12	Morimotoite	Ca_3	$TiFe^{2+}$	Si_3	O_{12}
13	*Na–Ti garnet*	Na_2Ca	Ti_2	Si_3	O_{12}
14	*Morimotoite-Mg*	Ca_3	$TiMg$	Si_3	O_{12}
15	*Morimotoite-Fe*	Fe_3^{2+}	$TiFe^{2+}$	Si_3	O_{12}
16	Majorite	Mg_3	$MgSi$	Si_3	O_{12}
17	*Sc-garnet*	Ca_3	Sc_2	Si_3	O_{12}
18	Goldmanite	Ca_3	V_2	Si_3	O_{12}
19	*Yamatoite*	Mn_3^{2+}	V_2	Si_3	O_{12}
20	Uvarovite	Ca_3	Cr_2	Si_3	O_{12}
21	Knorringite	Mg_3	Cr_2	Si_3	O_{12}
22	Spessartine	Mn_3^{2+}	Al_2	Si_3	O_{12}
23	Pyrope	Mg_3	Al_2	Si_3	O_{12}
24	Almandine	Fe_3^{2+}	Al_2	Si_3	O_{12}
25	Grossular	Ca_3	Al_2	Si_3	O_{12}
26	Andradite	Ca_3	Fe_2^{3+}	Si_3	O_{12}
27	Calderite	Mn_3^{2+}	Fe_2^{3+}	Si_3	O_{12}
28	*Skiagite*	Fe_3^{2+}	Fe_2^{3+}	Si_3	O_{12}
29	*Khoharite*	Mg_3	Fe_2^{3+}	Si_3	O_{12}

Mineral species names are in regular font, but hypothetical endmembers are in italics

Reydarfjordur, Iceland (inferred temperature of garnet formation 300–360 °C; Exley 1982; Mehegan et al. 1982), 8 are from Los Humeros, Mexico (measured temperatures of 300–325 °C; Martinez Serrano 2002), 6 are from Sabatini, Italy (measured temperatures 165–290 °C; Cavarretta and Tecce 1987), 5 are from Kilauea, Hawaii (present-day temperature 322 °C; Bargar et al. 1996), 4 are from Miravalles, Costa Rica, (measured temperature 255 °C; Milodowski et al. 1989), and 1 each are from Cerro Prieto, Mexico (present-day temperature 325 °C; Schiffman et al. 1985) and Mofete, Italy (measured temperature 324 °C; Balducci and Chelini 1992).

The sum of the mole fractions of endmember garnets calculated adopting the Locock's method and using the Microsoft Excel spreadsheet provided by Locock (2008) is higher than 0.9 in 192 cases, between 0.87 and 0.90 in 5 cases (2 from Reydarfjordur and 1 each from Reykjanes, Los Humeros, and Darajat), close to 0.81 in 2 cases (1 from Reykjanes and 1 from Darajat) and 0.53–0.58 in 2 cases from Darajat. These few samples with values of the sum of the mole fractions of endmember garnets significantly lower than 1 and correspondingly very high remainders might be affected by analytical problems but were not rejected to avoid loss of information.

Andradite and grossular are the most frequent and important endmembers of the considered hydrothermal garnets, with 195 and 139 concentrations $\neq 0$, varying from 0.02 to 99.39 mol% and from 0.07 to 81.59 mol%, respectively. Pyrope and spessartine are also common endmembers but less important, with 105 and 94 contents $\neq 0$, ranging from 0.04 to 13.92 mol% and from 0.03 to 10.80 mol%, respectively. Almandine is not so common, with 37 concentrations $\neq 0$, but varying from 0.01 to 84.69 mol%. The highest almandine concentrations (58.05–84.69 mol%) are found in eight garnets from Larderello rocks affected by contact metamorphism (Gianelli and Ruggieri 2002). Titanium is mainly accommodated in schorlomite-Al and morimotoite, with 77 and 49 contents $\neq 0$, ranging from 0.01 to 7.76 mol% and from 0.02 to 9.98 mol%, respectively, and subordinately in morimotoite-Mg and Na–Ti garnet, with 28 and 24 contents $\neq 0$, ranging from 0.06 to 5.00 mol% and from 0.02 to 1.83 mol%, respectively.

Moreover, the sum of the andradite and grossular mole fractions is relatively close to unity in most of the considered hydrothermal garnets. In fact, they are situated close to the dashed line corresponding to the equation $X_{Adr} = 1 - X_{Grs}$ in the binary diagram of Fig. 4.16a. The only exceptions are the eight almandine-rich garnets from Larderello (see above), which are situated relatively close to the origin in the binary diagram of Fig. 4.16a, having small contents of grossular and especially andradite. Authors of previous studies (e.g., Cavarretta et al. 1982, 1985; Exley 1982; Mehegan et al. 1982; Bird et al. 1984; McDowell and Elders 1983; Schiffman et al. 1985; Cavarretta and Tecce 1987; Milodowski et al. 1989; Balducci and Chelini 1992; Gianelli et al. 1998; Martinez-Serrano 2002; Marks et al. 2010; Herdianita 2012; Libbey and Williams-Jones 2016) already recognized the prevalence of andradite and grossular in most hydrothermal garnets from active geothermal systems, which are therefore called grandites.

Moreover, disregarding the eight almandine-rich garnets from Larderello, andradite prevails over grossular in 145 grandites, i.e., 75% of the 193 cases, whereas

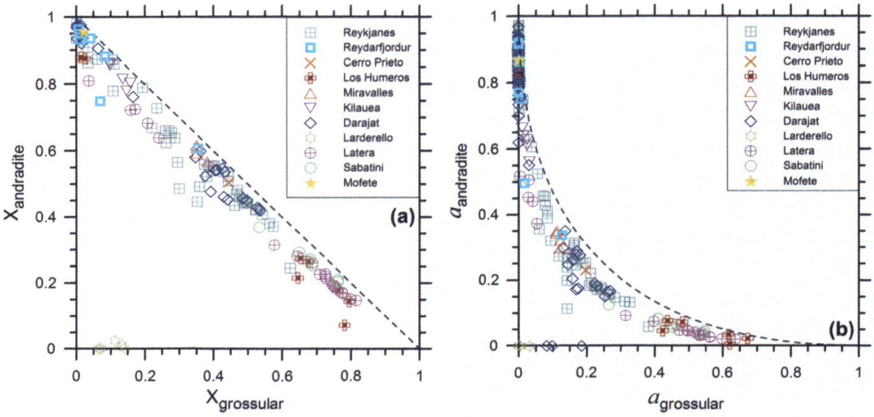

Fig. 4.16 Correlation diagrams between **a** the mole fractions of grossular and andradite and **b** the activities of grossular and andradite for the 173 garnet solid solutions coming from different geothermal systems (see legend). The dashed lines refer to the equations $X_{Adr} = 1 - X_{Grs}$ in **a** and $a_{Adr} = 1 + a_{Grs} - 2 \cdot \sqrt{a_{Grs}}$ in **b**

grossular dominates over andradite in the other 48 grandites, i.e., 25% of the 193 cases (Fig. 4.16a). Of these 48 grossular-rich garnets, 15 are from Latera, 13 are from Reykjanes, 8 are from Darajat, and 6 each are from Sabatini and Los Humeros. It should be noted that garnets are often zoned and exhibit wide intra-crystal compositional variations, with grossular-rich core and andradite-rich rim in some cases, and vice versa in other samples, that were attributed to changes in fluid chemistry in some previous studies. Contact metamorphism and metasomatic processes were also invoked to explain garnet compositions.

4.8.3 *The Activity of Grossular and Andradite in the Hydrothermal Garnet Solid Solutions*

Excluding the eight almandine-rich garnets from Larderello and the three samples from Darajat with very high remainders, the other 190 garnet solid solutions have andradite activity varying from 0.00539 to 0.972, with mean of 0.541, median of 0.625, and standard deviation of 0.340, whereas grossular activity ranges from $<2.27 \times 10^{-7}$ (in 22 cases) to 0.674, with average of 0.127, median of 0.0142, and standard deviation of 0.179.

The correlation plot of Fig. 4.16b shows that andradite and grossular activities of most garnets approximate the relation $a_{Adr} = 1 + a_{Grs} - 2 \cdot \sqrt{a_{Grs}}$ (dashed line), apart from the eight almandine-rich garnets from Larderello and the three samples from Darajat with very high remainders (see above).

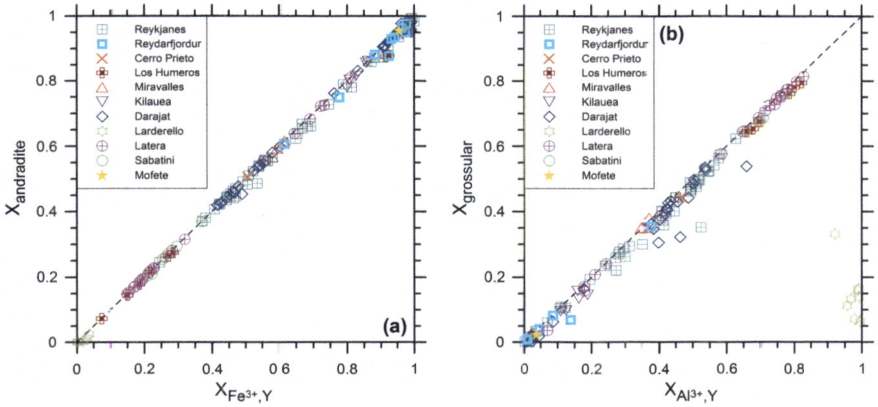

Fig. 4.17 Correlation diagrams between **a** the mole fraction of andradite and the mole fraction of trivalent iron on the octahedral sites and **b** the mole fraction of grossular and the mole fraction of aluminum on the octahedral sites for the 173 garnet solid solutions coming from different geothermal systems (see legend). The dashed lines refer to the equations $X_{Adr} = X_{Fe^{3+}_{Y}}$ in **a** and $X_{Grs} = X_{Al^{3+}_{Y}}$ in **b**

This relationships is obtained considering that $X_{Adr} \approx X_{Fe^{3+}_{Y}}$ and $X_{Grs} \approx X_{Al^{3+}_{Y}}$, which are reasonable approximations in most cases, as indicated by Fig. 4.17a, b, respectively. Consequently, Eqs. (4.54) and (4.55) simplify to $a_{Grs} \approx X^2_{Al^{3+}_{Y}} \approx X^2_{Grs}$ and $a_{Adr} \approx X^2_{Fe^{3+}_{Y}} \approx X^2_{Adr}$, respectively. Considering that $X_{Adr} \approx 1 - X_{Grs}$ (see above), it is a simple matter to obtain, by substitution and rearrangement, the relation written above linking andradite and grossular activities.

4.8.4 *The Thermodynamic Properties of Grossular and Andradite*

Helgeson et al. (1978) report the thermodynamic properties of grossular and andradite only, whereas almandine, pyrope, and spessartine are also included in the dataset of Holland and Powell (1998), and almandine, knorringite, majorite, pyrope, and spessartine are also comprised in the database of Holland and Powell (2011). The standard state molal thermodynamic properties, at 25 °C, 1 bar of grossular $[Ca_3Al_2Si_3O_{12}]$ and andradite $[Ca_3Fe^{3+}_2Si_3O_{12}]$ listed in these three compilations are shown in Table 4.16.

For grossular, the $\Delta G°_f$ and $\Delta H°_f$ of Holland and Powell (1998) differ from those of Helgeson et al. (1978) by 4214 and 4574 cal mol^{-1}, respectively, whereas the $\Delta G°_f$ and $\Delta H°_f$ listed by Holland and Powell (2011) differ from those of Helgeson et al. (1978) by 3946 and 4306 cal mol^{-1}, respectively.

Table 4.16 Standard state molal thermodynamic properties, at 25 °C, 1 bar of grossular and andradite from H78 = Helgeson et al. (1978), HP98 = Holland and Powell (1998), and HP11 = Holland and Powell (2011)

Name	ΔG°_f	ΔH°_f	S°	V°	a	b (×10^3)	c (×10^{-5})	d	References
	cal mol^{-1}	cal mol^{-1}	cal K^{-1} mol^{-1}	cm^3 mol^{-1}	cal K^{-1} mol^{-1}	cal K^{-2} mol^{-1}	cal K mol^{-1}	cal K$^{-0.5}$ mol^{-1}	
Grossular	−1,496,967	−1,583,397	60.87	125.30	104.017	17.013	−27.318		H78
Grossular	−1,501,181	−1,587,971	60.95	125.35	149.62	0	−13.813	−956.7	HP98
Grossular	−1,500,913	−1,587,703	60.95	125.35	149.62	0	−13.813	−956.7	HP11
Andradite	−1,297,479	−1,381,005	70.13	131.85	113.532	15.636	−30.889		H78
Andradite	−1,296,446	−1,378,607	76.00	132.04	152.63	0	−11.843	−953.4	HP98
Andradite	−1,296,569	−1,378,843	75.62	132.04	152.63	0	−11.843	−953.4	HP11

For andradite, the ΔG°_{f} and ΔH°_{f} of Holland and Powell (1998) differ from those of Helgeson et al. (1978) by 1033 and 2398 cal mol^{-1}, respectively, whereas the ΔG°_{f} and ΔH°_{f} listed by Holland and Powell (2011) differ from those of Helgeson et al. (1978) by 910 and 2162 cal mol^{-1}, respectively.

More recently, Dachs and Geiger (2019) investigated the thermodynamics of three grossular-rich natural solid solutions and three synthetic andradite-grossular solid solutions of composition $Ca_3(Al_{0.5}Fe^{3+}_{1.5})Si_3O_{12}$, $Ca_3(AlFe^{3+})Si_3O_{12}$, and $Ca_3(Al_{1.5}Fe^{3+}_{0.5})Si_3O_{12}$. Using the S° calorimetric value of 325.0 ± 2.0 J K^{-1} mol^{-1} (corresponding to 77.68 ± 0.48 cal K^{-1} mol^{-1}) for andradite (Geiger et al. 2018) and published phase-equilibrium results on three andradite-bearing reactions, they obtained the value of -5763.3 ± 1.5 kJ mol^{-1} (corresponding to -1377.462 ± 0.359 kcal mol^{-1}) for the standard enthalpy of formation of andradite. This ΔH°_{f} value is lower than those of Helgeson et al. (1978), Holland and Powell (1998), and Holland and Powell (2011) by 3543, 1145, and 1381 cal mol^{-1}, respectively.

Following the approach outlined in Sect. 2.2.1, it is advisable to consider the grossular/andradite solid solution with average activity of grossular, $a_{Grs} = 0.127$, and to calculate the Gibbs free energy and the thermodynamic equilibrium constant of its dissolution reaction. The results obtained considering the grossular thermodynamic data of Helgeson et al. (1978) are listed in Table 4.17 as a function of temperature together with the standard molal Gibbs free energies and the natural and decimal logarithms of the thermodynamic equilibrium constant of the dissolution reaction of pure grossular:

$$Ca_3Al_2Si_3O_{12} + 12\,H^{+} = 2\,Al^{3+} + 3\,Ca^{2+} + 3\,SiO_{2(aq)} + 6\,H_2O. \qquad (4.56)$$

Table 4.17 Standard molal Gibbs free energies and natural and decimal logarithms of the thermodynamic equilibrium constant of the dissolution reactions of both pure grossular and the grossular/andradite solid solution with average activity of grossular as a function of temperature

T(°C)	$\Delta_r G^{o}_{Grs}$ cal mol^{-1}	$\Delta_r G^{o}_{Grs,aa}$ cal mol^{-1}	ln K_{Grs}	ln $K_{Grs,aa}$	log K_{Grs}	log $K_{Grs,aa}$
0	−73,115	−71,995	134.6984	132.6348	58.4988	57.6026
25	−70,836	−69,613	119.5573	117.4938	51.9231	51.0269
60	−66,627	−65,261	100.6393	98.5757	43.7071	42.8109
100	−61,295	−59,765	82.6606	80.5970	35.3990	35.0029
150	−54,197	−52,462	64.4522	62.3887	27.9912	27.0950
200	−46,554	−44,614	49.5125	47.4490	21.5030	20.6068
250	−38,059	−35,914	36.6090	34.5455	15.3991	15.0029
300	−28,211	−25,861	24.7689	22.7054	10.7570	9.8608

4.9 Calcite

Anhydrous calcium carbonate occurs as three distinct polymorphs, i.e., calcite, aragonite, and vaterite, whose crystals have trigonal, orthorhombic, and hexagonal symmetry, respectively. Calcite is the thermodynamically stable polymorph at low pressures, aragonite is stable at high pressures (e.g., >3 kbar at 0 °C and >4.4 kbar at 100 °C; Crawford and Hoersch 1972 and references therein), whereas vaterite is unstable and is quickly converted into aragonite or calcite through dissolution and growth (Ogino et al. 1987).

Aragonite is a member of the isomorphous group of orthorhombic carbonates comprising strontianite [$SrCO_3$], witherite [$BaCO_3$], and cerussite [$PbCO_3$], in which Ca^{2+}, Sr^{2+}, Ba^{2+}, and Pb^{2+} have nine-fold coordination. Orthorhombic $BaCO_3$ and $SrCO_3$ invert to the trigonal calcite-type structure at temperatures of 806 and 912 °C, respectively, which are much higher than the range of geothermal interest (Chang 1965).

Calcite belongs to the isomorphous group of trigonal carbonates including magnesite [$MgCO_3$], siderite [$FeCO_3$], rhodochrosite [$MnCO_3$], smithsonite [$ZnCO_3$], sphaerocobaltite [$CoCO_3$], gaspeite [$NiCO_3$], and otavite [$CdCO_3$]. As required by the electroneutrality condition, these trigonal carbonates host divalent cations, the largest of which is Ca^{2+}, whose size represents the uppermost value for six-fold octahedral coordination (Reeder 1990). Complete miscibility occurs between calcite and otavite, whereas limited miscibility occurs between calcite and the trigonal carbonates of Mg, Fe, Mn, Co, and Ni. Nevertheless, significant amounts of Mg, Fe, Mn, Zn, Co, Ni, and Cd are likely to be scavenged from the aqueous solution by calcite precipitation (e.g., McIntire 1963; Marini et al. 2001). Similar considerations hold true for Sr, Ba, and Pb although their carbonates are expected to mix with aragonite instead of calcite.

4.9.1 The Composition of Calcite-Rich Trigonal Carbonates from Active Geothermal Systems

A total of 313 chemical analyses of solid solutions of calcite-rich trigonal carbonates sampled from deep geothermal wells at different depths and temperatures were compiled in this work. The weight percentages of relevant oxides (CaO, MgO, MnO, FeO, ZnO, SrO, BaO, and PbO) were converted into the corresponding moles and their sum was computed. A number of moles of CO_2 equal to the sum of the moles of the relevant oxides was assumed to be present in the mineral and the corresponding CO_2 weight percentage was computed accordingly. The chemical analyses with sum of weight percentages, including CO_2, between 92 and 104% were accepted, whereas the few analyses with sum of oxides <92 wt% were rejected.

Most of the accepted chemical analyses of calcite-rich trigonal carbonates, namely 251, are representative of the two geothermal systems of Darajat, Indonesia (N =

175, temperature 110–300 °C; Herdianita 2012) and Reykjanes, Iceland (N = 76; temperature 19–289 °C; Libbey and Williams-Jones 2016). Of the other 62 chemical analyses, 25 are from Milos, Greece (temperature 190 °C; Liakopoulos 1987), 16 are from Onikobe, Japan (temperature 110–225 °C; Seki et al. 1983), 10 are from Broadlands, New Zealand (temperature 230–290 °C; Lonker et al. 1990), 5 are from Cerro Prieto, Mexico (temperature 185–335 °C; Schiffman et al. 1985), 4 are from Pantelleria, Italy (temperature 85–120 °C; Fulignati et al. 1997), and 2 are from Newberry Caldera, Oregon (temperature 35–130 °C; Keith and Bargar 1988). The main statistical parameters for the mole fractions of calcite, magnesite, siderite, rhodochrosite, $SrCO_3$, $BaCO_3$, smithsonite, and $PbCO_3$ in the considered solid solutions of calcite-rich trigonal carbonates are listed in Table 4.18.

As expected, calcite is by far the main component of the solid solutions of interest, with mole fractions ranging between 0.705 and 1.000, average of 0.972 median of 0.988, and standard deviation of 0.043. Based on the average values, the second major component is rhodochrosite, with mole fractions ranging between 0.0000570 and 0.130, average of 0.0139, median of 0.00522, and standard deviation of 0.0214, and is followed by magnesite, siderite, $SrCO_3$, $PbCO_3$, smithsonite, and $BaCO_3$, in order of decreasing importance. To be noted that the number of available data, N, decreases in the following order: $CaCO_3 > MgCO_3 > MnCO_3 > FeCO_3 > SrCO_3 > PbCO_3 > BaCO_3 > ZnCO_3$.

4.9.2 The Carbonate Minerals Other Than Calcite from Active Geothermal Systems

Calcite is not the only carbonate mineral present in active geothermal systems. Although calcite is a common and important mineral at Ohaaki-Broadlands, occurring in all boreholes, siderite is also present (Browne and Ellis 1970). Siderite is uncommon, but it occurs in quantities of up to 10% in some Broadlands wells, at temperatures between 37 and 130 °C. At Salton Sea, siderite occurs at weakly higher temperatures, close to 135 °C, but it is absent at the depths/temperatures where base metal sulfides are found (Muffler and White 1969). At Wairakei, siderite is reported to be present at depth of only 27 m and low temperatures (Steiner 1953). At Newberry, siderite is abundant and is locally associated with magnesite and solid solutions of ankerite [$CaFe(CO_3)_2$] and dolomite [$CaMg(CO_3)_2$] at depths of 320 to 697 m, where measured temperature varies from 35 to 100 °C. It was probably precipitated after the emplacement of a sill, during a thermal event that caused alteration of Fe-sulfides (Keith and Bargar 1988). Both siderite and ankerite occur also at Darajat, where calcite is the most common hydrothermal mineral (Herdianita 2012).

Table 4.18 Main statistical parameters for the mole fractions of calcite, magnesite, siderite, rhodochrosite, $SrCO_3$, $BaCO_3$, smithsonite, and $PbCO_3$ in the solid solutions of calcite–rich trigonal carbonates from different active geothermal systems

	X $CaCO_3$	X $MgCO_3$	X $FeCO_3$	X $MnCO_3$	X $SrCO_3$	X $BaCO_3$	X $ZnCO_3$	X $PbCO_3$
N	313	292	252	279	68	34	12	57
Minimum	0.705	5.01E−05	1.42E−05	5.70E−05	9.93E−06	6.66E−06	2.39E−04	4.57E−06
Maximum	1.000	0.102	0.109	0.130	1.15E−02	9.90E−03	1.23E−03	1.21E−02
Mean	0.972	9.82E−03	7.86E−03	1.39E−02	1.60E−03	4.31E−04	7.50E−04	9.49E−04
Median	0.988	3.11E−03	2.97E−03	5.22E−03	9.20E−04	2.05E−04	7.23E−04	1.25E−04
Std. Dev.	0.043	1.69E−02	1.41E−02	2.14E−02	2.28E−03	1.30E−03	3.22E−04	2.50E−03

4.9.3 *The Thermodynamic Properties of Calcite*

Helgeson et al. (1978), Holland and Powell (1998), and Holland and Powell (2011) report the thermodynamic properties of calcite and aragonite of chemical formula $CaCO_3$ (Table 4.19). For calcite, the $\Delta G°_f$ and $\Delta H°_f$ of Holland and Powell (1998) differ from those of Helgeson et al. (1978) by 308 and 163 cal mol^{-1}, respectively, whereas the $\Delta G°_f$ and $\Delta H°_f$ listed by Holland and Powell (2011) differ from those of Helgeson et al. (1978) by 227 and 82 cal mol^{-1}, respectively.

For aragonite, the $\Delta G°_f$ and $\Delta H°_f$ of Holland and Powell (1998) differ from those of Helgeson et al. (1978) by 269 and 88 cal mol^{-1}, respectively, whereas the $\Delta G°_f$ and $\Delta H°_f$ listed by Holland and Powell (2011) differ from those of Helgeson et al. (1978) by 207 and 47 cal mol^{-1}, respectively.

These differences are much smaller than those generally observed for Al-silicates. Nevertheless, they determine some differences in the Gibbs free energy of calcite-aragonite conversion. For instance, this $\Delta G°_r$ value, at 25°C, 1 bar, is 225 cal mol^{-1} based on the data of Helgeson et al. (1978), 186 cal mol^{-1} according to Holland and Powell (1998), and 205 cal mol^{-1} based on the data of Holland and Powell (2011).

Since calcite generally occurs as a virtually pure phase (see Sect. 4.9.1), its activity can be assumed to be equal to one. Therefore, there is no need to correct the Gibbs free energy and the thermodynamic equilibrium constant of its dissolution reaction.

4.10 Quartz and Other Silica Minerals

The main crystalline polymorphs of silica stable at low pressures are quartz, cristobalite, and tridymite. Quartz and chalcedony (i.e., microcrystalline quartz with submicroscopic pores) are the most common hydrothermal minerals in most geothermal fields, if not all, and distribute in a wide temperature range, from 120 to 350 °C according to Henley and Ellis (1983). Above ~180 °C, aqueous solutions are saturated with respect to quartz, whereas at lower temperatures, silica activity is controlled by chalcedony solubility (e.g., Arnórsson et al. 1983a; Fournier 1991). Cristobalite and tridymite also occur as hydrothermal minerals, especially at relatively low temperatures, but are by far less common than quartz and chalcedony.

The crystal structure of all these silica polymorphs consists of a three-dimensional framework of SiO_4 tetraedra, which are linked together by sharing the corners with other tetraedra. In these lattices, therefore, each oxygen atom has two silicon atoms as nearest neighbors and each silicon atom is surrounded by four oxygen atoms (Deer et al. 2004). In spite of these similarities, each silica polymorph has its own crystal structure: α-quartz and β-quartz crystals have trigonal and hexagonal symmetry, respectively; α-cristobalite and β-cristobalite crystals have tetragonal and cubic symmetry, respectively, whereas tridymite occurs in seven distinct crystalline forms indicated by the symbols HP, LHP, OC, OS, OP, MC, and MX-1 (Pryde and Dove 1998).

Table 4.19 Standard state molal thermodynamic properties, at 25°C, 1 bar of calcite and aragonite from H78 = Helgeson et al. (1978), HP98 = Holland and Powell (1998), and HP11 = Holland and Powell (2011)

Name	ΔG°_f	ΔH°_f	S°	V°	a	b ($\times 10^3$)	c ($\times 10^{-5}$)	d	References
	cal mol^{-1}	cal mol^{-1}	cal K^{-1} mol^{-1}	cm^3 mol^{-1}	cal K^{-1} mol^{-1}	cal K^{-2} mol^{-1}	cal K mol^{-1}	cal $K^{-0.5}$ mol^{-1}	
Calcite	−270100	−288772	22.15	36.934	24.98	5.24	−6.2	–	H78
Calcite	−269792	−288609	22.11	36.890	33.68	1.202	−2.272	−205.2	HP98
Calcite	−269873	−288690	22.11	36.890	33.68	1.202	−2.272	−205.2	HP11
Aragonite	−269875	−288723	21.56	34.150	20.13	10.24	−3.340	–	H78
Aragonite	−269606	−288635	21.39	34.150	45.96	−0.729	2.748	−506.3	HP98
Aragonite	−269668	−288676	21.46	34.150	45.96	−0.729	2.748	−506.3	HP11

4.10.1 The Activity of Endmember Quartz in Hydrothermal Quartz

A total of 209 chemical analyses of quartz were compiled in this work. All these analyses are representative of quartz samples collected in the deep wells of the Reykjanes geothermal field, Iceland at different depths, from 150 to 2346 m, and temperatures, from 19 to 342 °C (Libbey and Williams-Jones 2016).

As expected, SiO_2 is by far the main component of these quartz samples, with mole fractions varying between 0.967 and 0.999, average of 0.993, median of 0.994, and standard deviation of 0.005. These statistical parameters indicate that hydrothermal quartz is a pure phase and its activity can be assumed to be equal to one. Strictly speaking this finding concerns the Reykjanes geothermal system only, but it may be considered a general conclusion, in spite of the lack of data for hydrothermal quartz from other geothermal systems.

4.10.2 The Thermodynamic Properties of the Quartz/Chalcedony Mechanical Mixture

Helgeson et al. (1978) report the thermodynamic properties of α- and β-quartz, α- and β-cristobalite, chalcedony, amorphous silica, and coesite. The latter mineral is disregarded in Table 4.20 being stable at very high pressures, outside the range of interest of geothermal systems. Holland and Powell (1998) and Holland and Powell (2011) report instead the thermodynamic properties of quartz, cristobalite, and tridymite (Table 4.20).

For quartz, the $\Delta G°_f$ and $\Delta H°_f$ of Holland and Powell (1998) differ from those of Helgeson et al. (1978) by 53 and 56 cal mol^{-1}, respectively, whereas the $\Delta G°_f$ and $\Delta H°_f$ listed by Holland and Powell (2011) are closer to those of Helgeson et al. (1978), with differences of 5 and 13 cal mol^{-1}, respectively. The opposite is true for cristobalite. In fact, the $\Delta G°_f$ and $\Delta H°_f$ of Holland and Powell (1998) deviate from those of Helgeson et al. (1978) by 6 and 206 cal mol^{-1}, respectively, whereas the $\Delta G°_f$ and $\Delta H°_f$ reported by Holland and Powell (2011) exhibit larger deviations from those of Helgeson et al. (1978), 114 and 636 cal mol^{-1}, respectively.

The methodological approach of Helgeson et al. (1978) for treating quartz is different from that adopted by Holland and Powell (1998) and Holland and Powell (2011). In fact, Helgeson et al. (1978) report the thermodynamic properties of α-quartz at 25 °C, 1 bar, the enthalpy, entropy, and volume of the α-β quartz transition at 575 °C, 1 bar (which are listed in the column $\Delta H°_f$, S°, and V° in Table 4.20 to avoid inserting three more columns), and the heat capacity coefficients of β-quartz. These are the data stored in SUPCRT92, which are used to compute the thermodynamic properties of quartz up to 2000 K. The same applies to other minerals that undergo phase transitions, such as high-albite, 7Å-clinochlore, and prehnite, among those of interest to us.

Table 4.20 Standard state molal thermodynamic properties, at 25 °C, 1 bar of silica polymorphs from H78 = Helgeson et al. (1978), HP98 = Holland and Powell (1998), and HP11 = Holland and Powell (2011)

Name	ΔG°_{f}	ΔH°_{f}	S°	V°	a	b ($\times 10^{3}$)	c ($\times 10^{-5}$)	d	References
	cal mol^{-1}	cal mol^{-1}	cal K^{-1} mol^{-1}	cm^{3} mol^{-1}	cal K^{-1} mol^{-1}	cal K^{-2} mol^{-1}	cal K mol^{-1}	cal K$^{-0.5}$ mol^{-1}	
α-Quartz	−204,646	−217,650	9.88	22.688	11.22	8.20	−2.70	–	H78
Transition	–	290	0.342	0.372	–	–	–	–	H78
β-Quartz	–	–	–	–	14.41	1.94	0.00	–	H78
Chalcedony	−204,276	−217,282	9.88	22.688	11.22	8.20	−2.70	–	H78
α-Cristobalite	−203,895	−216,755	10.372	25.74	13.98	3.34	−3.81	–	H78
β-Cristobalite	−203,290	−215,675	11.963	27.38	17.39	0.31	−9.89	–	H78
Am. Silica	−202,892	−214,568	14.34	29.00	5.93	47.2	−22.78	–	H78
Quartz	−204,699	−217,706	9.92	22.690	26.46	−1.240	0.000	−269.7	HP98
Quartz	−204,651	−217,663	9.90	22.690	22.20	−0.153	−1.709	−171.2	HP11
Cristobalite	−203,901	−216,549	11.11	26.100	23.40	−0.801	−1.521	−185.0	HP98
Cristobalite	−203,781	−216,119	12.16	27.450	17.38	0.312	−9.869	0.0	HP11
Tridymite	−204,037	−216,714	11.02	27.000	23.40	−0.801	−1.521	−185.0	HP98
Tridymite	−203,978	−216,797	10.54	28.000	17.90	0.741	−2.806	−56.6	HP11

In contrast, quartz is treated adopting a pressure-dependent Landau tricritical model by Holland and Powell (1998) and Holland and Powell (2011). In this way, there is no need to give separate thermodynamic data for α- and β-quartz. However, it is necessary to consider the pressure-dependent critical temperature and the excess volume, entropies, and energies. The properties given in the Holland and Powell data sets are the Landau critical temperature at 1 bar, $T_c^\circ = 847$ K (i.e., the α-β quartz transition within 1 K); the thermal expansion parameter $a^\circ = 0.65$ K^{-1} mol^{-1}; the bulk modulus at 298 K, $\kappa_{298} = 750$ kbar mol^{-1}; the maximum entropy of disorder, $S_{max} = 4.95$ J K^{-1} mol^{-1} (=1.183 cal K^{-1} mol^{-1}); and the maximum volume of disorder $V_{max} = 0.1188$ J bar^{-1} mol^{-1} (=1.188 cm^3 mol^{-1}). The approach of Holland and Powell allows calculation of mineral-fluids equilibria up to a pressure of at least 100 kbar, a value exceeding the maximum pressures present in geothermal systems by about three orders of magnitude.

Unlike quartz, α- and β-cristobalite are treated by Helgeson et al. (1978) as two distinct minerals. This different approach was probably adopted because the transition temperature of cristobalite is poorly defined.

The thermodynamic data of chalcedony, cristobalite, and amorphous silica reported by Helgeson et al. (1978) are consistent with those of quartz and were retrieved from solubility measurements at high temperatures by Walther and Helgeson (1977) in their study on the thermodynamic properties of aqueous silica as a function of temperature and pressure.

As shown in Sect. 5.2.2, the solubilities of quartz and chalcedony, at saturation temperatures and pressures, computed using the thermodynamic data of quartz and chalcedony from Helgeson et al. (1978) and those of aqueous SiO_2 from Shock et al. (1989), bracket the functions describing quartz solubilities according to different studies. Therefore, a mechanical mixture constituted by 50% chalcedony and 50% quartz was added to the SUPCRT92 database. At 25 °C, 1 bar, α-quartz and chalcedony have the same entropy, volume, and heat capacity coefficients, whereas the ΔG°_f and ΔH°_f values differ by 370 and 368 cal mol^{-1}, respectively, due to the distinct grain size of these two silica minerals. Therefore, the quartz/chalcedony mechanical mixture has the same entropy, volume, and heat capacity coefficients of the two endmember minerals, whereas its Gibbs free energy and enthalpy are different, with $\Delta G^\circ_f = -204{,}461$ cal mol^{-1} and $\Delta H^\circ_f = -217{,}466$ cal mol^{-1}.

Interpolation of the undissociated SiO_2 concentration data fixed by the quartz/chalcedony mechanical mixture leads to the following polynomial equation (c_{SiO_2} in mg/kg):

$$T(^\circ C) = 2.7195E-12 \cdot c_{SiO_2}^5 - 6.6268E-9 \cdot c_{SiO_2}^4 + 6.4341E-6 \cdot c_{SiO_2}^3 - 3.2559E-3 \cdot c_{SiO_2}^2 + 1.1233 \cdot c_{SiO_2} + 39.790 \quad (4.57)$$

which is valid from 100 to 350 °C, at water saturation pressures, and reproduces the equilibrium temperatures within 0.5 °C below 325 °C and within 5.5 °C above 325 °C.

4.11 Conclusive Considerations on Hydrothermal Minerals

Over 2200 chemical analyses of hydrothermal minerals collected at depth in different active geothermal systems were compiled, processed, and discussed in the previous sections of this chapter. In agreement with the outcomes of previous studies, it turned out that hydrothermal quartz, calcite, adularia, albite, anorthite, and laumontite are virtually pure solid phases and their activities can be assumed to be equal to one. In contrast, the other hydrothermal minerals of interest, i.e., white mica, chlorite, epidote, prehnite, wairakite, and garnet, are solid solutions of variable composition. Therefore, the activities of pertinent endmembers were computed, under the assumption of random mixing of atoms on energetically equivalent sites, obtaining the following results:

(i) In the considered 75 hydrothermal white mica solid solutions, muscovite activity varies from 0.007 to 0.875, but it is higher than 0.40 in most cases. Accordingly, average muscovite activity is 0.654, the median value is 0.688 and the standard deviation is relatively small, 0.156.

(ii) The selected 181 hydrothermal chlorites have highly different $Fe^{2+}/(Fe^{2+} + Mg)$ ratio, from 0.23 to 0.94, with compositions varying from Mg-rich values, relatively similar to clinochlore or sheridanite, to Fe^{2+}-rich values, close to ripidolite. The activity of the clinochlore endmember spans a very wide interval, from 4.39×10^{-7} to 0.250, with an average of 0.0553, a median of 0.0451 and a relatively large standard deviation of 0.0484.

(iii) Clinozoisite is the main component of the 436 hydrothermal epidote solid solutions of interest. Its activity varies from 0.523 to 0.888, with a mean of 0.703, a median of 0.706, and a small standard deviation of 0.057. According to the equilibrium model of Bird and Helgeson (1980), these hydrothermal epidote samples appear to be relatively ordered, but deviations from the equilibrium distribution are possible, due to occurrence of metastable states of substitutional order/disorder on the octahedral sites.

(iv) Prehnite prevails in most of the 127 selected prehnite/ferri-prehnite solid solutions. Its activity ranges from 0.373 to 0.998, with average of 0.763, median of 0.779, and a relatively small standard deviation of 0.154.

(v) Wairakite is the main endmember component of the 136 wairakite/analcime solid solutions under consideration. Its activity varies from 0.573 to 0.999, with mean of 0.891, median of 0.920, and a comparatively small standard deviation of 0.090.

(vi) The sum of the andradite and grossular mole fractions is relatively close to unity in most of the considered 201 hydrothermal garnets, which can be classified as grandites, apart from eight almandine-rich garnets from Larderello. Grossular is the main component in 25% of the cases only. Its activity ranges from $<2.27 \times 10^{-7}$ (in 22 cases) to 0.674, with average of 0.127, median of 0.0142, and relatively large standard deviation of 0.179.

The solid solutions with average activity of muscovite, 7Å-clinochlore, clinozoisite, prehnite, wairakite, and grossular were considered as points of reference.

The ΔG°_{f} and the log K of their dissolution reactions were computed and compared with the ΔG°_{f} and the log K of the dissolution reactions of the corresponding pure minerals.

The dilution of the endmember component of interest in a given solid phase increases its stability with respect to that of the pure mineral. An overall view of this effect is provided by the diagram of CO_2 fugacity versus temperature for the system SiO_2–Al_2O_3–CaO–H_2O–CO_2 (Fig. 4.18, modified from Giggenbach 1997), which was constructed assuming saturation with calcite and quartz/chalcedony at any f_{CO_2} and temperature.

In this diagram, the red lines depict the stability fields of the pure hydrothermal minerals, but the blue lines describe the stability fields of the garnet, epidote, wairakite/analcime and prehnite/ferri-prehnite solid solutions with average activities of the relevant endmembers, that is grossular, clinozoisite, wairakite, and prehnite, respectively. Both grids were prepared assuming that laumontite, kaolinite and pyrophyllite are pure phases. The comparison of the two grids shows what follows:

(i) The stability field of the garnet solid solution expands considerably with respect to that of pure grossular, at the expense of the stability field of prehnite. This significant effect is due to the substantial difference between the average activity

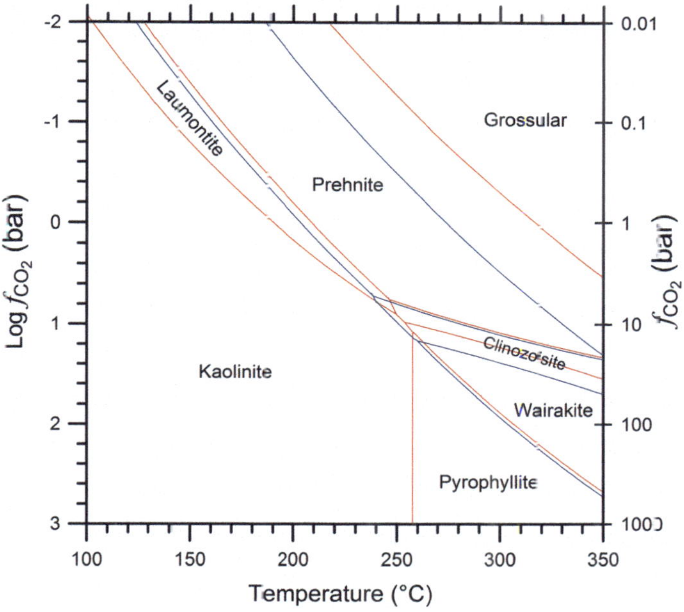

Fig. 4.18 Diagram of CO_2 fugacity versus temperature for the system SiO_2–Al_2O_3–CaO–H_2O–CO_2 (modified from Giggenbach 1997), prepared assuming saturation with quartz/chalcedony and calcite at any f_{CO_2} and temperature

of prehnite, 0.763, and the average activity of grossular, 0.127. To be noted that the following relation:

$$\log\ f_{CO_2,Prh-Grs} = \log\ K_{Cal} + \log\ K_{Prh} - \log\ K_{Grs}, \quad (4.58)$$

links the logarithm of CO_2 fugacity to the log K of the dissolution reactions of calcite, prehnite, and grossular, which are as follows:

$$Cal + 2\ H^+ = Ca^{2+} + CO_{2(g)} + H_2O \quad (4.59)$$

$$Prh + 3\ Qtz/Chc + 10\ H^+ = 2\ Ca^{2+} + 2\ Al^{3+} + 6\ H_2O \quad (4.60)$$

$$Grs + 3\ Qtz/Chc + 12\ H^+ = 3\ Ca^{2+} + 2\ Al^{3+} + 6\ H_2O \quad (4.61)$$

(ii) Also the stability field of the epidote solid solution exhibits a certain expansion in comparison to that of pure clinozoisite, especially to disadvantage of the stability field of wairakite and laumontite, whereas the clinozoisite-prehnite boundary experiences a small shift. This behavior is partly attributable to the fact that the average activity of clinozoisite, 0.703, is similar to that of prehnite, 0.763, but is significantly lower than those of wairakite, 0.891, and laumontite, which is assumed to be equal to 1. However, the differential expansion of the stability field of clinozoisite is also controlled, in part, by the stoichiometry of relevant reactions, as pointed out by the following relations:

$$\log\ f_{CO_2,Czo\text{-}Prh} = \log\ K_{Cal} + \log\ K_{Czo} - 1.5 \cdot \log\ K_{Prh} \quad (4.62)$$

$$\log\ f_{CO_2,Wrk\text{-}Czo} = \log\ K_{Cal} + 3 \cdot \log\ K_{Wrk} - 2 \cdot \log\ K_{Czo} \quad (4.63)$$

$$\log\ f_{CO_2,Lmt\text{-}Czo} = \log\ K_{Cal} + 3 \cdot \log\ K_{Lmt} - 2 \cdot \log\ K_{Czo} \quad (4.64)$$

linking the logarithm of CO_2 fugacity to the log K of the dissolution reactions of clinozoisite, wairakite, and laumontite, which are as follows:

$$Czo + 3\ Qtz/Chc + 13\ H^+ = 2\ Ca^{2+} + 3\ Al^{3+} + 7\ H_2O \quad (4.65)$$

$$Wrk + 4\ Qtz/Chc + 8\ H^+ = Ca^{2+} + 2\ Al^{3+} + 6\ H_2O \quad (4.66)$$

$$Lmt + 4\ Qtz/Chc + 8\ H^+ = Ca^{2+} + 2\ Al^{3+} + 8\ H_2O. \quad (4.67)$$

as well as by the log K of the dissolution reactions of calcite, Eq. (4.59), and prehnite, Eq. (4.60).

Since hydrothermal quartz, calcite, adularia, albite, anorthite, and laumontite occur as virtually pure solid phases, they can be involved without any particular worries and precautions in the theoretical, activity-based geothermometers and f_{CO_2}-indicators representing the aim of this work.

Also the use of white mica, epidote, prehnite, and wairakite for theoretical geothermometry and f_{CO_2} evaluation does not pose particular problems because these minerals exhibit limited compositional variations and, consequently, minor deviations are generally expected from their average activities.

The involvement of chlorite and garnet in theoretical geothermometers and f_{CO_2}-indicators is instead a risky business because these minerals show considerable compositional variations and, therefore, large deviations may occur from their average activities.

Moreover, it must be recalled that (i) hydrothermal feldspars present in active geothermal systems are most likely triclinic, fully ordered, albite and variably ordered adularia, from fully ordered microcline to completely disordered sanidine, with either triclinic or monoclinic symmetry (see Sect. 4.2.3) and (ii) microcline and sanidine have different thermodynamic properties (see Sect. 4.2.5).

The thermodynamic data of Helgeson et al. (1978) are used in this book because they resulted to be consistent with the monoclinic/triclinic transition temperatures of alkali feldspars whereas the thermodynamic data of Holland and Powell (1998) and Arnórsson and Stefánsson (1999) did not, as discussed in Sect. 4.2.4. Unfortunately, as shown in previous sections, there are large differences between the thermodynamic data of Helgeson et al. (1978) and those of Holland and Powell (1998, 2011), for most relevant solid phases, although these three databases are internally consistent. Owing to these large differences, it is meaningless to select data from distinct databases because it would be like mixing apples and oranges, as already recalled above.

In principle, the preparation of a common thermodynamic database for the whole scientific community would be highly desirable. In practice, it is a prohibitive challenge that is probably doomed to remain a chimera for a certain lapse of time, hopefully not too long.

References

Aagaard P, Helgeson HC (1983) Activity/composition relations among silicates and aqueous solutions: II. Chemical and thermodynamic consequences of ideal mixing of atoms on homological sites in montmorillonites, illites, and mixed-layer clays. Clay Clay Miner 31:207–217

Aagaard P, Jahren JS (1992) Diagenetic illite-chlorite assemblages in arenites. II. Thermodynamic relations. Clay Clay Miner 40:547

Akasaka M, Hashimoto H, Makino K, Hino R (2003) 57Fe Mössbauer and X-ray Rietveld studies of ferrian prehnite from Kouragahana, Shimane Peninsula, Japan. J Minerol Petrol Sci 98:31–40

Akizuki M (1987) Al, Si order and the internal texture of prehnite. Can Mineral 25:707–716

Albee AL (1962) Relationships between the mineral association, chemical composition and physical properties of the chlorite series. Am Mineral 47:851–870
Armbruster T, Gunter ME (2001) Crystal structures of natural zeolites. In: Bish DL, Ming DW (eds) Natural zeolites: occurrence, properties, applications. Reviews in mineralogy and geochemistry, vol 45, pp 1–67
Arnórsson S (1978) Major element chemistry of the geothermal sea-water at Reykjanes and Svartsengi. Iceland. Mineral. Mag. 42:209–220
Arnórsson S (1995) Geothermal systems in Iceland: structure and conceptual models—I. High-temperature areas. Geothermics 24:561–602
Arnórsson S, Gunnlaugsson E (1985) New gas geothermometers for geothermal exploration—calibration and application. Geochim Cosmochim Acta 49:1307–1325
Arnórsson S, Stefánsson A (1999) Assessment of feldspar solubility constants in water in the range of 0° to 350°C at vapor saturation pressures. Am J Sci 299:173–209
Arnórsson S, Gunnlaugsson E, Svavarsson H (1983a) The chemistry of geothermal waters in Iceland. III. Chemical geothermometry in geothermal investigations. Geochim Cosmochim Acta 47:567–577
Arnórsson S, Gunnlaugsson E, Svavarsson H (1983b) The chemistry of geothermal waters in Iceland. II. Mineral equilibria and independent variables controlling water compositions. Geochim Cosmochim Acta 47:547–566
Artioli G, Quartieri S, Deriu A (1995) Spectroscopic data on coexisting prehnite-pumpellyite and epidote-pumpellyite. Can Mineral 33:67–75
Badaut D, Decarreau A, Besson G (1992) Ferripyrophyllite and related Fe^{3+}-rich 2:1 clays in recent deposits of Atlantis II Deep, Red Sea. Clay Miner 27:227–244
Bailey SW (1988) Chlorites: structures and crystal chemistry. In: Bailey SW (ed) Hydrous Phyllosilicates (Exclusive of Micas). Reviews Mineralogy, vol 19, pp 347–403
Balducci S, Chelini W (1992) Hydrothermal equilibria in the active Mofete geothermal system, Phlegraean Fields, Naples, Italy. Acta Vulcanologica Marinelli 2:17–34
Ballantyne JM (1978) Hydrothermal alteration at the roosevelt hot springs thermal area, Utah: modal mineralogy, and geochemistry of sericite, chlorite, and feldspar from altered rocks. Thermal Power Company well Utah State 14-2. Topical Report, Dept. of Geology and Geophysics, Utah Univ., Salt Lake City, USA, 42 pp
Bargar KE, Beeson MH (1981) Hydrothermal alteration in research drill hole Y-2, Lower Geyser Basin, Yellowstone National Park, Wyoming. Am Mineral 66:473–490
Bargar KE, Beeson MH (1985) Hydrothermal alteration in research drill hole Y-3, Lower Geyser Basin, Yellowstone National Park, Wyoming. U.S. Geological Survey Professional Paper 1054-C, 23 pp
Bargar KE, Keith TEC, Trusdell FA, Evans SR, Sykes ML (1996) Hydrothermal alteration mineralogy of SOH drill holes, Kilauea East Rift Zone geothermal area, Hawaii. U.S. Geological Survey Open-File Report No. 96-0010, 75 pp
Baur WH, Joswig W, Kassner D, Hofmeister W (1990) Prehnite: structural similarity of the monoclinic and orthorhombic polymorphs and their Si/Al ordering. J Solid State Chem 86:330–333
Bird DK, Helgeson HC (1980) Chemical interaction of aqueous solutions with epidote-feldspar mineral assemblages in geologic systems—I. Thermodynamic analysis of phase relations in the system CaO-FeO-Fe_2O_3-Al_2O_3-SiO_2-H_2O-CO_2. Am J Sci 280:907–941
Bird DK, Norton DL (1981) Theoretical prediction of phase relations among aqueous solutions and minerals: Salton Sea geothermal system. Geochim Cosmochim Ac 45:1479–1494
Bird DK, Spieler AR (2004) Epidote in geothermal systems. In: Liebscher A, Franz G (eds) Epidotes, Reviews in Mineralogy and Geochemistry, vol 56, pp 235–300
Bird DK, Schiffman P, Elders WA, Williams AE, McDowell SD (1984) Calc-silicate mineralization in active geothermal systems. Econ Geol 79:671–695

Bird DK, Cho M, Janik CJ, Liou JG, Caruso LJ (1988) Compositional, order-disorder, and stable isotope characteristics of Al-Fe epidote, State 2-14 drill hole, Salton Sea geothermal system. J Geophys Res Sol Earth 93:13135–13144

Bish DL, Boak JM (2001) Clinoptilolite-heulandite nomenclature. In: Bish DL, Ming DW (eds) Natural zeolites: occurrence, properties, applications. Reviews in mineralogy and geochemistry, vol 45, pp 207–216

Bishop BP, Bird DK (1987) Variation in sericite compositions from fracture zones within the Coso Hot Springs geothermal system. Geochim Cosmochim Acta 51:1245–1256

Bourdelle F, Parra T, Chopin C, Beyssac O (2013) A new chlorite geothermometer for diagenetic to low-grade metamorphic conditions. Contrib Mineral Petrol 165:723–735

Boyce AJ, Fulignati P, Sbrana A (2003) Deep hydrothermal circulation in a granite intrusion beneath Larderello geothermal area (Italy): constraints from mineralogy, fluid inclusions and stable isotopes. J Volcanol Geoth Res 126:243–262

Brigatti MF, Guggenheim S (2002) Mica crystal chemistry and the influence of pressure, temperature, and solid solution on atomistic models. Rev Mineral Geochem 46:1–97

Browne PRL (1970) Hydrothermal alteration as an aid in investigating geothermal fields. Geothermics 2:564–570

Browne PRL (1977) Occurrence and hydrothermal alteration of diabase, Heber geothermal field, Imperial Valley, California. UCR/IGPP-77/9 Report, 61 pp

Browne PRL (1978) Hydrothermal alteration in active geothermal fields. Annu. Rev Earth Pl. Sc. 6:229–248

Browne PRL, Ellis AJ (1970) The Ohaaki-Broadlands hydrothermal area, New Zealand: mineralogy and related geochemistry. Am J Sci 269:97–131

Capuano RM, Cole DR (1982) Fluid-mineral equilibria in a hydrothermal system, Roosevelt Hot Springs, Utah. Geochim Cosmochim Acta 46:1353–1364

Carpenter MA, Putnis A (1985) Cation order and disorder during crystal growth: some implications for natural mineral assemblages. In: Thompson AB, Rubie DC (eds) Metamorphic reactions: kinetics, textures, and deformation. Springer, New York, pp 1–26

Cathelineau M (1988) Cation site occupancy in chlorites and illites as function of temperature. Clay Miner 23:471–485

Cathelineau M, Izquierdo G (1988) Temperature-composition relationships of authigenic micaceous minerals in the Los Azufres geothermal system. Contrib Mineral Petr 100:418–428

Cathelineau M, Nieva D (1985) A chlorite solid solution geothermometer—the Los Azufres (Mexico) geothermal system. Contrib Mineral Petr 91:235–244

Cavarretta G, Tecce F (1987) Contact metasomatic and hydrothermal minerals in the SH2 deep well, Sabatini volcanic district, Latium, Italy. Geothermics 16:127–145

Cavarretta G, Gianelli G, Puxeddu M (1980) Hydrothermal metamorphism in the Larderello geothermal field. Geothermics 9:297–314

Cavarretta G, Gianelli G, Puxeddu M (1982) Formation of authigenic minerals and their use as indicators of the physicochemical parameters of the fluid in the Larderello-Travale geothermal field. Econ Geol 77:1071–1084

Cavarretta G, Gianelli G, Scandiffio G, Tecce F (1985) Evolution of the Latera geothermal system II: metamorphic, hydrothermal mineral assemblages and fluid chemistry. J Volcanol Geoth Res 26:337–364

Černý P, Chapman R (1986) Adularia from hydrothermal vein deposits: extremes in structural state. Can Mineral 24:717–728

Chang LLY (1965) Subsolidus phase relations in the systems $BaCO_3$-$SrCO_3$, $SrCO_3$-$CaCO_3$, and $BaCO_3$-$CaCO_3$. J Geol 73:346–368

Chiodini G, Cioni R, Guidi M, Marini L (1991) Chemical geothermometry and geobarometry in hydrothermal aqueous solutions: a theoretical investigation based on a mineral-solution equilibrium model. Geochim Cosmochim Acta 55:2709–2727

Chipera SJ, Apps JA (2001) Geochemical stability of natural zeolites. Rev Mineral Geochem 45:117–161

Cho M, Maruyama S, Liou JG (1987) An experimental investigation of heulandite-laumontite equilibrium at 1000 to 2000 bar P fluid. Contrib Mineral Petrol 97:43–50

Cho M, Liou JG, Bird DK (1988) Prograde phase relations in the State 2-14 Well metasandstones, Salton Sea geothermal field, California. J Geophys Res Sol Earth 93:13081–13103

Coombs DS (1952) Cell size, optical properties and chemical composition of laumontite and leonhardite. Am Mineral 37:812–830

Coombs DS (1955) X-ray observations of wairakite and non-cubic analcime. Mineral Mag 30:699–708

Crawford WA, Hoersch LA (1972) Calcite-aragonite equilibrium from 50°C to 150°C. Am Mineral 57:995–998

Dachs E, Geiger CA (2019) Thermodynamic behaviour of grossular–andradite, $Ca_3(Al_xFe^{3+}_{1-x})_2Si_3O_{12}$, garnets: a calorimetric study. Eur J Mineral 31:443–451

De Caritat P, Hutcheon I, Walshe JL (1993) Chlorite geothermometry: a review. Clay Clay Miner 41:219–239

Deer WA, Howie RA, Zussman J (1982) Rock forming minerals. Orthosilicates, vol 1A, 2nd edn. Geological Society of London, 919 pp

Deer WA, Howie RA, Zussman J (2001) Rock-forming minerals. Framework silicates: Feldspars, vol 4A, 2nd edn. Geological Society of London, 972 pp

Deer WA, Howie RA, Wise WS, Zussman J (2004) Rock-forming minerals. Framework silicates: silica minerals, feldspathoids, and the zeolites, vol 4B, 2nd edn. Geological Society of London, 982 pp

Droop GTR (1987) A general equation for estimating Fe^{3+} concentrations in ferromagnesian silicates and oxides from microprobe analyses, using stoichiometric criteria. Mineral Mag 51:431–435

Ellis AJ (1970) Quantitative interpretation of chemical characteristics of hydrothermal systems. Geothermics 2:516–528

Ellis AJ, Mahon WAJ (1977) Chemistry and geothermal systems. Academic Press, 392 pp

Ernst WG (1963) Significance of phengitic micas from low-grade schists. Am Mineral 48:1357–1373

Exley RA (1982) Electron microprobe studies of Iceland Research Drilling Project high-temperature hydrothermal mineral geochemistry. J Geophys Res Sol Earth 87(B8):6547–6557

Fishman NS, Turner CE, Brownfield IK (1995) Authigenic albite in a Jurassic alkaline, saline lake deposit, Colorado Plateau. Evidence for early diagenetic origin. U.S. Geological Survey Bulletin 1808, Evolution of sedimentary basins—San Juan Basin, P1-P13

Flehmig W (1977) The synthesis of feldspars at temperatures between 0–80°C, their ordering behaviour and twinning. Contrib Mineral Petr 65:1–9

Foster MD (1962) Interpretation of the composition and a classification of the chlorites: U.S. Geological Survey Professional Paper 414-A, 33 pp

Fournier RO (1991) Water geothermometers applied to geothermal energy. In: D'Amore F (co-ordinator) Application of geochemistry in geothermal reservoir development. UNITAR, pp 37–69

Fowler AP, Zierenberg RA, Schiffman P, Marks N, Friðleifsson GÓ (2015) Evolution of fluid–rock interaction in the Reykjanes geothermal system, Iceland: evidence from Iceland Deep Drilling Project core RN-17B. J Volcanol Geoth Res 302:47–63

Franz G, Liebscher A (2004) Physical and chemical properties of the epidote minerals. An introduction. In: Liebscher A, Franz G (eds) Epidotes. Reviews in mineral and geochemistry, vol 56, pp 1–82

Fulignati P, Malfitano G, Sbrana A (1997) The Pantelleria caldera geothermal system: data from the hydrothermal minerals. J Volcanol Geoth Res 75:251–270

Geiger CA, Dachs E, Vielreicher NM, Rossman GR (2018) Heat capacity and entropy behavior of andradite: a multi-sample and—methodological investigation. Eur J Mineral 30:681–694

Gianelli G, Ruggieri G (2002) Evidence of a contact metamorphic aureole with high-temperature metasomatism in the deepest part of the active geothermal field of Larderello, Italy. Geothermics 31:443–474

Gianelli G, Mekuria N, Battaglia S, Chersicla A, Garofalo P, Ruggieri G, Manganelli M, Gebregziabher Z (1998) Water–rock interaction and hydrothermal mineral equilibria in the Tendaho geothermal system. J Volcanol Geoth Res 86:253–276

Giggenbach WF (1980) Geothermal gas equilibria. Geochim Cosmochim Acta 44:2021–2032

Giggenbach WF (1984) Mass transfer in hydrothermal alterations systems. Geochim Cosmochim Acta 48:2693–2711

Giggenbach WF (1988) Geothermal solute equilibria. derivation of Na-K-Mg-Ca geoindicators. Geochim Cosmochim Acta 52:2749–2765

Giggenbach WF (1997) The origin and evolution of fluids in magmatic-hydrothermal systems. In: Barnes HL (ed) Geochemistry of hydrothermal ore deposits, 3rd edn. Wiley, New York, pp 737–796

Gottschalk M (2004) Thermodynamic properties of zoisite, clinozoisite and epidote. In: Liebscher A, Franz G (eds) Epidotes, reviews in mineralogy and geochemistry, vol 56, pp 83–124

Guidi M, Marini L, Scandiffio G, Cioni R (1990) Chemical geothermometry in hydrothermal aqueous solutions: the influence of ion complexing. Geothermics 19:415–441

Guidotti CV, Sassi FP (2002) Constraints on studies of metamorphic K-Na white micas. Rev Mineral Geochem 46:413–448

Hedenquist JW, Browne PR (1989) The evolution of the Waiotapu geothermal system, New Zealand, based on the chemical and isotopic composition of its fluids, minerals and rocks. Geochim Cosmochim Acta 53:2235–2257

Helgeson HC, Aagaard P (1985) Activity/composition relations among silicates and aqueous solutions; I, Thermodynamics of intrasite mixing and substitutional order/disorder in minerals. Am J Sci 285:769–844

Helgeson HC, Delany JM, Nesbitt HW, Bird DK (1978) Summary and critique of the thermodynamic properties of rock-forming minerals. Am J Sci 278A, 229 p

Henley RW, Ellis AJ (1983) Geothermal systems ancient and modern: a geochemical review. Earth-Sci Rev 19:1–50

Herdianita NR (2012) The evolution of the Darajat geothermal system, West Java—Indonesia. Ph.D. Thesis, University of Auckland, New Zealand, 264 pp

Holland TJ (1989) Dependence of entropy on volume for silicate and oxide minerals: A review and predictive model. Am Mineral 74:5–13

Holland TJB, Powell R (1998) An internally consistent thermodynamic data set for phases of petrological interest. J Metamorph Geol 16:309–343

Holland TJB, Powell R (2011) An improved and extended internally consistent thermodynamic dataset for phases of petrological interest, involving a new equation of state for solids. J Metamorp Geol 29:333–383

Inoue A, Meunier A, Beaufort D (2004) Illite-smectite mixed-layer minerals in felsic volcaniclastic rocks from drill cores, Kakkonda, Japan. Clays Clay Miner 52:66–84

Inoue A, Kurokawa K, Hatta T (2010) Application of chlorite geothermometry to hydrothermal alteration in Toyoha geothermal system, southwestern Hokkaido, Japan. Resour Geol 60:52–70

Jahren JS, Aagaard P (1992) Diagenetic illite-chlorite assemblages in arenites. I. Chemical evolution. Clay Clay Miner 40:540

Kastner M, Siever R (1979) Low temperature feldspars in sedimentary rocks. Am J Sci 279:435–479

Kastner M, Waldbaum DR (1968) Authigenic albite from Rhodes. Am Mineral 53:1579–1602

Keith TE, Bargar KE (1988) Petrology and hydrothermal mineralogy of U.S. Geological Survey Newberry 2 drill core from Newberry Caldera, Oregon. J Geophys Res Sol Earth 93(B9):10174–10190

Keith TE, Muffler LP, Cremer M (1968) Hydrothermal epidote formed in the Salton Sea geothermal system, California. Am Mineral 53:1635–1644

Kiseleva I, Navrotsky A, Belitsky IA, Fursenko BA (1996) Thermochemistry and phase equilibria in calcium zeolites. Am Mineral 81:658–667

Kristmannsdóttir H (1979) Alteration of basaltic rocks by hydrothermal activity at 100–300°C. In: Mortland MM, Farmer VC (eds) International clay conference 1978. Developments in sedimentology, vol. 27. Elsevier, Amsterdam, pp 359–367

Kristmannsdóttir H, Tomasson J (1976) Zeolite zones in geothermal areas in Iceland. Orkustofnun Report OS-JHD-7649, 11 pp

Li G, Peacor DR, Coombs DS, Kawachi Y (1997) Solid solution in the celadonite family: The new minerals ferroceladonite, $K_2Fe_2{}^{2+}Fe_2{}^{3+}Si_8O_{20}(OH)_4$, and ferroaluminoceladonite, $K_2Fe_2{}^{2+}Al_2Si_8O_{20}(OH)_4$. Am Mineral 82:503–511

Liakopoulos A (1987) Hydrothermalisme et minéralisations métallifères de l'île de Milos (Cyclades, Grèce). Ph.D. Thesis Université de Paris VI, 276 pp

Libbey RB, Williams-Jones AE (2016) Compositions of hydrothermal silicates and carbonates as indicators of physicochemical conditions in the Reykjanes geothermal system, Iceland. Geothermics 64:15–27

Liebscher A (2004) Spectroscopy of epidote minerals. In: Liebscher A, Franz G (eds) Epidotes, Reviews in mineralogy and geochemistry, vol 56, pp 125–170

Liou JG (1970) Synthesis and stability relations of wairakite, CaAl2Si4O12 & #x22C5; 2H2O. Contrib Mineral Petrol 27:259–282

Liou JG (1971a) P-T stabilities of laumontite, wairakite, lawsonite, and related minerals in the system $CaAl_2Si_2O_8$-SiO_2-H_2O. J Petrol 12:379–411

Liou JG (1971b) Stilbite-laumontite equilibrium. Contrib Mineral Petr 31:171–177

Locock AJ (2008) An excel spreadsheet to recast analyses of garnet into end-member components, and a synopsis of the crystal chemistry of natural silicate garnets. Comput Geosci 34:1769–1780. Code available from server at http://www.iamg.org/CGEditor/index.htm

Lonker SW, Gerald JDF (1990) Formation of coexisting 1 M and 2 M polytypes in illite from an active hydrothermal system. Am Mineral 75:1282–1289

Lonker SW, Gerald JDF, Hedenquist JW, Walshe JL (1990) Mineral-fluid interactions in the Broadlands-Ohaaki geothermal system, New Zealand. Am J Sci 290:995–1068

Marini L, Canepa M, Cipolli F, Ottonello G, Vetuschi Zuccolini M (2001) Use of stream sediment chemistry to predict trace element chemistry of groundwater. A case study from the Bisagno valley (Genoa, Italy). J Hydrol 241:194–220

Marks N, Schiffman P, Zierenberg RA, Franzson H, Fridleifsson GÓ (2010) Hydrothermal alteration in the Reykjanes geothermal system: Insights from Iceland deep drilling program well RN-17. J Volcanol Geoth Res 189:172–190

Martin RF (1969) The hydrothermal synthesis of low albite. Contrib Mineral Petrol 23:323–339

Martinez-Serrano RG (2002) Chemical variations in hydrothermal minerals of the Los Humeros geothermal system, Mexico. Geothermics 31:579–612

Martinez-Serrano RG, Dubois M (1998) Chemical variations in chlorite at the Los Humeros geothermal system, Mexico. Clay Clay Miner 46:615–628

Mas A, Guisseau D, Mas PP, Beaufort D, Genter A, Sanjuan B, Girard JP (2006) Clay minerals related to the hydrothermal activity of the Bouillante geothermal field (Guadeloupe). J Volcanol Geoth Res 158:380–400

McCulloh TH, Frizzell VA, Stewart RJ, Barnes I (1981) Precipitation of laumontite with quartz, thenardite, and gypsum at Sespe Hot Springs, western Transverse Ranges, California. Clays Clay Miner 29:353–364

McDowell SD (1986) Composition and structural state of coexisting feldspars, Salton Sea geothermal field. Mineral Mag 50:75–84

McDowell SD, Elders WA (1980) Authigenic layer silicate minerals in borehole Elmore 1, Salton Sea geothermal field, California, USA. Contrib Mineral Petrol 74:293–310

McDowell SD, Elders WA (1983) Allogenic layer silicate minerals in borehole Elmore 1, Salton Sea geothermal field, California. Am Mineral 68:1146–1159

McIntire W (1963) Trace element partition coefficients- a review of theory and applications to geology. Geochim Cosmochim Acta 27:1209–1264

Mehegan JM, Robinson PT, Delaney JR (1982) Secondary mineralization and hydrothermal alteration in the Reydarfjordur drill core, eastern Iceland. J Geophys Res Scl Earth 87(B8):6511–6524
Milodowski AE, Savage D, Bath AH, Fortey NJ, Nancarrow PHA, Shepherd TJ (1989) Hydrothermal mineralogy in geothermal assessment: studies of Miravalles field, Costa Rica and experimental simulations of hydrothermal alteration. British Geological Survey Technical Report WE/89/63, 128 pp
Morad S (1978) Feldspars in sedimentary rocks. In: Sedimentology. Encyclopedia of earth science. Springer, Berlin
Muffler PLJ, White DE (1969) Active metamorphism of Upper Cenozoic sediments in the Salton Sea geothermal field and the Salton Trough, Southeastern California. Bull Geol Soc Am 80:157–182
Muramatsu Y, Doi N (2000) Prehnite as an indicator of productive fractures in the shallow reservoir, Kakkonda geothermal system, northeast Japan. J Miner Petrol Sci 95:32–42
Novak GA, Gibbs GV (1971) The crystal chemistry of the silicate garnets. Am Mineral 56:791–825
Ogino T, Suzuki T, Sawada K (1987) The formation and transformation mechanism of calcium carbonate in water. Geochim Cosmochim Acta 51:2757–2767
Openshaw RE, Hemingway BS, Robie RA, Waldbaum DR, Krupka KM (1976) The heat capacities at low temperatures and entropies at 298.15 K of low albite, analbite, microcline, and high sanidine. US Geol Surv J Res 4:195–204
Oscarson RL, Bargar KE (1996) Electron microprobe analyses of zeolite minerals from Neogene volcanic rocks in the Breitenbush-Austin Hot Springs area, Oregon. U.S. Geological Survey Open-File Report 96–41, 61 pp
Ottolini LP, Raffone N, Fridleifsson GÓ, Tonarini S, D'Orazio M, Gianelli G (2012) A geochemical investigation of trace elements in well RN-17 at Reykjanes geothermal system, SW-Iceland. In: IOP conference series: materials science and engineering, vol 32. IOP Publishing, pp 012020
Papike JJ, Zoltai T (1967) Ordering of tetrahedral aluminum in prehnite, Ca_2(Al, Fe+3)[Si3AlO10](OH)2. Am Mineral 52:974–984
Parsons I (2010) Feldspars defined and described: a pair of posters published by the Mineralogical Society.Sources and supporting information. Mineral Mag 74:529–551
Passaglia E, Sheppard RA (2001) The crystal chemistry of zeolites. In: Bish DL, Ming DW (eds) Natural zeolites: occurrence, properties, applications. Reviews in mineralogy and geochemistry, vol 45, pp 69–116
Patrier P, Beaufort D, Meunier A, Eymery JP, Petit S (1991) Determination of nonequilibrium ordering state in epidote from the ancient geothermal field of Saint Martin: application of Mössbauer spectroscopy. Am Mineral 76:602–610
Petrova VV (1970) Zeolites of the Paratunskii formation. In: Nabokc SI (ed) Mineralogy of the hydrothermal systems of Kamchatka and the Kurile Islands. Academy of Sciences USSR, Siberian Branch, pp 97–116 (in Russian)
Pirajino F (2009) Hydrothermal processes and mineral systems. Springer, Berlin, p 1250
Preisinger A (1965) Prehnit-ein neuer Schichtsilikattyp. Tscher Miner Petrog 10:491–504
Pryde AKA, Dove MT (1998) On the sequence of phase transitions in tridymite. Phys Chem Miner 26:171–179
Reeder R.J. (1990) Crystal chemistry of the rhombohedral carbonates. In: R. J. Reeder (Ed.), Carbonates: Mineralogy and chemistry. Rev. Mineral. Geochem., 11, 1–47
Reyes AG (1990) Petrology of Philippine geothermal systems and the application of alteration mineralogy to their assessment. J Volcanol Geoth Res 43:279–309
Ribbe PH (ed) (1983) Feldspar mineralogy. Mineralogical Society of America, Washington D.C. Reviews in mineralogy, vol 2, 362 pp
Rickwood PC (1968) On recasting analyses of garnet into end-member molecules. Contrib Mineral Petrol 18:175–198
Ruggieri G, Petrone CM, Gianelli G, Arias A, Henriquez ET (2006) Hydrothermal alteration in the Berlin geothermal field (El Salvador) new data and discussion on the natural state of the system. Period Mineral 75:293–312

Sanjuan B., Brach M., Lasne E. (2001) Bouillante geothermal fluid: mixing and water/rock interaction processes at 250°C. In: Cidu R. (Ed.), Water-Rock Interaction, WRI-10, International Symposium on Water-Rock Interaction, Jun 2001, Villasimius, Italy. A.A. Balkema, 2, 911–914

Sawaki T, Sasaki M, Fujimoto K, Takeno N (2001) Zinc-bearing actinolite from the Kakkonda geothermal system, Iwate Prefecture, northeastern Japan. Bull. Geol. Surv. Jpn 52:315–320

Schiffman P, Bird DK, Elders WA (1985) Hydrothermal mineralogy of calcareous sandstones from the Colorado River delta in the Cerro Prieto geothermal system, Baja California, Mexico. Mineral Mag 49:435–449

Seki Y, Onuki H, Okumura K, Takashima I (1969) Zeolite distribution in the Katayama geothermal area, Onikobe, Japan. Jpn J Geol Geogr 40:63–79

Seki Y, Liou JG, Guillemette R, Sakai H, Oki Y, Hirano T, Onuki H (1983) Investigation of geothermal systems in Japan I. Onikobe geothermal area. Hydroscience and Geotechnology Laboratory, Saitama University, Memoir No. 3, 206 pp

Shearer CK, Papike JJ, Simon SB, Davis BL, Laul JC (1988) Mineral reactions in altered sediments from the California State 2-14 well: Variations in the modal mineralogy, mineral chemistry and bulk composition of the Salton Sea Scientific Drilling Project core. J Geophys Res Sol Earth 93(B11):13104–13122

Shimazu M, Yajima J (1973) Epidote and wairakite in drill cores at the Hachimantai geothermal area, northeastern Japan. J Miner Petrol Sci 68:363–371

Shock EL, Helgeson HC, Sverjensky DA (1989) Calculation of the thermodynamic and transport properties of aqueous species at high pressures and temperatures: standard partial molal properties of inorganic neutral species. Geochim Cosmochim Acta 53:2157–2183

Smith JV, Brown WL (1988) Feldspar minerals. 1. Crystal structure, physical, chemical, and microtextural properties, 2nd edn. Springer, Berlin, 828 pp

Stefánsson A, Arnórsson S (2000) Feldspar saturation state in natural waters. Geochim Cosmochim Acta 64:2567–2584

Steiner A (1953) Hydrothermal rock alteration at Wairakei, New Zealand. Econ Geol 48:1–13

Steiner A (1955) Wairakite, the calcium analogue of analcime, a new zeolite mineral. Mineral Mag 30:691–698

Steiner A (1970) Genesis of hydrothermal K-feldspar (adularia) in an active geothermal environment at Wairakei, New Zealand. Mineral Mag 37:916–922

Sverjensky DA, Hemley JJ, D'Angelo WM (1991) Thermodynamic assessment of hydrothermal alkali feldspar-mica-aluminosilicate equilibria. Geochim Cosmochim Ac 55:989–1004

Teklemariam M, Battaglia S, Gianelli G, Ruggieri G (1996) Hydrothermal alteration in the Aluto-Langano geothermal field, Ethiopia. Geothermics 25:679–702

Thompson JB Jr (1969) Chemical reactions in crystals. Am Mineral 54:341–375

Thompson AB (1970) Laumontite equilibria and the zeolite facies. Am J Sci 269:267–275

Turekian KK, Wedepohl KH (1961) Distribution of the elements in some major units of the earth's crust. Geol Soc Am Bull 72:175–192

Velde B (1977) Clays and clay minerals in natural and synthetic systems. Developments in Sedimentology, vol 21. Elsevier, Amsterdam, p 217

Viereck LG, Griffin BJ, Schmincke HU, Pritchard RG (1982) Volcaniclastic rocks of the Reydarfjordur drill hole, eastern Iceland: 2. Alteration. J Geophys Res Sol Earth 87(B8):6459–6476

Walther JV, Helgeson HC (1977) Calculation of the thermodynamic properties of aqueous silica and the solubility of quartz and its polymorphs at high pressures and temperatures. Am J Sci 277:1315–1351

Weaver CE, Pollard LD (1973) The chemistry of clay minerals. Developments in sedimentology, vol 15. Elsevier, Amsterdam

Weissberg BG, Browne PR, Seward TM (1979) Ore metals in active geothermal systems. In: Barnes HL (ed) Geochemistry of hydrothermal ore deposits, 2nd edn. Wiley, New York, pp 738–780

Wheeler RS, Browne PRL, Rodgers KA (2001) Iron-rich and iron-poor prehnites from the Way Linggo epithermal Au-Ag deposit, southwest Sumatra, and the Heber geothermal field, California. Mineral Mag 65:397–406

Wiewióra A, Weiss Z (1990) Crystallochemical classifications of phyllosilicates based on the unified system of projection of chemical composition: II. The chlorite group. Clay Miner 25:83–92

Winchell AN (1933) Elements of optical mineralogy: an introduction to microscopic petrography. Part 2. Descriptions of minerals, with special reference to their optical and microscopic characters, 3rd edn. Wiley, New York, p 459

Zunić TB, Šćavničar S, Molin G (1990) Crystal structure of prehnite from Komiža. Eur J Mineral 2:731–734

Chapter 5
Traditional Water Geothermometers and f_{CO_2}-Indicators

Abstract Traditional water geothermometers and f_{CO_2}-indicators have been thoroughly reviewed in this chapter. The solubility of different silica minerals has been recalled, but the main focus has been placed on quartz solubility, as it plays a pivotal role in water geothermometry, underscoring the effects of grain size, pH, and salinity. The different Na-K geothermometric functions proposed by different authors have been examined. Since they are all plausible, there is not a unique Na-K geothermometer, but an infinite number of Na-K geothermometers which are controlled by the exchange reaction between low-albite and variably ordered adularia, from fully ordered maximum-microcline to completely disordered high-sanidine. Consequently, there is also an infinite number of the other cation geothermometers which are controlled by exchange reactions involving adularia. In particular, the Na-K-Ca geothermometer has been split in the three separate Na-K, Na-Ca, and K-Ca functions, following the suggestions of Tonani (1980). It turns out that the Na-Ca and K-Ca geothermometers work for waters equilibrated with an unspecified Ca-Al-silicate at relatively low f_{CO_2} values, whereas they do not work for waters equilibrated with calcite under comparatively high f_{CO_2} values. This implies that the Na^2/Ca and K^2/Ca ratios represent the basis not only for geothermometers but also for f_{CO_2}-indicators, which is somewhat different from the point of view of previous authors. Although Mg concentration decreases considerably with increasing temperature, the K-Mg and Na-Mg geothermometers suffer several limitations due to different reasons, including the variable composition of chlorites and illites and the varying order-disorder of adularia, impacting the K-Mg geothermometer much more than the Na-Mg geothermometer. Hence, some words of caution are needed on the use of the very popular Na-K-$Mg^{1/2}$ triangular plot.

5.1 General Aspects of Geothermometers and f_{CO_2}-Indicators

Geothermometers and f_{CO_2}-indicators (sometimes improperly called geobarometers) are simple functions in which the chemical data of geothermal water samples are

R. Cioni and L. Marini, *A Thermodynamic Approach to Water Geothermometry*, Springer Geochemistry, https://doi.org/10.1007/978-3-030-54318-1_5

inserted into to compute the temperature and f_{CO_2}, respectively, which are expected to be present in the zones of the geothermal aquifer where these geothermal water samples have presumably attained chemical equilibrium with hydrothermal minerals.

5.1.1 Basic Hypotheses of Geothermometry

As underscored by Fournier et al. (1974), "*There are many basic assumptions inherent in using geochemical indicators to estimate subsurface temperatures .... The usual assumptions are*:

1. *Temperature-dependent reactions occur at depth.*
2. *All constituents involved in a temperature-dependent reaction are sufficiently abundant (that is, supply is not a limiting factor).*
3. *Water-rock equilibration occurs at the reservoir temperature.*
4. *Little or no re-equilibration or change in composition occurs at lower temperatures as the water flows from the reservoir to the surface.*
5. *The hot water coming from deep in the system does not mix with cooler shallow ground water.*"

The f_{CO_2}-indicators rely on similar hypotheses. Often, the hypotheses 4 and 5 are not satisfied, especially for thermal springs. In fact, the geothermal liquid frequently mixes with shallow waters or separates a vapor phase or precipitates one or more mineral phases, during the ascent towards the surface. If so, the effects of these secondary processes have to be properly modeled for a correct application of geothermometers and f_{CO_2}-indicators as already recognized by Fournier and Truesdell (1974).

The purpose of this chapter is to review the traditional geothermometers and f_{CO_2}-indicators, both to provide a picture of the present state of the art and to underscore the needs for possible improvements.

5.1.2 Historical Overview

The first attempt to use silica as a water geothermometer was probably done by Bödvarsson (1960) and Bödvarsson and Palmason (1961) who adopted the concentration of dissolved SiO_2, c_{SiO_2}, of the thermal springs in the low temperature areas in Iceland as an indicator of aquifer temperature, T, using the simple relation (c_{SiO_2} in mg/kg; T in °C): $T = c_{SiO_2} - 25$. However, this can be considered a qualitative geothermometer at best.

Morey et al. (1962) investigated the solubility of quartz in water, in the temperature range from 25 to 300 °C, by means of laboratory experiments. They calibrated two quartz solubility relations, one at P_{H_2O} of 1000 atm, the other at water saturation pressure using the results of their own experiments and those of Kennedy (1950).

In the conclusive section of the paper, Morey et al. (1962) wrote that "*it is probable that the silica contents of many hot springs are controlled by equilibrium between dissolved silica and quartz at relatively shallow depths where temperatures are above* 150 °C. *The waters from many springs and geysers in Yellowstone National Park are unsaturated in silica with respect to silica gel at the vent temperatures,* 80°–95 °C. *However, these waters are saturated in silica with respect to the solubility of quartz at temperatures of* 175°–275 °C." For the first time, quartz solubility was used as a real geothermometer.

Mahon (1966) utilized the quartz solubility curves of Morey et al. (1962) and the dissolved silica concentration in waters discharged from the geothermal wells of Wairakei, New Zealand, suitably corrected for steam separation, to evaluate the temperatures present in the geothermal reservoir. Fournier and Rowe (1966) described the method for calculating aquifer temperatures from the silica concentration of high-discharge, boiling springs and wet-steam wells assuming equilibrium with quartz at depth. The quartz geothermometer was thus thoroughly established and tested in the 60s. Since then, dissolved silica in hot spring waters and geothermal well discharges has been extensively used as a tool to estimate aquifer temperature.

More or less at the same time, it was observed that the Na/K concentration ratio of thermal waters decreases with increasing temperature and tables and/or diagrams displaying the atomic Na/K ratios versus temperature were generated using both results of laboratory experiments and water compositions of geothermal wells and hot springs (White 1965, 1968; Ellis and Mahon 1967; Ellis 1970; Mercado 1970). The ion exchange reaction between coexisting Na- and K-feldspars was generally invoked to explain the Na/K ratio—temperature relation (e.g., White 1965; Ellis 1970). Nevertheless, the possible role of Na and K micas instead of one of the two feldspars or both of them was discussed by Fournier and Truesdell (1970). These authors underscored also the changes in the Na/K ratio during the ascent of thermal waters from the deep aquifer to the surface discharge due to mineral-solution reactions as well as the possible effects of complex ion formation.

The studies of hydrothermal alteration mineralogy (e.g., Browne 1970, 1978) and geochemical modeling of mineral-solution equilibria and irreversible mass transfer during water-rock interaction in high-temperature hydrothermal systems (e.g., Helgeson, 1968, 1969; Helgeson et al. 1969, 1970) provided further evidence on the role of temperature-dependent reactions involving hydrothermal minerals and aqueous species in controlling water geothermometers.

Based on this knowledge, the solubility of different SiO_2 polymorphs in water was used to calibrate silica geothermometers (e.g., Fournier and Rowe 1962; Fournier 1973; Arnórsson 1970, 1975; Fournier and Potter 1982a, b; Arnórsson et al. 1983a) and ion exchange reactions involving suitable hydrothermal minerals were invoked to derive or substantiate ionic solute geothermometers, such as the Na–K geothermometer (e.g., Fournier 1979; Arnórsson et al. 1983a; Giggenbach 1988), the Na–K–Ca geothermometer (Fournier and Truesdell 1973), and the K–Mg geothermometer (Giggenbach et al. 1983; Giggenbach 1988), as well as the K–Ca f_{CO_2}-indicator (Giggenbach 1984, 1988).

These simple functions have been used in a countless number of studies and are still widely used today although some limitations became apparent through the years. Finally, a theoretical justification for ionic solute geothermometers and f_{CO_2}-indicators was found by Guidi et al. (1990) and Chiodini et al. (1991) who elaborated on the early suggestion of Fournier and Truesdell (1970) concerning the possible influence of complex ion formation.

5.1.3 The Simple Form of Most Geothermometric Equations

Although polynomial equations were used for some geothermometers, such as the quartz solubility functions of Fournier and Potter (1982a, b), Von Damm et al. (1991), Manning (1994), Shibue (1996), and Gunnarsson and Arnórsson (2000) (see Sect. 5.2.2) and the Na–K geothermometer of Arnórsson (2000; see Sect. 5.3.1), most geothermometers are written in the simple form:

$$T(°C) = \frac{A}{\log K - B} - 273.15. \quad (5.1)$$

In Eq. (5.1), K is the thermodynamic equilibrium constant of the reaction controlling the geothermometer and the meaning of A and B constitutes the subject of this discussion. Suitable examples are the dissolution reactions of quartz and chalcedony:

$$\text{quartz} = SiO_{2(aq)} \quad (5.2)$$

$$\text{chalcedony} = SiO_{2(aq)} \quad (5.3)$$

and the albite-adularia exchange reaction:

$$\text{albite} + K^+ = \text{adularia} + Na^+. \quad (5.4)$$

The thermodynamic equilibrium constants of reactions (5.2), (5.3), and (5.4) can be written in the following simplified forms:

$$K_{Qtz} \cong m_{SiO_{2(aq)}} \quad (5.5)$$

$$K_{Chc} \cong m_{SiO_{2(aq)}} \quad (5.6)$$

$$K_{Ab-Adl} \cong m_{Na^+} / m_{K^+} \quad (5.7)$$

assuming that: (i) solid phases are pure (see Sect. 2.1.2), which is reasonable for silica minerals and alkali feldspars (see Sects. 4.2, and 4.10), (ii) the activity coefficient of dissolved silica is 1, which is realistic for neutral species (see Sect. 2.3.3), and (iii) the activity coefficients of Na^+ and K^+ ions are nearly equal, which is true (or nearly so) for ions of the same charge if the B-dot equation is used (see Sect. 2.3.3).

As already recalled in Sect. 2.1.4, the temperature dependence of the thermodynamic equilibrium constant K of any reaction is expressed by the van't Hoff relation, Eq. (2.22). Assuming that the standard isobaric heat capacity of the reaction, $\Delta C^{o}_{P,r}$, is equal to zero, Eq. (2.22) is easily integrated and rearranged, obtaining:

$$\log K_T = -\frac{\Delta H^o_r}{2.303 \cdot R \cdot T} + \log K_{T_r} + \frac{\Delta H^o_r}{2.303 \cdot R \cdot T_r}. \qquad (5.8)$$

The natural logarithm of the thermodynamic equilibrium constant and the standard Gibbs free energy of reaction, $\Delta G°_r$, are linked by Eq. (2.12), which can be written in the following form, for $T = T_r$ and adopting the decimal logarithm of K instead of its natural logarithm:

$$\log K_{T_r} = -\frac{\Delta G^o_r}{2.303 \cdot R \cdot T_r}. \qquad (5.9)$$

Let us now insert Eq. (5.9) into Eq. (5.8) and reorganize the obtained relation, recalling Eq. (2.1) rewritten for $T = T_r$:

$$\Delta G^o_r = \Delta H^o_r - T_r \cdot \Delta S^o_r. \qquad (5.10)$$

In this way, one finds:

$$\log K_T = -\frac{\Delta H^o_r}{2.303 \cdot R \cdot T} + \frac{\Delta S^o_r}{2.303 \cdot R}. \qquad (5.11)$$

Let us now solve Eq. (5.1) with respect to log K expressing the temperature in Kelvin degrees:

$$\log K = \frac{A}{T(K)} + B. \qquad (5.12)$$

Comparison of Eqs. (5.11) and (5.12) shows that:

$$A = -\frac{\Delta H^o_r}{2.303 \cdot R} \qquad (5.13)$$

$$B = \frac{\Delta S^o_r}{2.303 \cdot R}. \qquad (5.14)$$

In other terms, the linear relation between the logarithm of the thermodynamic equilibrium constant and the inverse of the absolute temperature implies that the standard enthalpy and entropy of reaction are constant. This findings are expected because: (i) $\Delta C_P{}^{\circ}{}_r$ was assumed equal to zero in the integration of the van't Hoff equation (see above) and (ii) $\Delta C_P{}^{\circ}{}_r$ is related to the enthalpy and entropy of reaction through the relations:

$$\left(\frac{\partial(\Delta H_r^o)}{\partial T}\right)_P = \Delta C_{P,r}^o \tag{5.15}$$

$$\left(\frac{\partial(\Delta S_r^o)}{\partial T}\right)_P T = \Delta C_{P,r}^o. \tag{5.16}$$

The condition $\Delta C_P{}^{\circ}{}_r = 0$ is usually true for isocoulombic reactions, that is for reactions having the same number of ionic species of the same charge on each side. In fact, reactions (5.2), (5.3), and (5.4), and similar reactions, are isocoulombic (Puigdomènech et al. 1997). Therefore, it is reasonable to assume that the log K of these reactions, and of similar reactions as well, depends linearly on the inverse of the absolute temperature. In other terms, it is permissible to adopt Eqs. (5.1) and (5.12) for the silica and Na–K geothermometers.

5.2 The Silica Geothermometers

According to Fournier (1991), in some geothermal reservoirs, well-crystalline quartz controls undissociated SiO_2 concentration at temperatures as low as 100 °C, probably because water has been in contact with rocks for comparatively long lapses of time, whereas chalcedony (i.e., microcrystalline quartz with sub-microscopic pores) governs undissociated SiO_2 concentration up to 180 °C in the portions of geothermal systems recently affected by fracturing. Chalcedony is more soluble than quartz because the small size of its microcrystals determines large surface energies.

According to Arnórsson et al. (1983a), quartz solubility governs undissociated SiO_2 concentration above about 180 °C, whereas geothermal waters circulating in aquifers of relatively low temperatures attain metastable equilibrium with chalcedony instead of quartz.

Other silica minerals, such as α-cristobalite, opal-CT, and amorphous silica, occur in geothermal systems and may control undissociated SiO_2 concentration especially at comparatively low temperatures and/or in acidic solutions.

5.2.1 *Constant-Enthalpy Relations Expressing the Solubility of Silica Minerals in Pure Water*

The logarithm of total dissolved silica concentration in pure water in equilibrium with vapor and a silica mineral (i.e., quartz, chalcedony, α-cristobalite, opal-CT, and amorphous silica) increases linearly with decreasing values of the absolute temperature inverse, as shown by the solid black lines in Fig. 5.1. In the same diagram, the dashed blue lines apply to waters that were initially in equilibrium with these silica minerals at the considered temperatures and have experienced single-step steam separation at 100 °C, or maximum steam loss.

The solid black straight lines in Fig. 5.1 correspond to the following geothermometric functions, which were proposed by Fournier (1977) and references therein and are valid in the temperature range 0–250 °C (T is temperature, in °C, and c_{SiO_2} is total dissolved silica concentration, in mg/kg):

$$T_{Qz,F} = \frac{1309}{5.19 - \log c_{SiO_2}} - 273.15, \quad \text{for quartz} \tag{5.17}$$

$$T_{Chc,F} = \frac{1032}{4.69 - \log c_{SiO_2}} - 273.15, \text{ for chalcedony} \tag{5.18}$$

$$T_{\alpha-Crs,F} = \frac{1000}{4.78 - \log c_{SiO_2}} - 273.15, \text{ for } \alpha\text{-cristobalite} \tag{5.19}$$

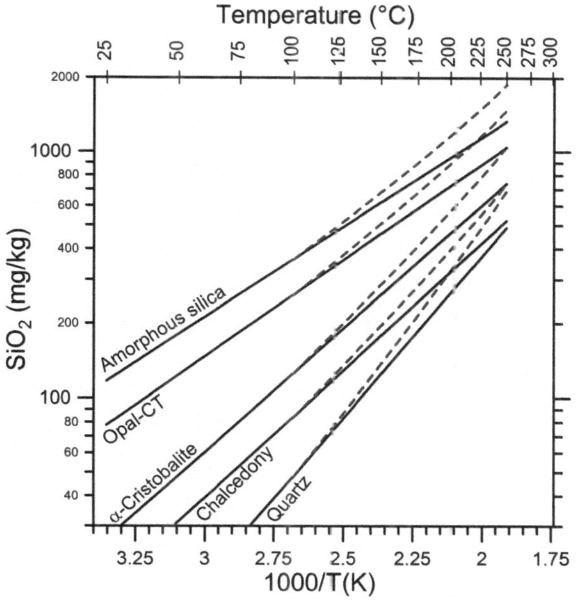

Fig. 5.1 Solid black lines indicate the solubility of several silica minerals in water, at the vapor pressure of the solution, as expressed by Eqs. (5.17)–(5.21). Dashed blue lines refer to the corresponding maximum steam loss relations, that is Eqs. (5.22)–(5.26). Equations (5.17)–(5.22) are from Fournier (1977) whereas Eqs. (5.23)–(5.26) were derived in this work

$$T_{Opl,F} = \frac{781}{4.51 - \log c_{SiO_2}} - 273.15, \text{ for opal - CT} \tag{5.20}$$

$$T_{A-SiO_2,F} = \frac{731}{4.52 - \log c_{SiO_2}} - 273.15, \text{ for amorphous silica.} \tag{5.21}$$

The plot of Fig. 5.1 and Eqs. (5.17)–(5.21) indicate that, among the silica minerals, amorphous silica has the highest solubilities, quartz has the lowest solubilities, while chalcedony, α-cristobalite, and opal-CT exhibit intermediate solubilities.[1]

Fournier (1977) and references therein proposed also the following maximum-steam-loss quartz geothermometer:

$$T_{Qz,msl,F} = \frac{1522}{5.75 - \log c_{SiO_2}} - 273.15. \tag{5.22}$$

Similar maximum-steam-loss geothermometric functions for the other silica minerals, corresponding to the dashed blue lines in Fig. 5.1, are as follows (this work):

$$T_{Chc,msl,F} = \frac{1227}{5.20 - \log c_{SiO_2}} - 273.15, \text{ for chalcedony} \tag{5.23}$$

$$T_{\alpha\text{-}Crs,msl,F} = \frac{1195}{5.29 - \log c_{SiO_2}} - 273.15, \text{ for } \alpha\text{-cristobalite} \tag{5.24}$$

$$T_{Opl,msl,F} = \frac{976}{5.02 - \log c_{SiO_2}} - 273.15, \text{ for opal - CT} \tag{5.25}$$

$$T_{Am.SiO_2,msl,F} = \frac{926}{5.03 - \log c_{SiO_2}} - 273.15, \text{ for amorphous silica.} \tag{5.26}$$

Based on data from geothermal wells in Iceland, Arnórsson et al. (1983a) performed an empirical calibration of the chalcedony geothermometer obtaining the following function:

$$T_{Chc,A} = \frac{1112}{4.91 - \log c_{SiO_2}} - 273.15, \tag{5.27}$$

which is valid from 25 to 180 °C. Arnórsson et al. (1983a) did not calibrate the quartz geothermometer, because their analytical data were inadequate, but recommended the use of the following function, in the temperature range 180–300 °C:

$$T_{Qz,A} = \frac{1164}{4.90 - \log c_{SiO_2}} - 273.15. \tag{5.28}$$

[1]Opal-CT was identified as β-cristobalite by Fournier (1977), but this erroneous identification was corrected in later papers.

Rimstidt (1997) presented results for quartz solubility measurements in pure water at 21, 50, 74, and 96 °C and 1 bar, in which equilibrium was attained from undersaturation. The duration of the two runs at 21 °C was 4917 days (~13.5 years) whereas the other runs lasted 54 to 210 days. Rimstidt's results indicate that quartz solubility at 25 °C is 11.0 ± 1.1 mg/kg, which is significantly higher than the results given by Eq. (5.17), 6.3 mg/kg, and Eq. (5.28), 9.9 mg/kg, although the latter should not be used at 25 °C.

Rimstidt (1997) reviewed the available experimental data for the temperature range 0–300 °C and obtained the following linear regression relationship (N = 71; $R^2 = 0.994$):

$$\mathrm{Log}\, m_{\mathrm{SiO_{2(aq)}}} = -1107.12(\pm 10.77) \cdot \mathrm{T}^{-1} - 0.0254(\pm 0.0247). \quad (5.29)$$

Equation (5.29) corresponds to the following geothermometric relation:

$$\mathrm{T_{Qz,R}} = \frac{1107.12}{4.753 - \log c_{\mathrm{SiO_2}}} - 273.15, \quad (5.30)$$

in which the uncertainties are disregarded.

Strictly speaking, undissociated silica concentration, $c_{\mathrm{SiO_{2(aq)}}}$, should appear in silica geothermometers instead of total dissolved silica concentration, $c_{\mathrm{SiO_2}}$, because the equilibrium condition between the aqueous solution and the considered silica mineral, $SiO_{2(s)}$, involves undissociated silica, $SiO_{2(aq)}$, as expressed by the following general reaction:

$$\mathrm{SiO_{2(s)} = SiO_{2(aq)}}. \quad (5.31)$$

Reactions (5.2) and (5.3) involve quartz and chalcedony, respectively, and are special cases of the general reaction (5.31). The thermodynamic equilibrium constant of reaction (5.31) can be considered nearly equal to $SiO_{2(aq)}$ molality:

$$\mathrm{K_{SiO_{2(s)}}} \cong m_{\mathrm{SiO_{2(aq)}}} \quad (5.32)$$

for the reasons already given above for Eqs. (5.5) and (5.6).

To investigate the relation between the concentrations of total dissolved silica and undissociated silica (or silicic acid) as a function of pH, we write the dissociation reaction of undissociated silica:

$$\mathrm{SiO_{2(aq)} + H_2O = H^+ + HSiO_3^-} \quad (5.33)$$

and the corresponding thermodynamic equilibrium constant:

$$\mathrm{K} = \frac{a_{\mathrm{HSiO_3^-}} \cdot a_{\mathrm{H^+}}}{a_{\mathrm{SiO_{2(aq)}}} \cdot a_{\mathrm{H_2O}}}. \quad (5.34)$$

Assuming that the aqueous solution is relatively dilute, then $a_{H_2O} = 1$. Hypothesizing that activity coefficients are equal or close to 1, and recalling that $a_{H^+} = 10^{-pH}$ and $K = 10^{-pK}$, Eq. (5.34) can be rearranged as:

$$\frac{m_{HSiO_3^-}}{m_{SiO_{2(aq)}}} = \frac{10^{-pK}}{10^{-pH}} = 10^{pH-pK} \tag{5.35}$$

Consequently, the following conditions hold true: $m_{HSiO_3^-} = m_{SiO_{2(aq)}}$ at pH = pK, $m_{HSiO_3^-} = 0.1 \cdot m_{SiO_{2(aq)}}$ at pH = pK − 1, $m_{HSiO_3^-} = 0.01 \cdot m_{SiO_{2(aq)}}$ at pH = pK − 2, and so on. For instance, since pK = 9.04 at 250 °C, it follows that $m_{SiO_{2(aq)}} = m_{HSiO_3^-}$ at pH = 9.04, $m_{HSiO_3^-} = 0.1 \cdot m_{SiO_{2(aq)}}$ at pH = 8.04, $m_{HSiO_3^-} = 0.01 \cdot m_{SiO_{2(aq)}}$ at pH = 7.04, and so on. Since, the pK of silicic acid dissociation varies from 10.3 to 8.8, in the 0–350 °C range (Fig. 5.2) and the pH of reservoir liquids is usually in the range 5–7, being constrained by water-rock reactions, in general, total dissolved SiO_2 concentration is nearly equal to undissociated SiO_2 concentration, under reservoir conditions.

However, loss of acid gases (CO_2 and H_2S) upon steam separation (boiling) increases significantly the pH of the liquid phase. Upon attainment of pH values higher than 7.8–9.3, depending on temperature, total dissolved silica concentration becomes significantly greater than undissociated silica concentration due to the presence of significant amounts of silicate ion produced through reaction (5.33). Therefore, it is advisable to insert undissociated silica concentration into silica geothermometers instead of total dissolved silica concentration, to get rid of these possible pH effects. The pH effect on quartz solubility was underscored *inter alias* by Crerar

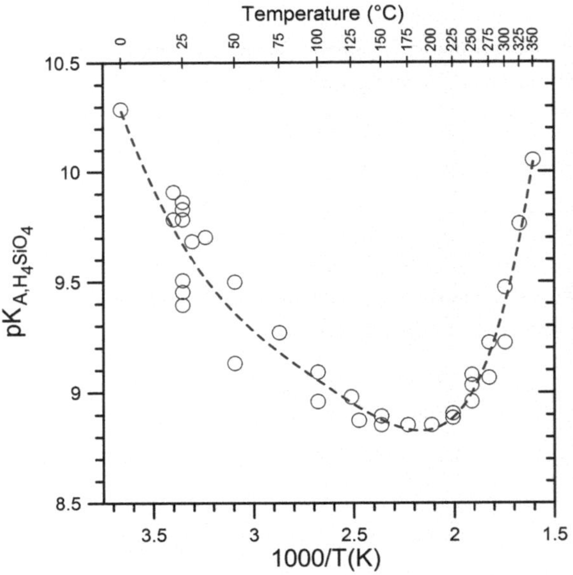

Fig. 5.2 Temperature dependence of the pK of the silicic acid dissociation (reaction 5.33), based on different experimental data (from Fleming and Crerar 1982 and references therein)

and Anderson (1971), who determined quartz solubility in the SiO_2–H_2O–NaOH system, and Arnórsson (1975), who discussed the increase in pH upon flashing due to extensive removal of dissolved CO_2 and H_2S from liquid water into the vapor phase.

The linear relation between the absolute temperature inverse and log c_{SiO_2} or log $c_{SiO_{2(aq)}}$, for all the silica geothermometers, is due to the fact that reaction (5.31) is isocoulombic, as already underscored in Sect. 5.1.3. Besides, based on Eq. (5.13):

(i) The slope of the quartz and chalcedony geothermometers of Fournier (1977), Eqs. (5.17) and (5.18), respectively, is consistent with $\Delta H°_r$ values of 5991 cal mol^{-1} and 4723 cal mol^{-1}, respectively, for the corresponding dissolution reactions.
(ii) The slope of the quartz and chalcedony geothermometric functions of Arnórsson et al. (1983a), Eqs. (5.27) and (5.28), respectively, is consistent with $\Delta H°_r$ values of 5327 cal mol^{-1} and 5089 cal mol^{-1}, respectively, for the dissolution reactions of these two silica minerals.
(iii) The slope of the quartz geothermometer of Rimstidt (1997), Eqs (5.30), is consistent with a $\Delta H°_r$ value of 5067 cal mol^{-1} for the dissolution reaction of quartz.

In Fig. 5.3, these enthalpies of quartz and chalcedony dissolution are compared with those predicted by SUPCRT92 (Johnson et al. 1992) as a function of temperature, at water saturation pressure, based on the thermodynamic properties of Helgeson et al. (1978). From 100 to 275 °C, the $\Delta H°_r$ values of the quartz and chalcedony geothermometers of Arnórsson et al. (1983a), the chalcedony geothermometer of Fournier (1977), and the quartz geothermometer of Rimstidt (1997) agree with the $\Delta H°_r$ values given by SUPCRT92 for quartz and chalcedony, 5150 ± 87 and 4782 ± 87 cal mol^{-1},

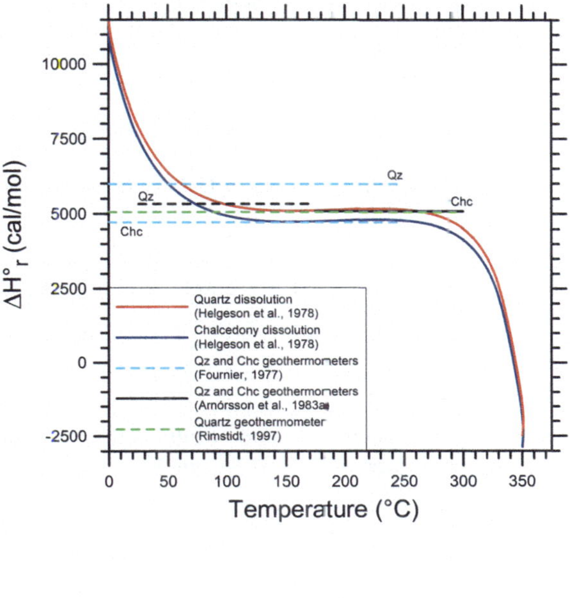

Fig. 5.3 Enthalpy of quartz and chalcedony dissolution predicted by SUPCRT92 as a function of temperature, at water saturation pressure, based on the thermodynamic properties of Helgeson et al. (1978) and $\Delta H°_r$ values consistent with the quartz and chalcedony geothermometers of Fournier (1977) and Arnórsson et al. (1983a) and the quartz geothermometer of Rimstidt (1997)

respectively, on average. In contrast, the $\Delta H°_r$ values of the quartz geothermometer of Fournier (1977) is higher than the $\Delta H°_r$ values of quartz dissolution computed using SUPCRT92 by 820 cal mol^{-1} on average.

To be noted that the $\Delta H°_r$ values of SUPCRT92 show limited variations in the temperature range 100–275 °C, whereas the SUPCRT92 enthalpies deviate progressively from the $\Delta H°_r$ values of silica geothermometers towards higher values below 100 °C and towards lower values above 275 °C. Therefore, based on SUPCRT92 predictions, the constant-enthalpy quartz and chalcedony geothermometers are expected to give increasingly wrong results both below 100 °C and above 275 °C.

5.2.2 Variable-Enthalpy Relations Expressing Quartz Solubility in Pure Water

Since the constant-enthalpy silica geothermometers do not describe properly the solubility of quartz and chalcedony below 100 °C and above 275 °C, variable-enthalpy relations were proposed by Fournier and Potter (1982a, b), Von Damm et al. (1991), Manning (1994), Shibue (1996), and Gunnarsson and Arnórsson (2000). The first four of these equations involve the density of H_2O or its molar volume and reproduce quartz solubility both in the liquid and vapor phases.

Fournier and Potter (1982a) used the experimental results obtained by different authors (Kennedy 1950; Morey and Hesselgesser 1951a; Wyart and Sabatier 1955; Khitarov 1956; Kitahara 1960; Van Lier et al. 1960; Morey et al. 1962; Siever 1962; Weill and Fyfe 1964; Anderson and Burnham 1965; Heitmann 1965; Stewart 1965, unpublished data; Semonova and Tsilkis, 1970; Crerar and Anderson 1971; Seward 1974; Hemley et al. 1980) at temperatures from 20 to 900 °C and at pressures up to 10 kbar, to express quartz solubility in pure water as a function of the absolute temperature and the specific volume of pure water, either liquid or gaseous, through the following relation:

$$\log m_{SiO_2} = A + B \cdot \log V + C \cdot (\log V)^2 \tag{5.36}$$

where V is the specific volume of H_2O ($cm^3\ g^{-1}$), and:

$$\begin{aligned} A &= -4.66206 + 0.0034063 \cdot T + 2179.7 \cdot T^{-1} \\ &\quad - 1.1292 \times 10^6 \cdot T^{-2} + 1.3543 \times 10^8 \cdot T^{-3} \\ B &= -0.0014180 \cdot T - 806.97 \cdot T^{-1} \\ C &= 3.9465 \times 10^{-4} \cdot T. \end{aligned}$$

Inserting the specific volume of pure water, at saturation temperatures and pressures, from Lemmon et al. (2017)[2] into Eq. (5.36), computed quartz solubility in the

[2]Data were obtained from the website https://webbook.nist.gov/chemistry/fluid/

liquid phase increases gradually with temperature, up to a maximum of 771.2 mg/kg at 339 °C, and decreases quickly above 360 °C, attaining a value close to 310 mg/kg at the critical temperature of pure water, 374 °C.[3] In contrast, the solubility of quartz in steam is less than 1 mg/kg below 270 °C and increases steeply close to the critical temperature, reaching 10.0 mg/kg at 322 °C, 30.7 mg/kg at 347 °C, 103.4 mg/kg at 368 °C, and 259.5 mg/kg at 373.9 °C.

In a separate scientific paper, Fournier and Potter (1982b) used the equation of Fournier and Potter (1982a) to compute the solubility of quartz in pure H_2O, at the vapor pressure of the solution, in both the liquid and gas (steam) phases, at 5 °C intervals up to the critical point. The obtained results are closely reproduced by the following polynomial function:

$$T_{Qz,FP} = C_1 + C_2 \cdot c_{SiO_2} + C_3 \cdot c_{SiO_2}^2 + C_4 \cdot c_{SiO_2}^3 + C_5 \cdot \log c_{SiO_2} \qquad (5.37)$$

in which $T_{Qz,FP}$ is in °C, $C_1 = -4.2198 \times 10^1$, $C_2 = 2.8831 \times 10^{-1}$, $C_3 = -3.6686 \times 10^{-4}$, $C_4 = 3.1665 \times 10^{-7}$, and $C_5 = 7.7034 \times 10^1$. The quartz geothermometer of Fournier and Potter (1982b), i.e., Eq. (5.37), can be applied to dilute aqueous solutions in the temperature range 20–330 °C.

As detailed in the next section, Von Damm et al. (1991) derived an equation describing quartz solubility in NaCl solutions of molality up to 0.5 m, which can be applied to pure water as well. Inserting the specific volume of pure water, at saturation temperatures and pressures, from Lemmon et al. (2017) into Eq. (5.41), calculated quartz solubility in liquid water increases gradually with temperature, up to a maximum of 739.0 mg/kg at 343 °C, and decreases rapidly above 360 °C, attaining a value close to 329 mg/kg at the critical temperature of pure water. In contrast, the solubility of quartz in steam is <1 mg/kg below 263 °C and increases steeply near the critical temperature, attaining 10.2 mg/kg at 319 °C, 30.0 mg/kg at 344 °C, 105.7 mg/kg at 367 °C, and 278.3 mg/kg at 373.9 °C.

Manning (1994) measured the solubility of quartz in pure water at pressures of 5 to 20 kbar and temperatures of 500 to 900 °C by means of crushed-quartz and single-crystal experiments. He considered the results of his experiments, those of Hemley et al. (1980) and Walther and Orville (1983) as well as the quartz solubilities computed at 25, 50, 100, and 150 °C, up to 5 kbar, by means of the equation of Fournier and Potter (1982a). Manning (1994) regressed all these data against T (in K) and ρ_{H_2O} (in g cm^{-3}) obtaining the following relation:

$$\log m_{SiO_2} = A(T) + B(T) \cdot \log \rho_{H_2O} \qquad (5.38)$$

where

$$A(T) = 4.2620 - \frac{5764.2}{T} + \frac{1.7513 \times 10^6}{T^2} - \frac{2.2869 \times 10^8}{T^3}$$

[3]The SiO_2 concentration at the critical point was computed by averaging the values of the liquid and vapor at 373.9 °C.

$$B(T) = 2.8454 - \frac{1006.9}{T} + \frac{3.5689 \times 10^5}{T^2}.$$

According to Manning (1994), Eq. (5.38) reproduces measured quartz solubilities up to 20 kbar and 900 °C and can be extrapolated to pressures higher than 20 kbar. Inserting the specific volume of pure water, at saturation temperatures and pressures, from Lemmon et al. (2017) into Eq. (5.38), quartz solubility in the liquid phase growths progressively with temperature, up to a maximum of 700 mg/kg at 341 °C, and decreases rapidly above 360 °C, attaining a value close to 263 mg/kg at the critical temperature of pure water. In contrast, the solubility of quartz in the gas phase is less than 1 mg/kg below 292 °C and rises suddenly near the critical temperature, reaching 10.3 mg/kg at 337 °C, 30.5 mg/kg at 356 °C, 104.1 mg/kg at 371 °C, and 215.3 mg/kg at 373.9 °C.

Shibue (1996) used 424 experimental data on quartz solubility in pure water from different studies (Kennedy 1950; Morey and Hesselgesser 1951a, b; Wyart and Sabatier 1955; Morey et al. 1962; Siever 1962; Weill and Fyfe 1964; Anderson and Burnham 1965, 1967; Crerar and Anderson 1971; Novgorodov 1975, 1977; Hemley et al. 1980; Ragnarsdottir and Walther 1983; Walther and Orville 1983; Manning 1994) and derived the following equation (T in K; P in bar; ρ_{H_2O} in g cm^{-3}):

$$\ln m_{SiO_2} = -1.51914 + 1.99970 \cdot \ln \rho_{H_2O} - \frac{2636.78}{T} + 0.00387766 \cdot T + \frac{0.0198582 \cdot P}{T} \quad (5.39)$$

which is valid up to 10 kbar and 900 °C. Inserting the density of pure water, at saturation temperatures and pressures, from Lemmon et al. (2017) into Eq. (5.39), quartz solubility in liquid water increments more and more with temperature, up to a maximum of 720.6 mg/kg at 341 °C, and decreases quickly above 360 °C, attaining a value close to 289 mg/kg at the critical temperature of pure water. In contrast, the solubility of quartz in steam is <1 mg/kg below 279 °C and increases abruptly approaching the critical temperature, attaining 10.3 mg/kg at 330 °C, 31.0 mg/kg at 352 °C, 108.3 mg/kg at 370 °C, and 239.5 mg/kg at 373.9 °C.

Gunnarsson and Arnórsson (2000) considered the experimental data on quartz solubility in water from several studies (Kennedy 1950; Fournier 1960; Kitahara 1960; Van Lier et al. 1960; Morey et al. 1962; Siever 1962; Crerar and Anderson 1971; Mackenzie and Gees 1971; Hemley et al. 1980; Rimstidt 1997; Gíslason et al. 1997) and proposed the following relation (T in K):

$$\log m_{SiO_2(aq)} = -34.188 + 197.47 \cdot T^{-1} - 5.851 \times 10^{-6} \cdot T^2 + 12.245 \cdot \log T. \quad (5.40)$$

Equation (5.40) is valid at water saturation pressure in the temperature range 0 to 350 °C.

All the functions expressing quartz solubility in pure water, both the constant-enthalpy relations presented in the previous section and the variable-enthalpy equations introduced in this section, are shown in the diagrams of SiO_2 concentration versus temperature of Fig. 5.4 and SiO_2 concentration (in logarithmic scale) versus the absolute temperature inverse of Fig. 5.5.

The solubility of quartz and chalcedony, at saturation temperatures and pressures, computed by means of the program SUPCRT92 (Johnson et al. 1992), based on the thermodynamic data of quartz and chalcedony from Helgeson et al. (1978) and those of aqueous SiO_2 from Shock et al. (1989), are also displayed in this diagram, together with the results of the polynomial Eq. (4.57) in Sect. 4.10.2, describing the solubility of a mechanical mixture constituted by 50% chalcedony and 50% quartz, called quartz/chalcedony in brief, at saturation temperatures and pressures.

Apart from the equation of Shibue (1996) below ~160 °C and the relations of Fournier (1977) and Fournier and Potter (1982a, b) below ~120 °C, all the functions describing quartz solubility in pure water are positioned between the solubility curves of quartz and chalcedony computed by means of SUPCRT92. This suggests that the differences between the experimental data of different authors might be ascribable,

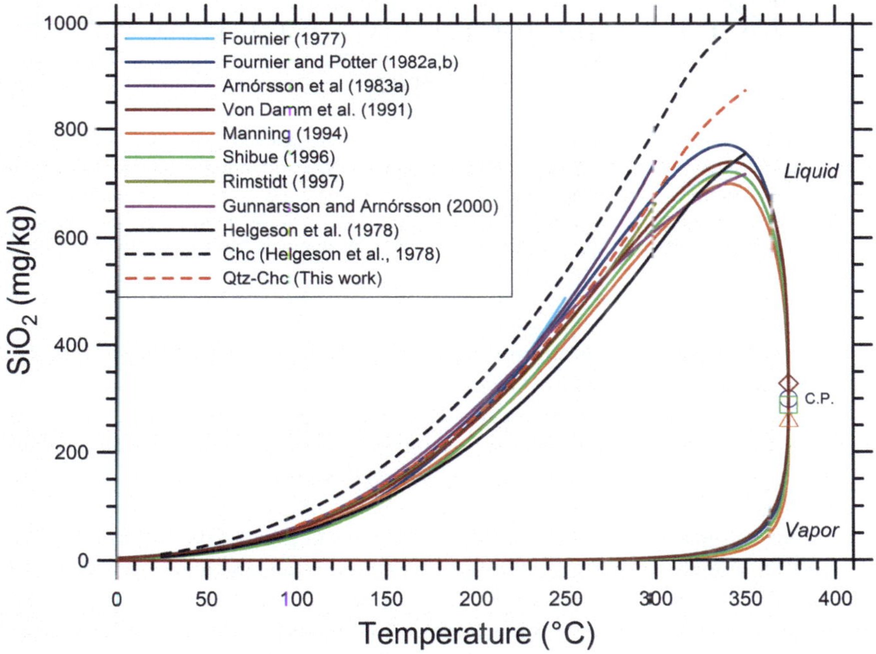

Fig. 5.4 Diagram of SiO_2 concentration versus temperature showing the constant-enthalpy and the variable-enthalpy equations expressing quartz solubility in pure water, at saturation temperatures and pressures. The blue circle, ruby red diamond, orange triangle, and green square indicate the SiO_2 concentration at critical point (C.P.) according to Fournier and Potter (1982a), Von Damm et al. (1991), Manning (1994), and Shibue (1996), respectively

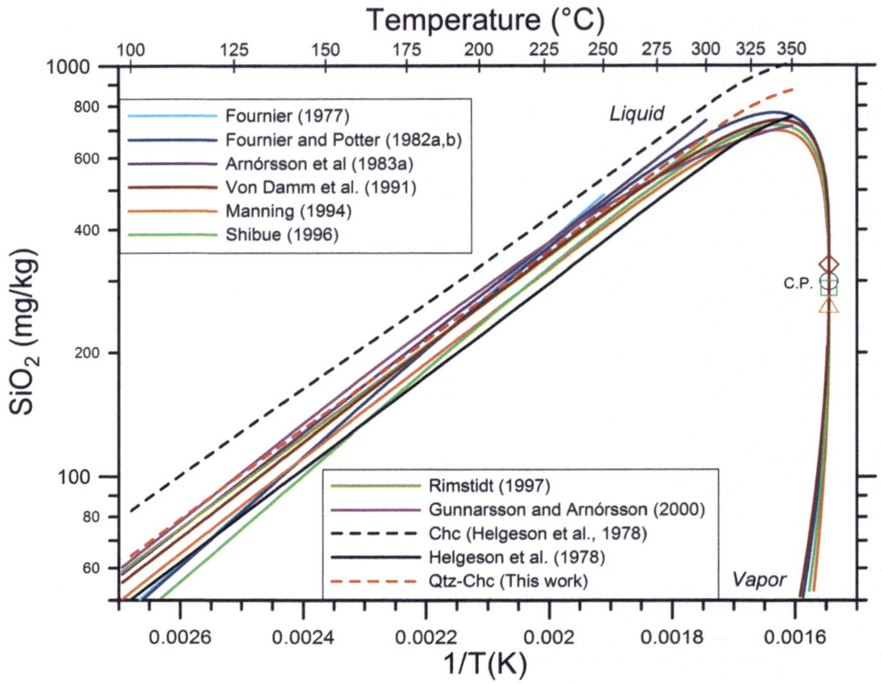

Fig. 5.5 Diagram of SiO_2 concentration (in logarithmic scale) versus absolute temperature inverse showing the constant-enthalpy and the variable-enthalpy equations describing quartz solubility in pure water, at saturation temperatures and pressures. The blue circle, ruby red diamond, orange triangle, and green square indicate the SiO_2 concentration at critical point (C.P.) according to Fournier and Potter (1982a), Von Damm et al. (1991), Manning (1994), and Shibue (1996), respectively

at least partly, to the different grain size of quartz crystals in equilibrium with the aqueous solution, as proposed by Azaroual et al. (1997) for the geothermal fluids of the Paris Basin. In fact, quartz microcrystals (i.e., chalcedony) are more soluble than quartz crystals, as already noted above. Accepting this explanation, it can be inferred that the quartz/chalcedony polynomial Eq. (4.57) describes the solubility of quartz crystals of average grain size, being representative of the central tendency.

The four variable-enthalpy relations of Fournier and Potter (1982a), Von Damm et al. (1991), Manning (1994), and Shibue (1996) reproduce the marked decrease in quartz solubility in the liquid phase and the steep increase in quartz solubility in the vapor phase occurring at temperatures approaching the critical point of pure water.

Moreover, the liquid and vapor branches of these four functions converge at the critical point, but there are considerable differences between the SiO_2 concentration value at the critical point, which varies from 263 to 329 mg/kg, as already noted above. In principle, SiO_2 concentrations in the liquid phase higher than the critical value correspond to two different temperatures of quartz equilibrium, one situated along the ascending part of the liquid branch of these four functions, the other positioned along the descending portion.

The other equations are characterized by a continuous increment of quartz solubility with increasing temperature and, consequently, they are unable to reproduce quartz solubility neither in the liquid phase at temperatures approaching the critical point nor in the vapor phase. Moreover, at temperatures higher than 250–275 °C there are large differences among these functions.

Consequently, the four variable-enthalpy equations of Fournier and Potter (1982a), Von Damm et al. (1991), Manning (1994), and Shibue (1996) can be applied up to the critical point of pure water, whereas the other relations cannot be used above temperatures varying from 250 to 350 °C, different from case to case.

5.2.3 Quartz Solubility in Salt Solutions

Fournier et al. (1982) measured the solubility of quartz in 2, 3, and 4 molal NaCl solutions at 350 °C and pressures varying from 180 to 500 bar using the gold bag hydrothermal apparatus of Seyfried et al. (1979). Quartz solubility in each of the considered NaCl solutions resulted to be greater than that in pure water over the whole pressure interval considered in the study, but the ratio between quartz solubility in NaCl solutions and in pure water decreases with increasing pressure. Fournier et al. (1982) underscored that the previous measurements of quartz solubility in NaCl solutions carried out by Kitahara (1960) are too high due to the high pH caused by reaction of elemental silver and NaCl with production of NaOH, AgCl, and H_2. According to Fournier et al. (1982), also quartz solubilities in NaCl solutions measured by Ganeyev (1975) might be affected by similar problems whereas quartz solubilities in NaCl solutions determined by Hemley et al. (1980) are reliable.

As anticipated in previous section, Von Damm et al. (1991) measured quartz solubility in seawater at temperatures of 200, 300, 363, 400, and 450 °C and pressures from 200 to 1000 bar performing the experiments in the modified Dickson apparatus of Seyfried et al. (1979). The 100 data thus obtained were merged with three data for 0.5 M NaCl (Hemley et al. 1980) and 391 for pure water from distinct sources (Kennedy 1950; Morey and Hesselgesser 1951b; Wyart and Sabatier 1955; Khitarov 1956; Kitahara 1960; Van Lier et al. 1960; Morey et al. 1962; Siever 1962; Weill and Fyfe 1964; Anderson and Burnham 1965; Heitmann 1965; Sommerfeld 1967; Semonova and Tsilkis 1970; Crerar and Anderson 1971; Ragnarsdottir and Walther 1983; Walther and Orville 1983). The whole dataset includes 494 quartz solubility data covering the temperature interval 45–900 °C and the pressure range <1–9860 bar. According to Von Damm et al. (1991), quartz solubility is described by the following equation:

$$\ln m_{SiO_2 \cdot bH_2O} = A + B \cdot \ln \rho_F + \left(C + D \cdot T^2\right) \cdot T^{-1} + E \cdot P \cdot T^{-} \quad (5.41)$$

in which ρ_F is density of the aqueous fluid in g cm^{-3}, P is pressure in bar, T is temperature in K, and coefficients are as follows: A = −2.32888, B = 1.79547, C = −2263.62, D = 0.00407350, and E = 0.0398808. In spite of the empirical derivation of Eq. (5.41), its terms are present in theoretical models describing quartz solubility. For instance: the coefficient A depends on the hydration number of dissolved silica, the weight fraction of free water, the activity coefficient of dissolved silica and water, as well as the effects of pressure on the activity coefficient of dissolved silica and water; the coefficient B is the hydration number of dissolved silica; the term $(C + D \cdot T^2)$ is equal to $-\Delta H/R$; the coefficient E is equal to $\Delta V/R$. Equation (5.41) may hold for ionic strengths higher than that of seawater, provided that coefficient A is suitably modified. However, Von Damm et al. (1991) did not test the model at high ionic strengths due to the lack of data.

Shibue (1996) used available experimental data to derive three empirical equations describing quartz solubility in pure water [i.e., Eq. (5.39) in previous section] as well as in $H_2O + CO_2$ and $H_2O + NaCl$ fluids. In particular, the relation for $H_2O + NaCl$ fluids was based on the experimentally measured quartz solubilities of Anderson and Burnham (1967), Novgorodov (1975), Hemley et al. (1980), and Fournier et al. (1982). For $H_2O + CO_2$ and $H_2O + NaCl$ fluids, the weight fraction of water in the solvent, F, the density of the aqueous fluid, ρ_F, and an empirical constant, Θ, were used as additional parameters. The ratio of quartz solubility in the aqueous fluid, $m_{SiO_2,F}$, to that in pure water, $m_{SiO_2,W}$, at the same P, T, was assumed to be expressed by the relation:

$$\ln\left(\frac{m_{SiO_2,F}}{m_{SiO_2,W}}\right) = \Theta \cdot \ln\left(\frac{\rho_F \cdot F}{\rho_{H_2O}}\right). \tag{5.42}$$

Densities of $H_2O + NaCl$ fluids were calculated by Shibue (1996) using the equations of Anderko and Pitzer (1993). The obtained equation describing quartz solubility in $H_2O + NaCl$ fluids is:

$$\ln m_{SiO_2,F} = A + B \cdot \ln \rho_F + C \cdot T^{-1} + D \cdot T + E \cdot P \cdot T^{-1} + \Theta \cdot \ln\left(\frac{\rho_F \cdot F}{\rho_{H_2O}}\right) \tag{5.43}$$

in which P is pressure in bar, T is temperature in K, densities are in g cm^{-3}, F is defined above, and coefficients are as follows: A = −1.51914, B = 1.99970, C = −2636.78, D = 0.00387766, E = 0.0198582, and Θ = 3.24740. The equation of Shibue (1996) for $H_2O + NaCl$ fluids differs from that of Von Damm et al. (1991) for the presence of an additional term, namely the last term on the right in Eq. (5.43), which is defined by Eq. (5.42) and does not appear in Eq. (5.41). Equation (5.43) applies to $H_2O + NaCl$ fluids with NaCl concentrations up to 33.6 wt%, at pressures from 0.2 to 2 kbar and temperatures from 200 to 700 °C.

To calculate the molality of aqueous SiO_2 in water-salt-CO_2 fluids in equilibrium with quartz, Akinfiev and Diamond (2009) proposed the following relation:

$$\log m_{SiO_2} = A(T) + B(T) \cdot \log \frac{18.0152}{V^*_{H_2O}} + 2 \cdot \log x_{H_2O} \quad (5.44)$$

in which A(T) and B(T) are the two polynomials of the equation of Manning (1994), i.e., Eq. (5.38), and x_{H_2O} and $V^*_{H_2O}$ are the mole fraction and the effective partial molar volume of H_2O in the fluid, respectively. In turn, $V^*_{H_2O}$ is calculated from the equation:

$$V_{mix} = x_{H_2O} \cdot V^*_{H_2O} + \sum x_s \cdot V_s \quad (5.45)$$

where V_{mix} is the molar volume of the fluid mixture (in $cm^3\ mol^{-1}$) and x_s and V_s are the mole fraction and the intrinsic volume of the s solute, respectively. For the binary system H_2O–NaCl, Eq. (5.45) simplifies as follows:

$$V_{mix} = x_{H_2O} \cdot V^*_{H_2O} + (1 - x_{H_2O}) \cdot V_{NaCl} \quad (5.46)$$

with the intrinsic volume of NaCl equal to $30.8 \pm 1.3\ cm^3\ mol^{-1}$. For pure water, the relation of Akinfiev and Diamond (2009), i.e., Eqn (5.44), becomes equal to that of Manning (1994) i.e., Eq. (5.38). The density model of Akinfiev and Diamond (2009) has been coded by Ronald Bakker into the program package LonerAP (Bakker 2003), which may be downloaded, free of charge, at the website: http://fluids.unileoben.ac.at/Computer.html.

We used the equations of Von Damm et al. (1991) and Akinfiev and Diamond (2009) to calculate quartz solubility in liquid seawater, at saturation temperatures and pressures, considering the molar volumes and densities of seawater from Bischoff and Rosenbauer (1985)[4]. Results are shown in Fig. 5.6, together with quartz solubility in pure water, at saturation temperatures and pressures, which was computed by means of the equations of Von Damm et al. (1991) and Manning (1994).

The two seawater curves mimic the corresponding pure water curves, apart from the flat portion close to the critical temperature of seawater, 408 °C (Bischoff 1991), which might be due to uncertainties in relevant data. Moreover, the two seawater curves depart progressively from the corresponding pure water curves, indicating that quartz becomes more and more soluble in seawater, compared to pure water, with increasing temperature. Some details are given below.

For instance, at 200 °C, quartz solubility increases by 6.4 mg/kg, from 259.3 mg/kg in pure water to 265.7 mg/kg in seawater, according to Von Damm et al. (1991), and by 5.4 mg/kg, from 237.7 mg/kg in pure water to 243.1 mg/kg in seawater, according to Akinfiev and Diamond (2009).

At 250 °C, quartz solubility growths by 24.1 mg/kg, from 436.4 mg/kg in pure water to 460.5 mg/kg in seawater, according to Von Damm et al. (1991), and by 23.0 mg/kg, from 404.8 mg/kg in pure water to 427.8 mg/kg in seawater, according to Akinfiev and Diamond (2009).

[4]The equation of Shibue (1996) was not used because it cannot be applied along the saturation curve of seawater.

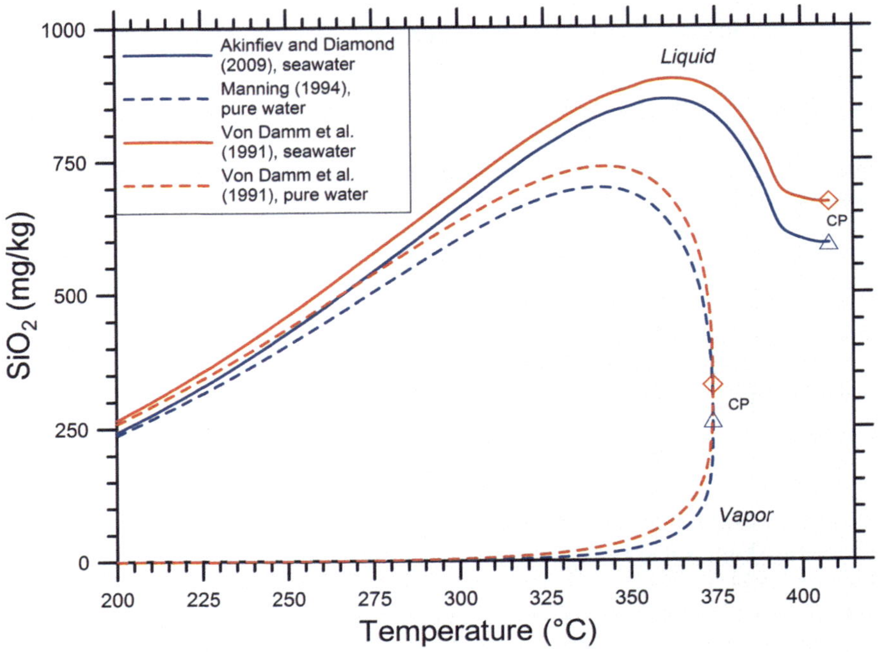

Fig. 5.6 Quartz solubility in liquid seawater, at saturation temperatures and pressures, computed by means of the equations of Von Damm et al. (1991) and Akinfiev and Diamond (2009), as a function of temperature. Quartz solubility in pure water, at saturation temperatures and pressures, computed by means of the equations of Von Damm et al. (1991) and Manning (1994) is also shown for comparison. The blue triangles and the red diamonds indicate the SiO_2 concentration at critical point (C.P.)

At 300 °C, quartz solubility increments by 59.3 mg/kg, from 636.7 mg/kg in pure water to 696.0 mg/kg in seawater, according to Von Damm et al. (1991), and by 59.4 mg/kg, from 602.0 mg/kg in pure water to 661.4 mg/kg in seawater, according to Akinfiev and Diamond (2009).

At 350 °C, quartz solubility increases by 153.0 mg/kg, from 732.6 mg/kg in pure water to 885.6 mg/kg in seawater, according to Von Damm et al. (1991), and by 160.8 mg/kg, from 689.7 mg/kg in pure water to 850.5 mg/kg in seawater, according to Akinfiev and Diamond (2009).

At 370 °C quartz solubility growths by 317.8 mg/kg, from 578.2 mg/kg in pure water to 896.0 mg/kg in seawater, according to Von Damm et al. (1991), and by 337.1 mg/kg, from 515.7 mg/kg in pure water to 852.8 mg/kg, according to Akinfiev and Diamond (2009).

Therefore, the increase in quartz solubility in seawater with respect to pure water is relatively similar in both cases, that is considering either the two curves of Von Damm et al. (1991) or the curves of Akinfiev and Diamond (2009) and Manning (1994), whereas the differences between the two seawater curves reflect essentially the differences between the two pure water curves, which are computed according to Von Damm et al. (1991) and Manning (1994), respectively.

5.2.4 The Preferred Silica Geothermometers

It is useful to compare the increase in quartz solubility in seawater with respect to pure water (see previous section) with the difference in solubility between quartz/chalcedony [polynomial Eq. (4.57), see Sect. 4.10.2] and quartz computed by means of the program SUPCRT92 (Johnson et al. 1992), using the thermodynamic data of quartz from Helgeson et al. (1978) and those of aqueous SiO_2 from Shock et al. (1989). All solubilities are at saturation temperatures and pressures.

At 200 °C, the difference in solubility between quartz/chalcedony and quartz, 47.7 mg/kg, is 7.5–8.8 times higher than the increase in quartz solubility in seawater with respect to pure water, 5.4–6.4 mg/kg.

At 250 °C, the difference in solubility between quartz/chalcedony and quartz, 72.8 mg/kg, is 3.0–3.2 times higher than the increase in quartz solubility in seawater with respect to pure water, 23.0–24.1 mg/kg.

At 300 °C, the difference in solubility between quartz/chalcedony and quartz, 101.6 mg/kg, is 1.7 times higher than the increase in quartz solubility in seawater with respect to pure water, 59.3–59.4 mg/kg.

At 350 °C, the situation inverts and the increase in quartz solubility in seawater with respect to pure water, 153.0–160.8 mg/kg becomes 1.3 times higher than the difference in solubility between quartz/chalcedony and quartz, 120.3 mg/kg.

Summing up, only at temperatures higher than 300 °C, the increment in quartz solubility caused by NaCl addition to pure water (~3.2 wt% corresponding to ~0.55 mol/kg) is more important than the increase in solubility due to the larger grain size of quartz crystal. Among the reservoir liquids considered in this work, the NaCl effect on quartz solubility is significant only for the high-salinity, high-temperature liquids of Salton Sea (with Cl concentrations in the range 116,000–162,000 mg/kg in most cases) and to, a lower degree, of Asal (with Cl concentrations of 66,800 to 74,600 mg/kg).

For these reasons, solubility of quartz/chalcedony was generally adopted, in this work, as preferred silica geothermometer for the neutral liquids of salinity significantly lower than the seawater value and temperature in the range 180–350 °C. Following Arnórsson et al. (1983a), the solubility of chalcedony was instead adopted for the same kind of liquids below 180 °C.

5.2.5 The Silica Versus Enthalpy Plot

As underscored by Fournier (1991), in data processing/interpretation, it is often advisable to use the diagram of silica versus enthalpy instead of the plot of silica versus temperature, because enthalpy can be involved in enthalpy balances (whereas this is not possible for temperature) and for other reasons given below. Therefore, the quartz solubility curve for pure water according to Fournier and Potter (1982a) was reported in the silica versus enthalpy plot of Fig. 5.7 by combining the silica concentrations given by Eq. (5.36), at saturation temperatures and pressures, with the enthalpies of pure water from the Steam Tables (Lemmon et al. 2017; see above) at the same conditions.

Besides, the curve expressing the solubility of quartz/chalcedony in pure liquid water [polynomial Eq. (4.57), see Sect. 4.10.2], again at saturation temperatures and pressures, was added to the diagram. As temperature increases above 313 °C, the quartz/chalcedony curve deviates more and more from the Fournier and Potter curve. In contrast, below 313 °C, the two curves are relatively close to each other, with differences in SiO_2 concentration lower than 20 mg/kg and average absolute deviation of 13 mg/kg.

In the silica versus enthalpy plot, the solubility of quartz in pure water (liquid and steam), at saturation temperatures and pressures, according to Fournier and

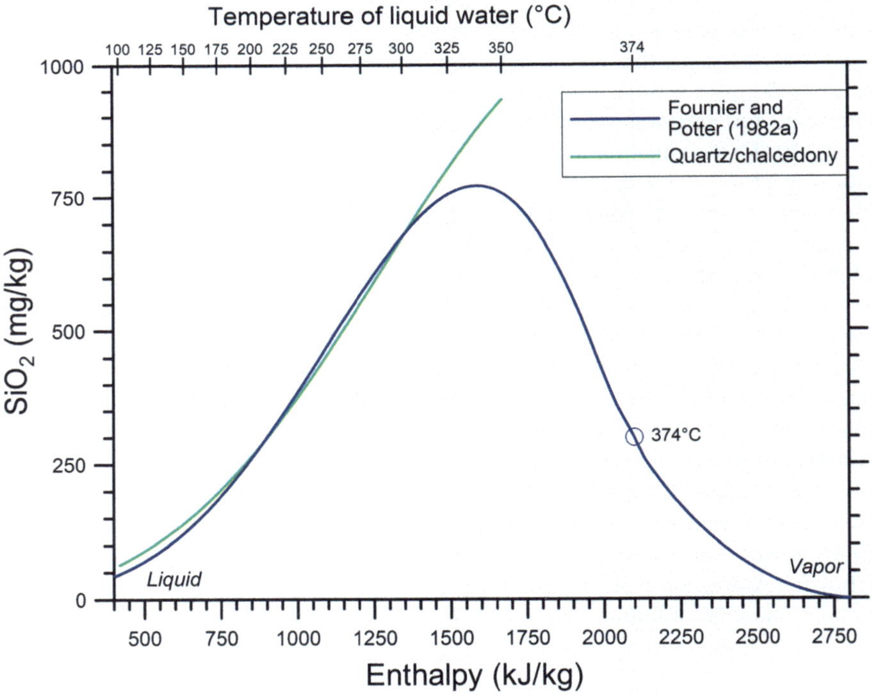

Fig. 5.7 Solubility of quartz in pure liquid water and in steam, at saturation temperatures and pressures, as a function of enthalpy (from Fournier and Potter 1982a). The blue circle indicates the SiO_2 concentration at critical point (C.P.). The solubility of quartz/chalcedony in pure liquid water, at saturation temperatures and pressures, is also shown

Potter (1982a), is represented by a bell-shaped symmetrical curve, which reaches its maximum of 771.2 mg/kg at 1587.4 kJ/kg, corresponding to 339 °C, and decreases with a further increase in enthalpy. Therefore, at a given enthalpy, there is only one value of dissolved silica, while at a given temperature there are two values of dissolved silica, one for the liquid phase and one for the steam phase (see Fig. 5.4). This is another good reason for using the diagram of silica versus enthalpy (Fig. 5.7) instead of the plot of silica versus temperature (Fig. 5.4).

Two possible quartz solubility curves in seawater, computed by means of the equations of Von Damm et al. (1991) and Akinfiev and Diamond (2009), and the related quartz solubility curves in pure water, computed using the functions of Von Damm et al. (1991) and Manning (1994), are shown in the silica versus enthalpy plot of Fig. 5.8.

Also the solubility of quartz in seawater is described by bell-shaped curves similar to that of pure water, apart from the flat part close to the critical temperature of

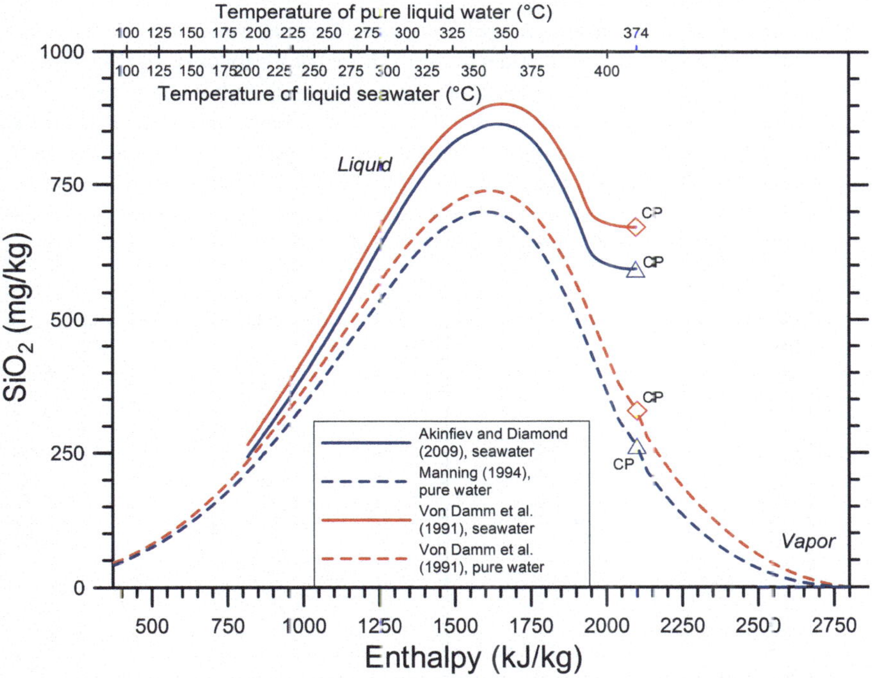

Fig. 5.8 Quartz solubility in liquid seawater, at saturation temperatures and pressures, computed using the equations of Von Damm et al. (1991) and Akinfiev and Diamond (2009), as a function of enthalpy. Quartz solubility in pure water, at saturation temperatures and pressures, calculated by means of the equations of Von Damm et al. (1991) and Manning (1994) is also shown for comparison. The blue triangles and the red diamonds indicate the SiO_2 concentration at critical point (C.P.)

seawater, 408 °C (Bischoff 1991), which might be due to uncertainties in pertinent data. As expected, the two seawater curves are situated above the two pure water curves because quartz solubility increases with increasing salt concentration, as already discussed in Sect. 5.2.3.

Quartz solubility curves in seawater can be of interest for high-salinity geothermal liquids. In most cases, however, the quartz solubility curve in pure water represents a suitable approximation and the basis for different geothermometric techniques, such as the silica mixing model (Truesdell and Fournier 1977), the silica boiling model (Fournier 1991), and the quartz geothermometry for wells with excess enthalpy (excess steam) discharges. These topics are treated in the next sections.

5.2.6 *The Silica Mixing Model*

Assuming conservation of enthalpy and conservative behavior for dissolved SiO_2, that is adopting the silica mixing model of Truesdell and Fournier (1977), the silica versus enthalpy plot reporting the quartz/chalcedony solubility curve for pure water can be used to determine the temperature of the thermal endmember involved in mixed water(s). In the ideal case, in which the cold endmember is known and the mixture is unaffected either by conductive heat losses or gain/loss of dissolved SiO_2, the intersection of the straight line drawn through the cold endmember and the mixed water with the quartz/chalcedony solubility curve gives the initial silica content and the enthalpy of the thermal endmember (Fig. 5.9 a).

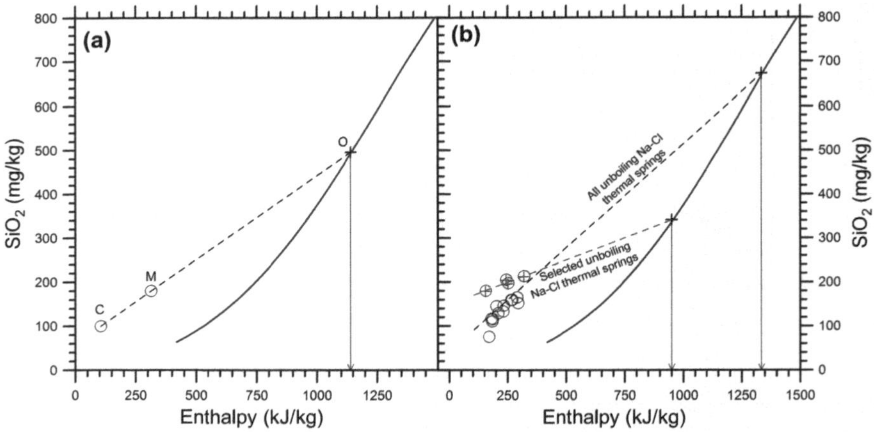

Fig. 5.9 Enthalpy-silica plot illustrating the application of the silica mixing model (Truesdell and Fournier 1977) to both **a** the ideal case in which the cold endmember C is known and the mixture M is unaffected either by conductive heat losses or gain/loss of dissolved SiO_2 and **b** the Na–Cl unboiling thermal springs of San Marcos, Guatemala (data from Marini et al. 1998)

This graphical exercise corresponds to solve a system constituted by the silica-enthalpy relation equivalent to the polynomial Eq. (4.57), see Sect. 4.10 2:

$$c_{SiO_2,O} = -20.195 + 0.21658 \cdot Ho - 2.9378 \times 10^{-4} \cdot H_0^2 + 7.2745 \times 10^{-7} \cdot Ho^3 - 2.5453 \times 10^{-10} \cdot Ho^4 \quad (5.47)$$

the enthalpy balance:

$$H_M = H_C \cdot (1 - x) + H_O \cdot x \quad (5.48)$$

and the mass balance:

$$c_{SiO_2,M} = c_{SiO_2,C} \cdot (1 - x) + c_{SiO_2,O} \cdot x. \quad (5.49)$$

In Eqs. (5.47)–(5.49), pedices O, C, and M refer to the reservoir water (i.e., the thermal endmember), the shallow water (i.e., the cold endmember), and the mixture, respectively, whereas x is the mass fraction of the thermal endmember in the mixture.

In real situations, mixed waters are more or less scattered due to conductive cooling and loss or (less likely) gain of dissolved SiO_2. Reconstruction of a reasonable mixing line requires use of linear regression through the most representative samples, which are presumably not affected (or least affected) by conductive cooling and loss (or gain) of dissolved SiO_2. In fact, if the mixtures are affected by heat losses through conduction, the enthalpy of the thermal endmember is overestimated, whereas if mixed waters are influenced by SiO_2 losses, the enthalpy of the thermal endmember is underestimated.

In the case of the Na–Cl unboiling thermal springs of San Marcos, Guatemala (data from Marini et al. 1998), the linear regression equation through all the samples intersects the quartz/chalcedony solubility curve at an enthalpy of 1335 kJ/kg, corresponding to a temperature of 298 °C (Fig. 5.9b). This temperature is too high for the thermal endmember, based on the results of other geothermometric techniques (Marini et al. 1998 and references therein). In contrast, the line drawn through the four springs less affected by SiO_2 loss (identified by the red crosses) leads to an estimated enthalpy of 950 kJ/kg for the thermal endmember. This enthalpy value corresponds to a temperature close to 220 °C, which is within the range indicated by the silica boiling model for the boiling springs present in the area (see next section).

5.2.7 The Silica Boiling Model

The silica versus enthalpy plot with the quartz/chalcedony solubility curve for pure water allows also a quick determination of the silica concentration and enthalpy of the initial liquid before boiling assuming (i) equilibrium between quartz/chalcedony and the aqueous solution under reservoir conditions, (ii) conservation of enthalpy during the travel of the liquid phase from the reservoir to the surface discharge point, which is a realistic assumption for the irreversible adiabatic process taking place in geothermal wells and high-flow rate natural geothermal manifestations, and (iii) occurrence of single-step vapor/liquid separation at surface conditions, i.e., 100 °C, 1.013 bar at sea level. Under these assumptions, the enthalpy and silica concentration of the initial liquid before boiling are given by the intersection of the line joining the boiled liquids and the steam point at 100 °C, 1.013 bar (code V) with the quartz/chalcedony solubility curve, as shown by the silica versus enthalpy plot for the boiling springs of San Marcos, Guatemala (Fig. 5.10).

This graphical exercise corresponds to solve a system constituted by the silica-enthalpy relation (5.47), the enthalpy balance:

$$H_O = H_L \cdot (1 - y) + H_V \cdot y, \tag{5.50}$$

and the mass balance:

$$c_{SiO_2,O} = c_{SiO_2,L} \cdot (1 - y) + c_{SiO_2,V} \cdot y, \tag{5.51}$$

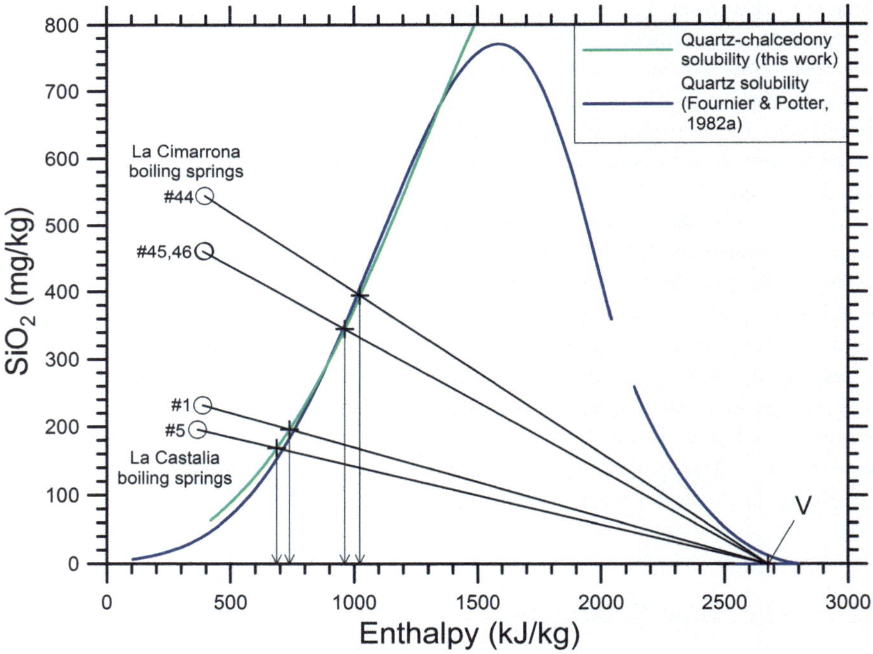

Fig. 5.10 Plot of silica versus enthalpy for the boiling springs of La Cimarrona and La Castalia, San Marcos area, Guatemala (circles, data from Marini et al., 1998), also showing both quartz solubility in pure water (from Fournier and Potter 1982a) and quartz/chalcedony solubility in pure water (this work), at saturation temperatures and pressures (see legend; modified from Fournier 1991)

where subscripts O, L, and V refer to the deep reservoir water, the separated liquid, and the separated vapor, respectively, whereas y is the mass steam fraction. At vapor/liquid separation conditions of 100 °C, 1.013 bar, $SiO_{2,V}$ can be assumed equal to zero and Eqs. (5.51) reduces to:

$$c_{SiO_2,O} = c_{SiO_2,L} \cdot (1 - y). \tag{5.52}$$

At these separation conditions, $H_L = 419.17$ kJ/kg and $H_V = 2675.6$ kJ/kg.

For La Cimarrona springs (# 44, 45, and 46) with $c_{SiO_2,L}$ of 461–545 mg/kg, H_O and $c_{SiO_2,O}$ turn out to be 961.1–1023.9 kJ/kg and 345.3–394.7 mg/kg, respectively, corresponding to quartz/chalcedony equilibrium temperatures of 224–237 °C. For La Castalia springs # 1 and #5, with $c_{SiO_2,L}$ of 232 and 196 mg/kg, the obtained H_O values are 737.8 and 686.4 kJ/kg and the resulting $c_{SiO_2,O}$ values are 196.4 and 168.8 mg/kg, corresponding to quartz/chalcedony equilibrium temperatures of 163 and 174 °C, respectively. These temperature values are moderately smaller than those proposed by Marini et al. (1998), i.e., 240 °C for La Cimarrona and 185 °C for La Castalia.

5.2.8 *Silica Geothermometry for Wells with Excess Enthalpy (Excess Steam) Discharges*

The excess enthalpy (excess steam) discharge is generally caused by the entry of excess steam into these wells, together with the reservoir liquid phase, although other models are possible as discussed by Arnórsson et al. (2007), see also Sect. 3.1.2. Since the excess steam dilutes the liquid, concentrations on a total-discharge basis decrease for non-volatile solutes and increase for gas species.

In many liquid-dominated systems, the excess steam comes from the vapor cap which is typically situated on top of the aquifer hosting the single liquid phase. Examples are Wairakei in New Zealand, Svartsengi in Iceland, and Olkaria in Kenya. In some systems, the steam cap is present in the natural state before the onset of exploitation, whereas in other systems, the steam cap develops due to pressure drawdown caused by the exploitation. The vapor cap is generally characterized by pressures and temperatures close to those of maximum enthalpy for saturated steam, 236 °C, 31.2 bar [5] (e.g., Ingebritsen and Sorey 1988; Arnórsson et al. 2007). Therefore, in the following discussion, the reservoir vapor is assumed to be at these P, T conditions and to have enthalpy of 2803.2 kJ/kg. Of course, the enthalpy of the reservoir vapor can be properly fixed if different data are available.

Let us assume also that liquid samples are collected from the weir box at atmospheric pressure. If so, the enthalpy of the separated steam is 2675.6 kJ/kg. Although the difference between these two enthalpies is less than 5%, it is important to consider

[5]The reason for these peculiar conditions in vapor-dominated systems were discussed by James (1968) and White et al. (1971) based on the pressure-enthalpy diagram for pure water.

the proper enthalpy values to minimize uncertainties on the quartz/chalcedony temperature of the reservoir fluid, which is hypothesized to be a pure saturated liquid phase.

The quartz/chalcedony geothermometric method is graphically presented in Fig. 5.11 for the sample collected from the Olkaria well OW-709 with discharge enthalpy of 1921 kJ/kg and SiO_2 concentration at the weir box of 649 mg/kg (Karingithi et al. 2010).

First, we plot the weir box sample on the silica versus enthalpy diagram associating the SiO_2 concentration to the enthalpy of the separated liquid phase, 419.17 kJ/kg (black triangle).

Second, we draw the black solid line connecting the separated liquid point and the separated steam point, with enthalpy of 2675.6 kJ/kg and SiO_2 concentration of 0 mg/kg (black square).

Third, we place the point representative of total discharge along this black solid line at enthalpy of 1921 kJ/kg, resulting in a SiO_2 concentration of 217.0 mg/kg (green circle).

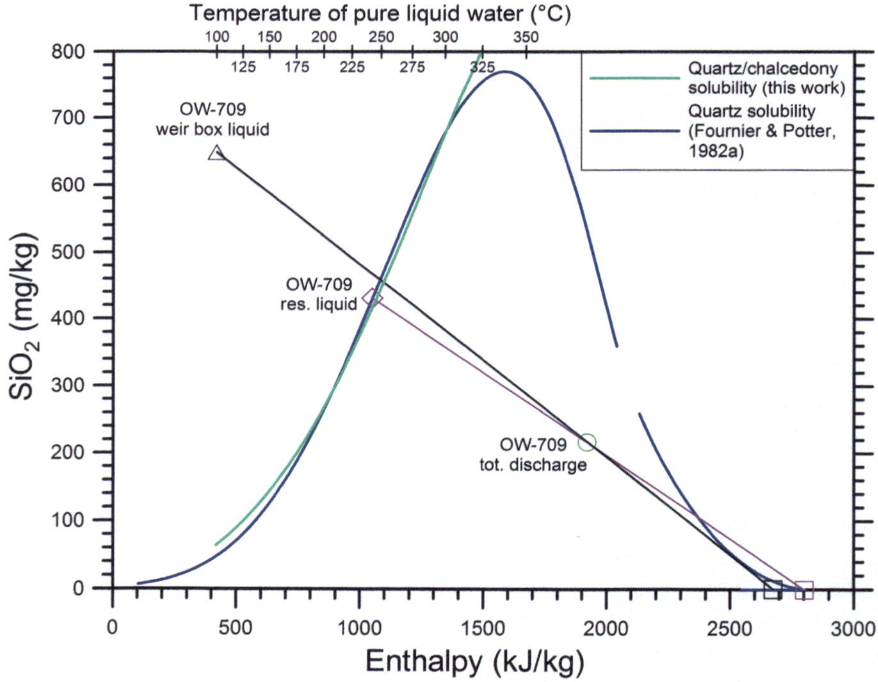

Fig. 5.11 Plot of silica versus enthalpy for the sample collected from the Olkaria well OW-709 (data from Karingithi et al. 2010), also showing both quartz solubility in pure water (from Fournier and Potter 1982a) and quartz/chalcedony solubility in pure water (this work), at saturation temperatures and pressures (see legend). The black square refers to the steam separated at atmospheric conditions. The pink square represents reservoir steam. See text for other explanations and discussion

Fourth, we draw the pink solid line from the reservoir steam point, with enthalpy of 2803.2 kJ/kg and SiO_2 concentration of 0 mg/kg (pink square) through the total discharge point (green circle). The intersection of this line and the quartz/chalcedony solubility curve (pink diamond) constrains the reservoir liquid enthalpy at 1064.18 kJ/kg and its silica concentration at 427.8 mg/kg, corresponding to a reservoir temperature of 245.5 °C. The mass fraction of reservoir vapor results to be 0.493.

Instead of this graphical exercise, we can first compute the SiO_2 concentration in total discharge, $c_{SiO_2,TD}$, using the following relation:

$$c_{SiO_2,TD} = c_{SiO_2,WB} \cdot (1 - y) \tag{5.53}$$

where $c_{SiO_2,WB}$ is the silica concentration in the separated liquid at the weir box and y is the mass steam fraction at the same atmospheric P,T conditions, which is calculated from total discharge enthalpy, H_{TD}:

$$y = \frac{H_{TD} - H_{L,WB}}{H_{V,WB} - H_{L,WB}}, \tag{5.54}$$

where indices L,WB and V,WB refer to the separated liquid at the weir box and the separated vapor at the weir box, respectively.

Second, we solve the system constituted by the silica-enthalpy relation (5.47), the silica mass balance:

$$c_{SiO_2,TD} = c_{SiO_2,RL} \cdot (1 - x) + c_{SiO_2,RV} \cdot x \tag{5.55}$$

and the enthalpy balance:

$$H_{TD} = H_{RL} \cdot (1 - x) + H_{RV} \cdot x \tag{5.56}$$

where subscripts RL and RV refer to the reservoir liquid and the reservoir vapor, respectively, and x stands for the mass fraction of reservoir vapor.

Table 5.1 lists the main results, including the mass fraction of reservoir vapor and the quartz/chalcedony temperature of the reservoir liquid, for all the Olkaria wells sampled by Karingithi et al. (2010), together with the input data used both in calculations and to prepare the silica versus enthalpy plot of Fig. 5.12.

Twenty five of the thirty samples have fraction of reservoir vapor varying from 0.1 and 0.9, whereas the remaining four samples have fraction of reservoir vapor varying between 0.095 and 0.001. Only ten samples have quartz/chalcedony temperatures, $T_{Qz/Chc}$, comparable with reported aquifer temperatures, T_R, within ± 5 °C. Thirteen samples have absolute differences between $T_{Qz/Chc}$ and T_R varying from 6 to 15 °C. Seven samples have even higher absolute differences between $T_{Qz/Chc}$ and T_R, ranging from 18 to 60 °C. Most of these differences are probably related to the definition of the aquifer temperature, which was computed by Karingithi et al. (2010) by taking the average of the temperatures given by the quartz and Na–K geothermometers.

Table 5.1 Results of the quartz/chalcedony geothermometer for the Olkaria wells sampled by Karingithi et al. (2010), most of which have high mass fractions of reservoir vapor, x. T_R is the reservoir temperature reported by the authors in their Table 4

Well	H_{TD}	$c_{SiO_2,WB}$	$y_{,WB}$	$c_{SiO_2,TD}$	H_{RL}	$c_{SiO_2,RL}$	x	$T_{Qz/Chc}$	T_R
	kJ/kg	mg/kg		mg/kg	kJ/kg	mg/kg		°C	°C
OW-02	1839	643	0.629	238.4	1066.15	429.5	0.445	245.9	251
OW-05	2599	624	0.966	21.2	764.10	211.5	0.900	180.4	240
OW-10	2531	773	0.936	49.5	950.50	337.2	0.853	221.4	262
OW-11	1894	597	0.654	206.8	1033.51	402.5	0.486	239.1	246
OW-15	2140	576	0.763	136.7	996.20	372.5	0.633	231.2	242
OW-16	1534	502	0.494	254.0	985.13	363.8	0.302	228.8	228
OW-19	1871	548	0.643	195.4	1002.45	377.5	0.482	232.5	229
OW-20	2541	778	0.940	46.4	940.50	329.7	0.859	219.3	254
OW-26	1881	657	0.648	231.4	1071.94	434.3	0.467	247.1	247
OW-28	2446	625	0.898	63.6	942.65	331.3	0.808	219.8	239
OW-29	2158	609	0.771	139.7	1014.68	387.2	0.639	235.1	243
OW-30i	2196	768	0.787	163.2	1099.77	458.0	0.644	252.8	262
OW-30ii	2196	701	0.787	149.0	1063.15	427.0	0.651	245.3	259
OW-10	2535	638	0.938	39.8	878.90	285.2	0.861	206.0	242
OW-15	1899	604	0.656	207.9	1037.67	405.9	0.488	239.9	246
OW-16	1384	573	0.428	328.0	1039.69	407.6	0.195	240.4	237
OW-19	1823	622	0.622	235.0	1054.04	419.4	0.440	243.4	253
OW-23	2191	653	0.785	140.2	1036.27	404.8	0.654	239.6	242
OW-25	2516	641	0.929	45.3	900.45	300.4	0.849	210.7	253
OW-202	1104	320	0.304	222.9	838.26	257.7	0.135	197.0	193
OW-301	1653	855	0.547	387.5	1195.28	541.7	0.285	272.0	262
OW-302	1234	744	0.361	475.3	1149.36	501.0	0.051	262.9	256
OW-304D	1672	364	0.555	161.9	867.10	277.1	0.416	203.4	190
OW-306	1037	551	0.274	400.1	1032.00	401.3	0.003	238.7	224
OW-709	1921	649	0.666	217.0	1064.18	427.8	0.493	245.5	245
OW-714	1303	739	0.392	449.5	1144.78	497.0	0.095	262.0	267
OW-719	1259	588	0.372	369.1	1052.87	418.4	0.118	243.1	241
OW-901	1854	529	0.636	192.6	990.32	367.9	0.476	229.9	220
OW-902	1108	477	0.305	331.4	976.50	357.1	0.072	227.0	209
OW-903	953	443	0.237	338.2	952.07	338.4	0.001	221.8	207

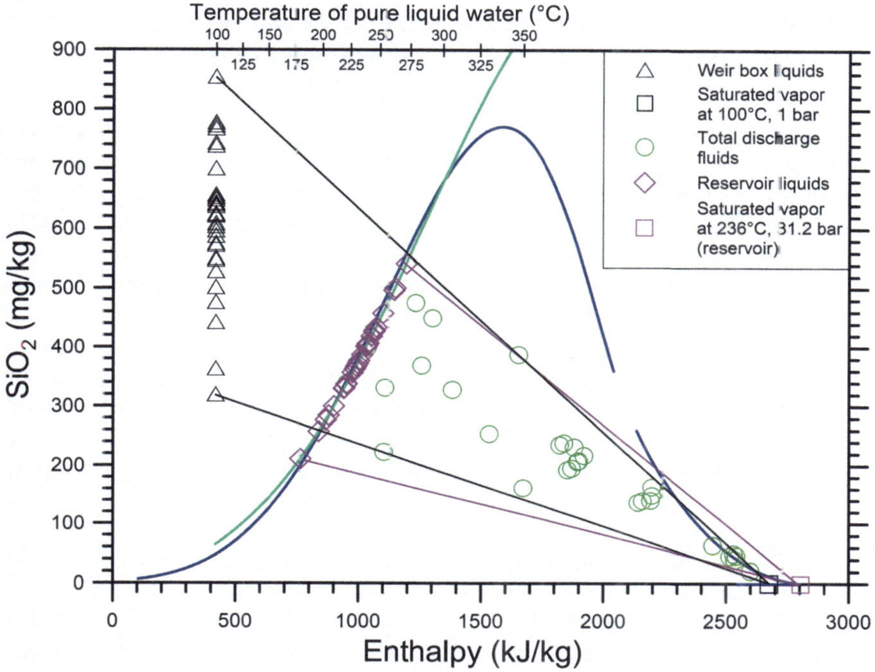

Fig. 5.12 Plot of silica versus enthalpy for all the samples collected from the Olkaria wells by Karingithi et al. (2010), also showing both quartz solubility in pure water (from Fournier and Potter 1982b; blue curves) and quartz/chalcedony solubility in pure water (this work; aquamarine curve), at saturation temperatures and pressures. The two solid black lines enclose the separated liquids collected at the weir box, whereas the two solid pink lines encompass the reservoir liquids. See text for other explanations and discussion

Finally, it must be underscored that the silica-enthalpy diagram and related calculations can be applied not only to wells with excess enthalpy discharges but also to wells discharging bi-phase fluids produced through boiling of single-phase reservoir liquids either unaffected by steam gain/loss or affected by steam loss. The well discharges unaffected by steam gain/loss are situated along the quartz/chalcedony solubility curve in the silica-enthalpy diagram, whereas the well discharges affected by steam loss, similar to the boiling springs, are positioned to the left of the quartz/chalcedony solubility curve having negative x values.

5.2.9 Relation Between Undissociated SiO_2 and Aquifer Temperature for the Selected Reservoir Liquids

The concentration of undissociated silica, $SiO_{2(aq)}$, of the 1013 selected reservoir liquids is contrasted with the aquifer temperature (or reservoir temperature or reference temperature) in the correlation diagrams of Fig. 5.13. The solubility of quartz/chalcedony (see Sect. 4.10.2), at saturation temperatures and pressures, is also shown in these diagrams, together with the solubilities of quartz and chalcedony, at saturation temperatures and pressures, constrained by the thermodynamic data of quartz and chalcedony from Helgeson et al. (1978) and of aqueous SiO_2 from Shock et al. (1989).

Most reservoir liquids are positioned along the solubility curves of quartz/chalcedony or chalcedony because their aquifer temperatures were assumed to be equal to either the quartz/chalcedony temperature given by Eq. (4.57) above 180 °C or the chalcedony temperature given by Eq. (5.18) at lower temperatures. This assumption was generally adopted owing to the lack of information on aquifer temperatures.

A number of reservoir liquids are positioned between the solubility curves of quartz and chalcedony, due to attainment of equilibrium with quartz crystals of different grain size, as suggested by Azaroual et al. (1997) for the geothermal fluids of the Paris Basin. Some samples are found slightly above (to the left of) the chalcedony solubility curve due to moderate supersaturation with this silica mineral or underestimation of the aquifer temperature.

All the reservoir liquids from Salton Sea and some from other fields, namely five from Los Azufres and three from Cerro Prieto (Fig. 5.13b), one from Fushime (Fig. 5.13c), four from Rotokawa (Fig. 5.13e), and two from Mofete (Fig. 5.13f) are located significantly below (to the right of) the solubility curve of quartz. These shifts are ascribable to either overestimation of the reference temperature or, more likely, loss of SiO_2 through precipitation of amorphous silica.

All previous findings are not surprising recalling that the rate of quartz dissolution/precipitation depends strongly on temperature and is comparatively fast at high temperatures and very slow at low temperatures (e.g., Marini 2006 and references therein). This explains why: (i) in the geothermal aquifers of medium to high temperature (generally > 180 °C), reservoirs liquids attain equilibrium with respect to quartz or quartz/chalcedony, after relatively long water-rock interaction processes and (ii) precipitation of quartz or quartz/chalcedony does not occur during the relatively fast upflow of the reservoir liquids, even though saturation with respect to these solid phases is largely exceeded.

In contrast, amorphous silica precipitates relatively fast when saturation is exceeded, thus representing an upper threshold on the silica concentrations attainable in many natural aqueous environments.

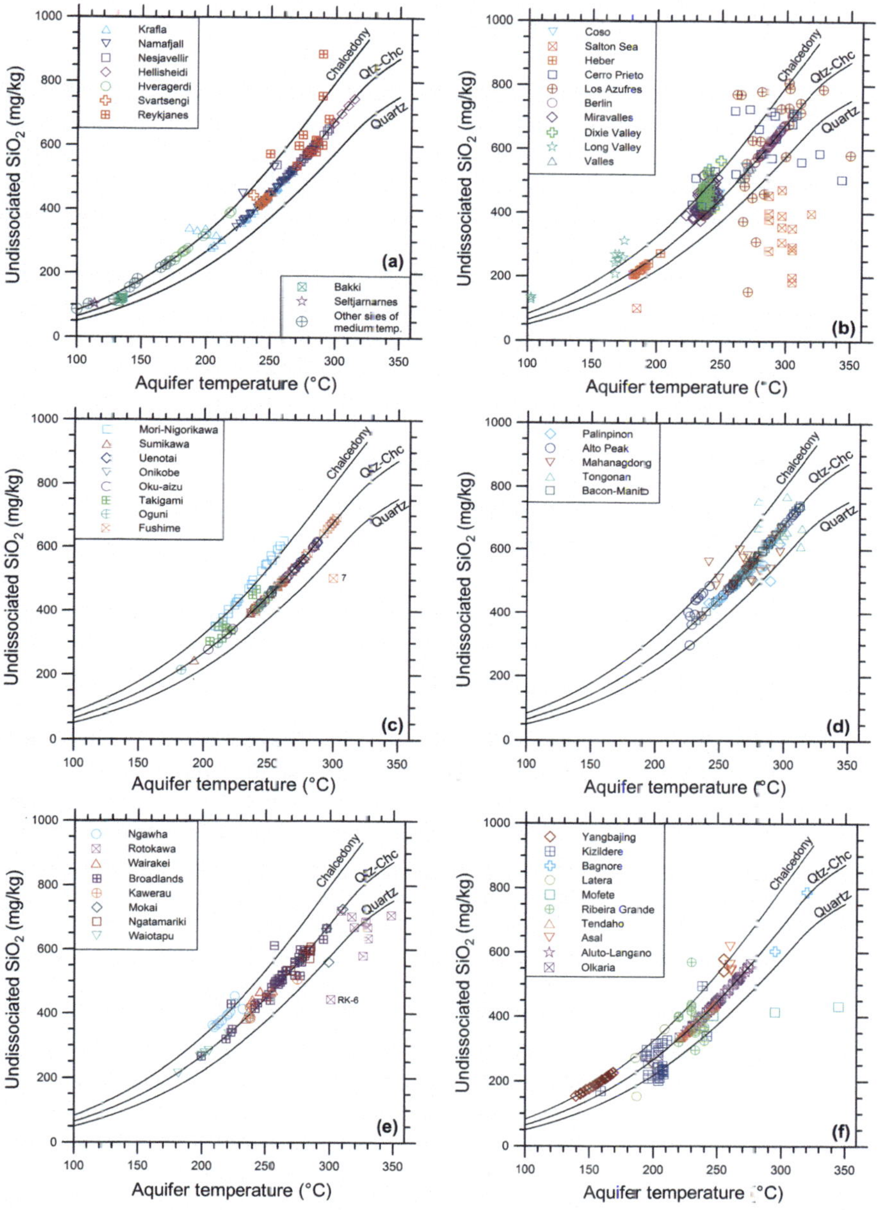

Fig. 5.13 Correlation diagrams of undissociated silica versus aquifer temperature for the selected reservoir liquids from **a** Iceland, **b** Northern-Central America, **c** Japan, **d** The Philippines, **e** New Zealand, and **f** miscellaneous geothermal systems. The solubility of quartz/chalcedony, at saturation temperatures and pressures, computed using the polynomial Eq. (4.57), is also shown in these plots, together with the solubilities of quartz and chalcedony, at saturation temperatures and pressures, constrained by the thermodynamic data of quartz and chalcedony from Helgeson et al. (1978) and of aqueous SiO_2 from Shock et al. (1989)

5.3 The Na–K Geothermometers

There are more than a few Na–K geothermometers in the literature as detailed in the next sections. These geothermometric equations were used in a large number of studies since the '60s and are still extensively used today. One of the main advantages of the Na–K geothermometers is that they are not affected by steam separation and are weakly influenced by mixing, provided that the Na and K concentrations of the diluting water are much lower than those of the liquid coming from the geothermal reservoir (e.g., Fournier 1981). According to several authors, the Na–K geothermometer is likely to re-adjust slower than the silica and K–Mg geothermometers in response to physical changes experienced by thermal waters and, therefore, deep Na/K-ratios are preserved upon cooling (see below for further details). We will show that alternative views on the behavior of Na–K geothermometers are possible.

5.3.1 The Na–K Geothermometric Functions Proposed by Different Authors

After the pioneering studies of White (1965, 1968), Ellis and Mahon (1967), Ellis (1970), and Mercado (1970), several geothermometric relations involving the Na/K ratio were proposed by different authors (e.g., Fournier and Truesdell 1973; Truesdell 1976; Fournier 1979; Tonani 1980; Arnórsson et al. 1983a; Giggenbach et al. 1983; Nieva and Nieva 1987; Giggenbach 1988; Kharaka and Mariner 1989; Michard 1990; Verma and Santoyo 1997; Santoyo and Díaz-González 2010; Simmons 2013).

Since the mid '70s, the van't Hoff equation integrated for constant standard enthalpy of reaction (see Sect. 5.1.3) was adopted as a common practice to express the Na–K geothermometers. This is an acceptable approximation because the Na–K geothermometers are controlled by ion-exchange isocoulombic reactions, irrespective of the involved minerals. Therefore, the $\Delta C_P{^\circ}_r$ is close to zero and the ΔH°_r is constant or nearly so (see Sect. 5.1.3).

As already recalled in Sect. 5.1.2, tables and diagrams showing the atomic Na/K ratios and temperatures were presented in the early studies, performed before the mid '70s. Therefore, we have transposed these tables and diagrams in the corresponding analytical forms, obtaining Na–K relations linear with respect to the absolute temperature inverse, which are directly comparable with the more recent Na–K functions.

All the Na–K geothermometers mentioned so far are reported in Table 5.2, in which two expressions are given for convenience of the reader, one including the molal concentrations of Na and K, the other comprising the concentrations of the two alkali metals in mg/kg, although the change from one formulation to the other is straightforward. All these Na–K geothermometric functions were calibrated on a purely empirical basis, utilizing both results of laboratory experiments and water compositions of geothermal boreholes, hot springs, and fluid inclusions.

Table 5.2 Considered Na–K geothermometers with linear dependence on the absolute temperature inverse. T is temperature in °C

References	Na–K function involving molal concentrations	Na–K function involving concentrations in mg/kg	Code	Eqn. #
White (1968)	$T = \frac{504.1}{\log\left(\frac{m_{Na}}{m_K}\right) - 0.0594} - 273.15$	$T = \frac{504.1}{\log\left(\frac{c_{Na}}{c_K}\right) + 0.1712} - 273.15$	Wh68	(5.57)
Ellis (1970)	$T = \frac{801.5}{\log\left(\frac{m_{Na}}{m_K}\right) + 0.5214} - 273.15$	$T = \frac{801.5}{\log\left(\frac{c_{Na}}{c_K}\right) + 0.752} - 273.15$	El70	(5.58)
Mercado (1970)	$T = \frac{640.5}{\log\left(\frac{m_{Na}}{m_K}\right) + 0.2411} - 273.15$	$T = \frac{640.5}{\log\left(\frac{c_{Na}}{c_K}\right) + 0.4717} - 273.15$	Me70	(5.59)
Fournier and Truesdell (1973)	$T = \frac{777}{\log\left(\frac{m_{Na}}{m_K}\right) + 0.47} - 273.15$	$T = \frac{777}{\log\left(\frac{c_{Na}}{c_K}\right) + 0.70} - 273.15$	FoTr73	(5.60)
Truesdell (1976)	$T = \frac{855.6}{\log\left(\frac{m_{Na}}{m_K}\right) + 0.6267} - 273.15$	$T = \frac{855.6}{\log\left(\frac{c_{Na}}{c_K}\right) + 0.8573} - 273.15$	Tr76	(5.61)
Fournier (1979)	$T = \frac{1217}{\log\left(\frac{m_{Na}}{m_K}\right) + 1.252} - 273.15$	$T = \frac{1217}{\log\left(\frac{c_{Na}}{c_K}\right) + 1.483} - 273.15$	Fo79	(5.62)
Tonani (1980)	$T = \frac{833}{\log\left(\frac{m_{Na}}{m_K}\right) + 0.55} - 273.15$	$T = \frac{833}{\log\left(\frac{c_{Na}}{c_K}\right) + 0.78} - 273.15$	To80	(5.63)
Arnórsson et al. (1983a), 25–250 °C	$T = \frac{933}{\log\left(\frac{m_{Na}}{m_K}\right) + 0.762} - 273.15$	$T = \frac{933}{\log\left(\frac{c_{Na}}{c_K}\right) + 0.993} - 273.15$	Ar83, LT	(5.64)
Arnórsson et al. (1983a), 250–350 °C	$T = \frac{1319}{\log\left(\frac{m_{Na}}{m_K}\right) + 1.468} - 273.15$	$T = \frac{1319}{\log\left(\frac{c_{Na}}{c_K}\right) + 1.699} - 273.15$	Ar83, HT	(5.65)
Giggenbach (1988)	$T = \frac{1390}{\log\left(\frac{m_{Na}}{m_K}\right) + 1.52} - 273.15$	$T = \frac{1390}{\log\left(\frac{c_{Na}}{c_K}\right) + 1.75} - 273.15$	Gi88	(5.66)

(continued)

Table 5.2 (continued)

References	Na–K function involving molal concentrations	Na–K function involving concentrations in mg/kg	Code	Eqn. #
Nieva and Nieva (1987)	$T = \frac{1178}{\log\left(\frac{m_{Na}}{m_K}\right)+1.239} - 273.15$	$T = \frac{1178}{\log\left(\frac{c_{Na}}{c_K}\right)+1.470} - 273.15$	NiNi87	(5.67)
Kharaka and Mariner (1989)	$T = \frac{1180}{\log\left(\frac{m_{Na}}{m_K}\right)+1.08} - 273.15$	$T = \frac{1180}{\log\left(\frac{c_{Na}}{c_K}\right)+1.31} - 273.15$	KhMa89	(5.68)
Michard (1990)	$T = \frac{1170}{\log\left(\frac{m_{Na}}{m_K}\right)+1.42} - 273.15$	$T = \frac{1170}{\log\left(\frac{c_{Na}}{c_K}\right)+1.65} - 273.15$	Mi90	(5.69)
Verma and Santoyo (1997)[a]	$T = \frac{1289}{\log\left(\frac{m_{Na}}{m_K}\right)+1.384} - 273.15$	$T = \frac{1289}{\log\left(\frac{c_{Na}}{c_K}\right)+1.615} - 273.15$	VeSa97	(5.70)
Santoyo and Díaz-González (2010)[b]	$T = \frac{876.3}{\log\left(\frac{m_{Na}}{m_K}\right)+0.6469} - 273.15$	$T = \frac{876.3}{\log\left(\frac{c_{Na}}{c_K}\right)+0.8775} - 273.15$	SaDi10	(5.71)
Simmons (2013)	$T = \frac{315}{\log\left(\frac{m_{Na}}{m_K}\right)-0.467} - 273.15$	$T = \frac{315}{\log\left(\frac{c_{Na}}{c_K}\right)-0.2364} - 273.15$	Si13	(5.72)

[a]Uncertainty of ±76 units on A and ±0.179 units on B
[b]Uncertainty of ±26.26 units on A and ±0.0508 units on B

The Na–K geothermometric function of White (1968) corresponds to curve B in his Fig. 2, in which other four curves are depicted. It is derived from Fig. 2 of Ellis and Mahon (1967) and is valid from 60 to 700 °C. Owing to this wide temperature range, this function has a relatively high uncertainty.

The Na–K geothermometer of Ellis (1970) is expected to work in a much smaller temperature range, from 175 to 300 °C, because the aluminum-silicate equilibria are not certain at temperatures lower than ca. 175–200 °C, according to the author. Ellis (1970) noted that "*with cooling of water, e.g., through boiling on rising towards the surface, the Na/K exchange reaction is slower to readjust than the silica equilibrium. A memory of deep temperatures is carried upwards, but the temperatures indicated by the silica and the Na/K method may disagree, the latter showing a higher temperature. The Na/K ratio can for example indicate whether or not higher temperatures exist beneath the depth of drilling (Mahon* 1970*). The Na/K method is a more reliable method for estimating deep temperatures from spring flows, but for drillhole discharges, the silica method is capable of giving more accurate temperatures at the production levels.*"

The Na–K geothermometric relation of Mercado (1970) was calibrated with the geothermal liquids discharged from ten Cerro Prieto wells, whose temperatures range from 130 to 375 °C.

Truesdell (1976) reports two Na–K geothermometric equations, one based on Ellis and White diagrams, the other from Fournier and Truesdell (1973). As underscored by Truesdell (1976) as well, the Na–K geothermometer is less affected by re-equilibration and near-surface dilution than are the silica geothermometers, but it fails at temperatures below 100–120 °C.

The Na–K geothermometric function of Fournier (1979) was calibrated using data from boreholes drilled in geothermal systems in a wide variety of geological environments and spanning a large temperature interval, from 80 to 340 °C. Fournier (1979) noted that most waters from aquifers with temperatures higher than 80–100 °C have Na/K ratios intermediate between those of the aqueous solutions in equilibrium with low-albite and maximum-microcline and those of the waters in saturation with high-albite and high-sanidine, based on the thermodynamic data of Helgeson (1969).

According to Fournier (1979), at temperatures greater than 100–150 °C, exchange of Na^+ and K^+ ions between coexisting alkali feldspars may govern the Na/K ratio in most natural waters. At temperatures lower than 100–150 °C, either equilibrium is not attained or other solid phases, such as clays, micas, and zeolites control dissolved Na and K. Consequently, the Na–K geothermometer usually works well for waters from aquifers with temperature higher than 150–200 °C, but overestimates reservoir temperatures for waters from environments where temperatures is less than 100 °C approximately.

Tonani (1980) proposed to use the Na–K geothermometer together with the Ca–Na and Ca–K functions[6] obtained from the Na–K–Ca geothermometer of Fournier and Truesdell (1973) because the simultaneous use of the three separate geothermometers (two of which are independent) has the advantage of allowing some degree

[6]These functions are presented in Sect. 5.4.

of internal cross-checking, which is impossible by using a linear combination of the geothermometers, like the Na–K–Ca geothermometer of Fournier and Truesdell (1973).

Arnórsson et al. (1983a) presented two distinct equations, one for temperatures from 25 to 250 °C, the other for higher temperatures, up to 350 °C. Arnórsson et al. (1983a) underscored that the Na/K ratios given by these two functions agree with those fixed by equilibrium coexistence of the aqueous solution with low-albite and maximum-microcline, based on the thermodynamic data of Helgeson (1969). In contrast, the thermodynamic data of Helgeson et al. (1978) give a much worse fit. Arnórsson et al. (1983a) presented also two temperature functions to be used with activities of Na^+ and K^+ ions. One is linear in 1/T and is valid from 25 to 250 °C. The other is a polynomial equation and can be used from 25 to 350 °C. They agree within few degrees with Eq. (5.64) and (5.65) and therefore are not considered any further here below.

Giggenbach (1988) calibrated a Na–K relationship (already reported by Giggenbach et al. 1983) based on the analytical data of a large number of deep well discharges, including those of Fournier (1979) and Fournier and Potter (1979). The maximum Na/K ratios at a given temperature were considered to constrain the Na–K geothermometric function, assuming that equilibrium Na/K ratios are approached from lower values, which are initially controlled by rock dissolution.

Nieva and Nieva (1987) devised a systematic, computerized procedure, taking into account the six distinct exchange reactions involving pair of cations (Na^+, K^+, Ca^{2+}, and Mg^{2+}) and some combinations of such reactions. The Na–K geothermometer of Nieva and Nieva (1987) applies to waters with total cation content (sum of the concentrations of $Na + K + Mg + Ca) > 8$ meq/L and $\%Mg \leq 3.5$.

The Na–K geothermometer of Kharaka and Mariner (1989) was developed to evaluate the temperatures at depth in sedimentary basins, which are generally in the range 30–200 °C. Moreover, these environments have water salinities and pressures usually much greater than those occurring in geothermal systems sustained by magmatic processes.

The empirical Na–K geothermometer of Michard (1990) was calibrated with waters interacting with granitic rocks at temperatures lower than 150 °C and should be used in the temperature range 50–150 °C. Aquifer temperatures were calculated by Michard (1990) using the log(Q/K) versus temperature diagram (i.e., multicomponent chemical geothermometry, see Sect. 2.3.2) and the silica, Na–K–Ca, and Na–K geothermometers. Therefore, the good correlation between the logarithm of the Na/K ratio and the absolute temperature inverse is partly biased by a circular argument, as recognized by the author. Michard (1990) underscored that his Na–K relation is similar to the temperature dependence of the equilibrium constant of the ion exchange reaction involving albite and adularia[7].

[7]The thermodynamic data used by Michard (1990) are given in the report EUR 8590 FR entitled "*Recueil de données thermodynamiques concernant les équilibres eaux-minéraux dans les réservoirs géothermaux*", which was prepared by Prof. Gil Michard for the European Commission in 1983. Unfortunately, we were not able to find this report.

The function proposed by Verma and Santoyo (1997) is a refinement of the Na–K geothermometer of Fournier (1979) and was obtained by a statistical procedure consisting of detection and rejection of outliers, ordinary linear regression, and error propagation theory. Santoyo and Díaz-González (2010) adopted a similar statistical approach, but used a different geochemometric method and considered a larger and more representative chemical database of geothermal fluids. Quite surprisingly, the enthalpy term turned out to be relatively low and similar to the values of the early studies.

The function of Simmons (2013) involves the Na/K activity ratio and is empirical, being a best fit function of the data from some geothermal fields of the United States, namely Beowave, Dixie Valley, Long Valley-Mammoth, Roosevelt Hot Springs, and Raft River. It applies in the temperature range 120–270 °C. As noted by Simmons (2013), a small interval of Na/K activity ratios corresponds to a large range of temperatures due to the small slope of this empirical equation, whose usefulness needs to be proven.

In addition to the Na–K geothermometers considered so far, it is worth recalling the Na–K function of Arnórsson (2000):

$$T = 733.6 - 770.551 \cdot \alpha + 378.189 \cdot \alpha^2 - 95.753 \cdot \alpha^3 + 9.544 \cdot \alpha^4, \quad (5.73)$$

in which T is temperature in °C and $\alpha = \log(m_{Na}/m_K)$. This Na–K function was calibrated theoretically, based on the equilibrium condition between the aqueous solution, pure low-albite and pure microcline, both of which were assumed to be fully ordered. Arnórsson (2000) obtained the thermodynamic properties of Na^+ and K^+ ions from Shock and Helgeson (1988) and those of low-albite and microcline from Arnórsson and Stefánsson (1999). However, the reliability of the thermodynamic data of alkali feldspar of Arnórsson and Stefánsson (1999) is questionable, as shown in Sect. 4.2.4.

The Na–K geothermometric relations obtained through use of artificial neural network (e.g., Bayram 2001; Can 2002; Serpen et al. 2009) are not considered here because they add little to the present discussion, although their performance is superior to that of the Na–K equation utilized to provide input data.

5.3.2 *Why so Many Empirical Na–K Geothermometers Were Derived in Previous Studies?*

Sixteen different empirical Na–K functions are listed in Table 5.2 and were briefly described in previous section. They might seem too many and, indeed, Arnórsson (1998) noted that "*it was in fashion that every geochemist studied the relationship between aquifer temperature and aqueous Na/K ratios in "his" wells and, as a product, came up with a new calibration. Careful assessment of experimental data has just recently resulted in a theoretical calibration of this geothermometer so all*

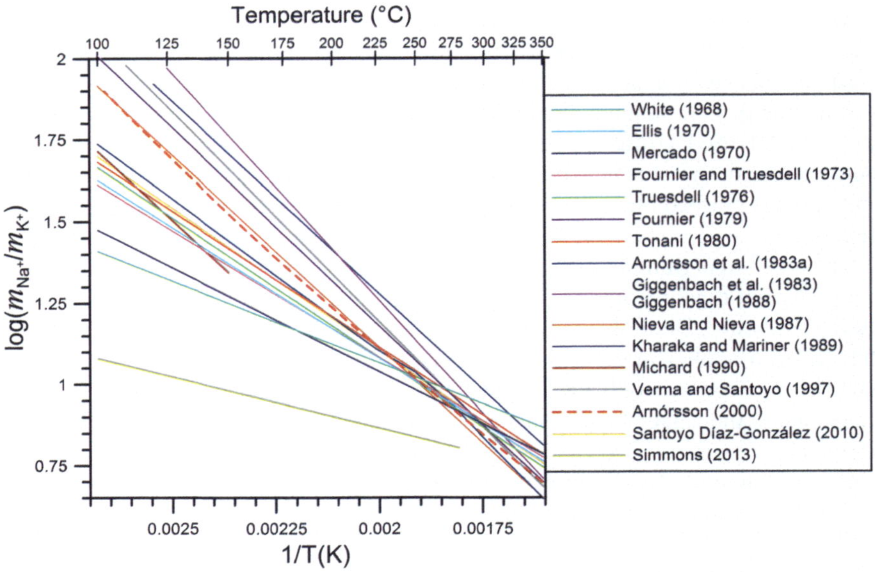

Fig. 5.14 Diagram of $\log(m_{Na}/m_K)$ versus the absolute temperature inverse showing the sixteen empirical Na–K geothermometers (see legend) and the theoretical Na–K function of Arnórsson (2000). To be noted that Arnórsson et al. (1983a) presented two distinct empirical Na–K equations, one for 25 < T < 250 °C, the other for 250 < T < 350 °C

the empirical calibrations can now be forgotten". Perhaps not every geochemist calibrated his/her own Na–K geothermometer, but certainly many geochemists did so. Can we really forget about all the empirical Na–K geothermometers or should we try to understand why so many Na–K functions have been proposed in previous studies?

To answer this questions let us examine the plot of $\log(m_{Na}/m_K)$ versus the absolute temperature inverse of Fig. 5.14, showing all the empirical Na–K geothermometric functions listed in Table 5.2 and the theoretical equation of Arnórsson (2000), i.e., Eq. (5.73).

If we showed this graph to a child, she/he would probably recognize pick-up sticks, a game in which a bundle of sticks are dropped as a loose bunch onto a table or the floor, jumbling into a random pile. Looking at the graph seriously, it can be noted that the scatter of the lines increases with decreasing temperature, whereas several lines converge at temperatures higher than ~250 °C. The function of Arnórsson (2000) passes more or less in the middle of the bundle, but it cannot be adopted as the best Na–K geothermometer, unless we are able to prove that the Na–K functions deviating from it are wrong because of one or more reasons which must be properly understood.

The presence of poorly equilibrated waters in the datasets chosen for calibrating the empirical Na–K geothermometers, especially the early ones, was invoked by some authors (e.g., Fournier 1979 and Giggenbach 1988) to reject the relationships characterized by low slopes. Among the three functions having the lowest slopes, those of White (1968) and Michard (1990) might actually be influenced by far-from-equilibrium aqueous solutions. However, the function of Mercado (1970) is constrained by the reservoir liquids tapped by ten Cerro Prieto wells with temperatures in the interval 130–375 °C, as noted in the previous section. Therefore, the influence of low-temperature, poorly equilibrated waters does not seem to be a convincing explanation, at least for the relation of Mercado (1970).

Further evidence on this important point is provided by the correlation diagrams of $\log(m_{Na}/m_K)$ versus the absolute temperature inverse of Fig. 5.15. In these plots, most of the 1013 selected reservoir liquids are found in the same area occupied by the sixteen empirical Na–K functions and the theoretical Na–K relation of Arnórsson (2000). Based on this good correspondence between the Na–K geothermometers and the selected reservoir liquids and assuming that the latter ones or at least most of them are representative of mineral-solution thermochemical equilibrium at aquifer temperatures of 100–350 °C, it can be concluded that all the Na–K geothermometric functions are plausible, because it is possible to choose a suitable set of reservoir liquids consistent with each geothermometer.

This implies that there is not a unique Na–K geothermometer, but an infinite number of Na–K geothermometers. The Na–K functions proposed so far, therefore, are not too many but are rather a small number compared to the infinite possibilities. They should not be forgotten but rather used to gain knowledge on the identity of the minerals governing the Na–K geothermometers, another key point representing the subject of the next section.

A diagram like that of Fig. 5.15, although reporting only the six Na–K geothermometers of Truesdell (1976), Fournier (1979), Tonani (1980), Arnórsson et al. (1983a), Giggenbach et al. (1983), and Nieva and Nieva (1987), was previously presented and discussed by Fournier (1991) who concluded that "*there is no single universally best Na/K geothermometer because one may give the correct temperature in one place and an erroneous temperature in another, depending on the particular mineral assemblage (and structural state of the minerals) with which the circulating water equilibrates.*"

5.3.3 The Hydrothermal Minerals Controlling the Na–K Geothermometers

Albites and adularias were suggested as the most probable candidates by several authors (e.g., White 1965; Ellis 1970; Fournier 1979; Arnórsson et al. 1983a; Giggenbach 1988; Michard 1990), although other solid phases, such as clays, micas, and zeolites were proposed to control dissolved Na and K concentrations, especially at low temperatures (e.g., Fournier and Truesdell 1970; Fournier 1979, 1991).

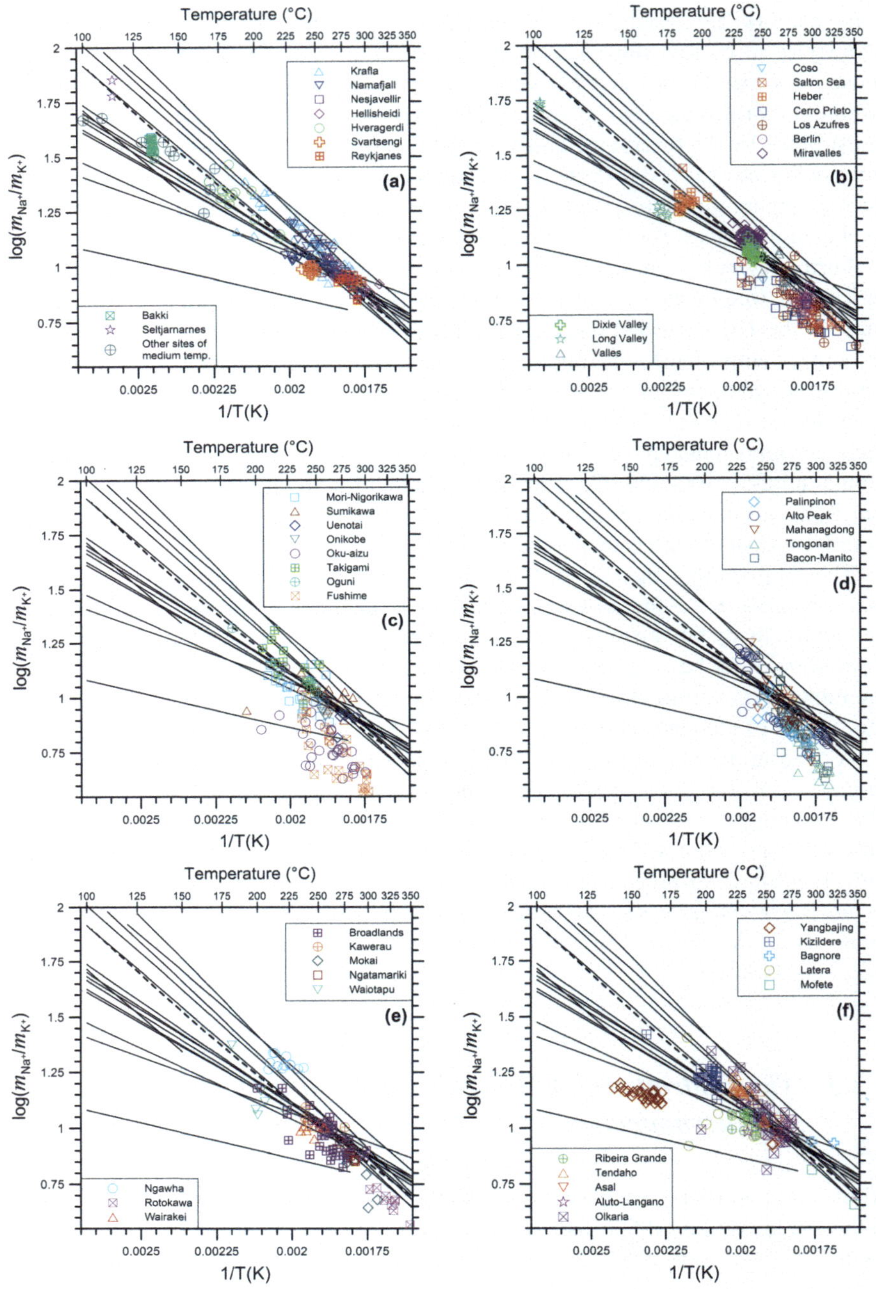

Fig. 5.15 Diagrams of log(m_{Na}/m_K) versus the absolute temperature inverse showing the selected reservoir liquids from **a** Iceland, **b** Northern-Central America, **c** Japan, **d** The Philippines, **e** New Zealand, and **f** miscellaneous geothermal systems as well as the sixteen empirical Na–K geothermometers (solid black lines) and the theoretical Na–K function of Arnórsson (2000; dashed black line)

To gain more information on the minerals governing the Na–K geothermometers, it is useful to compare the ΔH°_r and ΔS°_r values of the empirical Na–K functions, calculated from their slope and intercept using Eqs. (5.13) and (5.14), with the enthalpies and entropies of the Na–K exchange reactions involving the pertinent solid phases mentioned above. Since no clays and zeolites are included in the thermodynamic dataset of Helgeson et al. (1978), the thermodynamic properties of Na–montmorillonite, K-montmorillonite, Na–clinoptilolite, and K-clinoptilolite were obtained from Wolery et al. (2007) and these four minerals were added to the SUPCRT92 database. Then, SUPCRT92 was run to calculate the thermodynamic properties of the reactions of interest, as a function of temperature at water saturation pressure. The considered reactions are:

(i) low-albite/maximum-microcline (L–Ab/Mc), low-albite/high-sanidine (L–Ab/Sa), high-albite/maximum-microcline (H–Ab/Mc), and high-albite/high-sanidine (H–Ab/Sa), all described by:

$$NaAlSi_3O_8 + K^+ = KAlSi_3O_8 + Na^+. \tag{5.74}$$

(ii) paragonite/muscovite (Prg/Ms):

$$NaAl_3Si_3O_{10}(OH)_2 + K^+ = KAl_3Si_3O_{10}(OH)_2 + Na^+. \tag{5.75}$$

(iii) Na–clinoptilolite/K-clinoptilolite (Na–Cpt/K–Cpt):

$$\begin{aligned} &0.909\ \ Na_{1.1}Si_{4.9}Al_{1.1}O_{12}\cdot 3.5H_2O + K^+ \\ &\quad = 0.909\ K_{1.1}Si_{4.9}Al_{1.1}O_{12}\cdot 2.7H_2O + Na^+ + 0.727H_2O \end{aligned} \tag{5.76}$$

(iv) Na–montmorillonite/K-montmorillonite (Na–Mm/K–Mm):

$$\begin{aligned} &3.03Na_{0.33}Mg_{0.33}Al_{1.67}Si_4O_{10}(OH)_2 + K^+ \\ &\quad = 3.03K_{0.33}Mg_{0.33}Al_{1.67}Si_4O_{10}(OH)_2 + Na^+. \end{aligned} \tag{5.77}$$

The correlation plot of Fig. 5.16 shows that the ΔH°_r and ΔS°_r values of the sixteen empirical Na–K geothermometric equations or at least most of them are intermediate between the enthalpies and entropies of the low-albite/high-sanidine exchange reaction and those of the low-albite/maximum-microcline exchange reaction or, less likely, of the high-albite/high-sanidine exchange reaction. In contrast, the ΔH°_r and ΔS°_r values of the sixteen empirical Na–K geothermometric equations

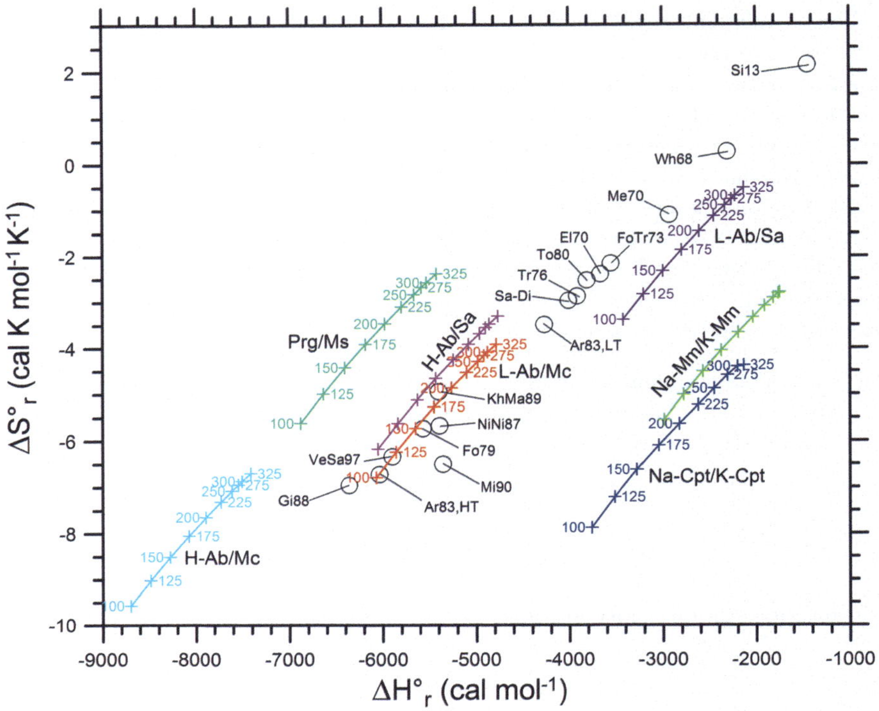

Fig. 5.16 Comparison of the temperature-independent standard enthalpies and entropies of the sixteen empirical Na–K geothermometers (open black circles), obtained from their slope and intercept, with the standard enthalpies and entropies of the Na–K exchange reactions low-albite/maximum-microcline (L–Ab/Mc), low-albite/high-sanidine (L–Ab/Sa), high-albite/maximum-microcline (H-Ab/Mc), high-albite/high-sanidine (H–Ab/Sa), paragonite/muscovite (Prg/Ms), Na-clinoptilolite/K-clinoptilolite (Na–Cpt/K–Cpt), and Na–montmorillonite/K-montmorillonite (Na–Mm/K–Mm), computed as a function of temperature at water saturation pressure using SUPCRT92

are at variance with the enthalpies and entropies of the Na–K exchange reactions comprising high-albite and maximum-microcline as well as micas, clinoptilolites, and montmorillonites.

5.3.4 Conclusive Remarks on the Na–K Geothermometers

Summing up, it appears that there is not a unique Na–K geothermometer, but that all the Na–K geothermometers proposed so far are plausible and that they are a small number compared to the infinite number of possible Na–K geothermometers, as already underscored by Fournier (1991).

Based on the $\Delta H°_r$ and $\Delta S°_r$ values of the sixteen empirical Na–K geothermometers of interest, it turns out that they are probably controlled by the exchange reactions low-albite/high-sanidine or low-albite/maximum-microcline or any "in-between" exchange reaction, involving low-albite and adularia with a degree of ordering intermediate between that of fully ordered maximum-microcline and that of completely disordered high-sanidine whereas the possible role of the high-albite/high-sanidine exchange reaction appears to be much less likely. These inferences are consistent with the indications given by the authigenic alkali feldspars occurring in sedimentary rocks and synthesized by means of hydroxide gels (see Sect. 4.2.3).

Furthermore, our conclusions represent a step forward with respect to previous findings of Bird and Norton (1981) and Bird and Helgeson (1981), who showed that the tetrahedral site distributions of Al^{3+} and Si^{4+} ions in coexisting alkali feldspars have a dramatic influence on the Na^+/K^+ activity ratio in the aqueous phase and on the temperatures given by Na–K geothermometers. According to Bird and Norton (1981) "*metastable tetrahedral ordering in alkali feldspars is a likely cause of many of the discrepancies found among alkali cation geothermometers and actual measured temperatures in geothermal systems*".

Accepting our conclusions, it follows that there is no reason to look for the universally best Na–K geothermometer, either empirical or theoretical, that can be applied everywhere because it does not exists. Rather than using the Na/K ratio as the basis for an infinite number of geothermometers, it is advisable to use it for obtaining different indications, changing completely the approach to water geothermometry. This discussion will be resumed in Chap. 6.

5.4 The Na–K–Ca Geothermometer

5.4.1 Formulation, Controlling Reactions, and Limitations/Problems of the Na–K–Ca Geothermometer

The Na–K–Ca geothermometer is one of the most popular and used empirical geothermometers, although it is affected by a certain number of limitations and problems. We consider the Na–K–Ca geothermometer immediately after the Na–K geothermometer because of the strict link between these two functions. Such a link was pointed out by Fournier and Truesdell (1973) who decided to introduce Ca in a revised and expanded Na–K geothermometer because of at least two distinct reasons. One reason is that the Na–K geothermometer does not yield reasonable results for Ca-rich waters. The other reason is that Ca^{2+} ion may have an effect upon the Na/K ratio of the aqueous solution as suggested by the reaction in which plagioclase is destructed and K-feldspar is produced with the simultaneous exchange of K^+ ion for both Na^+ and Ca^{2+} ions. Fournier and Truesdell (1973) postulated that "*aqueous*

Na–K–Ca relationships generally can be explained entirely in terms of silicate reactions even though the absolute quantity of aqueous Ca is controlled by the solubility of carbonate"[8]. According to Fournier and Truesdell (1973), the reactions involving silicate minerals and controlling Na–K–Ca relationships have net stoichiometry:

$$Na_{2/3}Ca_{1/6} - \text{solid} + K^+ = 1/6Ca^{2+} + 2/3Na^+ + K - \text{solid}, \qquad (5.78)$$

above 100 °C, and:

$$Ca_{2/3} - \text{solid} + K^+ + 1/3Na^+ = 2/3Ca^{2+} + KNa_{1/3} - \text{solid} \qquad (5.79)$$

below 100 °C. Fournier and Truesdell (1973) did not provide the temperature dependence of the log K of reactions (5.78) and (5.79), but only the following overall relation (concentrations in mg/kg):

$$T(°C) = \frac{1647}{\log(c_{Na}/c_K) + \beta \cdot [\log(\sqrt{c_{Ca}}/c_{Na} + 2.06] + 2.47} - 273.15, \qquad (5.80)$$

in which β may be equal to 1/3 or 4/3. To choose the proper value of β, Fournier and Truesdell (1973) suggested to proceed as follows: Express the concentrations of dissolved species in units of molality and calculate $\log(\sqrt{Ca}/Na)$. If $\log(\sqrt{Ca}/Na)$ is positive, compute the Na–K–Ca temperature for $\beta = 4/3$. The obtained Na–K–Ca temperature is correct, if it is <100 °C. Otherwise, recalculate the Na–K–Ca temperature for $\beta = 1/3$. However, β can be assumed equal to 1/3 for aquifer temperature >100 °C, as is the case of the 1013 reservoir liquids of interest.

Precipitation of calcite causes an overestimation of the equilibrium temperature obtained by means of the Na–K–Ca function, as already recognized by Fournier and Truesdell (1973). Mixing has limited effects on the Na–K–Ca geothermometer if the high-temperature water has salinity much greater than the low-temperature water and if the fraction of the geothermal water is higher than 0.2–0.3. Otherwise, the effects of mixing on the Na–K–Ca geothermometer should be taken into consideration, as discussed by Fournier (1981).

Fournier and Potter (1979) noted that many Mg-rich waters have high Na–K–Ca temperatures, well above 150 °C, which are unexpected considering that all high-temperature (>175 °C) waters discharged from boreholes drilled into active geothermal systems have low concentrations of Mg relative to the other dissolved cations. Therefore, based on the value of $R = [Mg/(Mg + Ca + K)] \cdot 100$, with concentrations expressed in equivalent units, Fournier and Potter (1979), proposed two different temperature corrections, Δt_{Mg} (in °C), that should be subtracted from

[8] Actually, this hypothesis is erroneous because Ca^{2+} activity cannot be constrained by both calcite and a Ca–Al-silicate at the same time, as hypothesized by Fournier and Truesdell (1973). In case of coexistence of calcite and a Ca–Al-silicate, calcite fixes the activity of the carbonate species of reference [either $CO_{2(aq)}$ or HCO_3^- or CO_3^{2-}], whereas the Ca–Al-silicate constrains the activity of Ca^{2+} ion. However, Fournier and Truesdell (1973) excluded calcite and CO_2 from reactions (5.78) and (5.79).

the Na–K–Ca temperature, T. For 5 < R < 50 and T > 70 °C:

$$\Delta t_{Mg} = 10.66 - 4.7415 \cdot R + 325.87 \cdot (\log R)^2 - 1.032 \times 10^5 \cdot \frac{(\log R)^2}{T} - 1.968 \times 10^7 \cdot \frac{(\log R)^2}{T^2} + 1.605 \times 10^7 \cdot \frac{(\log R)^3}{T^2}. \quad (5.81)$$

For 0.5 < R < 5:

$$\Delta t_{Mg} = -1.03 + 59.971 \cdot \log R + 145.05 \cdot (\log R)^2 - 36711 \cdot \frac{(\log R)^2}{T} - 1.67 \times 10^7 \cdot \frac{\log R}{T^2}. \quad (5.82)$$

Irrespective of the R value, the Mg correction should not be applied if Δt_{Mg} is negative. Moreover, Fournier and Potter (1979) suggested that Mg-rich waters may be unsuitable for the application of geothermometers because high Mg concentrations indicate the occurrence of water-rock reactions at relatively low temperatures, causing not only Mg acquisition but also possible changes in the concentrations of other dissolved constituents.

The behavior of the Na–K–Ca geothermometer has stimulated the interest of several authors, who approached this topic both theoretically and experimentally. Shikazono (1976) and Michard and Fouillac (1976) provided a thermodynamic interpretation of the Na–K–Ca geothermometer.

Shikazono (1976) underscored that in NaCl aqueous solutions in equilibrium with typical hydrothermal minerals at fixed temperature, the logarithms of Na^+, K^+, and Ca^{2+} molalities are linear functions of the logarithm of Cl^- molality. The linear functions involving alkali cations have slope 1, whereas those comprising alkali-earth cations have slope 2.

According to Michard and Fouillac (1976), above 100 °C, the Na–K–Ca geothermometer is controlled by the equilibration of the aqueous solution with K-feldspar and a plagioclase of nearly constant composition close to $Ab_{0.8}An_{0.2}$. In contrast, below 100 °C, water chemistry is assumed to be controlled by the relative dissolution kinetics of relevant minerals and the Na–K–Ca geothermometer would be chiefly explained by the increasing salinity with temperature.

Janecki et al. (1986) performed laboratory experiments at 260, 280, 300, 320, and 340 °C, reacting granite disks in 10 mmol/kg NaCl solutions and in mixed solutions containing 10 mmol/kg NaCl and 0.1 mmol/kg $CaCl_2$. Hydrothermal batch reactors were used for all the experiments, keeping total pressure at 272 bar, except one run that was performed in a flow-through reactor. According to Janecki et al. (1986), variations in solution chemistry closely follow the empirical Na–K–Ca geothermometric function although weak deviations were observed at 340 °C. Primary minerals composing the granite were etched. Authigenic minerals recognized at the end of the runs comprise minor quantities of Fe–, Mg–, Ca-rich clay minerals and a Ca-silicate (truscottite or a similar phase) at all temperatures, a Ca-zeolite (possibly faujasite) below 300 °C, feldspar overgrowths (apparently albitic) above 300 °C, and amphibole overgrowths at 340 °C.

Benjamin et al. (1983) performed a thorough review of the Na–K–Ca geothermometer. Through the least-squares fit of the Fournier and Truesdell data, they obtained the following revised Na–K–Ca geothermometric relations:

$$T(^{\circ}C) = \frac{22200}{64.2 - \log(Na/K) + 6.3 \cdot \log(\sqrt{Ca}/Na)} - 273.15 \quad (5.83)$$

$$T(^{\circ}C) = \frac{1416}{\log(Na/K) + 0.055 \cdot \log(\sqrt{Ca}/Na) + 1.69} - 273.15 \quad (5.84)$$

Equation (5.83) applies to $T \leq 100$ °C and it is of little interest for us, whereas Eq. (5.84) is valid for $T > 100$ °C and deserves more attention. It is consistent with a ΔH°_{r} of -6480 ± 260 cal mol^{-1} and a ΔS°_{r} of -7.74 ± 0.50 cal mol^{-1} K^{-1}. The error on the stoichiometric coefficient β is ± 0.053, whereas the error on the computed temperatures varies from 10 to 25 °C.

Benjamin et al. (1983) performed also experimental studies reacting separately two granodiorites and a basalt with distilled water, both in a fixed-temperature circulation system, at 200 and 300 °C, and in a controlled-temperature-gradient circulation system, at 72, 119, 161, 209, 270, and 310 °C, for 1440 h. Pressure was kept at 1/3 kbar in all the experiments. The duration of the fixed-temperature experiments was 6579 and 5816 h for those with the granodiorites and 3600 h for that with the basalt. Concentrations of Na, K, and Ca attained steady state after 2–4 weeks approximately, but cation ratios stabilized in a few days. Overgrowths of zeolites, usually Ca-rich, and clay minerals of variable composition were found over the primary minerals at the end of the runs. Based on their analytical results, Benjamin et al. (1983) computed a stoichiometric coefficient β of 0.98 ± 0.10 for the Na–K–Ca geothermometric function as well as a ΔH°_{r} of $-10{,}300 \pm 1100$ cal mol^{-1} and a ΔS°_{r} of -13.3 ± 1.7 cal mol^{-1} K^{-1} for its controlling reaction which was written as follows:

$$\begin{aligned} &K^{+} + 0.02NaAlSi_3O_8 + 0.49CaAl_2Si_2O_8 + 1.96SiO_2 \\ &= 0.02Na^{+} + 0.49Ca^{2+} + KAlSi_3O_8 \end{aligned} \quad (5.85)$$

adopting a stoichiometric coefficient of K^{+} ion equal to 1. Reaction (5.85) is essentially a Ca-K exchange reaction with a marginally involvement of Na^{+} ion. Benjamin et al. (1983) were forced to interpret their experimental results in this way, owing to the lack of thermodynamic data for zeolites and clay minerals, although they recognized that feldspar endmembers do not govern solution chemistry. To be noted that the ΔH°_{r} and ΔS°_{r} values and the stoichiometric coefficient β derived from the results of the hydrothermal experiments are significantly different from those obtained through the least-squares fit of the Fournier and Truesdell data for T > 100 °C. These differences suggest that the hydrothermal experiments did not reproduce the water-rock reactions controlling Na–K–Ca relations in natural geothermal systems.

Pope et al. (1987) performed another experimental study in which a powdered peralkaline rhyolite was reacted with distilled water in separate runs with 0.1 M NaCl and 0.01 M $NaHCO_3$ at different temperatures in the range 100–500 °C, mostly at 200 and 300 °C, maintaining total pressure at 1 kbar. Reactants were sealed inside gold capsules. Run durations were usually 2–3 weeks, with some longer experiments to control the attainment of equilibrium or steady-state conditions. No alteration mineral was detected by SEM and XRD analyses but the presence of alteration minerals in small quantities cannot be excluded. The Na–K–Ca geothermometer works well for the experiments with 0.1 M NaCl. In fact, the mean computed Na–K–Ca temperatures are 191 ± 14 °C and 297 ± 14 °C, at 200 and 300 °C, respectively. In contrast, the Na–K–Ca temperatures are lower than the experimental values by 10–50 °C for the experiments with 0.01 M $NaHCO_3$. These poor results are ascribable, at least partly, to the low Ca concentrations, which are close to detection limit at 200 °C and are generally below detection limit at 300 °C.

5.4.2 Performance of the Na–K–Ca Geothermometer for the Selected Reservoir Liquids

The temperature given by the Na–K–Ca geothermometer of Fournier and Truesdell (1973) for the 1013 selected reservoir liquids is contrasted with the aquifer temperature in the binary diagrams of Fig. 5.17.

Among the Icelandic geothermal systems (Fig. 5.17a), the average absolute deviation of the Na–K–Ca temperature from the reservoir temperature is: (i) 18 ± 10 (1σ) °C for the high-temperature (>175 °C), low-salinity reservoir liquids of Hellisheidi, Hveragerdi, Krafla, Namafjall, and Nesjavellir; (ii) 35 ± 12 °C for the high-temperature reservoir liquids of Reykjanes; (iii) 6 ± 3 °C for the high-temperature reservoir liquids of Svartsengi; (iv) 11 ± 5 °C for the medium-temperature (<175 °C) Icelandic reservoir liquids. The Na–K–Ca geothermometer has a good performance at Svartsengi and a very poor behavior at Reykjanes, although both systems are similar, being partly recharged by seawater. The Na–K–Ca geothermometer has a poor performance for the high-temperature (>175 °C), low-salinity reservoir liquids, whereas it works relatively well for the medium-temperature (<175 °C) Icelandic reservoir liquids. The Na–K–Ca temperature is higher than the aquifer temperature for most liquids from Hveragerdi and the medium-temperature systems. In contrast, the Na–K–Ca geothermometer underestimates the aquifer temperature for most reservoir liquids of Hellisheidi, Krafla, Namafjall, Nesjavellir, Reykjanes, and Svartsengi.

Among the geothermal systems of Northern and Central America (Fig. 5.17b), the average absolute deviation of the Na–K–Ca temperature from the aquifer temperature is: (i) 5 ± 3 (1σ) °C for Dixie Valley; (ii) 42 ± 6 °C for Long Valley; (iii) 14 ± 6 °C

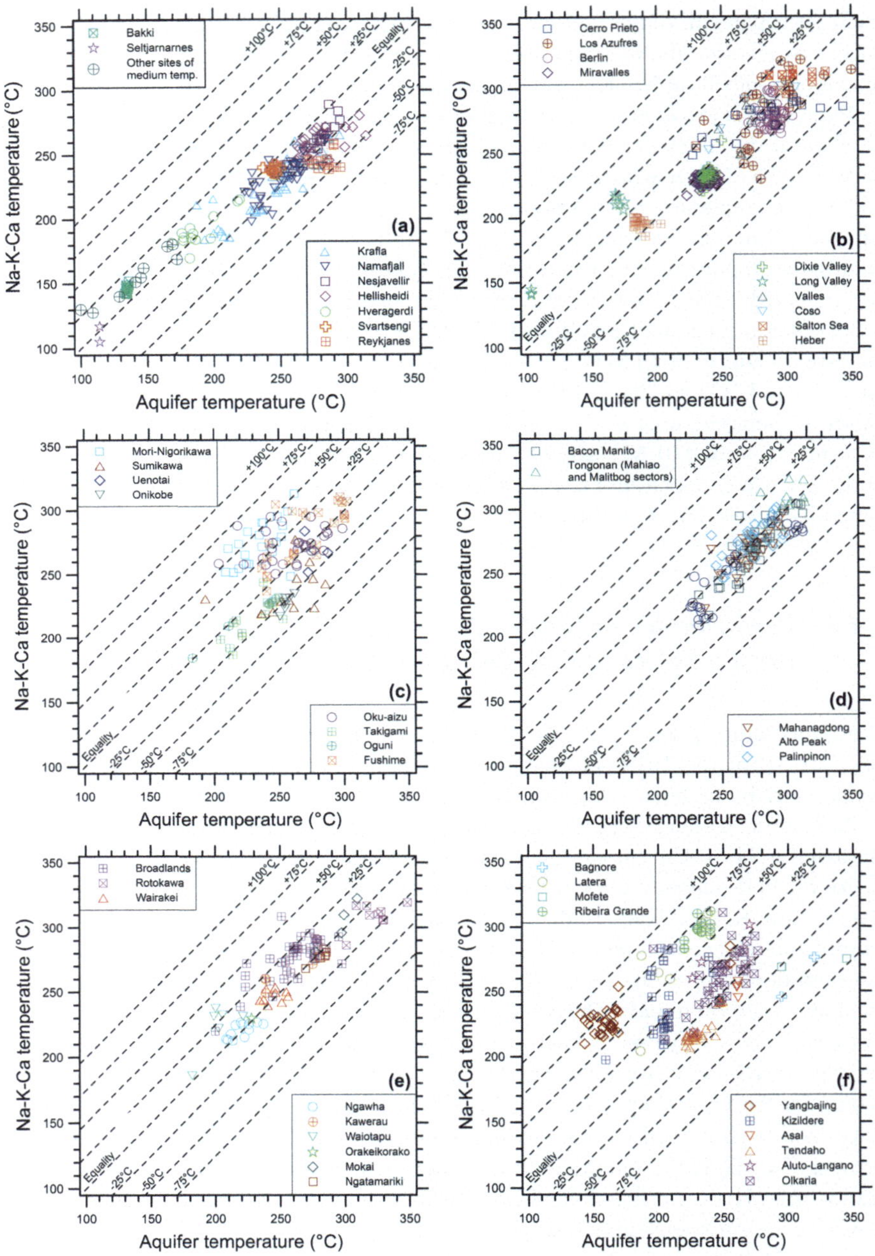

Fig. 5.17 Correlation diagrams of the temperature given by the Na–K–Ca geothermometer of Fournier and Truesdell (1973) versus the aquifer temperature for the selected reservoir liquids from **a** Iceland, **b** Northern-Central America, **c** Japan, **d** The Philippines, **e** New Zealand, and **f** miscellaneous geothermal systems

for Valles; (iv) 14 ± 8 °C for Coso; (v) 13 ± 7 °C for Salton Sea; (vi) 10 ± 5 °C for Heber; (vii) 17 ± 14 °C for Cerro Prieto; (viii) 19 ± 12 °C for Los Azufres; (ix) 14 ± 7 °C for Berlin; (x) 9 ± 4 °C for Miravalles. The Na–K–Ca geothermometer has a good performance at Dixie Valley and a relatively good performance at Miravalles and Heber, whereas its behavior is very poor at Long Valley and relatively poor in the other systems. The Na–K–Ca temperature overestimates considerably and systematically the aquifer temperature at Long Valley.

Among the geothermal systems of Japan (Fig. 5.17c), the average absolute deviation of the Na–K–Ca temperature from the reservoir temperature is: (i) 41 ± 12 (1σ) °C for Mori-Nigorikawa; (ii) 19 ± 4 °C for Sumikawa; (iii) 16 ± 5 °C for Uenotai; (iv) 27 ± 3 °C for Onikobe; (v) 18 ± 18 °C for Oku-aizu; (vi) 16 ± 10 °C for Takigami; (vii) 11 ± 7 °C for Oguni; (viii) 14 ± 14 °C for Fushime. The Na–K–Ca geothermometer has a very poor behavior at Mori-Nigorikawa and Onikobe and a poor performance is all the other geothermal systems Japan, apart from Oguni, to some extent. The Na–K–Ca temperature is higher than the reservoir temperature in most wells of Mori-Nigorikawa and in several wells of Oku-aizu and Fushime, whereas the reverse situation occurs frequently at Sumikawa, Uenotai, Onikobe, Takigami, and Oguni.

Among the geothermal systems of the Philippines (Fig. 5.17d), the average absolute deviation of the Na–K–Ca temperature from the aquifer temperature is: (i) 8 ± 7 °C for Bacon-Manito; (ii) 10 ± 9 °C for Tongonan; (iii) 8 ± 7 °C for Mahanagdong; (iv) 14 ± 10 °C for Alto Peak; (v) 10 ± 7 °C for Palinpinon. The Na–K–Ca geothermometer works well at Mahanagdong and Bacon-Manito and relatively well at Palinpinon and Tongonan. In contrast, its behavior is comparatively poor at Alto Peak, possibly due to the presence of feed zones with different characteristics as pointed out by Reyes et al. (1993). The Na–K–Ca temperatures are generally lower than aquifer temperatures, especially at Alto Peak, and to a lower extent at Mahanagdong and Bacon-Manito, whereas the opposite is true for Palinpinon and Tongonan.

Among the geothermal systems of New Zealand (Fig. 5.17e), the average absolute deviation of the Na–K–Ca temperature from the aquifer temperature is: (i) 5 ± 3 °C for Ngawha; (ii) 15 ± 8 °C for Kawerau; (iii) 22 ± 12 °C for Waiotapu; (iv) 7 ± 5 °C for Mokai; (v) 4 ± 2 °C for Ngatamariki; (vi) 18 ± 12 °C for Broadlands; (vii) 7 ± 4 °C for Wairakei; (viii) 15 ± 8 °C for Rotokawa. The Na–K–Ca geothermometer has a good performance at Ngawha, Mokai, Ngatamariki, and Wairakei, whereas its behavior is poor at Kawerau, Rotokawa, Broadlands, and especially at Waiotapu. The Na–K–Ca temperature overestimates significantly the aquifer temperature in most wells of Broadlands, Kawerau, and Waiotapu, whereas the opposite occurs at Rotokawa.

Among the miscellaneous geothermal systems (Fig. 5.17f), the average absolute deviation of the Na–K–Ca temperature from the aquifer temperature is: (i) 65 ± 15 °C for Yangbajing; (ii) 35 ± 23 °C for Kizildere; (iii) 59 ± 27 °C for Latera; (iv) 33 ± 34 °C for Mofete; (v) 66 ± 6 °C for Ribeira Grande; (vi) 10 ± 5 °C for Tendaho; (vii) 8 ± 5 °C for Asal; (viii) 35 ± 4 °C for Aluto-Langano; (ix) 16 ± 15 °C for Olkaria.

The Na–K–Ca geothermometer works well at Asal, relatively well at Tendaho, but it has a very poor performance at Olkaria, Mofete, Aluto-Langano, Kizildere and especially at Latera, Yangbajing, and Ribeira Grande. The Na–K–Ca temperature overestimates considerably the reservoir temperature at Yangbajing, Ribeira Grande, Kizildere, Latera, Aluto-Langano, and in most wells of Olkaria, whereas the Na–K–Ca temperature underestimates, to different extents, the aquifer temperature at Mofete, Bagnore, Asal and Tendaho.

Loss of Ca from the aqueous solution due to precipitation of calcite or other Ca-bearing minerals is a possible reason explaining why Na–K–Ca temperatures are higher than aquifer temperatures. Although there is no relation between the performance of the Na–K–Ca geothermometer and Ca concentration for the reservoir liquids of interest, loss of Ca is expected to have a much stronger impact on Ca-poor aqueous solutions, such as Na–HCO_3 waters, than on Ca-rich aqueous solutions, such as high-salinity Na–Cl waters. Of course, this problem affects not only the Na–K–Ca geothermometer, but also all the Ca-based geoindicators.

Na–K–Ca temperatures higher than aquifer temperatures could be explained also by the lack of re-equilibration of the Na–K–Ca system upon cooling of the aqueous solutions, owing to the slow kinetics of relevant reactions. This explanation could apply not only to spring waters but also to wells discharges, especially to wells producing from a secondary shallow reservoir, in which the geothermal waters, coming from a deeper and hotter reservoir, have a relatively short residence time. For instance, this could be the case of Yangbajing. However, further data on the kinetics of relevant re-equilibration processes are needed to convert this hypothesis in a convincing explanation.

The deviations of Na–K–Ca temperatures from aquifers temperatures could also be ascribable to the different constraints controlling the activity of Ca^{2+} ion in reservoir liquids, either saturation with calcite, at high f_{CO_2} values, or saturation with a Ca–Al–silicate, at low f_{CO_2} values. Further complications are caused by the possible role of different Ca–Al-silicates, such as laumontite and other zeolites at low-temperatures (indicatively up to 200 °C), epidote, prehnite, and wairakite at temperatures of 200–300 °C, as well as grandite garnet, anorthite, pyroxene, and amphibole, above 300 °C.

Participation of variably-ordered adularias (see Sect. 5.3.4) to the exchange reactions of interest, is another possible reason for the discrepancies between Na–K–Ca temperatures and aquifer temperatures. Finally, poor knowledge or choice of the aquifer temperature cannot be excluded, although it cannot be invoked for deviations higher than a few degrees to 10 °C approximately.

5.4.3 Na–Ca and K–Ca Geothermometers

As shown by Tonani (1980), the Na–K–Ca geothermometer can be split into three temperature functions involving the Na/K, $\sqrt{Ca}/Na$, and $\sqrt{Ca}/K$ ratios. The use of three separate geothermometric functions instead of the single Na–K–Ca geothermometer is advantageous because it is possible to check if the three computed temperatures are consistent with each other or not, although only two of these three relations are mutually independent (Tonani 1980). Also Arnórsson et al. (1983a) underscored that the separate Na/K and $\sqrt{Ca}/Na$ cation ratios composing the Na–K–Ca geothermometer are fixed by temperature alone, if the condition of overall chemical equilibrium between the geothermal waters and the hydrothermal mineral assemblage is attained.

Based on these findings and considering that there is an infinite number of Na–K geothermometers (see 5.3), it is permissible to suppose that there is also an infinite number of Na–Ca and K–Ca geothermometers. To avoid complicating too much the following discussion, we decided to take into account the sixteen empirical Na–K geothermometers with linear dependence on the absolute temperature inverse listed in Table 5.2 only. Moreover, we decided to write the geothermometric relations we are looking for in terms of the Na^2/Ca and K^2/Ca ratios instead of the $\sqrt{Ca}/Na$ and $\sqrt{Ca}/K$ ratios, respectively. To obtain the Na–Ca and K–Ca geothermometers of interest, first, β is assumed equal to 1/3 (consistent with T > 100 °C) and Eq. (5.80) is rewritten as (concentrations in mg/kg):

$$\log(c_{Na}/c_K) + 1/3 \cdot \log(\sqrt{c_{Ca}}/c_{Na}) = \frac{1647}{T + 273.15} - 3.1567. \quad (5.86)$$

Second, Eq. (5.86) is multiplied by 2 and is suitably rearranged obtaining:

$$\log(c_{Na}^2/c_{Ca}) = 6 \cdot \log(c_{Na}/c_K) - \frac{9882}{T + 273.15} + 18.9402. \quad (5.87)$$

Third, each Na–K empirical geothermometer listed in Table 5.2 is alternatively inserted into Eq. (5.87) and the relation is suitably reorganized, deriving the corresponding Na–Ca function. The related K–Ca geothermometer is then obtained either subtracting the Na–K relation multiplied by 2 from the Na–Ca equation:

$$\log(c_K^2/c_{Ca}) = \log(c_{Na}^2/c_{Ca}) - 2 \cdot \log(c_{Na}/c_K) \quad (5.88)$$

or through the relation:

$$\log(c_K^2/c_{Ca}) = 4 \cdot \log(c_{Na}/c_K) - \frac{9882}{T - 273.15} + 18.9402. \quad (5.89)$$

The results of these simple but tedious calculations are listed in Tables 5.3 and 5.4, where two relations are given, one for molal concentrations, the other for concentrations in mg/kg, although the conversion from one form to the other is straightforward. These Na–Ca and K–Ca geothermometers are expected to work at comparatively low f_{CO_2} values stabilizing one of the several possible Ca–Al-silicates (see above) and destabilizing calcite.

Table 5.3 Na–Ca functions obtained from the Na–K–Ca geothermometer and the sixteen empirical Na–K geothermometers

References	Na–Ca function (molal concentrations)	Na–Ca function (concentrations in mg/kg)	Code	Eqn. #
White (1968)	$T = \frac{6857.4}{13.793 - \log\left(\frac{m_{Na}^2}{m_{Ca}}\right)} - 273.15$	$T = \frac{6857.4}{17.913 - \log\left(\frac{c_{Na}^2}{c_{Ca}}\right)} - 273.15$	Wh68	(5.90)
Ellis (1970)	$T = \frac{5073}{10.308 - \log\left(\frac{m_{Na}^2}{m_{Ca}}\right)} - 273.15$	$T = \frac{5073}{14.428 - \log\left(\frac{c_{Na}^2}{c_{Ca}}\right)} - 273.15$	El70	(5.91)
Mercado (1970)	$T = \frac{6039}{11.990 - \log\left(\frac{m_{Na}^2}{m_{Ca}}\right)} - 273.15$	$T = \frac{6039}{16.110 - \log\left(\frac{c_{Na}^2}{c_{Ca}}\right)} - 273.15$	Me70	(5.92)
Fournier and Truesdell (1973)	$T = \frac{5220}{10.620 - \log\left(\frac{m_{Na}^2}{m_{Ca}}\right)} - 273.15$	$T = \frac{5220}{14.740 - \log\left(\frac{c_{Na}^2}{c_{Ca}}\right)} - 273.15$	FoTr73	(5.93)
Truesdell (1976)	$T = \frac{4748.4}{9.676 - \log\left(\frac{m_{Na}^2}{m_{Ca}}\right)} - 273.15$	$T = \frac{4748.4}{13.796 - \log\left(\frac{c_{Na}^2}{c_{Ca}}\right)} - 273.15$	Tr76	(5.94)
Fournier (1979)	$T = \frac{2580}{5.922 - \log\left(\frac{m_{Na}^2}{m_{Ca}}\right)} - 273.15$	$T = \frac{2580}{10.042 - \log\left(\frac{c_{Na}^2}{c_{Ca}}\right)} - 273.15$	Fo79	(5.95)
Tonani (1980)	$T = \frac{4884}{10.140 - \log\left(\frac{m_{Na}^2}{m_{Ca}}\right)} - 273.15$	$T = \frac{4884}{14.260 - \log\left(\frac{c_{Na}^2}{c_{Ca}}\right)} - 273.15$	To80	(5.96)
Arnórsson et al. (1983a), 25–250°C	$T = \frac{4284}{8.862 - \log\left(\frac{m_{Na}^2}{m_{Ca}}\right)} - 273.15$	$T = \frac{4284}{12.982 - \log\left(\frac{c_{Na}^2}{c_{Ca}}\right)} - 273.15$	Ar83, LT	(5.97)
Arnórsson et al. (1983a), 250–350°C	$T = \frac{1968}{4.626 - \log\left(\frac{m_{Na}^2}{m_{Ca}}\right)} - 273.15$	$T = \frac{1968}{8.746 - \log\left(\frac{c_{Na}^2}{c_{Ca}}\right)} - 273.15$	Ar83, HT	(5.98)

(continued)

Table 5.3 (continued)

References	Na–Ca function (molal concentrations)	Na–Ca function (concentrations in mg/kg)	Code	Eqn. #
Giggenbach (1988)	$T = \frac{1542}{4.320 - \log\left(\frac{m_{Na}^2}{m_{Ca}}\right)} - 273.15$	$T = \frac{1542}{8.440 - \log\left(\frac{c_{Na}^2}{c_{Ca}}\right)} - 273.15$	Gi88	(5.99)
Nieva and Nieva (1987)	$T = \frac{2814}{6.000 - \log\left(\frac{m_{Na}^2}{m_{Ca}}\right)} - 273.15$	$T = \frac{2814}{10.120 - \log\left(\frac{c_{Na}^2}{c_{Ca}}\right)} - 273.15$	NiNi87	(5.100)
Kharaka and Mariner (1989)	$T = \frac{2802}{6.960 - \log\left(\frac{m_{Na}^2}{m_{Ca}}\right)} - 273.15$	$T = \frac{2802}{11.080 - \log\left(\frac{c_{Na}^2}{c_{Ca}}\right)} - 273.15$	KhMa89	(5.101)
Michard (1990)	$T = \frac{2862}{4.920 - \log\left(\frac{m_{Na}^2}{m_{Ca}}\right)} - 273.15$	$T = \frac{2862}{9.040 - \log\left(\frac{c_{Na}^2}{c_{Ca}}\right)} - 273.15$	Mi90a	(5.102)
Verma and Santoyo (1997)	$T = \frac{2148}{5.130 - \log\left(\frac{m_{Na}^2}{m_{Ca}}\right)} - 273.15$	$T = \frac{2148}{9.250 - \log\left(\frac{c_{Na}^2}{c_{Ca}}\right)} - 273.15$	VeSa97	(5.103)
Santoyo and Díaz-González (2010)	$T = \frac{4624.2}{9.555 - \log\left(\frac{m_{Na}^2}{m_{Ca}}\right)} - 273.15$	$T = \frac{4624.2}{13.675 - \log\left(\frac{c_{Na}^2}{c_{Ca}}\right)} - 273.15$	SaDi10	(5.104)
Simmons (2013)	$T = \frac{7992}{16.238 - \log\left(\frac{c_{Na}^2}{c_{Ca}}\right)} - 273.15$	$T = \frac{7992}{20.359 - \log\left(\frac{c_{Na}^2}{c_{Ca}}\right)} - 273.15$	Si13	(5.105)

Table 5.4 K–Ca functions obtained from the Na–K–Ca geothermometer and the sixteen empirical Na–K geothermometers

References	K–Ca function (molal concentrations)	K–Ca function (concentrations in mg/kg)	Code	Eqn. #
White (1968)	$T = \frac{7865.6}{13.674 - \log\left(\frac{m_K^2}{m_{Ca}}\right)} - 273.15$	$T = \frac{7865.6}{18.255 - \log\left(\frac{c_K^2}{c_{Ca}}\right)} - 273.15$	Wh68	(5.106)
Ellis (1970)	$T = \frac{6676}{11.351 - \log\left(\frac{m_K^2}{m_{Ca}}\right)} - 273.15$	$T = \frac{6676}{15.932 - \log\left(\frac{c_K^2}{c_{Ca}}\right)} - 273.15$	El70	(5.107)
Mercado (1970)	$T = \frac{7320}{12.472 - \log\left(\frac{m_K^2}{m_{Ca}}\right)} - 273.15$	$T = \frac{7320}{17.053 - \log\left(\frac{c_K^2}{c_{Ca}}\right)} - 273.15$	Me70	(5.108)
Fournier and Truesdell (1973)	$T = \frac{6774}{11.559 - \log\left(\frac{m_K^2}{m_{Ca}}\right)} - 273.15$	$T = \frac{6774}{16.140 - \log\left(\frac{c_K^2}{c_{Ca}}\right)} - 273.15$	FoTr73	(5.109)
Truesdell (1976)	$T = \frac{6459.6}{10.930 - \log\left(\frac{m_K^2}{m_{Ca}}\right)} - 273.15$	$T = \frac{6459.6}{15.511 - \log\left(\frac{c_K^2}{c_{Ca}}\right)} - 273.15$	Tr76	(5.110)
Fournier (1979)	$T = \frac{5014}{8.427 - \log\left(\frac{m_K^2}{m_{Ca}}\right)} - 273.15$	$T = \frac{5014}{13.008 - \log\left(\frac{c_K^2}{c_{Ca}}\right)} - 273.15$	Fo79	(5.111)
Tonani (1980)	$T = \frac{6550}{11.239 - \log\left(\frac{m_K^2}{m_{Ca}}\right)} - 273.15$	$T = \frac{6550}{15.820 - \log\left(\frac{c_K^2}{c_{Ca}}\right)} - 273.15$	To80	(5.112)
Arnórsson et al. (1983a), 25–250 °C	$T = \frac{6150}{10.387 - \log\left(\frac{m_K^2}{m_{Ca}}\right)} - 273.15$	$T = \frac{6150}{14.968 - \log\left(\frac{c_K^2}{c_{Ca}}\right)} - 273.15$	Ar83, LT	(5.113)
Arnórsson et al. (1983a), 250–350 °C	$T = \frac{4606}{7.563 - \log\left(\frac{m_K^2}{m_{Ca}}\right)} - 273.15$	$T = \frac{4606}{12.144 - \log\left(\frac{c_K^2}{c_{Ca}}\right)} - 273.15$	Ar83, HT	(5.114)

(continued)

Table 5.4 (continued)

References	K–Ca function (molal concentrations)	K–Ca function (concentrations in mg/kg)	Code	Eqn. #
Giggenbach (1988)	$T = \frac{4322}{7.359 - \log\left(\frac{m_K^2}{m_{Ca}}\right)} - 273.15$	$T = \frac{4322}{11.940 - \log\left(\frac{c_K^2}{c_{Ca}}\right)} - 273.15$	Gi88	(5.115)
Nieva and Nieva (1987)	$T = \frac{5170}{8.479 - \log\left(\frac{m_K^2}{m_{Ca}}\right)} - 273.15$	$T = \frac{5170}{13.060 - \log\left(\frac{c_K^2}{c_{Ca}}\right)} - 273.15$	NiNi87	(5.116)
Kharaka and Mariner (1989)	$T = \frac{5162}{9.119 - \log\left(\frac{m_K^2}{m_{Ca}}\right)} - 273.15$	$T = \frac{5162}{13.700 - \log\left(\frac{c_K^2}{c_{Ca}}\right)} - 273.15$	KhMa89	(5.117)
Michard (1990)	$T = \frac{5202}{7.759 - \log\left(\frac{m_K^2}{m_{Ca}}\right)} - 273.15$	$T = \frac{5202}{12.340 - \log\left(\frac{c_K^2}{c_{Ca}}\right)} - 273.15$	Mi90a	(5.118)
Verma and Santoyo (1997)	$T = \frac{4726}{7.899 - \log\left(\frac{m_K^2}{m_{Ca}}\right)} - 273.15$	$T = \frac{4726}{12.480 - \log\left(\frac{c_K^2}{c_{Ca}}\right)} - 273.15$	VeSa97	(5.119)
Santoyo and Díaz-González (2010)	$T = \frac{6376.8}{10.849 - \log\left(\frac{m_K^2}{m_{Ca}}\right)} - 273.15$	$T = \frac{6376.8}{15.430 - \log\left(\frac{c_K^2}{c_{Ca}}\right)} - 273.15$	SaDi10	(5.120)
Simmons (2013)	$T = \frac{8622}{15.304 - \log\left(\frac{m_K^2}{m_{Ca}}\right)} - 273.15$	$T = \frac{8622}{19.886 - \log\left(\frac{c_K^2}{c_{Ca}}\right)} - 273.15$	Si13	(5.121)

In addition to the Na–Ca and K–Ca geothermometers derived from the Na–K–Ca geothermometer and the sixteen empirical Na–K geothermometers listed in Table 5.2, it is necessary to take into account the Na–Ca geothermometer of Arnórsson et al. (1983a) and the K–Ca geothermometer of Michard (1990), which were proposed as such.

Arnórsson et al. (1983a) presented a diagram of $\log(\sqrt{\mathrm{Ca}}/\mathrm{Na})$ versus the absolute temperature inverse, in which there is a large scatter of data points that was attributed to equilibrium with more than one mineral assemblage.

The $\sqrt{\mathrm{Ca}}/\mathrm{Na}$ geothermometric function of Arnórsson et al. (1983a) has a linear dependence on the absolute temperature inverse and corresponds to the following two relations identified by code Ar83b in Fig. 5.18:

$$\mathrm{T} = \frac{1120}{2.118 - \log(m_{\mathrm{Na}}^2/m_{\mathrm{Ca}})} - 273.15 \quad \mathrm{T} = \frac{1120}{6.238 - \log(c_{\mathrm{Na}}^2/c_{\mathrm{Ca}})} - 273.15. \tag{5.122}$$

The two associated K–Ca geothermometric functions can be computed by inserting the Na–Ca geothermometer defined by Eq. (5.122) and one of the two Na–K geothermometers of Arnórsson et al. (1983a), that is Eqs. (5.64) or (5.65), into Eq. (5.88). The K–Ca geothermometers thus obtained, identified by code Ar83b

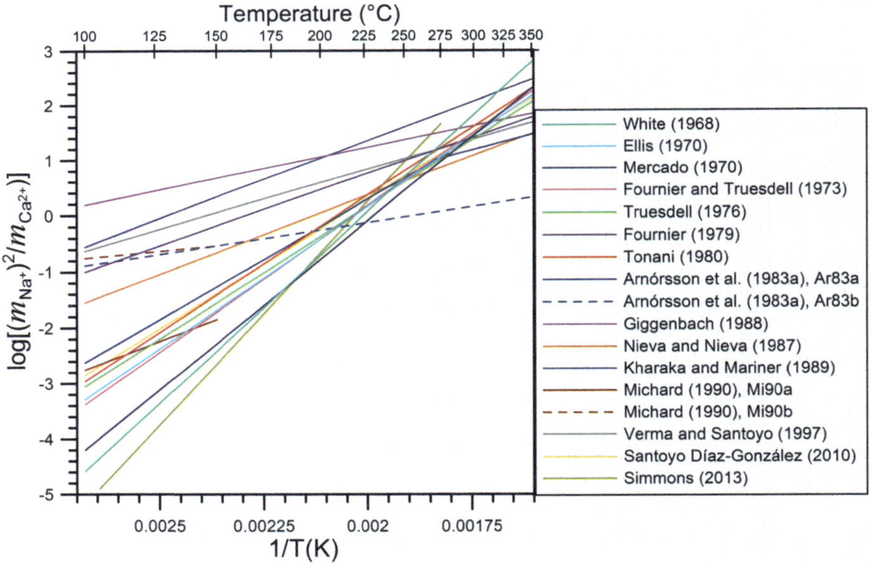

Fig. 5.18 Diagram of $\log(m_{\mathrm{Na}}^2/m_{\mathrm{Ca}})$ versus the absolute temperature inverse showing the sixteen Na–Ca geothermometers obtained from the sixteen empirical Na–K functions (see legend) and the Na–K–Ca geothermometer, as well as the Na–Ca geothermometers of Arnórsson et al. (1983a) and Michard (1990), indicated by codes Ar83b and Mi90b, respectively

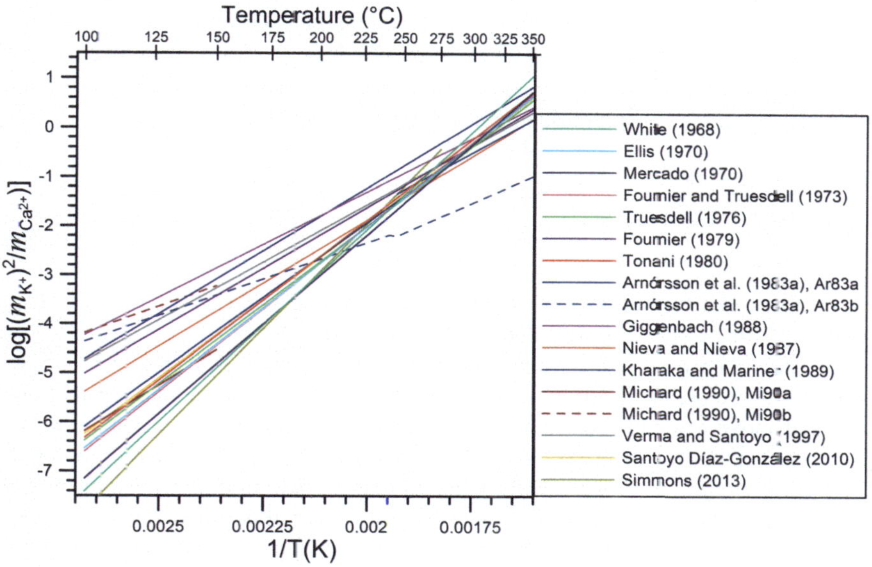

Fig. 5.19 Diagram of $\log(m_K^2/m_{Ca})$ versus the reciprocal of the absolute temperature showing the sixteen K–Ca geothermometers obtained from the sixteen empirical Na–K functions (see legend) and the Na–K–Ca geothermometer, as well as the K–Ca geothermometers of Arnórsson et al. (1983a) and Michard (1990), indicated by codes Ar83b and Mi90b, respectively

in Fig. 5.19, are:

$$T = \frac{2986}{3.643 - \log(m_K^2/m_{Ca})} - 273.15 \quad T = \frac{2986}{8.224 - \log(c_K^2/c_{Ca})} - 273.15 \tag{5.123}$$

$$T = \frac{3758}{5.055 - \log(m_K^2/m_{Ca})} - 273.15 \quad T = \frac{3758}{9.636 - \log(c_K^2/c_{Ca})} - 273.15. \tag{5.124}$$

Equation (5.123) is valid from 25 to 250 °C, whereas Eq. (5.124) applies in the temperature range 250–350 °C.

Based on the chemistry of several thermal waters interacting with granitic rocks at temperatures lower than 150 °C, Michard (1990) derived a K–Ca geothermometer, corresponding to the following two equations identified by code Mi90a in Fig. 5.19:

$$T = \frac{3030}{3.94 - \log(m_K^2/m_{Ca})} - 273.15 \quad T = \frac{3030}{8.52 - \log(c_K^2/c_{Ca})} - 273.15. \tag{5.125}$$

The related Na–Ca geothermometer can be obtained inserting the K–Ca and Na–K geothermometers of Michard (1990) defined by Eqs. (5.125) and (5.69), respectively, into Eq. (5.88). The Na–Ca geothermometer thus derived is expressed by the following two relations (identified by code Mi90a in Fig. 5.18):

$$T = \frac{690}{1.10 - \log(m_{Na}^{2}/m_{Ca})} - 273.15 \quad T = \frac{690}{5.22 - \log(c_{Na}^{2}/c_{Ca})} - 273.15. \tag{5.126}$$

Equations (5.125) and (5.126) should not be used above 150 °C, because their parent functions were calibrated below this threshold.

The plot of $\log(m_{Na}^{2}/m_{Ca})$ versus the absolute temperature inverse of Fig. 5.18 shows that there are large differences among the Na–Ca geothermometric functions discussed above, with most of them converging at temperatures higher than ~250 °C, apart from Eq. (5.102) from Michard (1990) and Eq. (5.122) from Arnórsson et al. (1983a), but diverging progressively with decreasing temperature. Since each pair of Na–Ca and Na–K functions is linked by Eq. (5.87), this graph reflects largely the shape of Fig. 5.14 with the Na–Ca geothermometers of relatively low slope corresponding to the Na–K geothermometers of comparatively high slope (e.g., Giggenbach 1988) and vice versa (e.g., Simmons 2013).

The diagram of $\log(m_{K}^{2}/m_{Ca})$ versus the absolute temperature reciprocal of Fig. 5.19 is similar to those of Figs. 5.14 and 5.18. To be noted that all the K–Ca geothermometers have slope higher than the corresponding Na–Ca geothermometers and, again, the K–Ca geothermometers of comparatively small slope correspond to the Na–K geothermometers of relatively large slope (e.g., Giggenbach 1988) and vice versa (e.g., Simmons 2013), since each couple of K–Ca and Na–K functions is connected by Eq. (5.89).

In the correlation diagrams of $\log(m_{Na}^{2}/m_{Ca})$ versus the absolute temperature inverse (Fig. 5.20) and of $\log(m_{K}^{2}/m_{Ca})$ versus the absolute temperature reciprocal (Fig. 5.21), most of the 1013 selected reservoir liquids are positioned within the same area occupied by the empirical Na–Ca and K–Ca functions, respectively. Not surprisingly, a similar distribution of data points was already noted by Fournier and Truesdell (1973) in the diagrams of log ($\sqrt{Ca}/Na$) versus the absolute temperature reciprocal and log ($\sqrt{Ca}/K$) versus the absolute temperature inverse.

Nevertheless, all or most reservoir liquids of Long Valley, Mori-Nigorikawa, Yangbajing, Kizildere, Ribeira Grande, Latera, and Aluto-Langano, as well as some reservoir liquids of Broadlands, Ngawha, and Olkaria are situated above the lines defined by the K–Ca geothermometers in Fig. 5.21 and most of these samples are positioned above the lines corresponding to the Na–Ca geothermometers in Fig. 5.20. These reservoir liquids have Na–K–Ca temperature significantly higher than aquifer temperature as discussed in Sect. 5.4.2 and shown by the diagrams of Fig. 5.17.

Again, these high Na^2/Ca and K^2/Ca log-ratios might be due to either precipitation of a Ca-bearing solid phase (typically calcite) or equilibration of the aqueous solutions

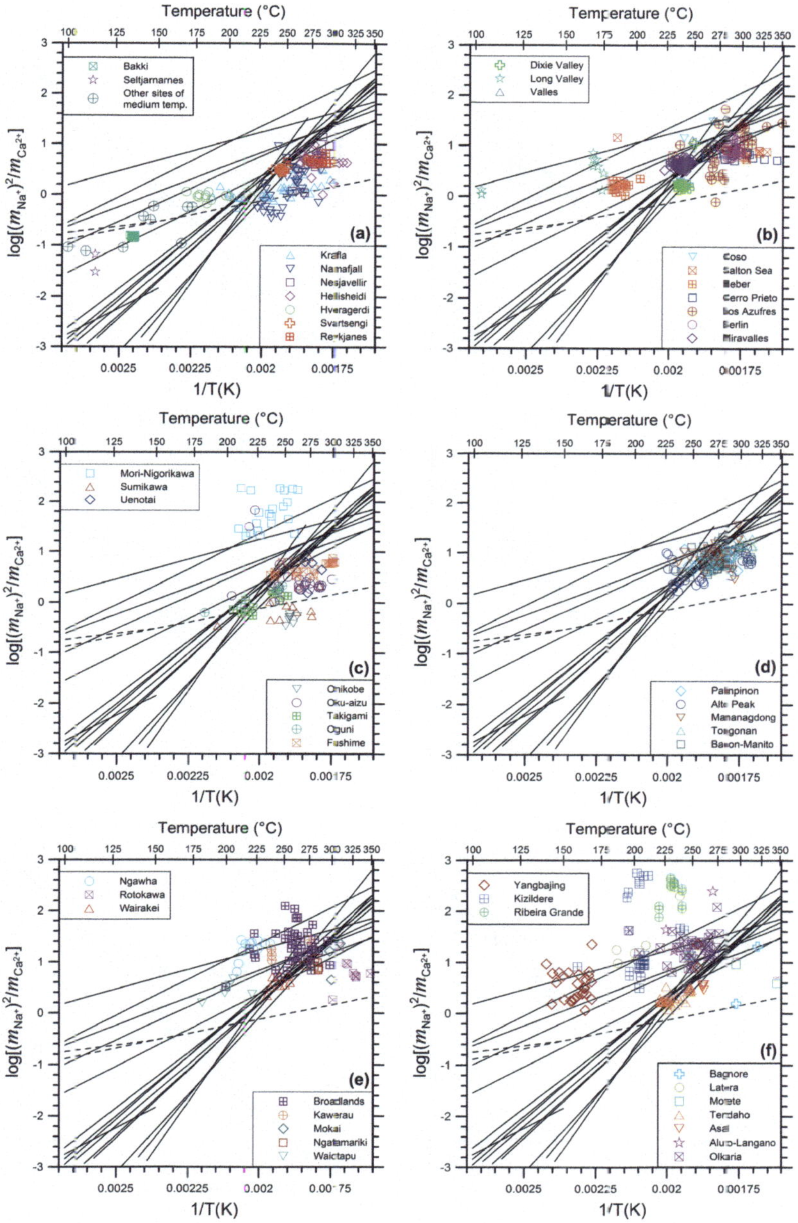

Fig. 5.20 Diagrams of $\log(m^2_{Na}/m_{Ca})$ versus the absolute temperature inverse showing the selected reservoir liquids from **a** Iceland, **b** Northern-Central America, **c** Japan, **d** The Philippines, **e** New Zealand, and **f** miscellaneous geothermal system, the sixteen Na–Ca geothermometers obtained from the sixteen empirical Na–K functions (see legend) and the Na–K–Ca geothermometer (solid black lines), as well as the Na–Ca functions of Arnórsson et al. (1983a) and Michard (1990), represented by dashed black lines

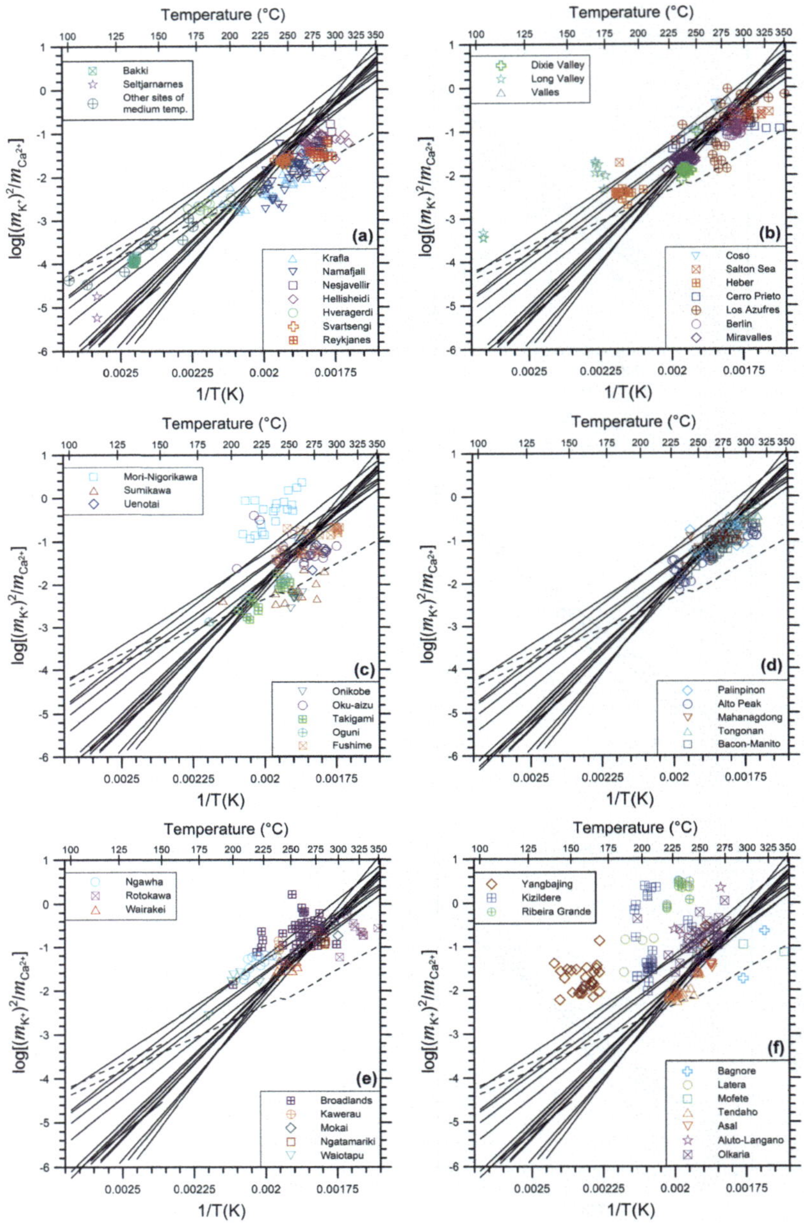

Fig. 5.21 Diagrams of $\log(m_{\mathrm{K}}^2/m_{\mathrm{Ca}})$ versus the absolute temperature inverse showing the selected reservoir liquids from **a** Iceland, **b** Northern-Central America, **c** Japan, **d** The Philippines, **e** New Zealand, and **f** miscellaneous geothermal, the sixteen K–Ca geothermometers obtained from the sixteen empirical Na–K functions (see legend) and the Na–K–Ca geothermometer (solid black lines), as well as the K-Ca functions of Arnórsson et al. (1983a) and Michard (1990), represented by dashed black lines

with calcite instead of one of the several possible Ca–Al-silicates, because of the relatively high f_{CO_2} values occurring in the geothermal reservoirs of provenance.

The Na^2/Ca and K^2/Ca log-ratios of the other reservoir liquids agree with those fixed by the Na–Ca and K–Ca geothermometers and, therefore, are probably controlled by equilibrium with one of the different Ca–Al-silicate minerals, because of the comparatively low f_{CO_2} values occurring in the related geothermal aquifers. These findings imply that the different Na–Ca and K–Ca geothermometric functions can be utilized, to some extent, to identify the governing Ca–Al-silicate minerals, as discussed in the next section.

5.4.4 The Hydrothermal Minerals Controlling the Na–Ca and K–Ca Geothermometers

Following the same approach adopted in Sect. 5.3.3 for the Na–K geothermometer, the $\Delta H°_r$ and $\Delta S°_r$ values of the empirical Na–Ca and K–Ca geothermometers, calculated from their slope and intercept using Eqs. (5.13) and (5.14), are compared with the standard enthalpies and entropies of the following Na–Ca exchange reactions:

$$2NaAlSi_3O_8 + 4H_2O + Ca^{2+} = CaAl_2Si_4O_{12} \cdot 4H_2O + 2SiO_2 + 2Na^+ \quad (5.127)$$

$$2NaAlSi_3O_8 + 0.25KAlSi_3O_8 + 0.5H_2O + Ca^{2+} = 0.25KAl_3Si_3O_{10}(OH)_2 + 0.5Ca_2Al_3Si_3O_{12}(OH) + 4.5SiO_2 + 2Na^+ \quad (5.128)$$

$$2\,NaAlSi_3O_8 + 0.5\,KAlSi_3O_8 + H_2O + Ca^{2+} = 0.5KAl_3Si_3O_{10}(OH)_2 + 0.5Ca_2Al_2Si_3O_{10}(OH)_2 + 4.5SiO_2 + 2Na^+ \quad (5.129)$$

$$2NaAlSi_3O_8 + 2H_2O + Ca^{2+} = CaAl_2Si_4O_{12} \cdot 2\,H_2O + 2SiO_2 + 2Na^+ \quad (5.130)$$

and of the following K–Ca exchange reactions:

$$2\,KAlSi_3O_8 + 4\,H_2O + Ca^{2+} = CaAl_2Si_4O_{12} \cdot 4\,H_2O + 2\,SiO_2 + 2\,K^+ \quad (5.131)$$

$$2.25\,KAlSi_3O_8 + 0.5\,H_2O + Ca^{2+} = 0.25\,KAl_3Si_3O_{10}(OH)_2 + 0.5\,Ca_2Al_3Si_3O_{12}(OH) + 4.5\,SiO_2 + 2\,K^+ \quad (5.132)$$

$$2.5\,KAlSi_3O_8 + H_2O + Ca^{2+} = 0.5\,KAl_3Si_3O_{10}(OH)_2 + 0.5\,Ca_2Al_2Si_3O_{10}(OH)_2 + 4.5\,SiO_2 + 2\,K^+ \quad (5.133)$$

$$2\,KAlSi_3O_8 + 2\,H_2O + Ca^{2+} = CaAl_2Si_4O_{12} \cdot 2\,H_2O + 2SiO_2 + 2\,K^+ \quad (5.134)$$

respectively. Laumontite [$CaAl_2Si_4O_{12} \cdot 4\ H_2O$] is involved in reactions (5.127) and (5.131), clinozoisite [$Ca_2Al_3Si_3O_{12}(OH)$] participates to reactions (5.128) and (5.132), prehnite [$Ca_2Al_2Si_3O_{10}(OH)_2$] takes part to reactions (5.129) and (5.133) and wairakite [$CaAl_2Si_4O_{12} \cdot 2\ H_2O$] plays a part in reactions (5.130) and (5.134). Among the other minerals involved in the Na–Ca and K–Ca exchange reactions of interest, maximum-microcline and high-sanidine were alternatively considered to represent K-feldspar, whereas the silica mineral was assumed to be quartz/chalcedony above 175 °C and chalcedony below 175 °C. The thermodynamic properties of these reactions were computed as a function of temperature at water saturation pressures using SUPCRT92.

The correlation diagram of Fig. 5.22 shows that the $\Delta H°_r$ and $\Delta S°_r$ values of most Na–Ca geothermometers are consistent with the standard enthalpies and entropies of the Na–Ca exchange reactions of interest computed using SUPCRT92. Only the Na–Ca functions derived from the Na–K geothermometers of White (1968), Mercado

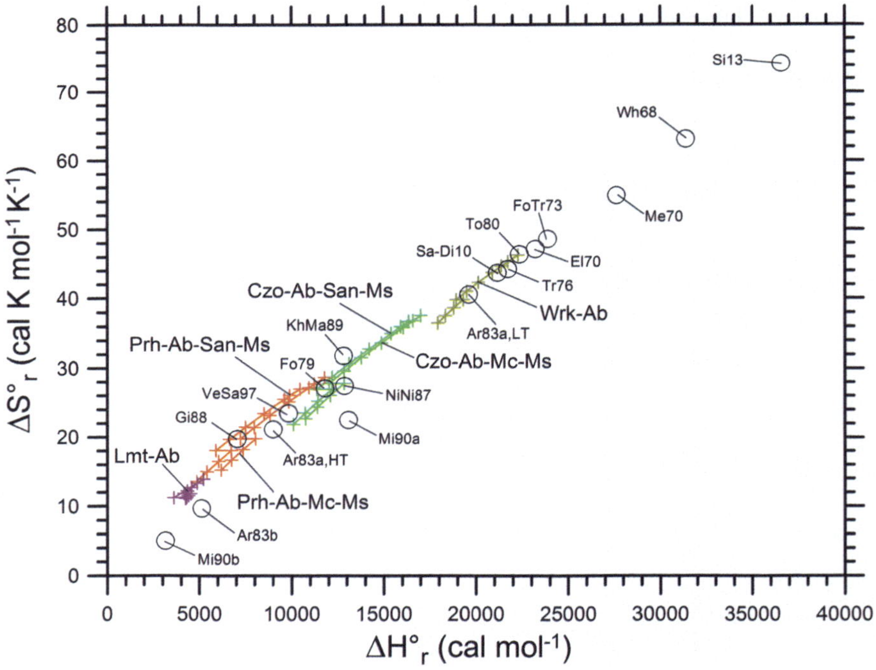

Fig. 5.22 Comparison of the temperature-independent $\Delta H°_r$ and $\Delta S°_r$ values of the empirical Na–Ca geothermometers (open black circles) with those of the Na–Ca exchange reactions involving laumontite (Lmt) or prehnite (Prh) or clinozoisite (Czo) or wairakite (Wrk), and other pertinent minerals, computed as a function of temperature at water saturation pressures using SUPCRT92

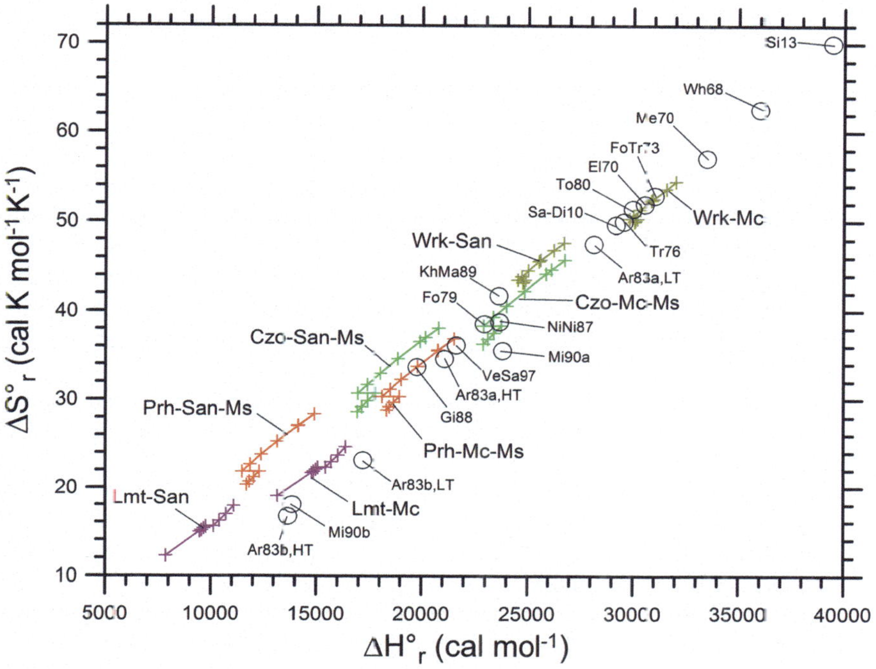

Fig. 5.23 Comparison of the temperature-independent ΔH°_r and ΔS°_r values of the empirical K–Ca geothermometers (open black circles) with those of the K–Ca exchange reactions involving laumontite (Lmt) or prehnite (Prh) or clinozoisite (Czo) or wairakite (Wrk), and other pertinent minerals, computed as a function of temperature at water saturation pressures using SUPCRT92

(1970), and Simmons (2013) have ΔH°_r and ΔS°_r values at variance with the standard enthalpies and entropies of the considered Na–Ca exchange reactions. Similar inferences can be drawn from the correlation diagram of Fig. 5.23, in which the temperature-independent standard enthalpies and entropies of the empirical K–Ca geothermometers, obtained from their slope and intercept, are contrasted with the ΔH°_r and ΔS°_r values of the K–Ca exchange reactions of interest, calculated as a function of temperature at water saturation pressures using SUPCRT92.

These findings have two implications. On the one hand, the Na–Ca and K–Ca geothermometers provide temperatures presumably controlled by the equilibrium condition between the aqueous solution of interest and a given Ca–Al-silicate, assuming that f_{CO_2} is low enough. However, it is impossible to identify the Ca–Al-silicate involved in the Na–Ca and K–Ca exchange reactions presumably controlling the Na–Ca and K–Ca geothermometers owing to their empirical nature, intrinsic and/or inherited from the parent functions, i.e., the Na–K–Ca geothermometer and a given Na–K geothermometer. On the other hand, it should be possible to calibrate theoretical Na–Ca and K–Ca geothermometers based on the exchange reactions considered in this section. This possibility will be explored in Chap. 8.

5.4.5 Conclusive Remarks on the Na–K–Ca Geothermometer

Summing up, the Na–Ca and K–Ca geothermometers obtained by splitting the Na–K–Ca function of Fournier and Truesdell (1973) as well as the Na–Ca geothermometer of Arnórsson et al. (1983a) and the K–Ca geothermometer of Michard (1990) work under favorable conditions, that is for waters presumably equilibrated with an unspecified Ca–Al-silicate at relatively low f_{CO_2} values. In contrast, these Na–Ca and K–Ca geothermometers do not work for waters equilibrated with calcite under comparatively high f_{CO_2} values. This implies that the Na^2/Ca and K^2/Ca ratios represent the basis not only for geothermometers but also for f_{CO_2}-indicators. Our view on this point is somewhat different from that of previous authors which is briefly summarized below.

Ellis (1970) pointed out that Ca is a potential CO_2-indicator for thermal waters, proposing a technique to evaluate deep f_{CO_2} values on the basis of Na and Ca concentrations.

As already noted in Sect. 5.4.1, Fournier and Truesdell (1973) recognized that calcite saturation controls Ca concentration, but they did not consider calcite and CO_2 in reactions (5.78) and (5.79), that were assumed to control the Na–K–Ca geothermometer by these authors.

Giggenbach (1988) summarized the results of several investigations which were carried out for providing both theoretical and experimental justifications to the Na–K–Ca geothermometer. He concluded that many problems in the use of this technique are connected with its sensitivity to differences in the CO_2 contents of geothermal fluids, especially at relatively low temperatures.[9] Therefore, Giggenbach (1988) suggested that Ca may be used to formulate a f_{CO_2}-indicator, representing the subject of the next section, rather than constituting the basis for a geothermometer.

This suggestion was later confirmed by Chiodini et al. (1991) based on the results of a mineral-solution equilibrium model (see Sect. 5.8).

5.5 The K–Ca f_{CO_2}-Indicator

5.5.1 Relevant Reactions, Derivation, and Limitations of the K–Ca f_{CO_2}-Indicator

According to Giggenbach (1988), it is likely that the initial CO_2 contents of deep geothermal fluids in areas affected by magmatic processes are externally controlled by variable contributions of CO_2-rich magmatic fluids and CO_2-poor meteoric waters.

[9]For instance, according to Paces (1975), the aqueous solution attains a steady state condition with the felsic rocks, in aquifers at temperatures below 75 °C and CO_2 partial pressures above 10^{-4} atm. Therefore, he suggested an empirical correction involving the aquifer CO_2 partial pressure to be applied to the Na–K–Ca temperature.

These deep geothermal fluids are expected to become reactive with respect to the conversion of Ca–Al-silicates to either calcite and clay minerals or calcite and K-mica. An example of the first process is the reaction:

$$2\,Ca_2Al_3Si_3O_{12}(OH) + 4\,CO_2 + 5\,H_2O = 3\,Al_2Si_2O_5(OH)_4 + 4\,CaCO_3 \quad (5.135)$$

in which the considered Ca–Al-silicate is clinozoisite. Reaction (5.135) involves the formation of kaolinite or similar minerals typical of acid alteration assemblages, such as dickite or pyrophyllite (Reyes 1990). An example of the second process is the reaction:

$$1.5\,KAlSi_3O_8 + Ca_2Al_3Si_3O_{12}(OH) + 2\,CO_{2(g)} + H_2O = 1.5\,KAl_3Si_3O_{10}(OH)_2 + 3\,SiO_2 + 2\,CaCO_3 \quad (5.136)$$

in which, again, the Ca–Al-silicate is clinozoisite. The temperature dependence of reaction (5.136) and similar reactions involving Ca–Al-silicates other than clinozoisite is described by the following equation (Giggenbach 1984; f_{CO_2} in bar, T in °C):

$$\log f_{CO_2} = 0.0168\,T - 3.78, \quad (5.137)$$

which is valid from 100 to 350 °C. The univariant reaction (5.136) involves two constituents of the full equilibrium hydrothermal mineral assemblage, K-mica and K-feldspar, and represents the calcite/Ca–Al-silicate boundary. For a given temperature, calcite is stable for f_{CO_2} higher than that given by Eq. (5.137), whereas the Ca–Al-silicate mineral (either laumontite, epidote or wairakite, depending on temperature) is stable for f_{CO_2} lower than that given by Eq. (5.137).

Arnórsson and Gunnlaugsson (1985) recognized that f_{CO_2} is governed, above ~230 °C, by a buffer comprising calcite, epidote, prehnite and quartz and Arnórsson (1985) proposed that the partial pressure of CO_2 is fixed by temperature-dependent mineral-solution equilibria as described by the following relation (P_{CO_2} in bar, T in K):

$$\log P_{CO_2} = -2.81 - 5012.7 \cdot T^{-1} - 0.00919 \cdot T + 6.464 \cdot \log T. \quad (5.138)$$

Equation (5.138) agrees closely with Eq. (5.137) from 200 to 250 °C, whereas the two functions diverge progressively the one from the other below 200 °C and above 250 °C with maximum deviations of about one order of magnitude at 100 and 350 °C. Similar temperature- f_{CO_2} (or P_{CO_2}) relations were also proposed by Giggenbach (1981), Arnórsson et al. (1983a), and Arnórsson and Gunnlaugsson (1985).

According to Giggenbach (1988), the conversion of K-feldspar to K-mica promoted by CO_2 with production of calcite and a silica mineral:

$$3\,KA1Si_3O_8 + CO_{2(g)} + Ca^{2+} + H_2O = CaCO_3 + KAl_3Si_3O_{10}(OH)_2 + 6SiO_2 + 2K^+, \quad (5.139)$$

is likely to occur over large portions of geothermal systems. The thermodynamic equilibrium constant of reaction (5.139) is equal to $\log(a^2_{K^+}/a_{Ca^{2+}}) - \log f_{CO_2}$ if the solid phases and water are assumed to be pure. Moreover, the log K of reaction (5.139) is practically temperature independent, being equal to -1.66 ± 0.15, in the temperature range 50–300 °C, based on the thermodynamic data of Bowers et al. (1984). Therefore, to a first approximation, the K^2/Ca concentration ratio can be assumed to be a f_{CO_2}-indicator, as described by the following equation (Giggenbach 1988; concentrations in mg/kg, f_{CO_2} in bar):

$$\log f_{CO_2} = \log(c^2_{K^+}/c_{Ca^{2+}}) - 3.0. \quad (5.140)$$

As underscored by Giggenbach (1988), the K–Ca f_{CO_2}-indicator is based on the hypothesis of chemical equilibrium attainment between the aqueous solution and all the solid phases participating to reaction (5.139). Therefore, the K–Ca f_{CO_2}-indicator can be applied to the comparatively high CO_2 fugacities at which calcite is stable and Ca–Al-silicate minerals are not, whereas it cannot be used at the relatively low CO_2 fugacities where the Ca–Al-silicate minerals are stable and calcite is not. The threshold between these two distinct situations is defined by Eqs. (5.137) or (5.138).

5.5.2 *Application of the K–Ca f_{CO_2}-Indicator to the Selected Reservoir Liquids*

Giggenbach (1988) used the K–Ca f_{CO_2}-indicator together with the K–Mg geothermometer because these two geo-indicators are expected to equilibrate with comparable speed and under comparable conditions. Since there are several doubts on the reliability of the K–Mg geothermometer (see Sect. 5.6), the results of the K–Ca f_{CO_2}-indicator for the 1013 selected reservoir liquids are plotted against the reference (aquifer) temperatures, instead of the K–Mg temperatures, in the diagrams of Fig. 5.24, also showing the approximate calcite/Ca–Al-silicate stability boundary or full equilibrium lines, represented by Eqs. (5.137) and (5.138).

The data points situated above the full equilibrium lines, such as all or most reservoir liquids of Long Valley (Fig. 5.24b), Mori-Nigorikawa (Fig. 5.24c), Ngawha, Kawerau, Broadlands (Fig. 5.24e), Yangbajing, Kizildere, Latera, Ribeira Grande, Aluto-Langano, and Olkaria (Fig. 5.24f) have computed CO_2 fugacities much higher than the full equilibrium value. Indeed, it is possible that the CO_2 fugacities of these reservoir liquids are externally controlled rather than internally fixed by mineral-solution equilibria.

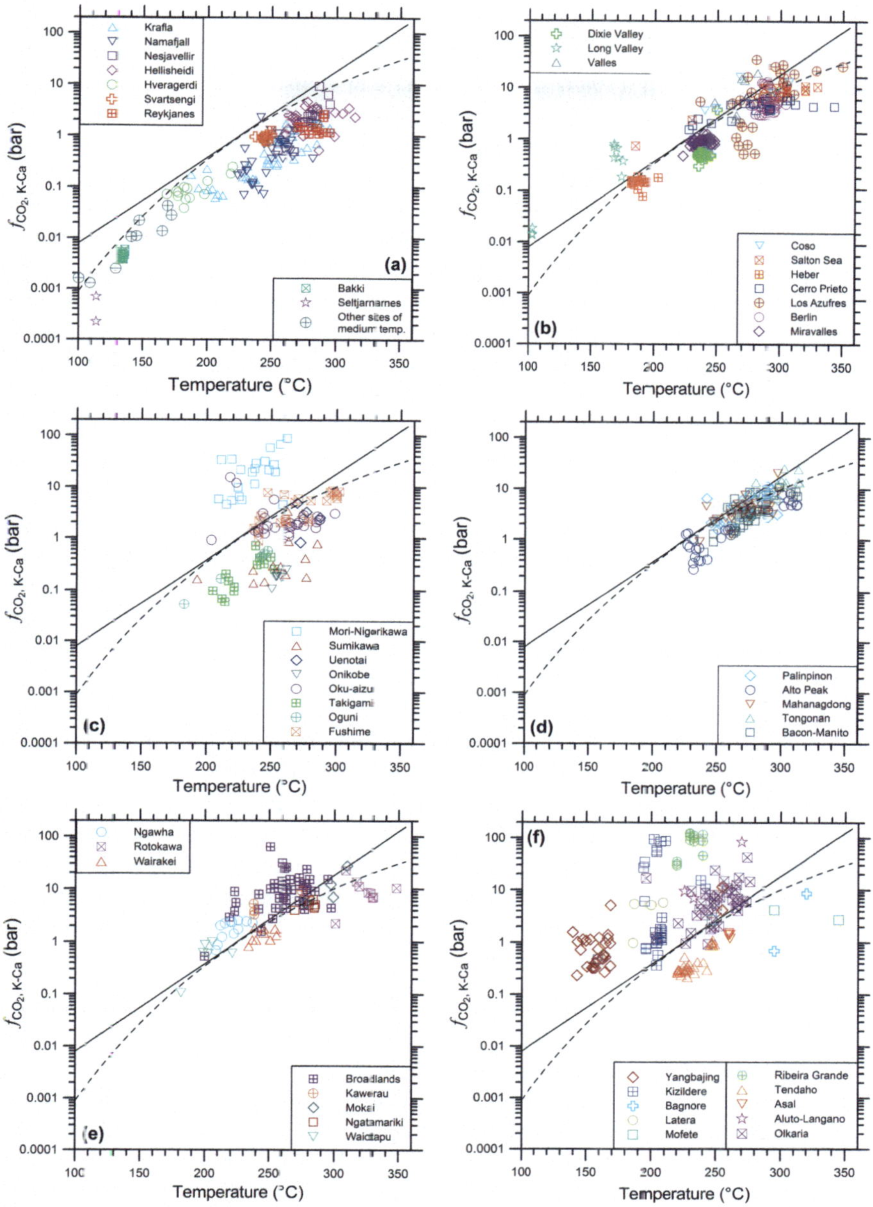

Fig. 5.24 Diagrams of the aquifer temperature versus the CO_2 fugacity computed by means of the K–Ca f_{CO_2}-indicator of Giggenbach (1988), Eq. (5.140), for the selected reservoir liquids from **a** Iceland, **b** Northern-Central America, **c** Japan, **d** The Philippines, **e** New Zealand, and **f** miscellaneous geothermal systems. The CO_2 fugacity spans six orders of magnitude and, therefore, is reported on a log scale. Also shown are the full equilibrium lines of Giggenbach (1984; solid black line), Eq. (5.137), and Arnórsson (1985; dashed black line), Eq. (5.138)

However, calcite scaling is known to occur in these systems, as reported for instance by Brown et al. (2013) for Long Valley, Yoshida (1991) for Mori-Nigorikawa, Sheppard (1984) for Ngawha, Christenson (1997) for Kawerau, Simmons and Christenson (1994) for Broadlands, Zheng et al. (2019) for Yangbajing, Tarcan et al. (2016) for Kizildere, Pieri et al. (1989) for Latera, Viveiros Pereira (2014) for Ribeira Grande, Tassew (2010) for Aluto-Langano, and Opondo (2015) for Olkaria. This process may cause a considerable decrease in Ca concentration and a related increase in both the K^2/Ca concentration ratio and in the computed CO_2 fugacity, which could overestimate the real value.

Further indications are given by the comparison of the CO_2 fugacities computed by means of the K–Ca f_{CO_2}-indicator and the corresponding CO_2 fugacities obtained through speciation calculations, generally carried out by means of the computer code WATCH, that is combining the analyses of the liquid and vapor phases separated at known pressure, temperature conditions. Such a comparison, performed through the log-log diagrams of Fig. 5.25 shows that:

(i) Most reservoir liquids of Long Valley (Fig. 5.25b), Ngawha (Fig. 5.25e) Kizildere, Aluto-Langano, and Latera (Fig. 5.25f) have CO_2 fugacities given by speciation calculations much higher than those given by the K–Ca f_{CO_2}-indicator and, therefore, they might have excess CO_2 with respect to the CO_2 concentration controlled by equilibrium coexistence of calcite, K-feldspar, K-mica and a silica mineral.

(ii) The reservoir liquids of Mori-Nigorikawa (Fig. 5.25c) and Kawerau (Fig. 5.25e) have CO_2 fugacities obtained through speciation calculations similar to those calculated using the K–Ca f_{CO_2}-indicator and, therefore, they might be close to equilibrium with calcite, K-feldspar, K-mica and a silica mineral.

(iii) Most reservoir liquids of Broadlands (Fig. 5.25e) Yangbajing, Ribeira Grande, and Olkaria (Fig. 5.24f) have CO_2 fugacities given by speciation calculations lower than those calculated using the K–Ca f_{CO_2}-indicator. Assuming that the CO_2 fugacities given by speciation calculations are reliable (which is far from certainty), the results of the K–Ca f_{CO_2}-indicator could overestimate the real values because of calcite precipitation.

Returning to consider the diagrams of Fig. 5.24, the sample points positioned below the full equilibrium line, such as the reservoir liquids from the high-temperature Icelandic geothermal systems (Fig. 5.24a), Berlin, Los Azufres (Fig. 5.24b), Fushime, Oku-aizu, Sumikawa, Uenotai (Fig. 5.24c), Alto Peak, Bacon-Manito, Mahanagdong, Palinpinon, and Tongonan (Fig. 5.24d), might indicate that comparatively low contributions of deep CO_2-rich fluids enter these geothermal aquifers from below, stabilizing the Ca–Al-silicates and destabilizing calcite. Alternatively, the low CO_2 concentrations of these reservoir liquids, or at least some of them, could be caused by aquifer boiling which is known to occur in these geothermal systems.

The log-log diagrams of the CO_2 fugacities computed by means of the K–Ca f_{CO_2}-indicator versus the corresponding CO_2 fugacities obtained through speciation calculations (Fig. 5.25) are useful for solving this dilemma. In fact:

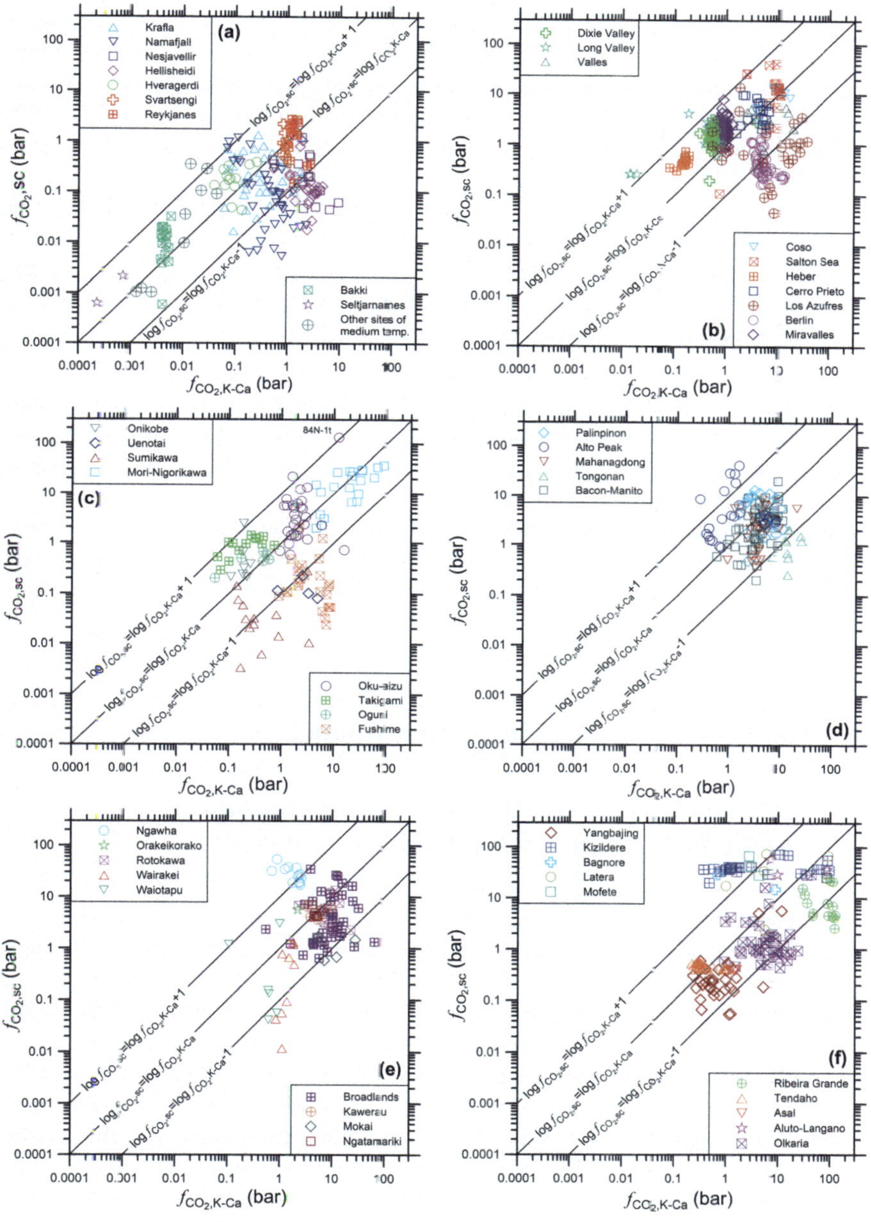

Fig. 5.25 Log-log diagrams of the CO_2 fugacities computed by means of the K–Ca f_{CO_2}-indicator versus the corresponding CO_2 fugacities obtained through speciation calculations, for the selected reservoir liquids from **a** Iceland, **b** Northern-Central America, **c** Japan, **d** The Philippines, **e** New Zealand, and **f** miscellaneous geothermal systems

(i) Most reservoir liquids of Reykjanes, Svartsengi, Hveragerdi, and Bakki (Fig. 5.25a), Oku-aizu (Fig. 5.25c), Alto Peak, Bacon-Manito, Mahanagdong, and Palinpinon (Fig. 5.25d) have CO_2 fugacities obtained through speciation calculations similar to or somewhat higher than those calculated through the K–Ca f_{CO_2}-indicator, suggesting that the occurrence of aquifer boiling is unlikely.

(ii) Several reservoir liquids of Krafla, Namafjall, Nesjavellir, Hellisheidi (Fig. 5.25a), Berlin, Los Azufres (Fig. 5.25b), Fushime, Sumikawa, Uenotai (Fig. 5.25c), and Tongonan (Fig. 5.25d), have CO_2 fugacities given by speciation calculations lower than those computed using the K–Ca f_{CO_2}-indicator, possibly because of aquifer boiling.

According to Giggenbach (1988), K and Ca concentrations can be used to evaluate reliable CO_2 fugacities only for sample points found near the full equilibrium line. However, it cannot be excluded that also these fluids might be affected by the processes mentioned above. Consequently, the results of the K–Ca f_{CO_2}-indicator have always a certain degree of uncertainty.

Further uncertainties are caused by the possible involvement of adularia of variable order-disorder degree in reactions (5.136) and (5.139) instead of fully ordered maximum-microcline as assumed by Giggenbach (1984, 1988). This discussion will be resumed in Chap. 8.

5.6 The K–Mg and Na–Mg Geothermometers

5.6.1 Relevant Reactions, Derivation, and Limitations of the K–Mg and Na–Mg Geothermometers

It was recognized long ago that the waters of high temperature (250–300 °C) and low salinity have exceptionally low Mg concentrations, typically 10–100 ppb only (Mahon 1965). Moreover, it was also understood that Mg concentrations are strongly dependent on temperature, and this relationship was attributed to equilibration of geothermal liquids with chlorites owing to their ubiquitous occurrence in the hydrothermal alteration parageneses. Ellis (1971) carried out laboratory experiments to investigate the chemistry of the aqueous solution in equilibrium with chlorite, calcite, and quartz, under controlled CO_2 partial pressures and to calibrate the temperature dependence of Mg concentration. Starting from these premises, Giggenbach (1988) assumed that the behavior of Mg in well-equilibrated geothermal waters is described by the following two reactions:

$$\begin{aligned} &0.8\,\text{muscovite} + 0.2\,\text{clinochlore} + 5.4\,\text{chalcedony} + 2\text{K}^{+} \\ &\quad = 2.8\text{K-feldspar} + 1.6\text{H}_2\text{O} + \text{Mg}^{2+} \end{aligned} \qquad (5.141)$$

and

$$\begin{aligned} &0.8\ \text{muscovite} + 0.2\ \text{clinochlore} + 5.4\ \text{chalcedony} + 2\text{Na}^+ \\ &\quad = 2\ \text{albite} + 0.8\text{K-feldspar} + 1.6\text{H}_2\text{O} + \text{Mg}^{2+}, \end{aligned} \tag{5.142}$$

which are written involving clinochlore and muscovite as endmember components of chlorites and illites, respectively. The corresponding equilibrium constants are:

$$\text{K}_{\text{Mg-K}} = \frac{a_{\text{Mg}^{2+}}}{a^2_{\text{K}^+}} \frac{a^{2.8}_{\text{Kfs}} \cdot a^{1.6}_{\text{H}_2\text{O}}}{a^{0.8}_{\text{Ms}} \cdot a^{0.2}_{\text{Clc}} \cdot a^{5.4}_{\text{Chc}}} \tag{5.143}$$

$$\text{K}_{\text{Mg-Na}} = \frac{a_{\text{Mg}^{2+}}}{a^2_{\text{Na}^+}} \frac{a^2_{\text{Ab}} \cdot a^{0.8}_{\text{Kfs}} \cdot a^{1.6}_{\text{H}_2\text{O}}}{a^{0.8}_{\text{Ms}} \cdot a^{0.2}_{\text{Clc}} \cdot a^{5.4}_{\text{Chc}}}. \tag{5.144}$$

The activities of H_2O, chalcedony, and alkali feldspars are close to unit, whereas the effects of clinochlore and muscovite incorporation into natural chlorites and illites was assessed by Giggenbach (1988) considering the data reported by Capuano and Cole (1982) for Roosevelt Hot Springs and by Walshe (1986) for a number of hydrothermal phyllosilicates.

The logarithm of clinochlore activity in the chlorites of interest varies from −4.7 to −1.3 with a mean of −2.2. The logarithm of muscovite activity in the considered illites ranges from −0.73 to −0.15 with an average of −0.33. Hence, the logarithm of the activity product AP $= a^{0.8}_{\text{Ms}} \cdot a^{0.2}_{\text{Clc}}$ ranges from −1.5 to −0.4 with a mean of −0.7. Instead of considering this average AP, Giggenbach (1988) arbitrarily assumed an AP value of +0.3 to derive the K–Mg geothermometer, obtaining the following equations, for concentrations c_i in mg/kg and m_i in mol/kg:

$$\text{T} = \frac{4410}{14.0 - \log(\text{c}^2_{\text{K}}/\text{c}_{\text{Mg}})} - 273.15 \quad \text{T} = \frac{4410}{9.2 - \log(m^2_{\text{K}}/m_{\text{Mg}})} - 273.15. \tag{5.145}$$

Giggenbach (1988) justified the choice of the AP value of +0.3 to adjust for the positions of high-temperature data points in a plot of log $(a_{\text{Mg}^{2+}}/a^2_{\text{K}^+})$ versus the absolute temperature inverse (Fig. 5b of Giggenbach 1988) and with the assumption that measured temperatures of low-temperature waters are likely to underestimate their actual mineral-solution equilibrium temperatures. Moreover, Giggenbach (1988) concluded that there is no reason to favor a theoretical over the empirical calibration, because of possible uncertainties in the thermodynamic properties adopted for K-feldspar, muscovite and clinochlore, that were obtained from Bowers et al. (1984). Summing up, the K–Mg geothermometer is empirical and its derivation leaves much to be desired. Equation (5.145) is slightly different from the relation previously published by Giggenbach et al. (1983) whose intercept (entropy term) is 13.95 instead of 14.0 for concentrations in mg/kg.

In addition to the temperature dependence of the Na/K and K^2/Mg concentration ratios, Giggenbach (1988) also computed the internally consistent temperature dependence of the Na^2/Mg concentration ratio inserting Eqs. (5.66) and (5.145) into the following simple relation, written for concentrations in mg/kg:

$$\log(c_{Na}^2/c_{Mg}) = \log(c_K^2/c_{Mg}) + 2 \cdot \log(c_{Na}/c_K). \quad (5.146)$$

Because of the infinite number of possible Na–K geothermometers (see chapter 5.3), there is also an infinite number of possible Na–Mg geothermometers, which are related to the K–Mg geothermometer of Giggenbach (1988) through Eq. (5.146). However, only the sixteen empirical Na–K geothermometers, with linear dependence on the absolute temperature inverse (see Table 5.2 in chapter 5.3), were alternatively inserted into Eq. (5.146) together with K–Mg geothermometer of Giggenbach (1988) expressed by Eq. (5.145) to keep the following discussion to a simple level. In this way, the sixteen Na–Mg geothermometers listed in Table 5.5 were obtained. Again, two relations are reported, one for concentrations in mg/kg, the other for molal concentrations, although the conversion from one form to the other is straightforward.

As part of their systematic, computerized procedure, Nieva and Nieva (1987) proposed the following Na–K–Mg relation:

$$T = \frac{11140}{18.30 + 6 \cdot \log(m_{Na}/m_K) + \log(m_{Mg}/m_{Na}^2)} - 273.15. \quad (5.163)$$

Inserting the Na–K geothermometer of Nieva and Nieva (1987), i.e. Eqn (5.67), into Eq. (5.163) and rearranging, one obtains the following Na–Mg geothermometric equation (indicated by code NiNi87b) for concentrations c_i in mg/kg and m_i in mol/kg:

$$T = \frac{4072}{10.866 - \log(m_{Na}^2/m_{Mg})} - 273.15,$$
$$T = \frac{4072}{15.203 - \log(c_{Na}^2/c_{Mg})} - 273.15. \quad (5.164)$$

The corresponding K–Mg geothermometer is then derived by inserting the Na–K and Na–Mg geothermometers of Nieva and Nieva (1987), that is Eqs. (5.67) and (5.164), in consistent units, into Eq. (5.146) and reorganizing:

$$T = \frac{6428}{13.344 - \log(m_K^2/m_{Mg})} - 273.15,$$
$$T = \frac{6428}{18.143 - \log(c_K^2/c_{Mg})} - 273.15. \quad (5.165)$$

Table 5.5 Na–Mg geothermometers obtained from the K–Mg geothermometer of Giggenbach (1988) and different Na–K geothermometers, as indicated

Na–K references	Na–Mg function involving molal concentrations	Na–Mg function involving concentrations in mg/kg	Code	Eqn. #
White (1968)	$T = \frac{3401.8}{9.32-\log\left(\frac{m_{Na}^2}{m_{Mg}}\right)} - 273.15$	$T = \frac{3401.8}{13.66-\log\left(\frac{c_{Na}^2}{c_{Mg}}\right)} - 273.15$	Wh68	(5.147)
Ellis (1970)	$T = \frac{2807}{8.16-\log\left(\frac{m_{Na}^2}{m_{Mg}}\right)} - 273.15$	$T = \frac{2807}{12.50-\log\left(\frac{c_{Na}^2}{c_{Mg}}\right)} - 273.15$	El70	(5.148)
Mercado (1970)	$T = \frac{3129}{8.72\;\log\left(\frac{m_{Na}^2}{m_{Mg}}\right)} - 273.15$	$T = \frac{3129}{13.06-\log\left(\frac{c_{Na}^2}{c_{Mg}}\right)} - 273.15$	Me70	(5.149)
Fournier and Truesdell (1973)	$T = \frac{2856}{8.26-\log\left(\frac{m_{Na}^2}{m_{Mg}}\right)} - 273.15$	$T = \frac{2856}{12.60-\log\left(\frac{c_{Na}^2}{c_{Mg}}\right)} - 273.15$	FoTr73	(5.150)
Truesdell (1976)	$T = \frac{2698.8}{7.95-\log\left(\frac{m_{Na}^2}{m_{Mg}}\right)} - 273.15$	$T = \frac{2698.8}{12.29-\log\left(\frac{c_{Na}^2}{c_{Mg}}\right)} - 273.15$	Tr76	(5.151)
Fournier (1979)	$T = \frac{1976}{6.70-\log\left(\frac{m_{Na}^2}{m_{Mg}}\right)} - 273.15$	$T = \frac{1976}{11.03-\log\left(\frac{c_{Na}^2}{c_{Mg}}\right)} - 273.15$	Fo79	(5.152)
Tonani (1980)	$T = \frac{2744}{8.10-\log\left(\frac{m_{Na}^2}{m_{Mg}}\right)} - 273.15$	$T = \frac{2744}{12.44-\log\left(\frac{c_{Na}^2}{c_{Mg}}\right)} - 273.15$	To80	(5.153)
Arnórsson et al. (1983a), 25–250 °C	$T = \frac{2544}{7.68-\log\left(\frac{m_{Na}^2}{m_{Mg}}\right)} - 273.15$	$T = \frac{2544}{12.01-\log\left(\frac{c_{Na}^2}{c_{Mg}}\right)} - 273.15$	Ar83, LT	(5.154)
Arnórsson et al. (1983a),250–350 °C	$T = \frac{1772}{6.27-\log\left(\frac{m_{Na}^2}{m_{Mg}}\right)} - 273.15$	$T = \frac{1772}{10.60-\log\left(\frac{c_{Na}^2}{c_{Mg}}\right)} - 273.15$	Ar83, HT	(5.155)

(continued)

Table 5.5 (continued)

Na–K references	Na–Mg function involving molal concentrations	Na–Mg function involving concentrations in mg/kg	Code	Eqn. #
Giggenbach (1988)	$T = \frac{1630}{6.16 - \log\left(\frac{m_{Na}^2}{m_{Mg}}\right)} - 273.15$	$T = \frac{1630}{10.50 - \log\left(\frac{c_{Na}^2}{c_{Mg}}\right)} - 273.15$	Gi88	(5.156)
Nieva and Nieva (1987)	$T = \frac{2054}{6.72 - \log\left(\frac{m_{Na}^2}{m_{Mg}}\right)} - 273.15$	$T = \frac{2054}{11.06 - \log\left(\frac{c_{Na}^2}{c_{Mg}}\right)} - 273.15$	NiNi87a	(5.157)
Kharaka and Mariner (1989)	$T = \frac{2050}{7.04 - \log\left(\frac{m_{Na}^2}{m_{Mg}}\right)} - 273.15$	$T = \frac{2050}{11.38 - \log\left(\frac{c_{Na}^2}{c_{Mg}}\right)} - 273.15$	KhMa89	(5.158)
Michard (1990)	$T = \frac{2070}{6.36 - \log\left(\frac{m_{Na}^2}{m_{Mg}}\right)} - 273.15$	$T = \frac{2070}{10.70 - \log\left(\frac{c_{Na}^2}{c_{Mg}}\right)} - 273.15$	Mi90	(5.159)
Verma and Santoyo (1997)	$T = \frac{1832}{6.43 - \log\left(\frac{m_{Na}^2}{m_{Mg}}\right)} - 273.15$	$T = \frac{1832}{10.77 - \log\left(\frac{c_{Na}^2}{c_{Mg}}\right)} - 273.15$	VeSa97	(5.160)
Santoyo and Díaz-González (2010)	$T = \frac{2657.4}{7.91 - \log\left(\frac{m_{Na}^2}{m_{Mg}}\right)} - 273.15$	$T = \frac{2657.4}{12.25 - \log\left(\frac{c_{Na}^2}{c_{Mg}}\right)} - 273.15$	SaDi10	(5.161)
Simmons (2013)	$T = \frac{3780}{10.14 - \log\left(\frac{m_{Na}^2}{m_{Mg}}\right)} - 273.15$	$T = \frac{3780}{14.47 - \log\left(\frac{c_{Na}^2}{c_{Mg}}\right)} - 273.15$	Si13	(5.162)

Based on the chemical characteristics of waters interacting with granitic lithotypes at temperatures lower than 150 °C, Michard (1990) derived a K–Mg geothermometer, which is defined by the following two relations (indicated by code Mi90b):

$$T = \frac{3000}{5.84 - \log(m_K^2/m_{Mg})} - 273.15,$$
$$T = \frac{3000}{10.64 - \log(c_K^2/c_{Mg})} - 273.15. \quad (5.166)$$

The corresponding Na–Mg geothermometer is obtained inserting the Na–K and K–Mg geothermometers of Michard (1990), that is Eqs. (5.69) and (5.166), in consistent units, into Eq. (5.146) and rearranging:

$$T = \frac{660}{3.00 - \log(m_{Na}^2/m_{Mg})} - 273.15, \quad T = \frac{660}{7.34 - \log(c_{Na}^2/c_{Mg})} - 273.15. \quad (5.167)$$

These two functions should not be used outside the calibration range, that is above 150 °C.

Two K–Mg geothermometric functions were proposed by Fournier (1991). However, the equations listed in Table 3 of Fournier (1991) are at variance with the corresponding lines reported in the diagram of $\log(K^2/Mg)$ versus the absolute temperature reciprocal, that is Fig. 9 of Fournier (1991). Relying on the diagram, the two K–Mg geothermometers of Fournier (1991) are defined by the following relations (indicated by code Fo91a and Fo91b) for concentrations c_i in mg/kg and m_i in mol/kg:

$$T = \frac{2680}{4.60 - \log(m_K^2/m_{Mg})} - 273.15$$
$$T = \frac{2680}{9.40 - \log(c_K^2/c_{Mg})} - 273.15, \quad (5.168)$$

$$T = \frac{6460}{13.30 - \log(m_K^2/m_{Mg})} - 273.15$$
$$T = \frac{6460}{18.10 - \log(c_K^2/c_{Mg})} - 273.15. \quad (5.169)$$

Equation (5.168) is valid for $\log(K^2/Mg) < 3.25$, corresponding to $T < 162.5$ °C, whereas Eq. (5.169) holds for $\log(K^2/Mg) > 3.25$, that is for $T > 162.5$ °C. The two related Na–Mg geothermometers are derived by inserting the Na–K geothermometer of Fournier (1979), Eq. (5.62), and each one of the two K–Mg geothermometers into Eq. (5.146), thus obtaining:

$$T = \frac{246}{2.097 - \log(m^2_{Na}/m_{Mg})} - 273.15,$$
$$T = \frac{246}{6.434 - \log(c^2_{Na}/c_{Mg})} - 273.15 \quad (5.170)$$

$$T = \frac{4026}{10.797 - \log(m^2_{Na}/m_{Mg})} - 273.15,$$
$$T = \frac{4026}{15.134 - \log(c^2_{Na}/c_{Mg})} - 273.15. \quad (5.171)$$

As shown by the diagram of $\log(m^2_K/m_{Mg})$ versus the absolute temperature inverse (Fig. 5.26), the high-slope K–Mg geothermometer of Fournier (1991), code Fo91b, is very close to the K–Mg function of Nieva and Nieva (1987), code NiNi87b. The K–Mg geothermometer of Giggenbach (1988) intersects these two functions at temperatures of 210–230 °C, while it progressively diverges from both relations at higher and lower temperatures. Of the two remaining K–Mg geothermometers, the low-slope function of Fournier (1991), code Fo91a, crosses the other relations, whereas that of Michard (1990), code Mi90b, is positioned somewhat above them.

The plot of $\log(m^2_{Na}/m_{Mg})$ versus the absolute temperature reciprocal (Fig. 5.27) shows that there are considerable differences among the twenty Na–Mg geothermometers, including those reported in Table 5.5, and Eqs. (5.164), (5.167), (5.170), and (5.171). The two low-slope relations of Michard (1990) and Fournier (1991), codes Mi90b and Fo91a, are limited below 150 and 160 °C, respectively. Most of the

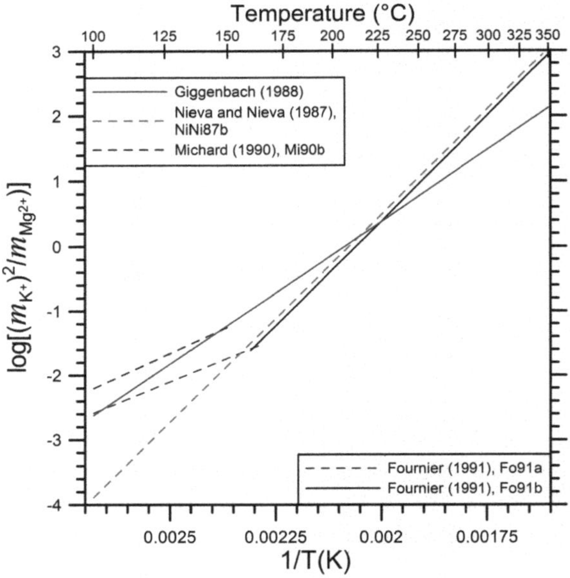

Fig. 5.26 Diagram of $\log(m^2_K/m_{Mg})$ versus the absolute temperature inverse showing the K–Mg geothermometers of Giggenbach (1988), Michard (1990), and Fournier (1991), as well as that obtained from the Na–K–Mg relation of Nieva and Nieva (1987)

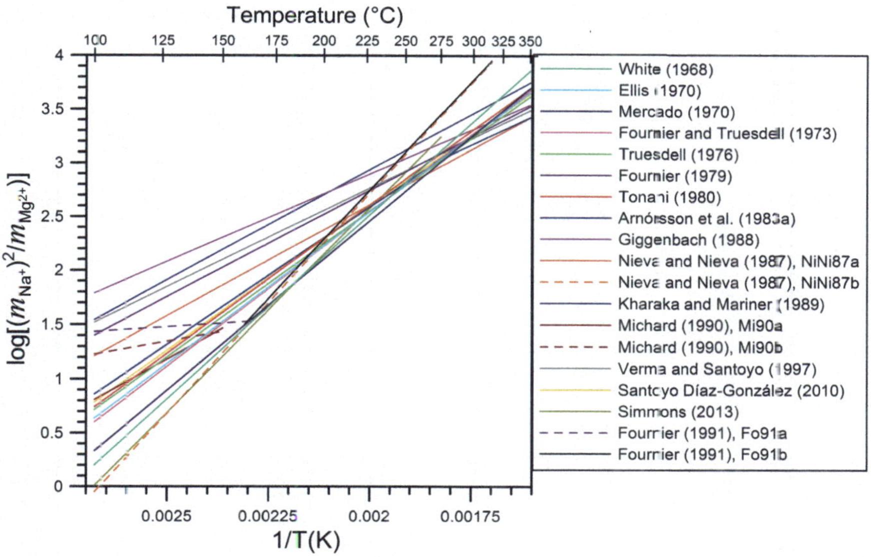

Fig. 5.27 Diagram of $\log(m_{Na}^{2}/m_{Mg})$ versus the absolute temperature inverse showing the sixteen Na–Mg geothermometers obtained from the K–Mg geothermometer of Giggenbach (1988) and the sixteen empirical Na–K functions (see legend) as well as the Na–Mg geothermometers related to the K–Mg geothermometers of Michard (1990) and Fournier (1991), and to the Na–K–Mg relation of Nieva and Nieva (1987)

other eighteen geothermometric functions converge at 225–275 °C, but diverge more and more at decreasing and increasing temperatures. As expected, the spread of the Na–Mg linear functions obtained from the K–Mg geothermometer of Giggenbach (1988) and the sixteen empirical Na–K functions is similar to that of the corresponding Na/K geothermometers (Fig. 5.14). To be noted that the Na–Mg geothermometers of comparatively low slope are derived from the Na–K geothermometers of relatively high slope (e.g., Giggenbach 1988) and vice versa (e.g., Simmons 2013). This is also expected, because each pair of Na–Mg and Na–K equations is interconnected by Eq. (5.146).

5.6.2 Application of the K–Mg and Na–Mg Geothermometers to the Selected Reservoir Liquids

In the correlation diagrams of $\log(m_{K}^{2}/m_{Mg})$ versus the absolute temperature reciprocal (Fig. 5.28) there is a large scatter of data points which are positioned both above and below the solid black lines representing the K–Mg geothermometric functions of Giggenbach (1988) and Fournier (1991), and that obtained from the Na–K–Mg

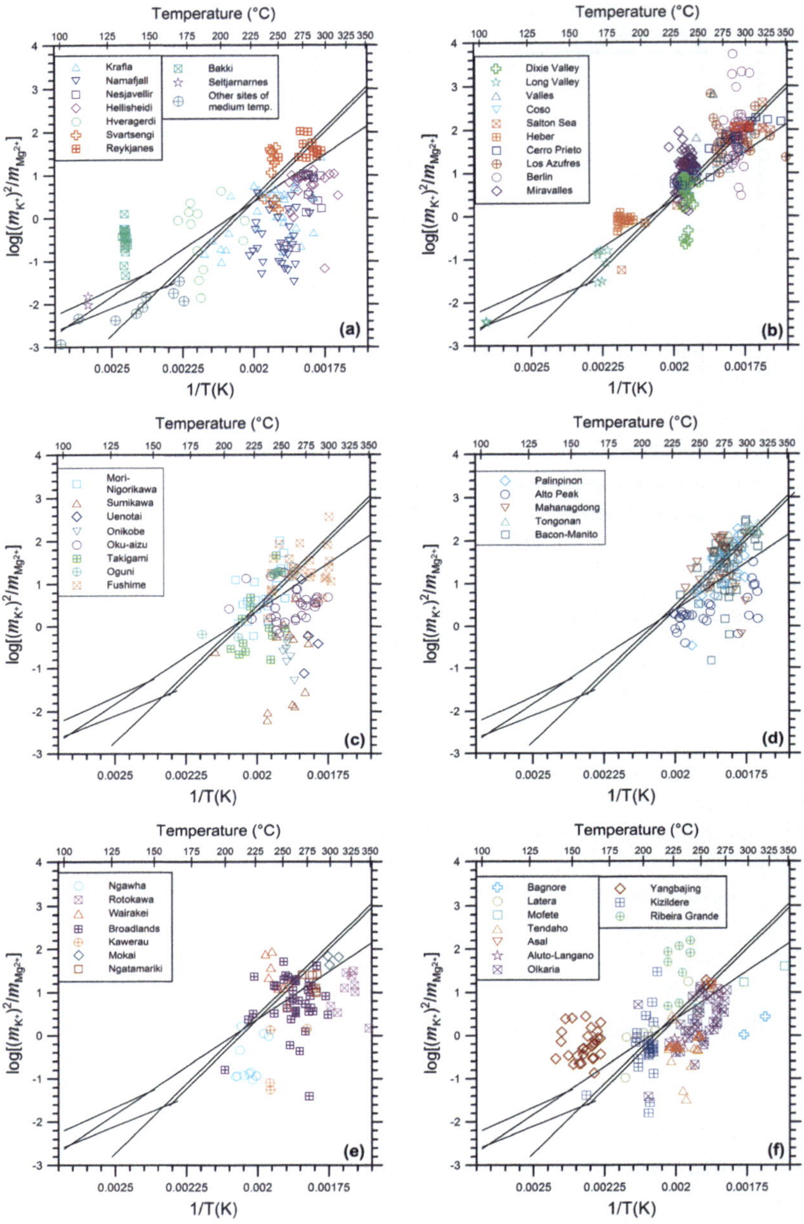

Fig. 5.28 Diagrams of $\log(m_{K}^{2}/m_{Mg})$ versus the absolute temperature inverse showing the selected reservoir liquids from **a** Iceland, **b** Northern-Central America, **c** Japan, **d** The Philippines, **e** New Zealand, and **f** miscellaneous geothermal systems as well as the K–Mg geothermometers of Giggenbach (1988), Michard (1990), Fournier (1991), and that obtained from the Na–K–Mg relation of Nieva and Nieva (1987), all represented by black solid lines

relation of Nieva and Nieva (1987). Most deviations of data points from these K–Mg geothermometers are less than ~ 1.5 log-units, but maximum shifts achieve ~2.5 log-units. Deviations of data points from the K-Mg geothermometers seem to be temperature independent. Also the binary plots of $\log(m^2_{Na}/m_{Mg})$ versus the absolute temperature reciprocal (Fig. 5.29) exhibit a considerable scatter of sample points which, again, are located both above and below the lines corresponding to the different Na–Mg geothermometric functions. The remarkable scatter of data points in the diagrams of Figs. 5.28 and 5.29 is probably due to different causes.

Low values of the (m^2_K/m_{Mg}) and (m^2_{Na}/m_{Mg}) ratios might be explained by the presence of excess magnesium due to mixing of reservoir liquids with Mg-rich external fluids coming from shallower and colder environments. Reservoir liquids with noticeable excess magnesium occur at Krafla, Namafjall, Nesjavellir, and Hellisheidi (Figs. 5.28a and 5.29a), Dixie Valley (Figs. 5.28b and 5.29b), Onikobe, Sumikawa, and Uenotai (Figs. 5.28c and 5.29c), Alto Peak (Figs. 5.28d and 5.29d), Broadlands, Kawerau, Ngawha, and Rotokawa (Figs. 5.28e and 5.29e), Bagnore, Kizildere, Olkaria, and Tendaho (Figs. 5.28f and 5.29f).

In contrast, high values of the (m^2_K/m_{Mg}) and (m^2_{Na}/m_{Mg}) ratios might be due to loss of magnesium, possibly because of Mg incorporation in the lattice of precipitating calcite or other solid phases. This process might be invoked for some reservoir liquids of Reykjanes, Svartsengi, Hveragerdi, and Bakki (Figs. 5.28a and 5.29a), Berlin, Los Azufres, Miravalles, and Valles (Figs. 5.28b and 5.29b), Mori-Nigorikawa, Fushime, Oguni, and Takigami (Figs. 5.28c and 5.29c), Bacon-Manito, Mahanagdong, Palinpinon, and Tongonan (Figs. 5.28d and 5.29d), Wairakei and Broadlands (Figs. 5.28e and 5.29e), Ribeira Grande, Kizildere, and Yangbajing (Figs. 5.28f and 5.29f).

High and low values of the (m^2_K/m_{Mg}) and (m^2_{Na}/m_{Mg}) ratios might be due to the presence of illites and especially chlorites of varying compositions, as documented by the chemical data of these hydrothermal solid phases coming from active geothermal systems, which are discussed in Sects. 4.3 and 4.4.

Still another cause might be the variable degree of ordering on the tetrahedral sites of adularia. Accepting that the differences among the Na–K geothermometers are due to the variable order-disorder degree of adularia involved in the exchange reaction with low-albite (see Sect. 5.3.4), also the differences among the K–Mg geothermometers and among the Na–Mg geothermometers could be ascribable to the same reason.

Since the stoichiometric coefficient of adularia is 0.8 in the exchange reaction (5.142) governing the Na–Mg geothermometer and 2.8 in the exchange reaction (5.141) controlling the K–Mg geothermometer, it is reasonable to hypothesize that the impact of variable ordering on the tetrahedral sites of adularia is more important on the K–Mg relation than on the Na–Mg function.

Finally, it is possible that in some systems or in some parts of a given geothermal system, the (m^2_K/m_{Mg}) and (m^2_{Na}/m_{Mg}) ratios are not controlled by chlorites, but by other Mg-bearing minerals, as discussed in the next section.

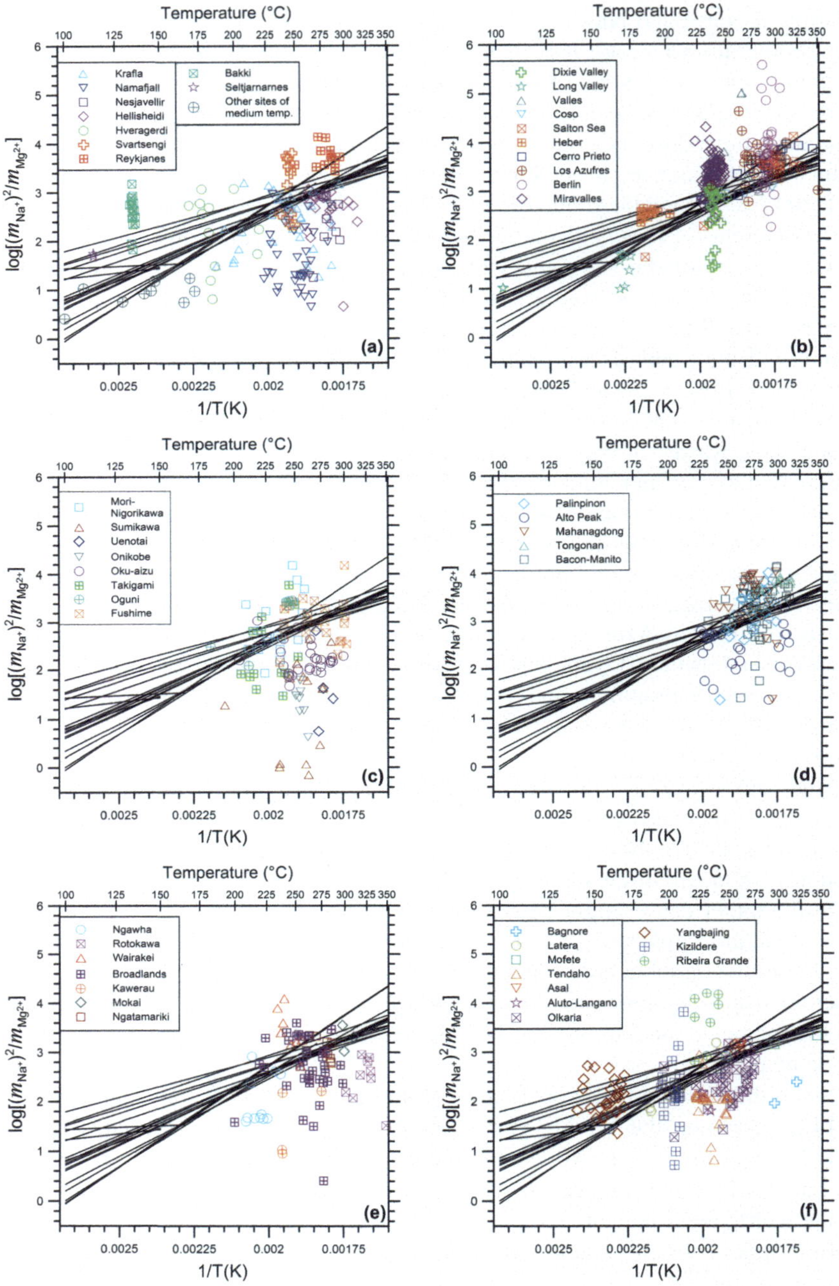

Fig. 5.29 Diagrams of $\log(m_{\mathrm{Na}}^2/m_{\mathrm{Mg}})$ versus the absolute temperature inverse showing the selected reservoir liquids from **a** Iceland, **b** Northern-Central America, **c** Japan, **d** The Philippines, **e** New Zealand, and **f** miscellaneous geothermal systems as well as different Na–Mg geothermometers (solid black lines)

5.6.2.1 The Hydrothermal Minerals Controlling the K–Mg and Na–Mg Geothermometers

As recalled above, Giggenbach (1988) assumed that the K–Mg and Na–Mg geothermometers are controlled by reactions (5.141) and (5.142), both involving clinochlore as Mg-bearing mineral. This is the only chlorite component present in the thermodynamic database of Helgeson et al. (1978), in the two distinct forms with basal spacing of 7 and 14 Å (Sect. 4.4.4).

Other Mg-bearing hydrothermal solid phases occur in active geothermal systems, such as biotite, clinopyroxene, and amphibole, generally above 275 °C, and dolomite, talc, and different clay minerals (e.g., smectites, vermiculites, and saponites), usually below 180 °C. Since no clays are included in the thermodynamic dataset of Helgeson et al. (1978), the thermodynamic properties of Na-montmorillonite, K-montmorillonite, and Mg-montmorillonite were taken from Wolery et al. (2007) and these three minerals were added to the SUPCRT92 database.

Then, the thermodynamic properties of the K–Mg and Na–Mg exchange reactions involving 7Å-clinochlore [Clc, $Mg_5Al_2Si_3O_{10}(OH)_8$], phlogopite [Phl, $KMg_3(AlSi_3)O_{10}(OH)_2$], diopside [Di, $CaMgSi_2O_6$], tremolite [Tr, $Ca_2Mg_5Si_8O_{22}(OH)_2$], ordered and disordered dolomite [Dol, $CaMg(CO_3)_2$], talc [Tlc, $Mg_3Si_4O_{10}(OH)_2$], and Mg-montmorillonite [Mg-Mnt, $Mg_{0.495}Al_{1.67}Si_4O_{10}(OH)_2$] were computed as a function of temperature at water saturation pressures using SUPCRT92. In addition to reaction (5.141), the following K–Mg exchange reactions were also considered:

$$3.33\,KAlSi_3O_8 + 1.33\,H_2O + Mg^{2+} = 0.33\,KMg_3(AlSi_3)O_{10}(OH)_2 + KAl_3Si_3O_{10}(OH)_2 + 6\,SiO_2 + 2\,K^+ \quad (5.172)$$

$$3.75\,KAlSi_3O_8 + 0.5\,Ca_2Al_3Si_3O_{12}(OH) + 1.5\,H_2O + Mg^{2+} = CaMgSi_2O_6 + 1.75\,KAl_3Si_3O_{10}(OH)_2 + 5.5\,SiO_2 + 2\,K^+ \quad (5.173)$$

$$3.3\,KAlSi_3O_8 + 0.2\,Ca_2Al_3Si_3O_{12}(OH) + 1.4\,H_2O + Mg^{2+} = 0.2\,Ca_2Mg_5Si_8O_{22}(OH)_2 + 1.3\,KAl_3Si_3O_{10}(OH)_2 + 5\,SiO_2 + 2\,K^+ \quad (5.174)$$

$$2.25\,KAlSiO_8 + 2\,CaCO_3 + 0.5\,H_2O + Mg^{2+} = CaMg(CO_3)_2 - 0.5\,Ca_2Al_3Si_3O_{12}(OH) + 0.25\,KAl_3Si_3O_{10}(OH)_2 + 4.5\,SiO_2 + 2\,K^+ \quad (5.175)$$

$$3\,KAlSi_3O_8 + 1.33\,H_2O + Mg^{2+} = 0.33\,Mg_3Si_4O_{10}(OH)_2 + KAl_3Si_3O_{10}(OH)_2 + 4.67\,SiO_2 + 2\,K^+ \quad (5.176)$$

$$6.06\,K_{0.33}Mg_{0.33}Al_{1.67}Si_4O_{10}(OH)_2 + Mg^{2+}$$

$$= 6.06\,Mg_{0.495}Al_{1.67}Si_4O_{10}(OH)_2 + 2\,K^+. \quad (5.177)$$

Similarly, in addition to reaction (5.142), the other Na–Mg exchange reactions taken into account are:

$$1.33\,KAlSi3O_8 + 2NaAlSi_3O_8 + 1.33H_2O + Mg^{2+} = 0.33KMg_3(AlSi_3)O_{10}(OH)_2 + KAl_3Si_{13}O_{10}(OH)_2 + 6SiO_2 + 2Na^+ \quad (5.178)$$

$$1.75\,KAlSi_3O_8 + 2NaAlSi_3O_8 + 0.5Ca_2Al_3Si_3O_{12}(OH) + 1.5H_2O + Mg^{2+} = CaMgSi_2O_6 + 1.75KAl_3Si_3O_{10}(OH)_2 + 5.5SiO_2 + 2Na^+ \quad (5.179)$$

$$1.3KAlSi_3O_8 + 2NaAlSi_3O_8 + 0.2Ca_2Al_3Si_3O_{12}(OH) + 1.4H_2O + Mg^{2+} = 0.2Ca_2Mg_5Si_8O_{22}(OH)_2 + 1.3KAl_3Si_3O_{10}(OH)_2 + 5SiO_2 + 2Na^+ \quad (5.180)$$

$$0.25\,KAlSi_3O_8 + 2\,NaAlSi_3O_8 + 2\,CaCO_3 + 0.5\,H_2O + Mg^{2+} = CaMg(CO_3)_2 + 0.5\,Ca_2Al_3Si_3O_{12}(OH) + 0.25\,KAl_3Si_3O_{10}(OH)_2 + 4.5\,SiO_2 + 2\,Na^+ \quad (5.181)$$

$$KAlSi_3O_8 + 2\,NaAlSi_3O_8 + 1.33\,H_2O + Mg^{2+} = 0.33\,Mg_3Si_4O_{10}(OH)_2 + KAl_3Si_3O_{10}(OH)_2 + 4.67\,SiO_2 + 2\,Na^+ \quad (5.182)$$

$$6.06\,Na_{0.33}Mg_{0.33}Al_{1.67}Si_4O_{10}(OH)_2 + Mg^{2+} = 6.06Mg_{0.495}Al_{1.67}Si_4O_{10}(OH)_2 + 2Na^+. \quad (5.183)$$

Clinozoisite was adopted to balance the calcium ions present in the lattice of diopside, tremolite, and dolomite. Calcite was assumed to take part to reactions (5.175) and (5.181) to balance the carbonate ions of dolomite. Apart from reactions (5.177) and (5.183), which comprise montmorillonites only, maximum-microcline and high-sanidine were alternatively involved in all the other reactions, in which the stoichiometric coefficients of alkali feldspars and silica minerals are higher than those of relevant Mg-bearing phases. Therefore, the contributions of Mg-bearing solid phases to the thermodynamic properties of all K–Mg and Na–Mg exchange reactions, apart from (5.177) and (5.183), are expected to be less important than those of alkali feldspars and silica minerals.

The $\Delta H°_r$ and $\Delta S°_r$ values of the empirical K–Mg geothermometers, calculated from their slope and intercept using Eqs. (5.13) and (5.14), are compared with the $\Delta H°_r$ and $\Delta S°_r$ values of the K–Mg exchange reactions of interest in the correlation diagram of Fig. 5.30. As expected, the $\Delta H°_r$ and $\Delta S°_r$ values of the K–Mg geothermometer of Giggenbach (1988) are consistent with the $\Delta H°_r$ and $\Delta S°_r$ values of

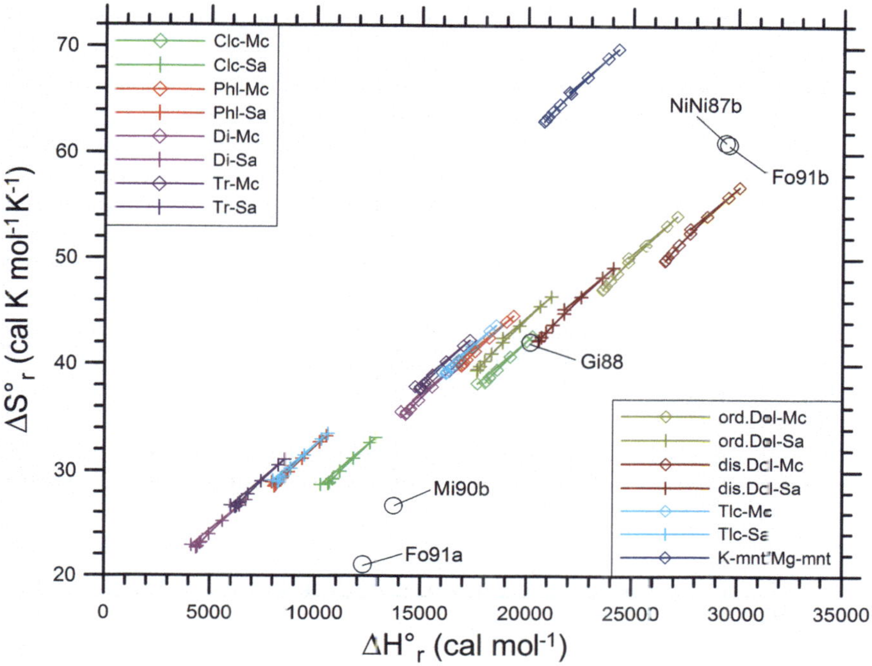

Fig. 5.30 Comparison of the temperature-independent ΔH°_r and ΔS°_r values of the empirical K–Mg geothermometers of Giggenbach (1988), Michard (1990), Fournier (1991), and that obtained from the Na–K–Mg relation of Nieva and Nieva (1987), all indicated by open black circles, with those of the K–Mg exchange reactions involving different hydrothermal minerals (see legend), computed as a function of temperature at water saturation pressures using SUPCRT92

the K–Mg exchange reaction (5.141) involving 7Å-clinochlore. However, the reactions comprising maximum-microcline together with phlogopite or talc or diopside or tremolite and those including high-sanidine and dolomite, ordered or disordered, might also control the K–Mg geothermometer of Giggenbach (1988), owing to the similar reaction entropies and the limited differences in reaction enthalpies. This finding is not surprising because of the small contributions of the different Mg-bearing solid phases to the thermodynamic properties of these exchange reactions. In contrast, the ΔH°_r and ΔS°_r values of the K–Mg geothermometer of Giggenbach (1988) are at variance with those of the reactions involving (i) K- and Mg-montmorillonites, (ii) high-sanidine together with 7Å-clinochlore or phlogopite or talc or diopside or tremolite, as well as (iii) maximum-microcline and dolomite, either disordered or ordered.

The high-slope K–Mg function of Fournier (1991), code Fo91b, and that obtained from Nieva and Nieva (1987), code NiNi87, have ΔH°_r values similar to those of the reactions including maximum-microcline and disordered dolomite, but higher ΔS°_r values. The low-slope K–Mg functions of both Michard (1990), code Mi90b, and Fournier (1991), code Fo91a, have (i) $\Delta S°_r$ values comparable with those of the exchange reactions comprising high-sanidine and diopside or tremolite, but much higher $\Delta H°_r$ values and (ii) $\Delta H°_r$ values similar to those of the exchange reactions involving high-sanidine and clinochlore, but much lower $\Delta S°_r$ values.

The $\Delta H°_r$ and $\Delta S°_r$ values of the empirical Na–Mg geothermometers, computed from their slopes and intercepts utilizing Eqs. (5.13) and (5.14), are compared with the $\Delta H°_r$ and $\Delta S°_r$ values of the considered Na–Mg exchange reactions in the binary diagram of Fig. 5.31. The low-slope Na–Mg functions of Michard (1990), code Mi90b, and Fournier (1991), code Fo91a, are at variance with all the considered exchange reactions. The high-slope Na–Mg function of Fournier (1991), code Fo91b, and the Na–Mg geothermometer obtained from the Na–K–Mg relation of Nieva and Nieva (1987), code NiNi87, have $\Delta H°_r$ and $\Delta S°_r$ values relatively similar to those of the reactions including maximum-microcline or high-sanidine and dolomite, either disordered or ordered.

The Na–Mg geothermometers derived from the K–Mg geothermometer of Giggenbach (1988) and different Na–K geothermometers distribute in the following two groups:

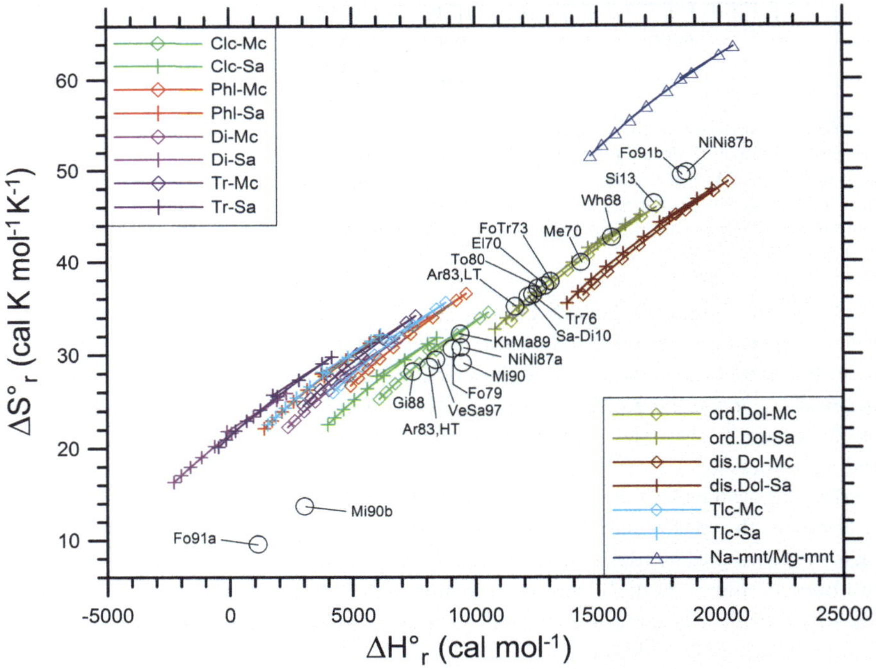

Fig. 5.31 Comparison of the temperature-independent $\Delta H°_r$ and $\Delta S°_r$ values of the empirical Na–Mg geothermometers (open black circles) with those of the Na–Mg exchange reactions involving different hydrothermal minerals, computed as a function of temperature at water saturation pressures using SUPCRT92

(i) Those related to the Na–K geothermometers of high slope (Fournier 1979; high-temperature function of Arnórsson et al., 1983a; Nieva and Nieva 1987; Kharaka and Mariner 1989; Giggenbach 1988; Michard 1990; Verma and Santoyo 1997) are consistent with the Na–Mg exchange reaction involving 7Å-clinochlore together with maximum-microcline or high-sanidine. The exchange reactions including phlogopite, diopside, talc, and tremolite have similar entropies but somewhat lower enthalpies.

(ii) Those linked to the Na–K geothermometers of low slope (White 1968; Ellis 1970; Mercado 1970; Fournier and Truesdell 1973; Truesdell 1975; Tonani 1980; low-temperature function of Arnórsson et al., 1983a; Santoyo and Díaz-González 2010; Simmons 2013) are consistent with the Na–Mg exchange reaction comprising ordered dolomite together with maximum-microcline or high-sanidine. The exchange reactions including disordered dolomite have comparable entropies and weakly higher enthalpies.

The ΔH°_{r} and ΔS°_{r} values of the reactions involving Na– and Mg-montmorillonites are at variance with those of all the Na–Mg geothermometers.

All in all, the K–Mg and Na–Mg geothermometers might be controlled by exchange reactions involving chlorites, as proposed by Giggenbach (1988), but the possible governing role of several other Mg-bearing solid phases cannot be excluded.

5.6.3 *The Na–K–Mg$^{0.5}$ Triangular Diagram*

Combining the fast-responding K–Mg geothermometer with the slowly re-equilibrating Na–K geothermometer by means of a triangular plot, Giggenbach (1988) suggested a graphical method (Fig. 5.32) to assess the degree of attainment of water-rock equilibrium.

In this Na–K–Mg$^{1/2}$ triangular diagram, the Na–K and K–Mg systems are represented by two sets of lines of constant Na/K and K/Mg$^{1/2}$ ratios, radiating from the Mg$^{1/2}$ and Na vertices, respectively. Since at each value of the Na/K and K/Mg$^{1/2}$ ratios corresponds a unique temperature value, each of these lines is an isotherm. The intersection points of the Na–K and K–Mg isotherms referring to the same temperature delineate the so-called "full equilibrium" curve comprising the compositions of waters in equilibrium with the mineral phases controlling both geothermometers.

The compositions of waters generated through isochemical dissolution of average crustal rocks, also shown in this triangular plot, delineate a rock dissolution area, which is far away from the full equilibrium curve.

Neutral chloride reservoir liquids encountered by deep geothermal wells generally plot along or close to the full equilibrium curve, at temperatures slightly higher than those physically measured in these wells. Corresponding neutral chloride spring waters generally plot in the field of partially equilibrated waters, being found below the full equilibrium curve and somewhat shifted towards the Mg$^{1/2}$ vertex. This might

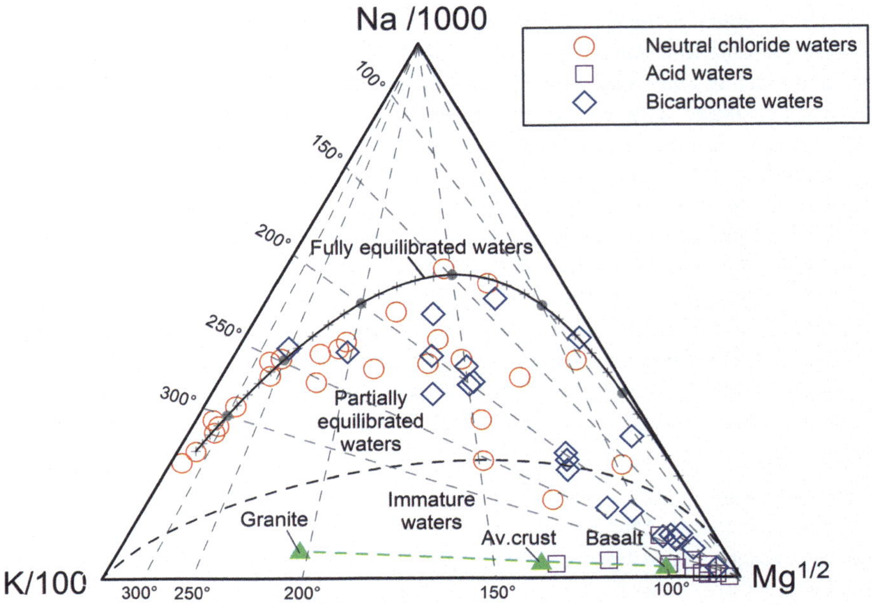

Fig. 5.32 Na–K–$Mg^{1/2}$ triangular diagram (modified from Giggenbach 1988) showing different types of geothermal waters (see legend), the compositions expected for rock dissolution (green closed triangles), the line of fully equilibrated waters, and the fields of both partially equilibrated waters and immature waters

be due to either Mg acquisition upon cooling, proceeding faster than Na acquisition, or to mixing of thermal waters with Mg-rich cold shallow waters.

Also some bicarbonate waters attain partial equilibrium, whereas other bicarbonate waters and the acid waters are situated close to the $Mg^{1/2}$ vertex. These are the so-called "immature waters", which provide unreliable Na–K temperatures, whereas their K–Mg temperatures may still be valid, at least for not too acid aqueous solutions.

This Na–K–$Mg^{1/2}$ plot has become very popular in geochemical investigations because it allows: (i) the immediate distinction between waters suitable or unsuitable for the application of ionic solute geothermometers, (ii) the assessment of deep equilibrium temperatures, and (iii) the straightforward evaluation of re-equilibration and/or mixing effects on a large number of water samples.

However, as already pointed out by Fournier (1991), some words of caution are needed, because different full equilibrium curves can be defined based on distinct Na–K and K–Mg geothermometric functions. As shown in the Na–K–$Mg^{1/2}$ triangular diagram of Fig. 5.33, these full equilibrium curves occupy a large part of the field of the waters in partial equilibrium defined by Giggenbach (1988) and some of them extend in the field of the immature waters towards the $Mg^{1/2}$ vertex. The consequence of this spread of full equilibrium curves is that a given water may be assumed to be either in full equilibrium or in partial equilibrium or even immature, depending

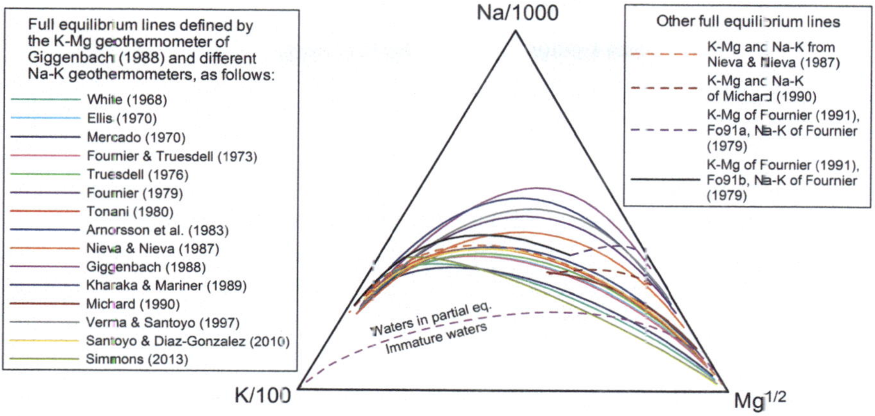

Fig. 5.33 Na–K–$Mg^{1/2}$ triangular diagram showing different "full equilibrium" curves defined by various Na–K and K–Mg geothermometers (see legend) as well as the line separating the field of the waters in partial equilibrium from the field of the immature waters, according to Giggenbach (1988)

on the Na–K and K–Mg of reference, as already pointed out by Fournier (1991). Alternatively, assuming that all the curves are actually representative of the full equilibrium conditions, the Na–K–$Mg^{1/2}$ triangular diagram of Fig. 5.33 suggests that a different approach to Na–K–Mg geothermometry is probably needed.

5.7 Other Ionic Solute Geothermometers

5.7.1 *The Li-Based Geothermometers*

Fouillac and Michard (1981) proposed two empirical Na–Li geothermometers calibrated considering a large number of geothermal waters. One relation applies to the waters with Cl < 0.2 M:

$$T = \frac{1000}{\log(m_{Na}/m_{Li}) + 0.38} - 273.15. \tag{5.184}$$

The other equation is valid for Cl > 0.3 M:

$$T = \frac{1195}{\log(m_{Na}/m_{Li}) - 0.19} - 273.15. \tag{5.185}$$

Fouillac and Michard (1981) identified also the following two functions relating temperature and Li molality:

$$T = \frac{2258}{1.44 - \log m_{Li}} - 273.15 \quad \text{for Cl} < 0.2\text{M} \tag{5.186}$$

$$T = \frac{1436}{0.61 - \log m_{Li}} - 273.15. \quad \text{for Cl} > 0.3\text{M} \tag{5.187}$$

More or less at the same time, a Na–Li geothermometer was also developed by Kharaka and coworkers (Kharaka et al. 1982; concentrations C_i in mg/L)

$$T = \frac{1590}{\log(c_{Na} / c_{Li}) + 0.779} - 273.15. \tag{5.188}$$

Considering the chemical characteristics of thermal waters interacting with granitic rocks at temperatures <150 °C, Michard (1990) derived another Li–Na geothermometer:

$$T = \frac{1040}{\log(m_{Na} / m_{Li}) + 0.43} - 273.15 \tag{5.189}$$

which is similar to Eq. (5.184).

Kharaka and coworkers assumed that the Mg-to-Li ratio may be a sensitive indicator of temperature because of both the decrease of Mg concentration with increasing temperature and the concurrent increase of Li concentration. Based on these premises, Kharaka et al. (1985) developed an Mg–Li geothermometer that can be applied up to 350 °C (concentrations C_i in mg/L):

$$T = \frac{1900}{\log(c_{Mg}^{0.5} / c_{Li}) + 4.67} - 273.15. \tag{5.190}$$

The use of the $c_{Mg}^{0.5} / c_{Li}$ ratio for water geothermometry was reconsidered by Kharaka and Mariner (1989) who derived two Mg–Li geothermometers (concentrations C_i in mg/L). One was calibrated using the whole database and applies up to 350 °C:

$$T = \frac{2200}{\log(c_{Mg}^{0.5} / c_{Li}) + 5.47} - 273.15. \tag{5.191}$$

The other was calibrated utilizing oil-field waters alone and is similar to Eq. (5.190):

$$T = \frac{1910}{\log(c_{Mg}^{0.5} / c_{Li}) + 4.63} - 273.15. \tag{5.192}$$

In our opinion, lithium geothermometers should be used with caution for at least two reasons. One is that lithium has virtually conservative (mobile) behavior in geothermal systems, although it may be incorporated into secondary quartz and chlorite, as found by Goguel (1983). In fact, owing to the conservative behavior of lithium, Giggenbach (1991) suggested to use it as tracer, together with rubidium and cesium and/or together with boron and chloride, to investigate the origin of geothermal fluids.

The second reason is that even admitting that Li has compatible behavior and its activity is therefore fixed by equilibrium with an unspecified solid solution[10], the Li content of this unspecified solid solution is expected to be small and highly variable from system to system and even within different portions of the same system. Therefore, the activity of the Li-endmember in the solid solution taking part to the exchange reactions supposedly governing the Na–Li and Mg–Li geothermometers cannot be properly constrained.

Actually, Fouillac and Michard (1981) recognized that "*it is unlikely that the Li temperature relationship is related to a chemical equilibrium between water and a lithium mineral. Lithium minerals seem to be very rare in hydrothermal environments.* In fact, they considered that "*lithium behaves in hot waters as a "soluble" element, i.e. it is not co-precipitated with any secondary mineral, except perhaps near-surface clays.*" Fouillac and Michard (1981) attributed the temperature—lithium relation to the increase in rock dissolution with increasing temperature coupled with the uptake of lithium in weathering products, such as clay minerals.

Accepting these explanations of Fouillac and Michard (1981), it follows that the development of a Na–Li geothermometer on a thermodynamic basis, by referring to a suitable exchange reaction, is meaningless. Therefore, Li-based geothermometers will not be considered any further in this work.

5.7.2 The "Auxiliary Geothermometers" of Michard (1990)

In addition to the Na–K, Ca–K, Mg–K, and Na–Li geothermometers which have been already discussed above, Michard (1990) calibrated other "auxiliary geothermometers" involving the Rb/Na, Cs/Na, Sr/K^2, Fe/K^2, and Mn/K^2 ratios as well as the F·K and $W \cdot K^2$ products for waters interacting with granitic rocks at temperatures lower than 150 °C. Michard (1990) suggested that these auxiliary geothermometers could be alternatively used to compute the concentrations of Rb, Cs, Sr, Fe, Mn, F, and W, for known temperature and concentrations of Na and K. However, Michard (1990) recognized that regional variations can be important for trace alkali elements, i.e., Li, Rb, and Cs. "*As these elements are governed by partition processes and not by equilibria with different solid phases, regional variations must be related to variations in the trace elements content of the granite.*" Similar to the Li-based geothermometers,

[10]We recall that the term compatible indicates the components whose solubility is limited by incorporation in authigenic minerals (Arnórsson et al. 1983b).

also those involving Rb, Cs, Sr, Fe, Mn, and W, will not be considered any further in this work. Discussion of F-based geoindicators will be resumed in the next section.

5.7.3 The Ca–Mg and SO_4–F Theoretical Geothermometers for Thermal Waters from Carbonate-Evaporite Reservoirs

Marini et al. (1986) proposed a Ca–Mg geothermometer and a SO_4–F geothermometer specific for waters coming from carbonate-evaporite aquifers. Both geothermometers were calibrated using the thermodynamic data of suitably selected exchange reactions. The Ca–Mg geothermometer is based on the exchange reaction involving calcite and dolomite:

$$2CaCO_3 + Mg^{2+} = CaMg(CO_3)_2 + Ca^{2+} \quad (5.193)$$

whereas the SO_4–F geothermometer is constrained by the exchange reaction between anhydrite and fluorite:

$$CaSO_4 + 2F^- = CaF_2 + SO_4^{2-}. \quad (5.194)$$

The thermodynamic equilibrium constant of reactions (5.193) and (5.194) are nearly equal to the free ion molality ratios $m_{Ca^{2+}} / m_{Mg^{2+}}$ and $m_{SO_4^{2-}} / m_{F^-}^2$, respectively, assuming that the solid phases are pure, the activity coefficients of Ca^{2+} and Mg^{2+} ions cancel out, and the activity coefficient of $SO_4{}^{2-}$ ion is nearly equal to the squared activity coefficient of F^- ion.

A major problem in the application of the Ca–Mg geothermometer is the variable degree of order-disorder of dolomite, as discussed by several authors (e.g., Marini et al. 1986; Chiodini et al. 1995; Hyeong and Capuano 2001; Zhiqian et al. 2007; Vespasiano et al. 2014; Blasco et al. 2018, 2019).

Another major problem is caused by ion complexing (Guidi et al. 1990). In fact, due to the formation of ion pairs of Ca^{2+} ion (e.g., $CaSO_4{}^\circ$, $CaCO_3{}^\circ$, $CaHCO_3{}^+$, $CaCl^+$, $CaCl_2{}^\circ$, and CaF^+) and Mg^{2+} ion (e.g., $MgSO_4{}^\circ$, $MgCO_3{}^\circ$, $MgHCO_3{}^+$, $MgCl^+$, and MgF^+), the total molality ratio m_{Ca} / m_{Mg} deviates significantly from the free ion molality ratio $m_{Ca^{2+}} / m_{Mg^{2+}}$ especially in Cl-poor solutions. Similarly, owing to the formation of ion complexes of $SO_4{}^{2-}$ ion (e.g., $NaSO_4{}^-$, $KSO_4{}^-$, $KHSO_4{}^\circ$, $CaSO_4{}^\circ$, and $MgSO_4{}^\circ$) and F^- ion (e.g., NaF°, CaF^+, and MgF^+), the total molality ratio m_{SO_4} / m_F^2 departs considerably from the free ion molality ratio $m_{SO_4^{2-}} / m_{F^-}^2$, especially in Cl-rich solutions.

To eliminate the impact of ion complexing, the Ca–Mg and SO_4–F geothermometers and other theoretical geoindicators were recalibrated by Chiodini et al. (1995) for waters coming from carbonate-evaporite aquifers of medium-low temperature

Table 5.6 Theoretical geoindicators for aqueous solutions coming from carbonate-evaporite aquifers of medium-low temperature (50–150 °C; from Chiodini et al. 1995)

$pH = 6.268 - 0.5564 \cdot \log P_{CO_2} - 98.77/T - 0.0005 \cdot \log \Sigma_{eq}$	(5.195)
$\log m_{Ca} = -3.809 + 727.29/T + 0.1279 \cdot \log P_{CO_2} + 0.2097 \cdot \log \Sigma_{eq}$	(5.196)
$\log m_{Mg} = -6.926 + 1707.1/T + 0.0888 \cdot \log P_{CO_2} + 0.1397 \cdot \log \Sigma_{eq}$	(5.197)
$\log m_{HCO_3} = -4.819 + 0.4570 \cdot \log P_{CO_2} + 970.29/T + 0.07323 \cdot \log \Sigma_{eq}$	(5.198)
$\log m_{SO_4} = -4.832 + 1145.9/T + 0.3118 \cdot \log \Sigma_{eq} - 0.07306 \cdot \log P_{CO_2}$	(5.199)
$\log m_F = -2.805 - 325.9/T + 0.1127 \cdot \log \Sigma_{eq} - 0.02174 \cdot \log P_{CO_2}$	(5.200)
$\log (m_{Ca}/m_{Mg}) = 3.117 - 979.8/T + 0.03904 \cdot \log P_{CO_2} + 0.07003 \cdot \log \Sigma_{eq}$	(5.201)
$\log (m_{HCO_3}/m_F) = -2.014 + 0.4787 \cdot \log P_{CO_2} + 1296.2/T - 0.03943 \cdot \log \Sigma_{eq}$	(5.202)
$\log (m_{SO_4}/m_F^2) = -0.7782 + 1797.7/T + 0.08653 \cdot \log \Sigma_{eq} - 0.02963 \cdot \log P_{CO_2}$	(5.203)
$\log (m_{HCO_3}^2/m_{SO_4}) = -4.807 + 0.9871 \cdot \log P_{CO_2} + 794.8/T - 0.1655 \cdot \log \Sigma_{eq}$	(5.204)

(50–150 °C). First, the total concentrations of Ca, Mg, SO_4, F, and HCO_3 were computed for aqueous solutions in equilibrium with calcite, disordered dolomite[11], anhydrite, and fluorite at temperatures between 0 and 150 °C, P_{CO_2} in the 0.1 to 100 bar range, and total concentrations of conservative components, Na and Cl, varying between 0.0001 and 0.3 mol/kg. These calculations were performed using the mineral-solution equilibrium model of Guidi et al. (1990) and Chiodini et al. (1991), representing the subject of the next section. Second, a stepwise multiple regression analysis was performed to describe the dependence of the total concentration of each chemical component and relevant concentration ratios upon 1/T(K), log P_{CO_2} and log Σ_{eq} (Σ_{eq} is total ionic salinity in eq/kg). The obtained theoretical geoindicators are reported in Table 5.6, indicating that: (i) the concentrations of Ca, Mg, SO_4, and F as well as the $SO_4/(F)^2$ and Ca/Mg ratios depend mainly on temperature and, therefore, can be used to formulate geothermometers, (ii) the $(HCO_3)^2/SO_4$ ratio and pH are mainly controlled by P_{CO_2} and, consequently, can be utilized as basis for P_{CO_2}-indicators (iii) the HCO_3/F ratio and HCO_3 concentrations are functions of both P_{CO_2} and temperature.

More recently, Vespasiano et al. (2014) developed geothermometric functions specific for the thermal aquifers discharging at the Luigiane Spa, near Guardia Piemontese (Calabria, Italy). These thermal aquifers are hosted in carbonate-evaporite rocks affected by low-grade metamorphic processes, which are presumably responsible of the order-disorder state of dolomite. The Ca–Mg and SO_4–F geothermometric functions obtained for these waters are (concentrations c_i in mg/kg):

$$T(K) = \frac{896.8}{3.408 - \log(c_{Ca}/c_{Mg})} - 273.15 \quad (5.205)$$

[11] Disordered dolomite was considered because ordered dolomite provided results at variance with the analytical data of the thermal springs of Central Italy considered by Chiodini et al. (1995).

$$T(K) = \frac{1700.4}{\log(c_{SO_4} / c_F^2) + 3.171} - 273.15. \quad (5.206)$$

Summing up, the studies on the geothermometry of thermal waters coming from carbonate-evaporite aquifers have stimulated the investigation of the ion complexing influence on geothermometers and P_{CO_2}-indicators, which is further considered in the next section.

5.8 The Influence of Ion Complexing on Geothermometers and f_{CO_2}-Indicators

5.8.1 Theoretical Approach

The concentrations of compatible dissolved components can be predicted specifying the identity of the minerals presumably in equilibrium with the aqueous solution (one mineral for each compatible component) as well as the temperature and the concentration of chloride and other conservative components, if any. The predictability of the concentrations of compatible dissolved components was recognized by several studies, such as those of Shikazono (1978), Michard et al. (1981), and Arnórsson et al. (1983b).

Building on the outcomes of these studies, Guidi et al. (1990) and Chiodini et al. (1991) investigated the influence of ion complexing on geothermometers and f_{CO_2}-indicators using a mineral-solution equilibrium model which has the same computational structure of the model adopted by Michard et al. (1981). The following aqueous species are considered in the model: $H_2O°$, H^+, OH^-, Na^+, $NaCl°$, $NaSO_4{}^-$, $NaCO_3{}^-$, $NaF°$, $NaOH°$, K^+, $KCl°$, $KSO_4{}^-$, $KHSO_4°$, Ca^{2+}, $CaSO_4°$, $CaCO_3°$, $CaHCO_3{}^+$, CaF^+, $CaOH^+$, Mg^{2+}, $MgSO_4°$, $MgCO_3°$, $MgHCO_3{}^+$, MgF^+, $MgOH^+$, $H_2CO_3°$, $HCO_3{}^-$, $CO_3{}^{2-}$, $SO_4{}^{2-}$, F^-, Cl^-, $H_4SiO_4°$, $H_3SiO_4{}^-$, $Al(OH)_4^-$, Al^{3+}, $Al(OH)^{2+}$, $Al(OH)_2^+$. The model finds the roots of a polynomial equation of H^+ activity, which is obtained considering: the electroneutrality equation, the thermodynamic constant of the dissolution reactions of relevant mineral phases and gaseous CO_2, and the thermodynamic constant of the dissociation reactions of complex aqueous species. In the light of the available information on high-temperature geothermal systems:

(i) Na^+, K^+, Ca^{2+}, Mg^{2+}, $Al(OH)_4^-$, $H_4SiO_4°$, $SO_4{}^{2-}$, F^-, and $HCO_3{}^-$ were chosen as compatible dissolved components of interest, whereas only Cl^- was taken into account among conservative constituents.
(ii) The hydrothermal mineral assemblage was assumed to be made up of low-albite, K-feldspar, either a Ca–Al-silicate[12] and/or calcite, depending on the

[12]Laumontite up to 200 °C, clinozoisite in the 200–280 °C interval, and wairakite at higher temperatures.

f_{CO_2} computed by Eq. (5.137), clinochlore, muscovite, quartz, anhydrite, and fluorite.

(iii) In general, f_{CO_2} was considered an externally controlled parameter fixing the activity of HCO_3^- ion, apart from the cases of Ca–Al-silicate/calcite coexistence. In these particular cases, f_{CO_2} is fixed by temperature, calcite governs the activity of HCO_3^- ion, and the Ca–Al-silicate controls the activity of Ca^{2+} ion.

The composition of the aqueous solution was computed at temperatures ranging from 150 to 300 °C, at steps of 50 °C, f_{CO_2} varying from 0.1 to 100 bar, every half log-unit, and total chloride molality from 0.0003 to 1, every half log-unit. Calculated total concentrations are in satisfactory agreement with those observed in different high-temperature geothermal systems.

Then, the logarithms of total concentrations of compatible constituents and their log-ratios were treated as dependent variables in multiple stepwise regression analysis, in which 1/T(K), log f_{CO_2} and log Σ_{eq} (Σ_{eq} is total ionic salinity in eq/kg) were adopted as independent variables.

5.8.2 *Complexing in Hydrothermal Aqueous Solutions and Related Effects*

To simplify the discussion, in this section, we examine only the chemical speciation of a hydrothermal aqueous solution in equilibrium with low-albite, maximum-microcline, clinozoisite, calcite, 7 Å-clinochlore, muscovite, quartz, anhydrite, and fluorite at temperature of 250 °C, CO_2 fugacity fixed by coexistence of calcite and clinozoisite, and total Cl concentration varying from 0.003 to 3 mol/kg.

We have performed a new series of calculations by means of the computer code EQ3NR, version 7.2b (Wolery 1992). Reference was made to the COM thermodynamic database, which includes a large number of solid, aqueous, and gaseous species, whose thermodynamic properties are mostly derived from SUPCRT92 (Johnson et al. 1992 and references therein). The aqueous species considered in speciation computations are listed in Table 5.7. In spite of some differences in the considered aqueous species and in some thermodynamic data, results are comparable with those of Guidi et al. (1990) and Chiodini et al. (1991).

As discussed by Guidi et al. (1990) and Chiodini et al. (1991), ion association in hydrothermal aqueous solutions has different effects on dissolved components, mainly depending on the charge of relevant ions, as summarized below.

Sodium and potassium (Fig. 5.34). Although the concentrations of the Cl complexes $NaCl_{(aq)}$ and $KCl_{(aq)}$ increase with total Cl molality, the free ions Na^+ and K^+ are always the dominant species of dissolved sodium and potassium. Therefore, total (analyzed) Na and K concentrations are always representative of the concentrations of free ions.

Magnesium and calcium (Fig. 5.35). The free ion Ca^{2+} is the dominant species for

Table 5.7 Aqueous species considered in speciation computations

	Al^{3+}	Ca^{2+}	H^+	HCO_3^-	K^+	Mg^{2+}	Na^+
F^-	$Al(OH)_2^+$	$CaCO_{3(aq)}$	$HCl_{(aq)}$	$CO_{2(aq)}$	$KCl_{(aq)}$	$MgCO_{3(aq)}$	$NaAlO_{2(aq)}$
Cl^-	AlO_2^-	$CaCl^+$	$HF_{(aq)}$	CO_3^{2-}	$KHSO_{4(aq)}$	$MgCl^+$	$NaCO_3^-$
SO_4^{2-}	$HAlO_{2(aq)}$	$CaCl_{2(aq)}$	HF_2^-		KSO_4^-	MgF^+	$NaCl_{(aq)}$
$SiO_{2(aq)}$		CaF^+	HSO_4^-			$MgHCO_3^+$	$NaF_{(aq)}$
OH^-		$CaHCO_3^+$	$HSiO_3^-$			$MgSO_{4(aq)}$	$NaHCO_{3(aq)}$
		$CaSO_{4(aq)}$					$NaHSiO_{3(aq)}$
							$NaOH_{(aq)}$

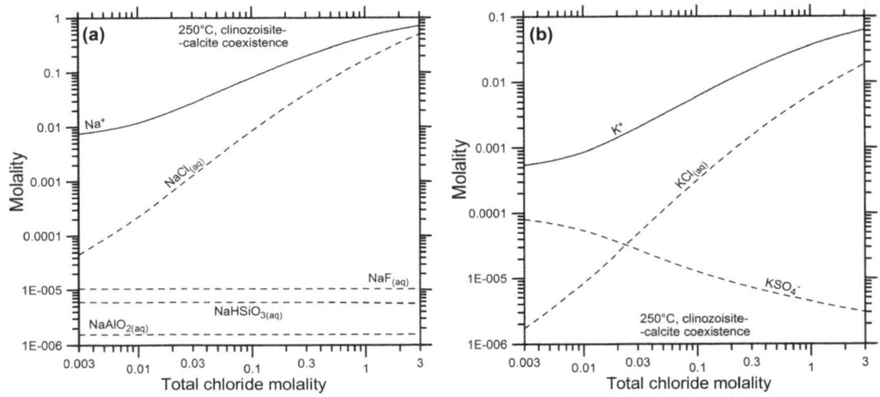

Fig. 5.34 Distribution of dissolved species of **a** sodium and **b** potassium in a hydrothermal aqueous solution in equilibrium with low-albite, maximum-microcline, clinozoisite, calcite, 7Å-clinochlore, muscovite, quartz, anhydrite, and fluorite at 250 °C as a function of total Cl molality

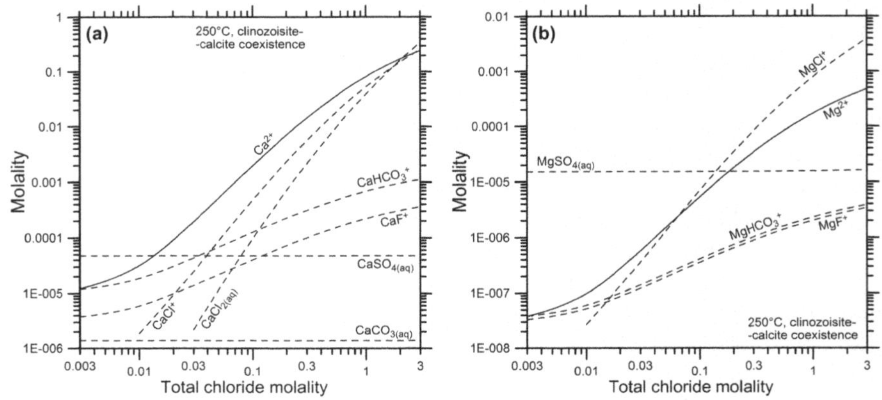

Fig. 5.35 Distribution of dissolved species of **a** calcium and **b** magnesium in a hydrothermal aqueous solution in equilibrium with low-albite, maximum-microcline, clinozoisite, calcite, 7Å-clinochlore, muscovite, quartz, anhydrite, and fluorite at 250 °C as a function of total Cl molality

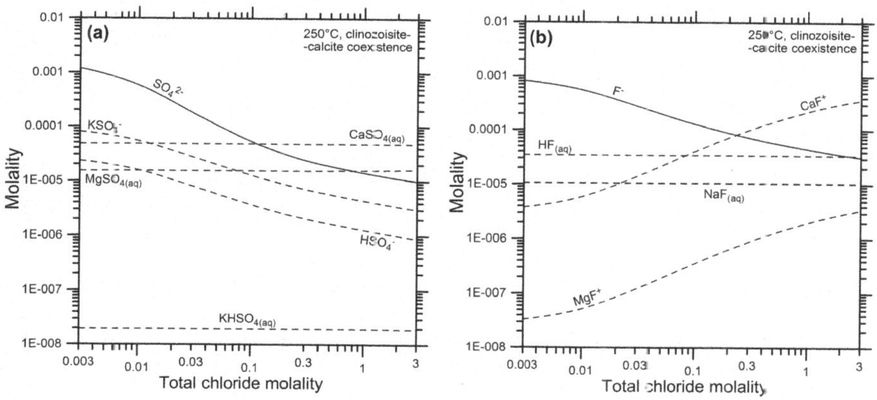

Fig. 5.36 Distribution of dissolved species of (a) sulfate and (b) fluoride in a hydrothermal aqueous solution in equilibrium with low-albite, maximum-microcline, clinozoisite, calcite, 7Å-clinochlore, muscovite, quartz, anhydrite, and fluorite at 250 °C as a function of total Cl molality

total Cl concentrations higher than 0.012 mol/kg and lower than 2 mol/kg, that is in the range 425–70,900 mg/kg. The Cl complexes $CaCl_2°$ and $CaCl^+$ dominate at high total Cl concentrations. In contrast, the $CaSO_4°$ ion pair prevails over other Ca-species at low total Cl concentrations. The free ion Mg^{2+} is never the main dissolved Mg species, because the $MgSO_4°$ aqueous complex dominates at total Cl concentrations lower than 0.13 mol/kg (4600 mg/kg) and the $MgCl^+$ ion pair prevails in Cl-rich solutions. Nevertheless, the relative proportion of Mg^{2+} ion increases with total Cl molality.

Sulfate and fluoride (Fig. 5.36). Free ions SO_4^{2-} and F^- are the prevailing species at total Cl concentrations lower than 0.1 mol/kg (3500 mg/kg) and 0.25 mol/kg (8900 mg/kg), respectively, while the $CaSO_4°$ and CaF^+ ion pairs are the dominant species at total Cl molalities higher than these values. Therefore total (analyzed) sulfate and fluoride concentrations are representative of the concentrations of free ions only at total Cl concentrations lower than 0.1 mol/kg (3500 mg/kg) and 0.25 mol/kg (8900 mg/kg), respectively.

Carbonate species (Fig. 5.37a). The main dissolved carbonate species is aqueous CO_2 (or carbonic acid, H_2CO_3) at all total Cl molalities. Similar to what has been observed for sulfate and fluoride, free HCO_3^- ion is the main carbonate species at total Cl concentrations lower than 0.18 mol/kg (6400 mg/kg), whereas the ion complex $CaHCO_3^+$ prevails at total Cl molalities above this value.

Proton donors and proton (Fig. 5.37b). Bicarbonate is the main acid (proton donor) at all total Cl molalities. The concentration of H^+ ion increases by almost 2 orders of magnitude with increasing total Cl concentrations, from 0.22 μmol/kg at total Cl concentration of 0.003 mol/kg to 13 μmol/kg at total Cl concentration of 3 mol/kg. Proton concentration is buffered by coexistence of aluminum-silicate minerals as already pointed out by Ellis (1970). For, instance, referring to the simplified system albite-muscovite-quartz-aqueous solution, if the concentration of Cl^-

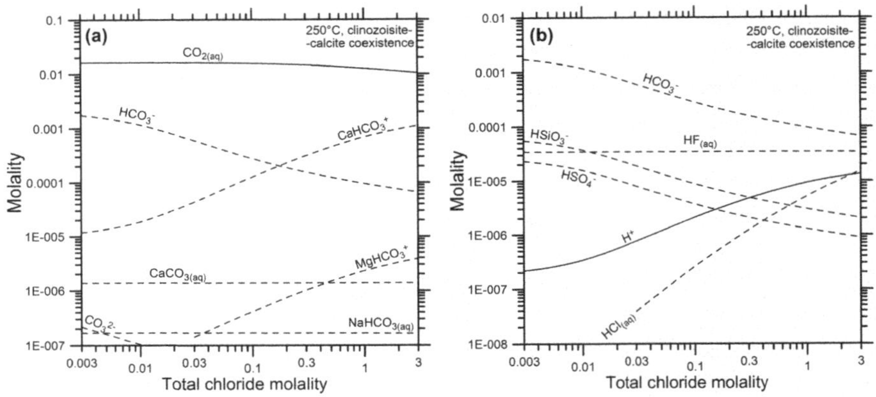

Fig. 5.37 Distribution of **a** carbonate species and **b** proton donors and H^+ in a hydrothermal aqueous solution in equilibrium with low-albite, maximum-microcline, clinozoisite, calcite, 7Å-clinochlore, muscovite, quartz, anhydrite, and fluorite at 250 °C as a function of total Cl molality

ion increases, the concentrations of Na^+ and K^+ ions have to increase accordingly, to maintain the electroneutrality condition. In turn, this increment in alkali concentrations determines an increase in H^+ molality owing to the equilibrium constraint dictated by the following reaction:

$$3\,\text{albite} + K^+ + 2H^+ = \text{muscovite} + 6\,\text{quartz} + 3Na^+. \tag{5.207}$$

Effects of ion complexing. Due to the effects of ion complexing, total (analytical) molalities of compatible cations (Na^+, K^+, Ca^{2+} and Mg^{2+}) and compatible anions ($SO_4{}^{2-}$, F^-, and $HCO_3{}^-$) diverge, to variable extents, from the molalities of free ions, which are uniquely fixed by mineral-solution equilibrium, at a given temperature or a given temperature, f_{CO_2} condition.

The ratios of total molalities exhibit variable deviations from the corresponding ratios of free ion activities, which are uniquely fixed by temperature. These effects are minor on the Na/K ratio because the total (analyzed) concentrations of both Na and K are representative of free ion concentrations, at all T–f_{CO_2}–Cl conditions (see above), and activity coefficients of Na^+ and K^+ ions are similar (Fig. 5.38a).

In contrast, the K^2/Mg total molality ratio deviates strongly from the free ion activity ratio (Fig. 5.38b), due to the differences in the speciation of K, which is dominated by free K^+ ion all T–f_{CO_2}–Cl conditions, and Mg, which is controlled by SO_4 and Cl complexes while the free Mg^{2+} ion is always present in lower concentrations (see above). The deviations of the total molality ratio from the free ion activity ratio are particularly important at low total Cl molality where the ion complexing effects are extremely important on Mg.

Also the K^2/Ca and Na^2/Ca total molality ratios exhibit considerable deviations from the corresponding free ion activity ratios (Fig. 5.39), because ion complexing is

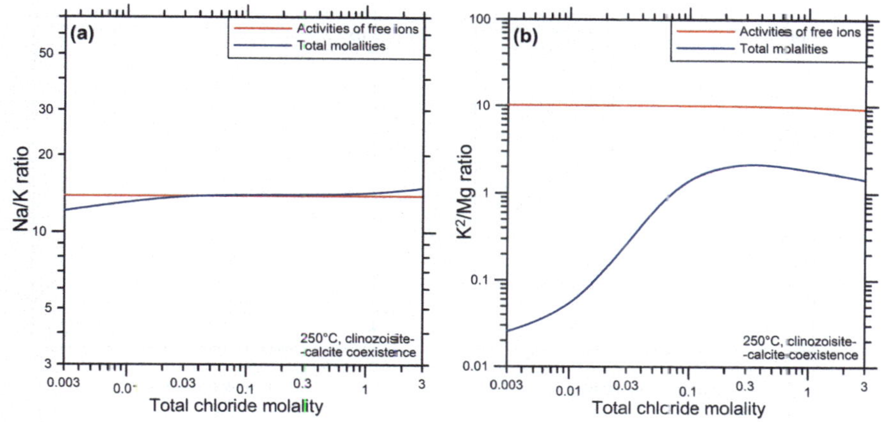

Fig. 5.38 Log-log diagrams of **a** Na/K and **b** K^2/Mg total molality ratios and free ion activity ratios as a function of total Cl molality in a hydrothermal aqueous solution in equilibrium with low-albite, maximum-microcline, clinozoisite, calcite, 7Å-clinochlore, muscovite, quartz, anhydrite, and fluorite at 250 °C

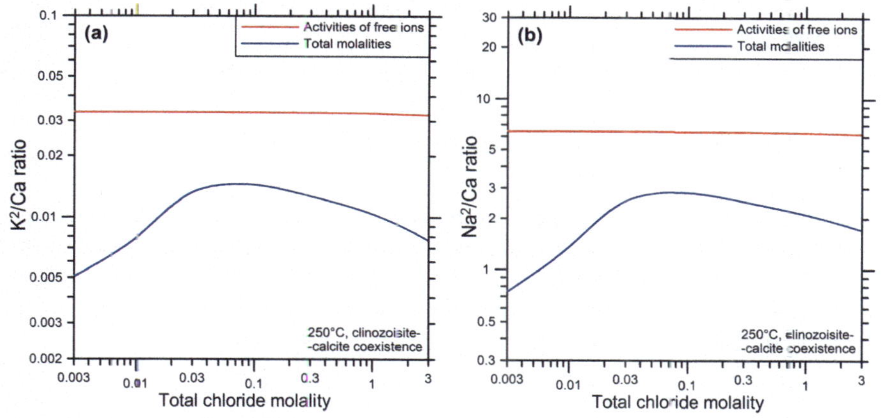

Fig. 5.39 Log-log diagrams of **a** K^2/Ca and **b** Na^2/Ca total molality ratios and free ion activity ratios as a function of total Cl molality in a hydrothermal aqueous solution in equilibrium with low-albite, maximum-microcline, clinozoisite, calcite, 7Å-clinochlore, muscovite, quartz, anhydrite, and fluorite at 250 °C

more important for Ca, especially at total Cl concentrations lower than 0.012 mol/kg and higher than 2 mol/kg, than for the two alkali metals (see above).

The SO_4/F^2 and SO_4/HCO_3^2 total molality ratios depart from the related free ion activity ratios as shown in Fig. 5.40. Nevertheless, these deviations are different from those of the ratios between mono- and bivalent cations, because SO_4, F, and HCO_3 have similar speciation, with the free ion prevailing at relatively low total Cl

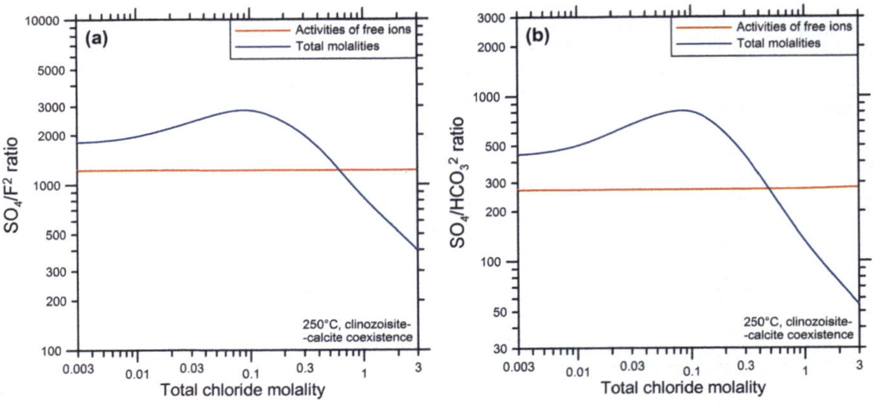

Fig. 5.40 Log-log diagrams of **a** SO_4/F^2 and **b** SO_4/HCO_3^2 total molality ratios and free ion activity ratios as a function of total Cl molality in a hydrothermal aqueous solution in equilibrium with low-albite, maximum-microcline, clinozoisite, calcite, 7Å-clinochlore, muscovite, quartz, anhydrite, and fluorite at 250 °C

molalities and the Ca-complexes dominating at comparatively high total Cl molalities (see above).

5.8.3 The Theoretical Geoindicators of Chiodini et al. (1991)

Chiodini et al. (1991) treated the logarithms of the total molalities of compatible components and their log-ratios as dependent variables in multiple stepwise regression analysis, in which 1/T(K), log f_{CO_2} and log Σ_{eq} were considered as independent variables. The final regression equations thus obtained, as well as a relation for pH, are reported in Tables 5.8 and 5.9.

Table 5.8 pH and logarithms of the total molalities of the compatible constituents dissolved in aqueous solutions in equilibrium with the neutral-pH hydrothermal mineral assemblage as a function of temperature, P_{CO_2} and ionic salinity (from Chiodini et al. 1991)

$\log m_{Na} = -0.527 + 0.982 \cdot \log \Sigma eq + 78.9/T + 0.0119 \cdot \log P_{CO_2}$	(5.208)
$\log m_{K} = +0.458 + 0.944 \cdot \log \Sigma eq - 1014/T + 0.0346 \cdot \log P_{CO_2}$	(5.209)
$\log m_{Ca} = -1.116 + 1.302 \cdot \log \Sigma eq - 0.390 \cdot \log P_{CO_2} - 443/T$	(5.210)
$\log m_{Mg} = -8.188 + 0.912 \cdot \log \Sigma eq + 2156/T + 0.356 \cdot \log P_{CO_2}$	(5.211)
$\log m_{HCO_3} = -10.004 + 0.904 \cdot \log P_{CO_2} + 3223/T - 0.489 \cdot \log \Sigma eq$	(5.212)
$\log m_{SO_4} = -12.503 + 4604/T + 0.501 \cdot \log P_{CO_2} - 0.626 \cdot \log \Sigma eq$	(5.213)
$\log m_{F} = -5.069 + 798/T + 0.127 \cdot \log P_{CO_2} - 0.0886 \cdot \log \Sigma eq$	(5.214)
$pH = +1.757 - 0.822 \cdot \log \Sigma eq + 1846/T - 0.0171 \cdot \log P_{CO_2}$	(5.215)

Table 5.9 Logarithms of the total molality ratios of compatible constituents dissolved in aqueous solutions in equilibrium with the neutral-pH hydrothermal mineral assemblage as a function of temperature, P_{CO_2} and ionic salinity (from Chiodini et al. 1991)

$\log(m_{Na}/m_K) = -0.985 + 1093/T + 0.0384 \cdot \log\Sigma eq - 0.0228 \cdot \log P_{CO_2}$	(5.216)
$\log(m_K^2/m_{Mg}) = +9.104 - 4184/T + 0.975 \cdot \log\Sigma eq - 0.286 \cdot \log P_{CO_2}$	(5.217)
$\log(m_{SO_4}/m_F^2) = -2.364 + 3007/T - 0.449 \cdot \log\Sigma eq + 0.247 \cdot \log P_{CO_2}$	(5.218)
$\log(m_K^2/m_{Ca}) = +2.031 + 0.459 \cdot \log P_{CO_2} + 0.585 \cdot \log\Sigma eq - 1584/T$	(5.219)
$\log(m_{Ca}/m_{Mg}) = +7.073 - 0.746 \cdot \log P_{CO_2} - 2600/T + 0.390 \cdot \log\Sigma eq$	(5.220)
$\log(m_{HCO_3}^2/m_{SO_4}) = -7.506 + 1.307 \cdot \log P_{CO_2} + 1843/T - 0.352 \cdot \log\Sigma eq$	(5.221)
$\log(m_{HCO_3}/m_F) = -4.935 + 0.777 \cdot \log P_{CO_2} + 2425/T - 0.401 \cdot \log\Sigma eq$	(5.222)
$\log(m_{Na}^2/m_{Mg}) = +7.134 + 1.052 \cdot \log\Sigma eq - 1999/T - 0.332 \cdot \log P_{CO_2}$	(5.223)
$\log(m_{Na}^2/m_{Ca}) = +0.061 + 0.662 \cdot \log\Sigma eq + 0.414 \cdot \log P_{CO_2} + 601/T$	(5.224)

Inspection of Tables 5.8 and 5.9 shows that:

(i) The Na/K, K^2/Mg, and SO_4/F^2 ratios as well as SO_4 and F molalities are mainly controlled by temperature and are therefore potential geothermometers.
(ii) The K^2/Ca, Ca/Mg, HCO_3/F, and $(HCO_3)^2/SO_4$ ratios as well as HCO_3 molality are largely controlled by f_{CO_2} and are therefore potential f_{CO_2}-indicators.
(iii) The Na^2/Mg and Na^2/Ca ratios, the molalities of cations, and pH are mainly constrained by total ionic salinity and are therefore less suitable, in principle, as either geothermometers or f_{CO_2}-indicators.

5.8.4 *The Lesson Learned and the Way Forward*

Summing up, the studies of Guidi et al. (1990) and Chiodini et al. (1991) found a theoretical justification for the ionic solute geothermometers and f_{CO_2}-indicators, which were originally derived on a purely empirical basis, and this is certainly a valuable result. The functions reported in Tables 5.8 and 5.9 can be used to investigate the behavior of each solute upon mixing and re-equilibration, as done by Marini et al. (1998) for the thermal waters of the San Marcos area, Guatemala.

However, the approach adopted by Chiodini et al. (1991), consisting in computing the speciation of fully-equilibrated geothermal aqueous solutions and processing the obtained results through multiple regression analysis with the ultimate aim to obtain geothermometers and f_{CO_2}-indicators involving the total concentrations of the compatible constituents, is probably not the best strategy. In fact, it is more correct to carry out speciation calculations on the water samples of interest obtaining the activities of the free ions of compatible constituents that are subsequently introduced into functions expressing the temperature dependence of the thermodynamic equilibrium constants of suitable exchange reactions (at water saturation pressures). In essence, the log K of these reactions are adopted as benchmarks in this second approach, which is therefore rigorous, whereas the same log K are manipulated in the first approach, a practice which leaves much to be desired, also considering that the rigmarole of multiple regression analysis introduces far from negligible approximations. Therefore, we decided to develop and investigate theoretical geothermometers and f_{CO_2}-indicators based on ion activity ratios, representing the subject of the next chapters of this book.

References

Akinfiev NN, Diamond LW (2009) A simple predictive model of quartz solubility in water–salt–CO_2 systems at temperatures up to 1000 °C and pressures up to 1000 MPa. Geochim Cosmochim Ac 73:1597–1608

Anderko A, Pitzer KS (1993) Equation-of-state representation of phase equilibria and volumetric properties of the system NaCl–H_2O above 573 K. Geochim Cosmochim Ac 57:1657–1680

Anderson G, Burnham CW (1965) The solubility of quartz in super-critical water. Am J Sci 263:494–511

Anderson GM, Burham CW (1967) Reactions of quartz and corundum with aqueous chloride and hydroxide solutions at high temperatures and pressures. Am J Sci 265:12–27

Arnórsson S (1970) Geochemical studies of thermal waters in the southern lowlands of Iceland. Geothermics 2:547–552

Arnórsson S (1975) Application of the silica geothermometer in low temperature hydrothermal areas in Iceland. Am J Sci 275:763–784

Arnórsson S (1985) The use of mixing models and chemical geothermometers for estimating underground temperatures in geothermal systems. J Volcanol Geoth Res 23:299–335

Arnórsson S (1998) Geothermal geochemistry. Historical overview, present status, future outlook. In: UNU-GTP 20th anniversary workshop, pp 179–185

Arnórsson S (2000) The quartz and Na/K geothermometers. I. New thermodynamic calibration. In: Proceedings of the world geothermal congress 2000, Kyushu-Tohoku, Japan, pp 929–934

Arnórsson S, Gunnlaugsson E (1985) New gas geothermometers for geothermal exploration—calibration and application. Geochim Cosmochim Ac 49:1307–1325

Arnórsson S, Stefánsson A (1999) Assessment of feldspar solubility constants in water in the range of 0° to 350 °C at vapor saturation pressures. Am J Sci 299:173–209

Arnórsson S, Gunnlaugsson E, Svavarsson H (1983a) The chemistry of geothermal waters in Iceland. III. Chemical geothermometry in geothermal investigations. Geochim Cosmochim Ac 47:567–577

Arnórsson S, Gunnlaugsson E, Svavarsson H (1983b) The chemistry of geothermal waters in Iceland. II. Mineral equilibria and independent variables controlling water compositions. Geochim Cosmochim Ac 47:547–566

Arnórsson S, Stefánsson A, Bjarnason JO (2007) Fluid-fluid interactions in geothermal systems. Rev Mineral Geochem 65:259–312

Azaroual M, Fouillac C, Matray JM (1997) Solubility of silica polymorphs in electrolyte solutions. II. Activity of aqueous silica and solid silica polymorphs in deep solutions from the sedimentary Paris Basin. Chem Geol 140:167–179

Bakker RJ (2003) Package FLUIDS 1. Computer programs for analysis of fluid inclusion data and for modelling bulk fluid properties. Chem Geol 194:3–23

Bayram AF (2001) Application of an artificial neural network model to a Na–K geothermometer. J Volcanol Geoth Res 112:75–81

Benjamin T, Charles R, Vidale R (1983) Thermodynamic parameters and experimental data for the Na–K–Ca geothermometer. J Volcanol Geoth Res 15:167–186

Bird DK, Helgeson HC (1981) Chemical interaction of aqueous solutions with epidote-feldspar mineral assemblages in geologic systems. II. Equilibrium constraints in metamorphic/geothermal processes. Am J Sci 281:576–614

Bird DK, Norton DL (1981) Theoretical prediction of phase relations among aqueous solutions and minerals: Salton Sea geothermal system. Geochim Cosmochim Ac 45:1479–1494

Bischoff JL (1991) Densities of liquids and vapors in boiling NaCl–H_2O solutions: a PVTX summary from 300° to 500 °C. Am J Sci 291:309–338

Bischoff JL, Rosenbauer RJ (1985) An empirical equation of state for hydrothermal seawater (3.2 percent NaCl). Am J Sci 285:725–763

Blasco M, Gimeno MJ, Auqué LF (2018) Low temperature geothermal systems in carbonate-evaporitic rocks: mineral equilibria assumptions and geothermometrical calculations. Insights from the Arnedillo thermal waters (Spain). Sci Total Environ 615:526–539

Blasco M, Auqué LF, Gimeno MJ, Acero P, Gómez J, Asta MP (2019) Mineral equilibria and thermodynamic uncertainties in the geothermometrical characterisation of carbonate geothermal systems of low temperature. The case of the Alhama-Jaraba system (Spain). Geothermics 78:170–182

Bödvarsson G (1960) Exploration and exploitation of natural heat in Iceland. B Volcanol 23:241–250

Bödvarsson G, Palmason G (1961) Exploration of subsurface temperature in Iceland. In: Geothermal energy. Proceedings of I: UN conference new sources energy, Rome 1961, vol 2, pp 91–98

Bowers TS, Jackson KJ, Helgeson HC (1984) Equilibrium activity diagrams for coexisting minerals and aqueous solutions at pressures and temperatures to 5 kb and 600 °C. Springer, 397pp

Brown ST, Kennedy BM, De Paolo DJ, Hurwitz S, Evans WC (2013) Ca, Sr, O and D isotope approach to defining the chemical evolution of hydrothermal fluids: example from Long Valley, CA, USA. Geochim Cosmochim Ac 122:209–225

Browne PRL (1970) Hydrothermal alteration as an aid in investigating geothermal fields. Geothermics 2:564–570

Browne PRL (1978) Hydrothermal alteration in active geothermal fields. Annu Rev Earth Pl Sci 6:229–248

Can I (2002) A new improved Na/K geothermometer by artificial neural networks. Geothermics 31:751–760

Capuano RM, Cole DR (1982) Fluid-mineral equilibria in a hydrothermal system, Roosevelt Hot Springs, Utah. Geochim Cosmochim Ac 46:1353–1364

Chiodini G, Cioni R, Guidi M, Marini L (1991) Chemical geothermometry and geobarometry in hydrothermal aqueous solutions: a theoretical investigation based on a mineral-solution equilibrium model. Geochim Cosmochim Ac 55:2709–2727

Chiodini G, Frondini F, Marini L (1995) Theoretical geothermometers and P_{CO2} indicators for aqueous solutions coming from hydrothermal systems of medium-low temperature hosted in carbonate-evaporite rocks. Application to the thermal springs of the Etruscan Swell, Italy. Appl Geochem 10:337–346

Christenson BW (1997) Kawerau geothermal field: geochemical structure of the reservoir and its response to exploitation. Geoth Res T 21:17–24

Crerar DA, Anderson GM (1971) Solubility and solvation reactions of quartz in dilute hydrothermal solutions. Chem Geol 8:107–122

Ellis AJ (1970) Quantitative interpretation of chemical characteristics of hydrothermal systems. Geothermics 2:516–528

Ellis AJ (1971) Magnesium ion concentrations in the presence of magnesium chlorite, calcite, carbon dioxide, quartz. Amer. J. Sci. 271:481–489

Ellis AJ, Mahon WAJ (1967) Natural hydrothermal systems and experimental hot water/rock interactions (part II). Geochim Cosmochim Ac 31:519–538

Fleming BA, Crerar DA (1982) Silicic acid ionization and calculation of silica solubility at elevated temperature and pH. Application to geothermal fluid processing and reinjection. Geothermics 11:15–29

Fouillac C, Michard G (1981) Sodium/lithium ratios in water applied to geothermometry of geothermal reservoirs. Geothermics 10:55–70
Fournier RO (1960) Solubility of quartz in water in the temperature interval from 25 to 300 °C. Geol Soc Am Bull 71:1867–1868
Fournier RO (1973) Silica in thermal waters: laboratory and field investigations. In: Proceedings international symposium on hydrogeochemistry and biogeochemistry, Tokyo, pp 122–139
Fournier RO (1977) Chemical geothermometers and mixing models for geothermal systems. Geothermics 5:41–50
Fournier RO (1979) A revised equation for the Na/K geothermometer. Geotherm Res Counc Trans 5:1–16
Fournier RO (1981) Application of water geochemistry to geothermal exploration and reservoir engineering (Chapter 4). In: Rybach L, Muffler LJP (eds) Geothermal systems: principles and case histories. Wiley, New York, pp 109–143
Fournier RO, Potter RW II (1982a) An equation correlating the solubility of quartz in water from 25° to 900°C at pressures up to 10,000 bars. Geochim Cosmochim Ac 46: 1969–1973
Fournier RO, Potter RW II (1982b) A revised and expanded silica (quartz) geothermometer. Geot Res Council Bull 11:3-12 (see also correction in vol. 12, page 32)
Fournier RO (1991) Water geothermometers applied to geothermal energy. Appl. Geochem. Geotherm. Reservoir Dev 37–69 (F. D'Amore, co-ordinator), UNITAR
Fournier RO, Potter RW (1979) Magnesium correction to the Na–K–Ca chemical geothermometer. Geochim Cosmochim Ac 43:1543–1550
Fournier RO, Rowe JJ (1962) The solubility of cristobalite along the three-phase curve, gas plus liquid plus cristobalite. Am Mineral 47:897–902
Fournier RO, Rowe JJ (1966) Estimation of underground temperatures from the silica content of water from hot springs and wet-steam wells. Am J Sci 264:685–697
Fournier RO, Truesdell AH (1970) Chemical indicators of subsurface temperature applied to hot spring waters of Yellowstone National Park, Wyoming, USA. Geothermics 2:529–535
Fournier RO, Truesdell AH (1973) An empirical Na–K–Ca geothermometer for natural waters. Geochim Cosmochim Ac 37:1255–1275
Fournier RO, White DE, Truesdell AH (1974) Geochemical indicators of subsurface temperature. Part I. Basic assumptions. J Res US Geol Surv 2:259–262
Fournier RO, Rosenbauer RJ, Bischoff JL (1982) The solubility of quartz in aqueous sodium chloride solution at 350 °C and 180 to 500 bars. Geochim Cosmochim Ac 46:1975–1978
Ganeyev IG (1975) Solubility and crystallization of silica in chloride. Doklady Acad Nauk SSSR 221:248–250 (citation from Shibue, 1996)
Giggenbach WF (1981) Geothermal mineral equilibria. Geochim Cosmochim Ac 45:393–410
Giggenbach WF (1984) Mass transfer in hydrothermal alterations systems. Geochim Cosmochim Ac 48:2693–2711
Giggenbach WF (1988) Geothermal solute equilibria. Derivation of Na–K–Mg–Ca geoindicators. Geochim Cosmochim Ac 52:2749–2765
Giggenbach WF (1991a) Chemical techniques in geothermal exploration. In Appl Geochem Geotherm Reservoir Dev 119–144 (F. D'Amore, co-ordinator), UNITAR
Giggenbach WF, Gonfiantini R, Jangi BL, Truesdell AH (1983) Isotopic and chemical composition of Parbati Valley geothermal discharges, NW-Himalaya, India. Geothermics 12:199–222
Gíslason SR, Heaney PJ, Oelkers EH, Schott J (1997) Kinetic and thermodynamic properties of moganite, a novel silica polymorph. Geochim Cosmochim Ac 61:1193–1204
Goguel RL (1983) The rare alkalies in hydrothermal alteration at Wairakei and Broadlands geothermal fields, NZ. Geochim Cosmochim Ac 47:429–437
Guidi M, Marini L, Scandiffio G, Cioni R (1990) Chemical geothermometry in hydrothermal aqueous solutions: the influence of ion complexing. Geothermics 19:415–441
Gunnarsson I, Arnórsson S (2000) Amorphous silica solubility and the thermodynamic properties of $H_4SiO_4^{\circ}$ in the range of 0° to 350 °C at P_{sat}. Geochim Cosmochim Ac 64:2295–2307

Heitmann HG (1965) Die Löslichkeit von Kieselsäure in Wasser und Wasserdampf. Glastech. Ber.-Glass 38:41–54

Helgeson HC (1968) Evaluation of irreversible reactions in geochemical processes involving minerals and aqueous solutions: I. Thermodynamic relations. Geochim Cosmochim Acta 32:853–877

Helgeson HC (1969) Thermodynamics of hydrothermal systems at elevated temperatures and pressures. Am J Sci 267:729–804

Helgeson HC, Brown TH, Nigrini A, Jones TA (1970) Calculation of mass transfer in geochemical processes involving aqueous solutions. Geochim Cosmochim Acta 34:569–592

Helgeson HC, Delany JM, Nesbitt HW, Bird DK (1978) Summary and critique of the thermodynamic properties of rock-forming minerals. Am J Sci 278A:229p

Helgeson HC, Garrels RM, Mackenzie FT (1969) Evaluation of irreversible reactions in geochemical processes involving minerals and aqueous solutions: II. Applications. Geochim Cosmochim Acta 33:455–481

Hemley JJ, Montoya JW, Marinenko JW, Luce RW (1980) Equilibria in the system Al_2O_3–SiO_2–H_2O and some general implications for alteration/mineralization processes. Econ Geol 75:210–228

Hyeong K, Capuano RM (2001) Ca/Mg of brines in Miocene/Oligocene clastic sediments of the Texas Gulf Coast: buffering by calcite/disordered dolomite equilibria. Geochim Cosmochim Ac 65:3065–3080

Ingebritsen SE, Sorey ML (1988) Vapor-dominated zones within hydrothermal systems: Evolution and natural state. J Geophys Res -Sol Ea 93(B11):13635–13655

James R (1968) Wairakei and Larderello: geothermal power systems compared. New Zeal. J Sci 11:706–719

Janecky DR, Charles RW, Bayhurst GK, Benjamin TM (1986) Physicochemical basis of the Na–K–Ca geothermometer. Los Alamos National Lab. Report No. LA-10806-MS, 10pp

Johnson JW, Oelkers EH, Helgeson HC (1992) SUPCRT 92: a software package for calculating the standard molal thermodynamic properties of minerals, gases, aqueous species, and reactions from 1 to 5000 bars and 0 to 1000 °C. Comput Geosci 18:899–947

Karingithi CW, Arnórsson S, Grönvold K (2010) Processes controlling aquifer fluid compositions in the Olkaria geothermal system, Kenya. J Volcanol Geoth Res 196:57–76

Kennedy GC (1950) A portion of the system silica-water. Econ Geol 45:629–653

Kharaka YK, Mariner RH (1989) Chemical geothermometers and their application to formation waters from sedimentary basins. In: Naeser ND, McCollon TH (eds) Thermal history of sedimentary basins. Springer, New York, pp 99–117

Kharaka YK, Lico MS, Law LM (1982) Chemical geothermometers applied to formation waters, Gulf of Mexico and California basins. Am Assoc Petr Geol B 66:588

Kharaka YK, Specht DJ, Carothers WW (1985) Low to intermediate subsurface temperatures calculated by chemical geothermometers. Am Assoc Petr Geol B 69:273

Khitarov NI (1956) The 400 °C isotherm for the system H_2O-SiO_2 at pressures up to 2,000 kg/cm^2. Geochem Int 1956:55–61

Kitahara S (1960) The solubility of quartz in water at high temperatures and high pressures. Rev Phys Chem Jpn 30:109–114

Lemmon EW, McLinden MO, Friend DG (2017) Thermophysical properties of fluid systems. In: Linstrom PJ, Mallard WG (eds) NIST chemistry webbook, NIST standard reference database number 69, National Institute of Standards and Technology, Gaithersburg MD. http://webbook.nist.gov/

Mackenzie FT, Gees R (1971) Quartz: synthesis at earth-surface conditions. Science 173:533–535

Mahon WAJ (1965) Calcium and magnesium in the natural thermal waters of New Zealand. New Zeal J Sci 8:66–78

Mahon WAJ (1966) Silica in hot water discharged from drillholes at Wairakei, New Zealand. New Zeal J Sci 9:135–144

Mahon WAJ (1970) Chemistry in the exploration and exploitation of hydrothermal systems. Geothermics 2:1310–1322
Manning CE (1994) The solubility of quartz in H_2O in the lower crust and upper mantle. Geochim Cosmochim Ac 58:4831–4839
Marini L (2006) Geological sequestration of carbon dioxide: Thermodynamics, kinetics, and reaction path modeling. In: Developments in geochemistry, vol 11. Elsevier, Amsterdam, 453pp
Marini L, Chiodini G, Cioni R (1986) New geothermometers for carbonate-evaporite geothermal reservoirs. Geothermics 15:77–86
Marini L, Cioni R, Guidi M (1998) Water chemistry of San Marcos area, Guatemala. Geothermics 27:331–360
Mercado S (1970) High activity hydrothermal zones detected by Na/K, Cerro Prieto, Mexico. Geothermics 2:1367–1376
Michard G (1990) Behaviour of major elements and some trace elements (Li, Rb, Cs, Sr, Fe, Mn, W, F) in deep hot waters from granitic areas. Chem Geol 89:117–134
Michard G, Fouillac C (1976) Remarques sur le geothermometre Na–K–Ca. J Volcanol Geoth Res 1:297–304
Michard G, Fouillac C, Grimaud D, Denis J (1981) Une méthode globale d'estimation des températures des resérvoirs alimentant les sources thermales. Exemple du Massif Central Français. Geochim Cosmochim Ac 45:1199–1207
Morey GW, Hesselgesser JM (1951a) The solubility of quartz and some other substances in superheated steam at high pressures. Am Soc Mech Eng Trans 73:864–875
Morey GW, Hesselgesser JM (1951b) The solubility of some minerals in superheated steam at high pressures. Econ Geol 46:821–835
Morey GW, Fournier RO, Rowe JJ (1962) The solubility of quartz in water in the temperature interval from 25° to 300 °C. Geochim Cosmochim Ac 26:1029–1043
Nieva D, Nieva R (1987) Developments in geothermal energy in Mexico—part twelve. A cationic geothermometer for prospecting of geothermal resources. Heat Recov Syst CHP 7:243–258
Novgorodov PG (1975) Solubility of quartz in H_2O-CO_2 mixtures at 700 °C and pressures of 3 and 5 kbar. Geochem Int 12:122–126
Novgorodov PG (1977) On the solubility of quartz in $H_2O + CO_2$ and H_2O + NaCl at 700 °C and 1.5 kb pressure. Geochem Int 14:191–193
Opondo KM (2015) Carbonate scale formed in well OW-202 in Olkaria central field, Kenya. In: Proceedings world geothermal congress 2015, Melbourne, Australia, 19–25 Apr 2015, 10pp
Paces T (1975) A systematic deviation from Na–K–Ca geothermometer below 75 °C and above 10^{-4} atm P_{CO2}. Geochim Cosmochim Ac 39:541–544
Pieri S, Sabatelli F, Tarquini B (1989) Field testing results of downhole scale inhibitor injection. Geothermics 18:249–257
Pope LA, Hajash A, Popp RK (1987) An experimental investigation of the quartz, Na–K, Na–K–Ca geothermometers and the effects of fluid composition. J Volcanol Geoth Res 31:151–161
Puigdomènech I, Rard JA, Plyasunov AV, Grenthe I (1997) Temperature corrections to thermodynamic data and enthalpy calculations. In: I. Grenthe and I. Puigdomènech (Eds.) Modelling in Aquatic Chemistry, NEA-OECD Publications, Paris 427–493
Ragnarsdóttir KV, Walther JV (1983) Pressure sensitive "silica geothermometer" determined from quartz solubility experiments at 250 °C. Geochim Cosmochim Acta 47:941–946
Reyes AG (1990) Petrology of Philippine geothermal systems and the application of alteration mineralogy to their assessment. J Volcanol Geoth Res 43:279–309
Reyes AG, Giggenbach WF, Saleras JRM, Salonga ND, Vergara MC (1993) Petrology and geochemistry of Alto Peak, a vapor-cored hydrothermal system, Leyte Province, Philippines. Geothermics 22:479–519
Rimstidt JD (1997) Quartz solubility at low temperatures. Geochim Cosmochim Ac 61:2553–2558
Santoyo E, Díaz-González L (2010) A new improved proposal of the Na/K geothermometer to estimate deep equilibrium temperatures and their uncertainties in geothermal systems. In: Proceedings world geothermal congress 2010, Bali, Indonesia, 25–29 Ap 2010, 9pp

Semenova AI, Tsiklis DS (1970) Solubility of silicon dioxide in steam at high pressures and temperatures. Russ J Phys Chem 44:1420–1422

Serpen G, Palabiyik Y, Serpen U (2009) An artificial neural network model for Na/K geothermometer. In: Proceedings, 34th workshop on geothermal reservoir engineering, Stanford University, Stanford, California, 9–11 Feb 2009, SGP-TR-187, 12pp

Seward TM (1974) Determination of the first ionization constant of silicic acid from quartz solubility in borate buffer solutions to 350 C. Geochim Cosmochim Ac 38:1651–1664

Seyfried WE, Gordon PC, Dickson FW (1979) A new reaction cell for hydrothermal solution equipment. Am Mineral 64:646–649

Sheppard DS (1984) Fluid chemistry of Ngawha reservoir. Proceedings of the 6th New Zealand geothermal workshop, pp 151–154

Shibue Y (1995) Empirical expressions of quartz solubility in H_2O, $H_2O + CO_2$, and $H_2O + NaCl$ fluids. Geochem J 30:339–354

Shikazono N (1976) Thermodynamic interpretation of Na–K–Ca geothermometer in the natural water system. Geochem J 10:47–50

Shikazono N (1978) Possible cation buffering in chloride-rich geothermal waters. Chem Geol 23:239–254

Shock EL, Helgeson HC (1988) Calculation of the thermodynamic and transport properties of aqueous species at high pressures and temperatures: correlation algorithms for ionic species and equation of state predictions to 5 kb and 1000 °C. Geochim Cosmochim Ac 52:2009–2036

Shock EL, Helgeson HC, Sverjensky DA (1989) Calculation of the thermodynamic and transport properties of aqueous species at high pressures and temperatures: standard partial molal properties of inorganic neutral species. Geochim Cosmochim Ac 53:2157–2183

Siever R (1962) Silica solubility, 0-200 °C, and the diagenesis of siliceous sediments. J Geol 70:127–150

Simmons SF (2013) Fluid-mineral equilibria in Great Basin geothermal systems: implications for chemical geothermometry. In: Proceedings of the 38th workshop on geothermal reservoir engineering, Stanford University, Stanford, California, February 11–13, 2013, SGP-TR-198, 8 pp

Simmons SF, Christenson BW (1994) Origins of calcite in a boiling geothermal system. Am J Sci 294:361–400

Sommerfeld RA (1967) Quartz solution reaction: 400°–500 °C, 1000 bars. J Geophys Res 72(16):4253–4257

Tarcan G, Özen T, Gemici Ü, Çolak M, Karamanderesi İH (2016) Geochemical assessment of mineral scaling in Kızıldere geothermal field, Turkey. Environ Earth Sci 75:1317 (19pp)

Tassew M (2010) Maintenance and operational experience gained by operating the Aluto Langano geothermal pilot power plant. In: Proceedings world geothermal congress 2010, Bali, Indonesia, 25–29 Apr 2010, 4pp

Tonani F (1980) Some remarks on the application of geochemical techniques in geothermal exploration. In: Proceedings of 2nd symposium advances in European geothermal research, Strasbourg, pp 428–443

Truesdell A.H. (1976) Summary of section III—Geochemical techniques in exploration. In Proceedings of 2nd UN symposium on the development and use of geothermal resources, vol 1, pp 53–79

Truesdell AH, Fournier RO (1977) Procedure for estimating the temperature of a hot-water component in a mixed water by using a plot of dissolved silica versus enthalpy. U.S. Geol. Surv. J. Res. 5:49–52

Van Lier J, Bruyn PD, Overbeek JTG (1960) The solubility of quartz. J Phys Chem-US 64:1675–1682

Verma SP, Santoyo E (1997) New improved equations for Na/K, Na/Li and SiO_2 geothermometers by outlier detection and rejection. J Volcanol Geoth Res 79:9–23

Vespasiano G, Apollaro C, Muto F, Dotsika E, De Rosa R, Marini L (2014) Chemical and isotopic characteristics of the warm and cold waters of the Luigiane Spa near Guardia Piemontese (Calabria, Italy) in a complex faulted geological framework. Appl Geochem 41:73–88

Viveiros PV (2014) Calcium carbonate scaling control in geothermal well PV8 in Sao Miguel, Azores, combining chemical inhibition and mechanical reaming. UNU-GTP Rep 33:721–744

Von Damm KL, Bischoff JL, Rosenbauer RJ (1991) Quartz solubility in hydrothermal seawater: An experimental study and equation describing quartz solubility for up to 0.5 M NaCl solutions. Am J Sci 291:977–1007

Walshe JL (1986) A six-component chlorite solid solution model and the conditions of chlorite formation in hydrothermal and geothermal systems. Econ Geol 81:681–703

Walther JV, Orville PM (1983) The extraction-quench technique for determination of the thermodynamic properties of solute complexes: application to quartz solubility in fluid mixtures. Am Mineral 68:731–741

Weill DF, Fyfe WS (1964) The solubility of quartz in H_2O in the range 1000–4000 bars and 400–550 °C. Geochim Cosmochim Ac 28:1243–1255

White DE (1965) Saline waters of sedimentary rocks. In: Fluids in subsurface environments pp 342–366

White DE (1968) Environments of generation of some base-metal ore deposits. Econ Geol 63:301–335

White DE, Muffler LJP, Truesdell AH (1971) Vapor-dominated hydrothermal systems compared with hot-water systems. Econ Geol, 66:75–97

Wolery TJ (1992) EQ3NR, a computer program for geochemical aqueous speciation-solubility calculations: theoretical manual, user's guide and related documentation (version 7.0). Report UCRL-MA-110662 PT III. Lawrence Livermore National Laboratory, Livermore

Wolery TW, Jove-Colon C (2007) Qualification of thermodynamic data for geochemical modeling of mineral-water interactions in dilute systems. Sandia National Laboratories Report ANL-WIS-GS-000003 REV 01

Wyart J, Sabatier G (1955) Nouvelles mesures de la solubilité du quartz, de la tridymite et de la cristobalite dans l'eau sous pression au-dessus de la temperature critique. CR Hebd Acad Sci 240:1905–1907

Yoshida Y (1991) Geochemistry of the Nigorikawa geothermal system, southwest Hokkaido, Japan. Geochem J 25:203–222

Zheng X, Duan C, Xia B, Jiang Y, Wen J (2019) Hydrogeochemical modeling of the shallow thermal water evolution in Yangbajing geothermal field, Tibet. J Earth Sci 30:870–878

Zhiqian G, Bingsong Y, Xingyun L (2007) The equilibrium between diagenetic calcites and dolomites and its impact on reservoir quality in the sandstone reservoir of Kela 2 gas field. Prog Nat Sci 17:1051–1058

Chapter 6
The Activity-Based Theoretical Na–K Geoindicators

Abstract Accepting that a single universally-working Na-K geothermometer does not exist and the Na-K geothermometers are controlled by the exchange reaction involving fully-ordered low-albite and variably-ordered adularia, from fully ordered maximum-microcline to completely disordered high-sanidine, the Na/K activity ratio of the aqueous solution is expected to be a function of both temperature and the ordering parameter of adularia, Z. However, the two limiting Na-K geothermometers involving low-albite and either high-sanidine (with Z = 0) or maximum-microcline (with Z = 1) give equilibrium temperatures differing by 125 ± 3 °C on average, in the range 100–350 °C, and have little usefulness. We used Multiple Regression Analysis to try to predict the ordering parameter of adularia based on the chemical characteristics of the 1013 selected reservoir liquids, but it turned out that Z is chiefly controlled by Na, K, and SiO_2. This is a disappointing result being affected by a circular argument. Therefore, for the moment, it is advisable to utilize the Na/K log activity ratio to estimate the ordering parameter Z of the hydrothermal adularia in hypothetical equilibrium with the considered aqueous solutions and to use this information in other geothermometers representing the subject of the following chapters.

We have already recalled that several previous authors (e.g., White 1965; Ellis 1970; Fournier 1979; Arnórsson et al. 1983; Giggenbach 1988; Michard 1990) suggested that the activities of Na^+ and K^+ ions in reservoir liquids are probably controlled by the exchange reaction (5.74) between hydrothermal alkali feldspars, which is rewritten here for convenience:

$$NaAlSi_3O_8 + K^+ = KAlSi_3O_8 + Na^+. \tag{6.1}$$

The possible involvement in reaction (6.1) of solid phases other than the alkali feldspars can be excluded comparing the enthalpies and entropies of several Na–K exchange reactions with those of Na–K geothermometers, as discussed in Sect. 5.3.3.

R. Cioni and L. Marini, *A Thermodynamic Approach to Water Geothermometry*, Springer Geochemistry, https://doi.org/10.1007/978-3-030-54318-1_6

6.1 The Log K of the Na–K Exchange Reactions Between Hydrothermal Alkali Feldspars

Since hydrothermal Na- and K-feldspars are pure or relatively pure minerals (see Sect. 4.2.2), their activities can be considered to be close to unity and the equilibrium constant of reaction (6.1) can be simplified as follows:

$$K = \frac{a_{K-\text{feldspar}} \cdot a_{Na^+}}{a_{Na-\text{feldspar}} \cdot a_{K^+}} \cong \frac{a_{Na^+}}{a_{K^+}}. \tag{6.2}$$

Equation (6.2) indicates that the Na^+/K^+ activity ratio of reservoir liquids is expected to be a simple temperature function, thus supporting the use of the Na^+/K^+ activity ratio or, to a first approximation, of the Na/K concentration ratio for geothermometry. However, the different structural state of the alkali feldspars and Al-Si order-disorder on their tetrahedral sites complicate this simple picture, making unlikely the existence of a unique Na–K geothermometer applicable everywhere, as underscored by Bird and Norton (1981) and already noted in Sect. 5.3.4.

Although the reasons for choosing the thermodynamic data of the alkali feldspars given by Helgeson et al. (1978) to derive the activity-based Na–K theoretical indicators of interest were thoroughly explained in Sect. 4.2.3, it is interesting to make a brief digression to see how the choice of thermodynamic properties of the alkali feldspars other than those of Helgeson et al. (1978) impacts the temperature dependence of the Na/K log activity ratio.

First, the thermodynamic equilibrium constant of the exchange reactions involving the four endmember alkali feldspars, that is the log K values of the exchange reactions between low-albite and maximum-microcline, low-albite and high-sanidine, high-albite and maximum-microcline, high-albite and high-sanidine were calculated by means of SUPCRT92 considering the thermodynamic data of Helgeson et al. (1978). Results of these SUPCRT92 calculations are shown in Fig. 6.1. Extrapolation of the lines for these four exchange reactions above 350 °C shows that the low-albite/maximum-microcline line intersects the low-albite/high-sanidine line at 500 °C and the high-albite/maximum-microcline line crosses the high-albite/high-sanidine line at the same temperature, which is the microcline-sanidine transition temperature consistent with the thermodynamic data of Helgeson et al. (1978) for these two K-feldspars, as discussed in Sect. 4.2.4. In other terms, the equality of the log K of the low-albite/high-sanidine and low-albite/maximum-microcline exchange reactions at 500 °C and the equality of the log K of the high-albite/high-sanidine and high-albite/maximum-microcline exchange reactions at the same temperature are not fortuitous coincidences, but the consequences of the condition $\Delta G^{\circ}_{f,Sa} = \Delta G^{\circ}_{f,Mc}$ at 500 °C.

Second, the thermodynamic equilibrium constants of the four exchange reactions involving the four endmember alkali feldspars were calculated using SUPCRT92 taking into account the thermodynamic properties of alkali feldspars of Holland and Powell (1998). Results are displayed in Fig. 6.2. Comparison of Figs. 6.1 and 6.2

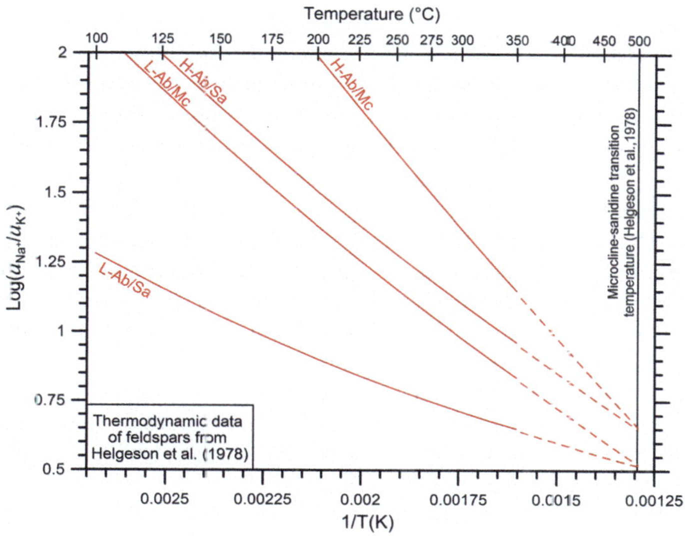

Fig. 6.1 Na/K log activity ratios fixed by low-albite/maximum-microcline (L-Ab/Mc), low-albite/high-sanidine (L-Ab/Sa), high-albite/maximum-microcline (H-Ab/Mc), and high-albite/high-sanidine (H-Ab/Sa) coexistence (red lines, solid below 350 °C, dashed above 350 °C), all computed considering the thermodynamic data of alkali feldspars of Helgeson et al. (1978)

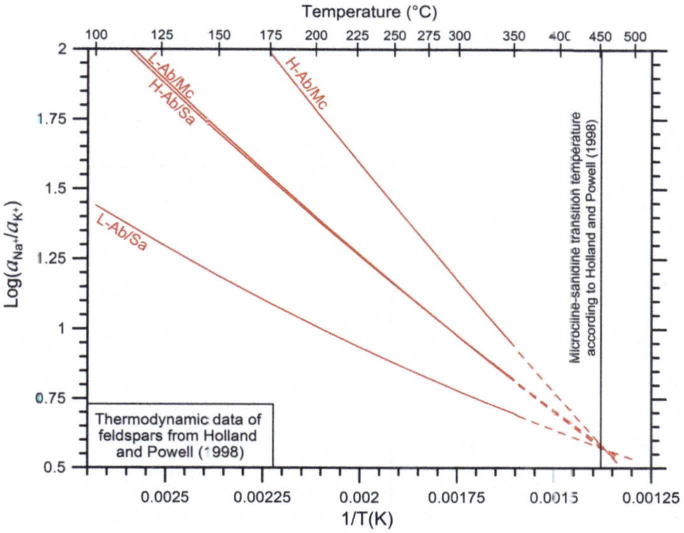

Fig. 6.2 Na/K log activity ratios fixed by low-albite/maximum-microcline (L-Ab/Mc), low-albite/high-sanidine (L-Ab/Sa), high-albite/maximum-microcline (H-Ab/Mc), and high-albite/high-sanidine (H-Ab/Sa) coexistence (red lines, solid below 350 °C, dashed above 350 °C), all computed considering the thermodynamic data of alkali feldspars of Holland and Powell (1998)

shows that the low-albite/maximum-microcline line in Fig. 6.2 is practically superimposed onto the same line in Fig. 6.1, whereas the low-albite/high-sanidine line in Fig. 6.2 is moderately shifted upwards with respect to the same line in Fig. 6.1. These small differences in the position of the low-albite/maximum-microcline and low-albite/high-sanidine lines in Figs. 6.1 and 6.2 reflect the relatively limited difference, ~50 °C, between the sanidine/microcline transition temperature consistent with the thermodynamic data adopted by Holland and Powell (1998) for the two K-feldspars, 452 °C, and the sanidine/microcline transition temperature of Helgeson et al. (1978), 500 °C. To be noted also that the high-albite/maximum-microcline and high-albite/high-sanidine lines in Fig. 6.2 are somewhat shifted downwards compared to the corresponding lines in Fig. 6.1, but also these two lines intersect at the sanidine/microcline transition temperature of 452 °C.

Third, the log K values of the exchange reactions involving the four endmember alkali feldspars were computed by means of SUPCRT92 considering the thermodynamic data of alkali feldspars of Arnórsson and Stefánsson (1999). The outcomes of these calculations are presented in Fig. 6.3.

In Fig. 6.3, the low-albite/high-sanidine line intersects the steeper low-albite/maximum-microcline line at 210 °C and the high-albite/high-sanidine line crosses the steeper high-albite/maximum-microcline line at the same temperature, which is the sanidine/microcline transition temperature consistent with the thermodynamic data of Arnórsson and Stefánsson (1999) for the two K-feldspars, as shown in Sect. 4.2.4.

Summing up, the comparison of Figs. 6.1, 6.2, and 6.3 shows that the sanidine/microcline transition temperature constrains the temperature—log K relations

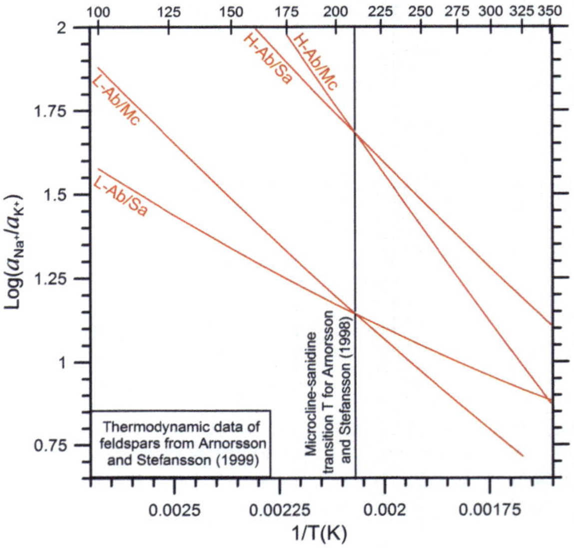

Fig. 6.3 Na/K log activity ratios fixed by low-albite/maximum microcline (L-Ab/Mc), low-albite/high-sanidine (L-Ab/Sa), high-albite/maximum-microcline (H-Ab/Mc), and high-albite/high-sanidine (H-Ab/Sa) coexistence (red solid lines), all computed considering the thermodynamic data of alkali feldspars of Arnórsson and Stefánsson (1999)

of the exchange reactions between alkali feldspars. Therefore, it is of utmost importance to rely on thermodynamic data of alkali feldspars which are consistent with the correct sanidine/microcline transition temperature, like those of Helgeson et al. (1978). Another good reason for choosing the thermodynamic data of Helgeson et al. (1978), not only for the alkali feldspars but also for all the other solid phases, is the high level of internal consistency of this thermodynamic database as already recalled in Sect. 2.2. Moreover, the shifts in the thermodynamic data of the Al-bearing solid phases with respect to other datasets do not affect the thermodynamic properties of the reactions of interest to us, owing to Al conservation in the solid phases.

Resuming the main discussion, it must be underscored that the hydrothermal feldspars participating to reaction (6.1) are most likely triclinic, fully-ordered low-albite and variably-ordered adularia, from fully ordered microcline to completely disordered sanidine, with either triclinic or monoclinic symmetry, as discussed in Sects. 4.2.3 and 5.3.3. Accepting this hypothesis, the Na/K activity ratio is expected to be a function of both temperature and the ordering parameter of adularia, Z, which is equal to 1 for maximum-microcline and assumes the value of 0 for high-sanidine. Therefore, the logarithm of the thermodynamic equilibrium constant of reaction (6.1), involving low-albite and adularia with variable ordering parameter Z, was computed using SUPCRT92 considering the thermodynamic data of alkali feldspars of Helgeson et al. (1978) for temperatures varying from 0 to 350 °C at steps of 25 °C and for Z values ranging from 0 to 1 at steps of 0.1 Z units. Results are reported in Table 6.1.

The computed log K values are nearly equal to the Na^+/K^+ activity ratio and are reproduced with acceptable approximations by the following two equations (T in °C):

$$\begin{aligned} Z = & \left(2.1911 \cdot 10^{-7} \cdot T^3 - 8.9451 \cdot 10^{-5} \cdot T^2 + 2.0023 \cdot 10^{-2} \cdot T - 0.11043\right) \\ & \cdot \log(a_{Na^+}/a_{K^+}) - 1.1269 \cdot 10^{-7} \cdot T^3 + 4.2393 \cdot 10^{-5} \cdot T^2 \\ & - 7.8201 \cdot 10^{-3} \cdot T - 1.1045 \end{aligned} \quad (6.3)$$

$$\begin{aligned} & 1000/(T + 273.15) = \\ & \left(-0.919194 \cdot Z^4 + 2.759529 \cdot Z^3 - 3.361149 \cdot Z^2 + 2.209325 \cdot Z - 0.791765\right) \\ & \cdot \left[\log(a_{Na^+}/a_{K^+})\right]^2 + \left(1.5035 \cdot Z^4 - 4.717177 \cdot Z^3 + 6.31701 \cdot Z^2 - 5.16263 \cdot Z + 3.215553\right) \\ & \cdot \log(a_{Na^+}/a_{K^+}) - 0.51088 \cdot Z^4 + 1.632504 \cdot Z^3 - 2.275505 \cdot Z^2 \\ & + 2.011206 \cdot Z - 0.147528 \end{aligned} \quad (6.4)$$

which are valid from 100 to 350 °C and can be alternatively used depending on what we know and what we want to compute.

Table 6.1 Logarithm of the thermodynamic equilibrium constant of the exchange reaction (6.1) involving low-albite and adularia with ordering parameter Z varying from 1 to 0 at steps of 0.1 units, as a function of temperature and the Z ordering parameter of adularia

T(°C)	Z = 1	Z = 0.9	Z = 0.8	Z = 0.7	Z = 0.6	Z = 0.5	Z = 0.4	Z = 0.3	Z = 0.2	Z = 0.1	Z = 0
0.01	3.4941	3.3564	3.2188	3.0811	2.9434	2.8057	2.668	2.5303	2.3926	2.2549	2.1172
25	3.0398	2.9199	2.7999	2.68	2.5601	2.4402	2.3203	2.2003	2.0804	1.9605	1.8406
50	2.6647	2.5598	2.4549	2.35	2.2451	2.1402	2.0353	1.9304	1.8255	1.7206	1.6157
75	2.3530	2.2609	2.1689	2.0769	1.9849	1.8928	1.8008	1.7088	1.6168	1.5247	1.4327
100	2.0919	2.0110	1.9301	1.8492	1.7683	1.6874	1.6066	1.5257	1.4448	1.3639	1.2830
125	1.8714	1.8003	1.7291	1.6580	1.5869	1.5157	1.4446	1.3735	1.3023	1.2312	1.1600
150	1.6839	1.6213	1.5588	1.4963	1.4337	1.3712	1.3086	1.2461	1.1835	1.1210	1.0584
175	1.5232	1.4683	1.4134	1.3585	1.3036	1.2487	1.1937	1.1388	1.0839	1.0290	0.9741
200	1.3846	1.3365	1.2884	1.2403	1.1923	1.1442	1.0961	1.0480	0.9999	0.9518	0.9038
225	1.2641	1.2221	1.1802	1.1382	1.0963	1.0544	1.0124	0.9705	0.9285	0.8866	0.8446
250	1.1583	1.1219	1.0855	1.0491	1.0127	0.9763	0.9399	0.9035	0.8671	0.8307	0.7943
275	1.0643	1.0329	1.0016	0.9702	0.9389	0.9075	0.8762	0.8449	0.8135	0.7822	0.7508
300	0.9802	0.9533	0.9266	0.8999	0.8731	0.8464	0.8196	0.7929	0.7662	0.7394	0.7127
325	0.9049	0.8823	0.8598	0.8372	0.8147	0.7922	0.7697	0.7471	0.7246	0.7021	0.6796
350	0.8388	0.8200	0.8013	0.7827	0.7640	0.7454	0.7267	0.7081	0.6895	0.6708	0.6522

Z = 1 corresponds to maximum-microcline and Z = 0 corresponds to high-sanidine. P = 1 bar for T < 100 °C; $P = P_{sat}$ for T ≥ 100 °C. Calculations performed using SUPCRT92 considering the thermodynamic data of alkali feldspars of Helgeson et al. (1978)

6.2 The Na^+/K^+ Log Activity Ratio of the Selected Reservoir Liquids

Let us now consider the binary plots in which the Na^+/K^+ log activity ratio is contrasted with the aquifer temperature inverse for the selected reservoir liquids of Iceland (Fig. 6.4a), Northern and Central America (Fig. 6.4b), Japan (Fig. 6.5a), the Philippines (Fig. 6.5b), New Zealand (Fig. 6.6a), and miscellaneous systems (Fig. 6.6b).

The Na^+/K^+ log activity ratios fixed by low-albite/maximum-microcline, low-albite/high-sanidine, high-albite/maximum-microcline, and high-albite/high-sanidine equilibrium coexistence, as well as by equilibrium co-occurrence of low-albite and adularia with ordering parameter Z varying from 0 to 1 at steps of 0.1 units, all computed considering the thermodynamic data of alkali feldspars of Helgeson et al. (1978) are also shown in these diagrams for comparison.

In these binary plots, 950 of the 1013 reservoir liquids (corresponding to 93.8% of the total) are found between the lines of low-albite/maximum microcline and low-albite/high-sanidine equilibrium coexistence, 19 samples only (1.9% of the total) are located above the low-albite/maximum microcline line and 45 samples only (4.4% of the total) are situated below the low-albite/high-sanidine line. Moreover, the 64 samples positioned outside the field of low-albite/adularia equilibrium co-occurrence exhibit small or relatively small deviations.

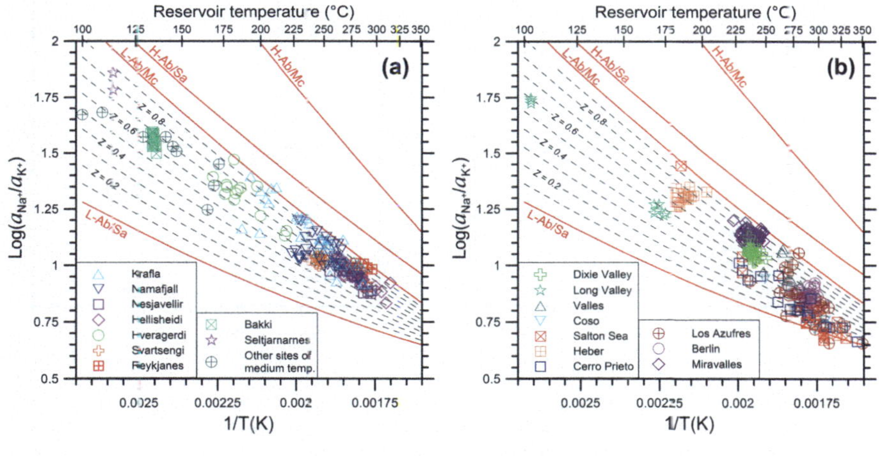

Fig. 6.4 Na^+/K^+ log activity ratios of the selected reservoir liquids of (a) Iceland and (b) North and Central America as a function of aquifer temperature. Also shown are the Na^+/K^+ log activity ratios fixed by low-albite/maximum-microcline (L-Ab/Mc), low-albite/high-sanidine (L-Ab/Sa), high-albite/maximum-microcline (H-Ab/Mc), and high-albite/high-sanidine (H-Ab/Sa) equilibrium coexistence (red solid lines), as well as by equilibrium co-occurrence of low-albite and adularia with ordering parameter Z varying from 0 to 1 at steps of 0.1 units (dashed black lines)

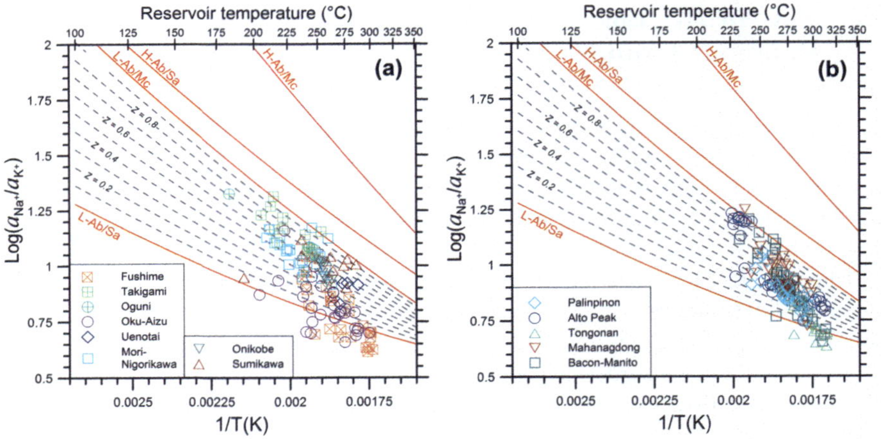

Fig. 6.5 Na^+/K^+ log activity ratios of the selected reservoir liquids of **a** Japan and **b** The Philippines as a function of aquifer temperature. Also shown are the Na^+/K^+ log activity ratios fixed by low-albite/maximum-microcline (L-Ab/Mc), low-albite/high-sanidine (L-Ab/Sa), high-albite/maximum-microcline (H-Ab/Mc), and high-albite/high-sanidine (H-Ab/Sa) equilibrium coexistence (red solid lines), as well as by equilibrium co-occurrence of low-albite and adularia with ordering parameter Z varying from 0 to 1 at steps of 0.1 units (dashed black lines)

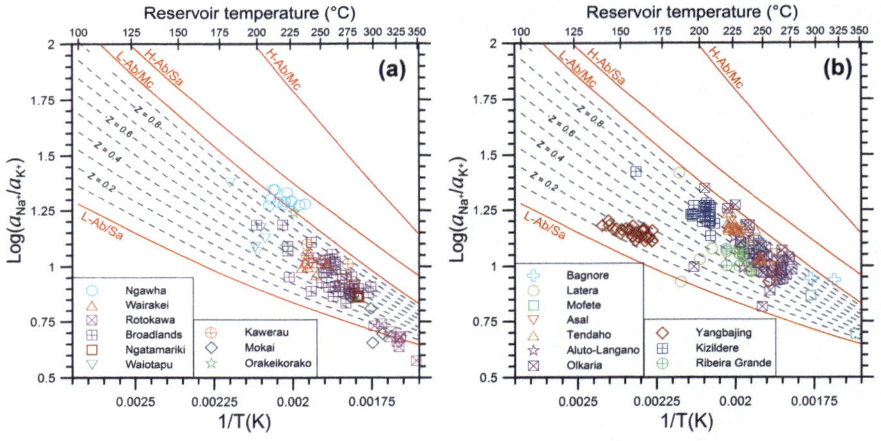

Fig. 6.6 Na^+/K^+ log activity ratios of the selected reservoir liquids of **a** New Zealand and **b** miscellaneous systems as a function of aquifer temperature. Also shown are the Na^+/K^+ log activity ratios fixed by low-albite/maximum-microcline (L-Ab/Mc), low-albite/high-sanidine (L-Ab/Sa), high-albite/maximum-microcline (H-Ab/Mc), and high-albite/high-sanidine (H-Ab/Sa) equilibrium coexistence (red solid lines), as well as by equilibrium co-occurrence of low-albite and adularia with ordering parameter Z varying from 0 to 1 at steps of 0.1 units (dashed black lines)

Of the 19 samples with Na^+/K^+ log activity ratios weakly higher than those fixed by low-albite/maximum microcline equilibrium coexistence, 2 are from Krafla (Fig. 6.4a), 1 is from Los Azufres (Fig. 6.4b), 2 are from Takigami, 1 is from Mori-Nigorikawa (Fig. 6.5a), 1 each are from Mahanagdong, Bacon Manito, and Alto Peak (Fig. 6.5b), 8 are from Ngawha (Fig. 6.6a), 1 is from Bagnore and 1 is from Olkaria (Fig. 6.6b). These small deviations are attributable to distinct reasons, including the undersaturation with completely ordered adularia[1], overestimation of aquifer temperature, underestimation of K^+ ion activity, and overestimation of Na^+ ion activity. The 12 reservoir liquids of Ngawha and the 2 reservoir liquids of Bagnore are unique due to the high concentrations of dissolved B (760–1026 mg/kg at Ngawha; 299 and 459 mg/kg at Bagnore) and NH_4 (69–293 mg/kg at Ngawha, 438 and 569 mg/kg at Bagnore). Since B^{3+} can be hosted in the T-sites of alkali feldspars substituting for Al^{3+} ion and NH_4^+ ion can occupy the M-sites of alkali feldspars replacing for Na^+ and K^+ ions (see Sect. 4.2.1), the hydrothermal alkali feldspars occurring in the Ngawha and Bagnore geothermal reservoirs might be different from pure albite and adularia. If so, the activity of albite and adularia would be different from unity, and should be properly considered in Eq. (6.2). Unfortunately, we were not able to find chemical analyses of the hydrothermal alkali feldspars present in the Ngawha and Bagnore geothermal aquifers to further investigate this matter.

Of the 45 samples with Na^+/K^+ log activity ratios slightly lower than those fixed by low-albite/high-sanidine equilibrium co-occurrence, 5 are from Salton Sea, 1 is from Los Azufres (Fig. 6.4b), 12 each are from Fushime and Oku-Aizu (Fig. 6.5a), 5 are from Tongonan, 4 are from Bacon-Manito (Fig. 6.5b), 4 are from Rotokawa and 2 are from Mokai (Fig. 6.6a). These limited deviations might be due to different causes, such as supersaturation with fully disordered adularia, underestimation of aquifer temperature, overestimation of K^+ ion activity, and underestimation of Na^+ ion activity. Furthermore, given the high temperatures and high salinities of some of these aqueous solutions, firstly those of Salton Sea and secondly those of Fushime and Oku-Aizu, the activities of Na^+ and K^+ ions might be affected by uncertainties in speciation calculations.

Most reservoir liquids of Northern-Central America (Fig. 6.4b), Japan (Fig. 6.5a), The Philippines (Fig. 6.5b), and New Zealand (Fig. 6.6a)are characterized by a general decrease in Z with increasing temperature in spite of the considerable scatter of sample points. In contrast, most reservoir liquids of Iceland have Z values ranging from 0.5 to 1 irrespective of temperature (Fig. 6.4a).

All in all, the spread of sample points in the binary plots of Figs. 6.4, 6.5 and 6.6 is in accordance with the expectations based on the characteristics of authigenic feldspars occurring in sedimentary rocks and to some extent in low-temperature hydrothermal veins as well as of K-feldspars and albites synthesized in hydroxide gels (Sect. 4.2.3). This suggests that the Na^+/K^+ log activity ratio of reservoir liquids is usually controlled by equilibrium coexistence of fully ordered low-albite and adularia with ordering parameter Z varying from 0 to 1. This inference is supported by a large

[1] Saturation with low-albite was assumed in speciation calculation to fix Al concentration.

number of data above ~200 °C, but is probably valid also in the temperature range 100–200 °C, in spite of the lower number of available data.

6.3 The Ordering Parameter Z of Hydrothermal Adularia in Hypothetical Equilibrium with the Selected Reservoir Liquids

The ordering parameter Z of hydrothermal adularia in hypothetical equilibrium with each reservoir liquid of interest was computed by inserting its Na^+/K^+ log activity ratio and reservoir temperature into Eq. (6.3). Considering together the 1013 selected reservoir liquids, Z has a mean of 0.559, a median of 0.612, and a standard deviation of 0.261. These statistical parameters were calculated forcing to 1 the 19 values of Z higher than 1 and forcing to 0 the 45 values of Z lower than 0. The main statistical parameters of Z for the selected reservoir liquids, divided according to the geothermal field of provenance, are listed in Table 6.2, whereas their mean and median values of Z are contrasted in the binary diagram of Fig. 6.7, in which the error bars correspond to ± 1 standard deviation.

Most geothermal systems have mean value of the ordering parameter Z of hydrothermal adularia in the range 0.347–0.808. The hydrothermal adularia of Bagnore and Ngawha has higher mean Z values, 0.908 and 0.977, respectively. These high means are expected on the basis of the unusual fluid chemistry of these two sites (see above). In contrast, the hydrothermal adularia of Cerro Prieto, Yangbajing, Mokai, Salton Sea, Fushime, Oku-aizu, Tongonan, and Rotokawa has lower average Z values, in the interval 0.047–0.239. The low Z values of hydrothermal adularia in hypothetical equilibrium with the Salton Sea reservoir liquids agree with previous findings of Bird and Helgeson (1981) who recognized that these aqueous solutions are close to equilibrium with an hydrothermal assemblage including high-sanidine, which has $Z = 0$.

For the Na–K geothermometers of Fournier (1979), Arnórsson et al. (1983), Giggenbach (1988), and Arnórsson (2000), which are probably those most widely used, we have computed the ordering parameter Z of hydrothermal adularia as a function of temperature by inserting the Na/K total molality ratio and the corresponding temperature into Eq. (6.3), assuming that the Na/K total molality ratio is equal to the Na^+/K^+ activity ratio. The calculated ordering parameter Z of adularia varies as follows, as shown by the diagram of Fig. 6.8: (i) 0.90 at 100–150 °C, 0.87 at 200 °C, 0.77 at 250 °C, 0.59 at 300 °C, and 0.26 at 350 °C for the function of Fournier (1979); (ii) 0.56 at 100 °C, 0.61 at 150 °C, 0.64 at 200 °C, 0.62 or 0.68 at 250 °C, 0.45 at 300 °C, and 0.00 at 348 °C for the two geothermometers of Arnórsson et al. (1983); (iii) 0.95 at 250 °C, 0.72 at 300 °C, and 0.32 at 350 °C for the geothermometer of Giggenbach (1988); (iv) 0.79 at 100 °C, 0.75 at 150 °C, 0.70 at 200 °C, 0.62 at 250 °C, 0.49 at 300 °C, and 0.24 at 350 °C for the theoretical function of Arnórsson (2000).

Table 6.2 Main statistical parameters for the ordering parameter Z of hydrothermal adularia in hypothetical equilibrium with the reservoir liquids of interest divided according to the field of provenance

Field	Country	N	Minimum	Maximum	Median	Mean	Std.Dev.
Hellisheidi	Iceland	27	0.539	0.949	0.698	0.699	0.080
Hveragerdi	Iceland	13	0.623	0.977	0.702	0.728	0.106
Krafla	Iceland	36	0.437	1.000	0.846	0.795	0.159
Namafjall	Iceland	37	0.460	0.958	0.686	0.704	0.148
Nesjavellir	Iceland	15	0.497	0.754	0.642	0.637	0.080
Reykjanes	Iceland	13	0.621	0.952	0.770	0.779	0.118
Svartsengi	Iceland	12	0.522	0.656	0.593	0.590	0.049
Iceland, medium T	Iceland	29	0.416	0.865	0.668	0.658	0.089
Dixie Valley	USA	35	0.525	0.776	0.627	0.627	0.048
Long Valley	USA	10	0.415	0.603	0.483	0.502	0.065
Valles	USA	7	0.478	0.920	0.580	0.648	0.172
Coso	USA	5	0.006	0.683	0.397	0.347	0.253
Salton Sea	USA	23	0.000	0.962	0.082	0.161	0.222
Heber	USA	16	0.585	0.916	0.715	0.717	0.090
Cerro Prieto	Mexico	19	0.036	0.539	0.193	0.239	0.137
Los Azufres	Mexico	26	0.000	1.000	0.365	0.370	0.242
Berlin	El Salvador	55	0.195	0.749	0.443	0.441	0.117
Miravalles	Costa Rica	105	0.673	0.968	0.801	0.808	0.055
Mori Nigorikawa	Japan	21	0.336	1.000	0.549	0.578	0.169
Sumikawa	Japan	14	0.052	0.928	0.657	0.623	0.240
Uenotai	Japan	4	0.221	0.639	0.550	0.490	0.185
Onikobe	Japan	7	0.438	0.654	0.506	0.540	0.083
Oku-aizu	Japan	31	0.000	0.735	0.074	0.138	0.184
Oguni	Japan	6	0.640	0.743	0.730	0.717	0.039
Takigami	Japan	13	0.421	1.000	0.728	0.749	0.166
Fushime	Japan	25	0.000	0.602	0.037	0.146	0.181
Palinpinon	The Philippines	29	0.065	0.745	0.351	0.378	0.164
Alto Peak	The Philippines	27	0.067	1.000	0.497	0.604	0.278
Tongonan	The Philippines	11	0.000	0.295	0.011	0.064	0.105
Mahanagdong	The Philippines	28	0.006	1.000	0.569	0.554	0.216
Bacon-Manito	The Philippines	51	0.000	1.000	0.434	0.444	0.252
Ngawha	New Zealand	12	0.852	1.000	1.000	0.977	0.045
Kawerau	New Zealand	4	0.515	0.838	0.669	0.673	0.142

(continued)

Table 6.2 (continued)

Field	Country	N	Minimum	Maximum	Median	Mean	Std.Dev.
Waiotapu	New Zealand	6	0.340	0.805	0.521	0.525	0.166
Ngatamariki	New Zealand	7	0.432	0.586	0.458	0.488	0.066
Broadlands + Orakeikorako	New Zealand	42	0.217	0.978	0.557	0.553	0.169
Mokai	New Zealand	4	0.000	0.366	0.172	0.177	0.205
Wairakei	New Zealand	9	0.434	0.675	0.585	0.563	0.098
Rotokawa	New Zealand	9	0.000	0.165	0.031	0.047	0.057
Bagnore	Italy	2	0.817	1.000	0.908	0.908	0.130
Latera	Italy	5	0.000	0.923	0.419	0.406	0.337
Mofete	Italy	3	0.197	0.807	0.541	0.515	0.306
Asal	Djibouti	4	0.545	0.826	0.651	0.668	0.117
Tendaho	Ethiopia	36	0.609	0.950	0.792	0.771	0.087
Aluto-Langano	Ethiopia	3	0.408	0.837	0.686	0.644	0.217
Olkaria	Kenya	42	0.064	1.000	0.675	0.673	0.199
Yangbajing	China	32	0.127	0.415	0.204	0.214	0.067
Kizildere	Turkey	28	0.544	0.838	0.702	0.714	0.078
Ribeira Grande	Azores, Portugal	15	0.343	0.638	0.532	0.517	0.084

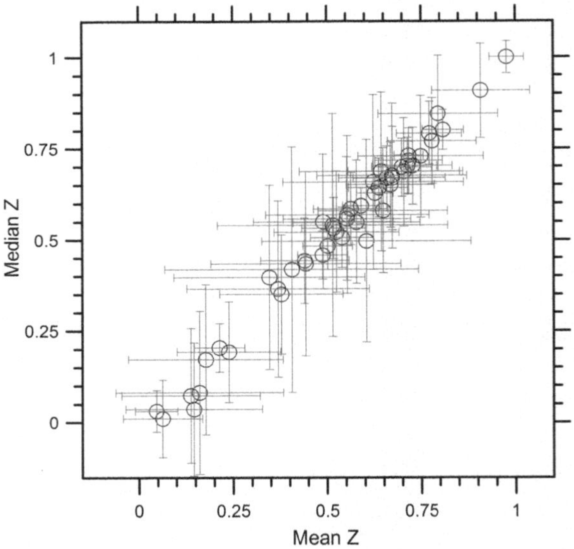

Fig. 6.7 Correlation plot between the mean and median values of the ordering parameter Z of hydrothermal adularia in hypothetical equilibrium with the considered reservoir liquids divided according to the field of provenance. The error bars correspond to ±1 standard deviation

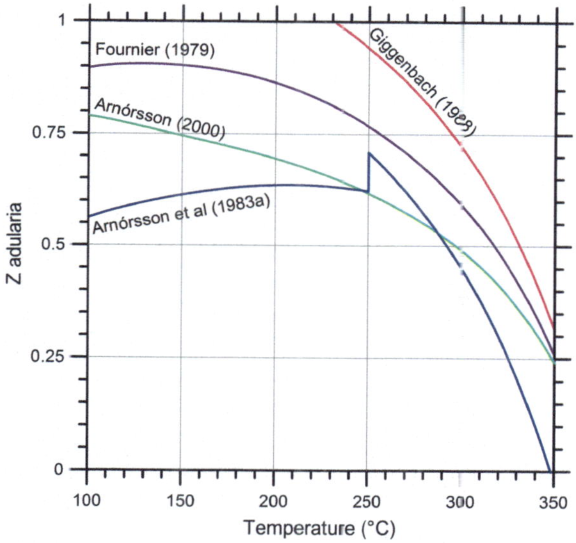

Fig. 6.8 Ordering parameter Z of hydrothermal adularia involved in the Na–K geothermometers of Fournier (1979), Arnórsson et al. (1983), Giggenbach (1988), and Arnórsson (2000), as a function of temperature

In spite of the considerable differences among these four Na–K geothermometers, they exhibit a common decrease in Z with increasing temperature, thus mimicking the general trend described by most data in the correlation diagrams of Figs. 6.4, 6.5 and 6.6.

6.4 The Theoretical Activity-Based Na–K Geothermometers Involving the Ordering Parameter of Adularia

In Figs. 6.4, 6.5 and 6.6 most considered reservoir liquids, presumably representative of mineral-solution equilibrium, are found above the line fixed by equilibrium coexistence of low-albite and high-sanidine and below the line constrained by equilibrium co-occurrence of low-albite and maximum-microcline. Consequently, the two corresponding geothermometric functions can be used to calculate the minimum and maximum Na–K equilibrium temperatures, respectively. These two theoretical activity-based Na–K geothermometers are described by the following two polynomial equations[2], in which $\beta = \log(a_{Na^+}/a_{K^+})$:

[2]In Figs. 6.4, 6.5 and 6.6, pronounced deviations from linearity can be observed for all the functions outlining the temperature dependence of the Na^+/K^+ log activity ratio, not only those fixed by low-albite/maximum-microcline and low-albite/high-sanidine equilibrium coexistence, but also those constrained by equilibrium co-occurrence of low-albite and adularia with ordering parameter

$$T_{Na-K(Z=0)}(^{\circ}C) = 300.04 \cdot \beta^4 - 1736.7 \cdot \beta^3 + 3813.9 \cdot \beta^2 - 3923.2 \cdot \beta + 1711.6 \quad (6.5)$$

$$T_{Na-K(Z=1)}(^{\circ}C) = 12.909 \cdot \beta^4 - 126.80 \cdot \beta^3 + 488.60 \cdot \beta^2 - 955.91 \cdot \beta + 875.63. \quad (6.6)$$

Equations (6.4) and (6.5) are applicable in the range 25–350 °C and the uncertainty in the computed temperature varies from 0.1 to 1 °C. Nevertheless, use of Eqs. (6.5) and (6.6) is not advisable below 100 °C, and sometimes even below 150 °C, owing to the marked decrease in the rate of the dissolution/precipitation reactions of alkali feldspars with decreasing temperature (e.g., Palandri and Kharaka 2004, Marini 2006, Bandstra et al. 2008 and references therein). Regardless of these considerations on reaction kinetics, the usefulness of Eqs. (6.4) and (6.5) is very limited because the Na–K-temperatures computed for $Z = 0$ and $Z = 1$ differ by 125 ± 3 °C on average, in the range 100–350 °C.

Therefore, it would be desirable to adopt a different approach to Na–K geothermometry, such as that consisting in the following two steps:

1. Calculation of the ordering parameter of adularia presumably in equilibrium with the aqueous solution of interest, based on either the logarithm of the total concentration of dissolved components (e.g., B, SiO_2, Na, K, Mg, Ca, SO_4, Cl, ΣCO_2, and ΣH_2S) or the logarithm of the activity of the related ionic or neutral species (e.g., $H_3BO_{3,aq}$, $SiO_{2,aq}$, Na^+, K^+, Mg^{2+}, Ca^{2+}, SO_4^{2-}, Cl^-, HCO_3^-, and HS^-), assuming that the degree of ordering of adularia is somehow related to water chemistry, as suggested by the experimental synthesis of alkali feldspars in hydroxide gels (see Sect. 4.2.3).
2. Use of the Z value obtained from water chemistry and the Na^+/K^+ activity ratio to compute the Na–K temperature of the aqueous solution of interest by means of Eq. (6.4).

Several attempts were done to find a suitable relation for predicting Z from water chemistry. Use of Factor Analysis (FA) followed by Multiple Regression Analysis (MRA) was explored to avoid multicollinearity issues, but the obtained relations resulted to be poor predictors due to the low R-squared values.

Therefore, MRA only was utilized, in spite of the high inter-associations among the independent variables. Different MRA runs were performed considering the selected reservoir liquids either altogether or in separate groups, but it turned out that Z is chiefly controlled by Na, K, and SiO_2 in all these attempts. This is a disappointing result being affected by a circular argument as explained, in more detail, referring to the following equation:

$$Z = 1.97264 - 2.24782 \cdot \log a_{K^+} + 2.19929 \cdot \log a_{Na^+} + 1.77758 \cdot \log a_{SiO_{2(aq)}}, \quad (6.7)$$

Z varying from 0 to 1 at steps of 0.1 units. Consequently, the use of polynomial equations of temperature, instead of linear functions of the absolute temperature inverse, is advisable (if not mandatory) for obtaining the desired Na–K activity-based theoretical geothermometers.

which was obtained considering the 1013 samples altogether in MRA. Equation (6.7) has standard errors of 0.057203 on the intercept, 0.033916 on the coefficient of $\log a_{K^+}$, 0.036593 on the coefficient of $\log a_{Na^+}$, and 0.038055 on the coefficient of $\log a_{SiO_{2(aq)}}$. Equation (6.7) has multiple R of 0.9138, multiple R-squared of 0.8351, and adjusted multiple R-squared of 0.8346. The standard error on the predicted Z values is 0.114, which is a relatively high value. For instance, for $\log(a_{Na^+}/a_{K^+}) = 1$, the corresponding error on the computed Na–K temperature is of 11–18 °C.

However, it must be noted that the coefficients of $\log a_{Na^+}$ and $\log a_{K^+}$ are positive and negative, respectively, but have almost the same absolute value, whereas $\log a_{SiO_{2(aq)}}$ is a proxy of temperature. Thus, Eq. (6.7) expresses the dependence of Z on $\log(a_{Na^+}/a_{K^+})$ and temperature and, therefore, is affected by a circular argument because Z was previously computed by inserting $\log(a_{Na^+}/a_{K^+})$ and temperature into Eq. (6.3). In short words, Eq. (6.7) and similar relations are meaningless.

6.5 Final Considerations on the Use of the Na/K-Activity Ratio

To conclude this chapter, it is worth repeating that a single universally-working Na–K geothermometer does not exist, regardless of whether it is calibrated empirically or it is based on the solid pillars of thermodynamics. This is due to the different degree of Al–Si order-disorder on the tetrahedral sites of alkali feldspars, as already pointed out by Bird and Norton (1981).

For this reason, Na–K geothermometry must necessarily imply the ordering parameter of adularia, Z. The two limiting Na–K geothermometers, involving low-albite and either high-sanidine or maximum-microcline, as proxies of completely disordered adularia (with Z = 0) and totally ordered adularia (with Z = 1), respectively, give equilibrium temperatures differing by 125 ± 3 °C on average, in the range 100–350 °C, and have little usefulness. The attempts that we carried out to predict Z on the basis of water chemistry were unsuccessful, but it is possible that the available data are inadequate and that this matter might be clarified by further researches.

All in all, for the moment, it is probably advisable to use the Na/K log activity ratio to estimate the ordering parameter Z of the hydrothermal adularia in hypothetical equilibrium with the aqueous solution of interest and to use this information in other geothermometric functions that are presented and discussed in the following chapters.

References

Arnórsson S (2000) The quartz and Na/K geothermometers. I. New thermodynamic calibration. In: Proceedings of the world geothermal congress 2000. Kyushu-Tohoku, Japan, pp 929–934

Arnórsson S, Stefánsson A (1999) Assessment of feldspar solubility constants in water in the range of 0–350 °C at vapor saturation pressures. Am J Sci 299:173–209

Arnórsson S, Gunnlaugsson E, Svavarsson H (1983) The chemistry of geothermal waters in Iceland. III. Chemical geothermometry in geothermal investigations. Geochim Cosmochim Ac 47:567–577

Bandstra JZ, Buss HL, Campen RK, Liermann LJ, Moore J, Hausrath EM, Navarre-Sitchler A, Jang J-H, Brantley SL (2008) Appendix: Compilation of mineral dissolution rates. In: Brantley SL, Kubicki JD, White AF (eds) Kinetics of water-rock interaction, Springer, New York, pp 737–823

Bird DK, Norton DL (1981) Theoretical prediction of phase relations among aqueous solutions and minerals: Salton Sea geothermal system. Geochim Cosmochim Ac 45:1479–1494

Ellis AJ (1970) Quantitative interpretation of chemical characteristics of hydrothermal systems. Geothermics 2:516–528

Fournier RO (1979) A revised equation for the Na/K geothermometer. Geotherm Res Counc Trans 5:1–16

Giggenbach WF (1988) Geothermal solute equilibria. Derivation of Na–K–Mg–Ca geoindicators. Geochim Cosmochim Ac 52:2749–2765

Helgeson HC, Delany JM, Nesbitt HW, Bird DK (1978) Summary and critique of the thermodynamic properties of rock-forming minerals. Am J Sci 278A:229

Holland TJB, Powell R (1998) An internally consistent thermodynamic data set for phases of petrological interest. J Metamorph Geol 16:309–343

Marini L (2006) Geological sequestration of carbon dioxide: Thermodynamics, kinetics, and reaction path modeling. Developments in geochemistry, 11. Elsevier, Amsterdam, p 453

Michard G (1990) Behaviour of major elements and some trace elements (Li, Rb, Cs, Sr, Fe, Mn, W, F) in deep hot waters from granitic areas. Chem Geol 89:117–134

Palandri JL, Kharaka YK (2004) A compilation of rate parameters of water-mineral interaction kinetics for application to geochemical modeling. US Geological Survey, Open File Report 2004–1068

White DE (1965) Saline waters of sedimentary rocks. In Fluids in Subsurface Environments. Symp Amer Assoc Petroleum Geologists 342–366

Chapter 7
The Activity-Based Theoretical K–Mg and Na–Mg Geoindicators

Abstract Assuming that geothermal reservoir liquids are in equilibrium with clinochlore, muscovite, adularia, albite, and a silica mineral (Giggenbach, 1988), their K^2/Mg- and Na^2/Mg-activity ratios are controlled by the aquifer temperature, T, the order-disorder degree of hydrothermal adularia, Z, and the mineral activity product, MAP, involving the activity of clinochlore in hydrothermal chlorite and the activity of muscovite in hydrothermal illite. Knowing Z and T, it is possible to obtain MAP. For the reservoir liquids with negative MAP values, the K^2/Mg- and Na^2/Mg-activity ratios can be used to calculate the minimum activities of clinochlore in chlorites and of muscovite in illites, which have limited importance for practical purposes. For the reservoir liquids with positive MAP values, the K^2/Mg- and Na^2/Mg-activity ratios can be utilized to compute the minimum concentrations of excess (non-equilibrium) Mg and, consequently, to identify and quantify the processes controlling excess Mg, such as cooling and mixing, which may lead to Mg-silicate (e.g., smectite) scaling. This is probably the most valuable information provided by the activity-based theoretical K-Mg and Na-Mg geoindicators. The K-Mg and Na-Mg exchange reactions involving other Mg-minerals are not explored to develop other K-Mg and Na-Mg geoindicators due to the restricted temperature intervals in which they are expected to be effective and the limitations caused by the varying characteristics of the relevant Mg-bearing minerals.

As already discussed in Sect. 5.6, Giggenbach (1988) assumed that (i) the K–Mg geothermometer is controlled by coexistence, at equilibrium, of clinochlore (Clc), muscovite (Ms), chalcedony (Chc), and adularia (Adl), as expressed by reaction (5.141), and (ii) the Na–Mg geothermometer is governed by co-occurrence, at equilibrium, of the same hydrothermal minerals plus albite (Ab), as indicated by reaction (5.142). Although the K–Mg and Na–Mg geothermometers might be controlled by exchange reactions involving Mg-bearing solid phases other than chlorites, as discussed in Sect. 5.6.3, let us consider reactions (5.141) and (5.142), which are rewritten here reversing their direction for convenience:

R. Cioni and L. Marini, *A Thermodynamic Approach to Water Geothermometry*, Springer Geochemistry, https://doi.org/10.1007/978-3-030-54318-1_7

$$2.8KAlSi_3O_8 + Mg^{2+} + 1.6H_2O = \\ 0.2Mg_5Al_2Si_3O_{10}(OH)_8 + 0.8KAl_3Si_3O_{10}(OH)_2 + 5.4SiO_2 + 2K^+, \tag{7.1}$$

$$0.8KAlSi_3O_8 + 2NaAlSi_3O_8 + Mg^{2+} + 1.6H_2O = \\ 0.2Mg_5Al_2Si_3O_{10}(OH)_8 + 0.8KAl_3Si_3O_{10}(OH)_2 + 5.4SiO_2 + 2Na^+. \tag{7.2}$$

7.1 The Log K of the K–Mg and Na–Mg Exchange Reactions

The thermodynamic equilibrium constants of reactions (7.1) and (7.2) are the inverse of Eqs. (5.143) and (5.144), that is:

$$K_{K-Mg} = \frac{a_{Clc}^{0.2} \cdot a_{Ms}^{0.8} \cdot a_{Chc}^{5.4} \cdot a_{K^+}^{2}}{a_{Adl}^{2.8} \cdot a_{Mg^{2+}} \cdot a_{H_2O}^{1.6}} \tag{7.3}$$

$$K_{Na-Mg} = \frac{a_{Clc}^{0.2} \cdot a_{Ms}^{0.8} \cdot a_{Chc}^{5.4} \cdot a_{Na^+}^{2}}{a_{Adl}^{0.8} \cdot a_{Ab}^{2} \cdot a_{Mg^{2+}} \cdot a_{H_2O}^{1.6}} \tag{7.4}$$

Since adularia, albite and chalcedony are usually pure or relatively pure minerals, their activities can be assumed to be equal to 1. Also the activity of water does not deviates significantly from unity, if the ionic strength of the aqueous solution is lower than ~1 mol/kg. Therefore, Eqs. (7.3) and (7.4) can be generally simplified and rearranged as follows:

$$\frac{a_{K^+}^2}{a_{Mg^{2+}}} \cong \frac{K_{K-Mg}}{a_{Clc}^{0.2} \cdot a_{Ms}^{0.8}}. \tag{7.5}$$

$$\frac{a_{Na^+}^2}{a_{Mg^{2+}}} \cong \frac{K_{Na-Mg}}{a_{Clc}^{0.2} \cdot a_{Ms}^{0.8}}. \tag{7.6}$$

Equations (7.5) and (7.6) indicate that the K^2/Mg- and Na^2/Mg-activity ratios of geothermal liquids are expected to depend not only on the equilibrium temperature but also on the mineral activity product $a_{Clc}^{0.2} \cdot a_{Ms}^{0.8}$, which is indicated by the acronym MAP in the subsequent discussion. In turn, the activities of clinochlore and muscovite depend on the crystallo-chemical characteristics of hydrothermal chlorites and illites, respectively. Therefore, it is worth to recall that:

1. The 181 hydrothermal chlorites considered in Sect. 4.4.3 have activity of the clinochlore endmember varying between 4.39×10^{-7} and 0.250, with an average of 0.0553, a median of 0.0451 and a standard deviation of 0.0484.
2. The 75 hydrothermal illites taken into account in Sect. 4.3.2 have activity of the muscovite endmember ranging between 0.007 and 0.875, with an average of 0.654, a median of 0.688 and a standard deviation of 0.156.
3. Based on the average a_{Ms} and a_{Clc} values, log MAP has a mean of −0.399 and is expected to vary from −0.674 to −0.270 considering a range of $\pm 1\sigma$ around the mean a_{Ms} and a_{Clc} values.

Further complications are posed by Al–Si order-disorder on the tetrahedral sites of adularia as discussed in Sects. 4.2.3 and 5.3.3. Therefore, an adularia with ordering parameter Z varying from 0 (i.e., high-sanidine) to 1 (i.e., maximum-microcline) at steps of 0.1 units was alternatively assumed to be involved in reactions (7.1) and (7.2).

Another point to be considered is the basal spacing of clinochlore. Based on the pertinent discussion in Sect. 4.4.4, 7Å-clinochlore was considered to take part to reactions (7.1) and (7.2)

Either the quartz/chalcedony mechanical mixture, above 175 °C, or chalcedony, below 175 °C, were assumed to participate to reactions (7.1) and (7.2), whose thermodynamic properties were calculated by means of SUPCRT92, considering the thermodynamic data of solid phases of Helgeson et al. (1978), for temperatures varying from 0 to 350 °C, at steps of 25 °C, and for values of the ordering parameter Z of adularia varying from 0 to 1, at steps of 0.1 units.

The computed log $K_{K\text{–}Mg}$ values (Table 7.1) are a function of the absolute temperature inverse and the ordering parameter Z of adularia as defined by the following two relations:

$$\log K_{K-Mg} = \left(\frac{-1621.2}{T} + 2.0798\right) \cdot Z - \frac{105,276}{T^2} - \frac{2019.8}{T} + 5.6865 \quad (7.7)$$

$$\log K_{K-Mg} = \left(\frac{-1621.2}{T} + 2.0798\right) \cdot Z + \frac{450,637}{T^2} - \frac{4280.5}{T} + 8.4530 \quad (7.8)$$

Equation (7.7) applies from 0 to 175 °C, whereas Eq. (7.8) is valid from 175 to 325 °C.

The calculated log $K_{Na\text{–}Mg}$ values (Table 7.2) depend on the absolute temperature inverse and the ordering parameter Z of adularia as described by the following two equations:

$$\log K_{Na-Mg} = \left(\frac{-463.264}{T} + 0.5944\right) \cdot Z + \frac{104,027}{T^2} - \frac{1646.0}{T} + 5.7524 \quad (7.9)$$

$$\log K_{Na-Mg} = \left(\frac{-462.976}{T} + 0.5938\right) \cdot Z + \frac{752,484}{T^2} - \frac{4407.2}{T} + 9.1807 \quad (7.10)$$

Table 7.1 Logarithm of the thermodynamic equilibrium constant of reaction (7.1) involving pure clinochlore, muscovite, adularia, and chalcedony (for T < 175 °C) or quartz/chalcedony (for T > 175 °C), log $K_{K\text{–}Mg}$, as a function of temperature and the ordering parameter Z of adularia

T(°C)	Z = 1	Z = 0.9	Z = 0.8	Z = 0.7	Z = 0.6	Z = 0.5	Z = 0.4	Z = 0.3	Z = 0.2	Z = 0.1	Z = 0
0.01	−6.9744	−6.5888	−6.2033	−5.8178	−5.4322	−5.0467	−4.6612	−4.2757	−3.8901	−3.5046	−3.1191
25	−5.6303	−5.2946	−4.9588	−4.6230	−4.2872	−3.9514	−3.6157	−3.2799	−2.9441	−2.6083	−2.2726
50	−4.5075	−4.2138	−3.9200	−3.6263	−3.3326	−3.0389	−2.7452	−2.4515	−2.1578	−1.8641	−1.5704
75	−3.5590	−3.3013	−3.0437	−2.7860	−2.5283	−2.2707	−2.0130	−1.7553	−1.4976	−1.2400	−0.9823
100	−2.7477	−2.5213	−2.2948	−2.0683	−1.8419	−1.6154	−1.3889	−1.1624	−0.9360	−0.7095	−0.4830
125	−2.0447	−1.8455	−1.6463	−1.4472	−1.2480	−1.0488	−0.8496	−0.6504	−0.4512	−0.2520	−0.0528
150	−1.4278	−1.2526	−1.0775	−0.9024	−0.7272	−0.5521	−0.3770	−0.2019	−0.0267	0.1484	0.3235
175	−0.8796	−0.7258	−0.5721	−0.4183	−0.2646	−0.1108	0.0429	0.1967	0.3505	0.5042	0.6580
175	−0.3924	−0.2386	−0.0849	0.0689	0.2226	0.3764	0.5301	0.6839	0.8376	0.9914	1.1451
200	0.0746	0.2093	0.3439	0.4786	0.6132	0.7478	0.8825	1.0171	1.1518	1.2864	1.4210
225	0.5004	0.6179	0.7353	0.8528	0.9702	1.0877	1.2051	1.3226	1.4400	1.5575	1.6749
250	0.8951	0.9971	1.0990	1.2009	1.3028	1.4047	1.5066	1.6085	1.7104	1.8123	1.9142
275	1.2663	1.3542	1.4420	1.5298	1.6175	1.7053	1.7931	1.8808	1.9686	2.0564	2.1441
300	1.6139	1.6890	1.7639	1.8388	1.9137	1.9885	2.0634	2.1383	2.2131	2.2880	2.3629
325	1.9215	1.9849	2.0480	2.1110	2.1741	2.2371	2.3002	2.3633	2.4263	2.4894	2.5524
350	2.1348	2.1875	2.2397	2.2919	2.3441	2.3963	2.4485	2.5007	2.5529	2.6051	2.6573

P = 1 bar for T < 100 °C; P = P_{sat} for T ≥ 100 °C. Calculations performed using SUPCRT92 considering the thermodynamic data of minerals of Helgeson et al. (1978)

Table 7.2 Logarithm of the thermodynamic equilibrium constant of reaction (7.2), for pure clinochlore, muscovite, adularia, low-albite, and chalcedony (for T < 175 °C) or quartz/chalcedony (for T > 175 °C), log K_{Na-Mg}, as a function of the temperature and the ordering parameter Z of adularia

T(°C)	Z = 1	Z = 0.9	Z = 0.8	Z = 0.7	Z = 0.6	Z = 0.5	Z = 0.4	Z = 0.3	Z = 0.2	Z = 0.1	Z = 0
0.01	0.0139	0.1241	0.2342	0.3444	0.4545	0.5647	0.6748	0.7850	0.8951	1.0053	1.1154
25	0.4492	0.5452	0.6411	0.7370	0.8330	0.9289	1.0248	1.1208	1.2167	1.3127	1.4086
50	0.8219	0.9058	0.9897	1.0736	1.1575	1.2414	1.3254	1.4093	1.4932	1.5771	1.6610
75	1.1469	1.2205	1.2942	1.3678	1.4414	1.5150	1.5886	1.6623	1.7359	1.8095	1.8831
100	1.4360	1.5007	1.5654	1.6301	1.6948	1.7595	1.8242	1.8889	1.9536	2.0183	2.0830
125	1.6981	1.7550	1.8119	1.8689	1.9258	1.9827	2.0396	2.0965	2.1534	2.2103	2.2672
150	1.9400	1.9901	2.0401	2.0901	2.1402	2.1902	2.2403	2.2903	2.3403	2.3904	2.4404
175	2.1669	2.2108	2.2547	2.2987	2.3426	2.3865	2.4304	2.4744	2.5183	2.5622	2.6062
175	2.6540	2.6980	2.7419	2.7858	2.8298	2.8737	2.9176	2.9615	3.0055	3.0494	3.0933
200	2.8438	2.8823	2.9208	2.9593	2.9977	3.0362	3.0747	3.1131	3.1516	3.1901	3.2285
225	3.0286	3.0622	3.0957	3.1293	3.1628	3.1964	3.2300	3.2635	3.2971	3.3306	3.3642
250	3.2117	3.2408	3.2699	3.2990	3.3282	3.3573	3.3864	3.4155	3.4446	3.4737	3.5028
275	3.3950	3.4201	3.4452	3.4702	3.4953	3.5204	3.5455	3.5705	3.5956	3.6207	3.6458
300	3.5743	3.5957	3.6171	3.6385	3.6599	3.6813	3.7027	3.7241	3.7455	3.7669	3.7883
325	3.7313	3.7495	3.7675	3.7855	3.8035	3.8215	3.8395	3.8576	3.8756	3.8936	3.9116
350	3.8123	3.8274	3.8423	3.8572	3.8721	3.8871	3.9020	3.9169	3.9318	3.9467	3.9616

P = 1 bar for T < 100 °C; P = P_{sat} for T ≥ 100 °C. Calculations performed using SUPCRT92 considering the thermodynamic data of minerals of Helgeson et al. (1978)

Equation (7.9) is valid from 0 to 175 °C, whereas Eq. (7.10) applies from 175 to 325 °C.

7.2 The $(K^+)^2/Mg^{2+}$ and $(Na^+)^2/Mg^{2+}$ Log Activity Ratios of the Selected Reservoir Liquids

The K^2/Mg log-activity ratios given by Eqs. (7.7) and (7.8) are shown in the diagrams of Figs. 7.1, 7.2, and 7.3, whereas the Na^2/Mg log-activity ratios given by Eqs. (7.9) and (7.10) are displayed in the graphs of Figs. 7.4, 7.5, and 7.6. In all these plots, the K^2/Mg and Na^2/Mg log-activity ratios are represented as a function of the absolute temperature inverse, from 100 to 350 °C, for the average log MAP value of −0.399 and for different values of the ordering parameter of adularia, namely Z = 1 (i.e., maximum-microcline), Z = 0 (i.e., high-sanidine), as well as Z = 0.559, Z = 0.820, and Z = 0.298, corresponding to the average value, the average value $+1\sigma$, and the average value -1σ, respectively, of the hydrothermal adularia in hypothetical equilibrium with the considered geothermal liquids (see Sect. 6.3). In all these diagrams are also displayed the K^2/Mg and Na^2/Mg log-activity ratios, as a function of 1/T(K), for both log MAP of −0.674, Z = 0 and log MAP of −0.270, Z = 1, to bracket most compositional variations of hydrothermal chlorites and illites. These theoretical K^2/Mg and Na^2/Mg log-activity ratios suggest that:

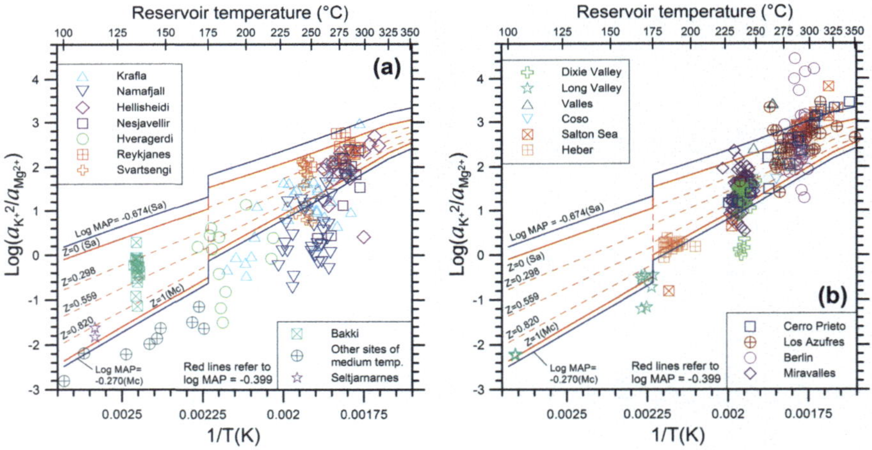

Fig. 7.1 K^2/Mg-log activity ratios of the selected reservoir liquids from the geothermal systems of **a** Iceland and **b** North-Central America as a function of aquifer temperature. Also shown are the K^2/Mg-log activity ratios fixed by equilibrium coexistence of clinochlore, muscovite, either chalcedony (below 175 °C) or quartz/chalcedony (above 175 °C), and adularia with ordering parameter Z = 1 and 0 (red solid lines), Z = 0.820, 0.559, and 0.298 (red dashed lines), all for log MAP = −0.399, as well as for log MAP = −0.674 and Z = 0 and for log MAP = −0.270 and Z = 1 (blue solid lines)

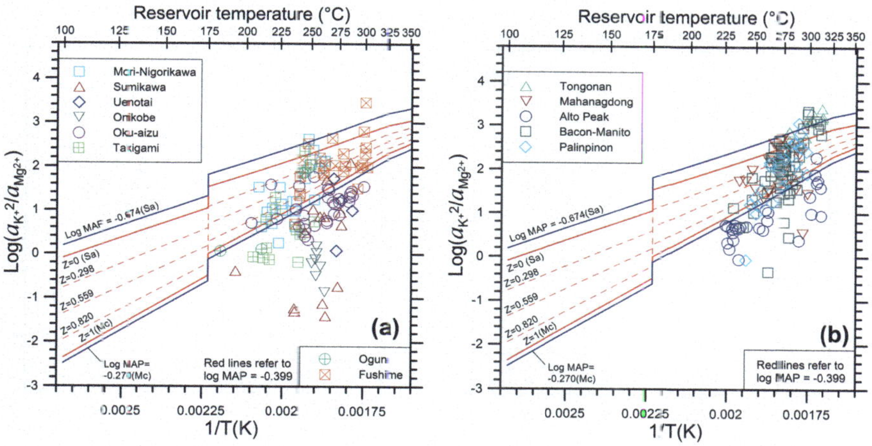

Fig. 7.2 K^2/Mg-log activity ratios of the selected reservoir liquids from the geothermal systems of **a** Japan and **b** The Philippines as a function of aquifer temperature. Also shown are the K^2/Mg-log activity ratios fixed by equilibrium coexistence of clinochlore, muscovite, either chalcedony (below 175 °C) or quartz/chalcedony (above 175 °C), and adularia with ordering parameter Z = 1 and 0 (red solid lines), Z = 0.820, 0.559, and 0.298 (red dashed lines), all for log MAP = −0.399, as well as for log MAP = −0.674 and Z = 0 and for log MAP = −0.270 and Z = 1 (blue solid lines)

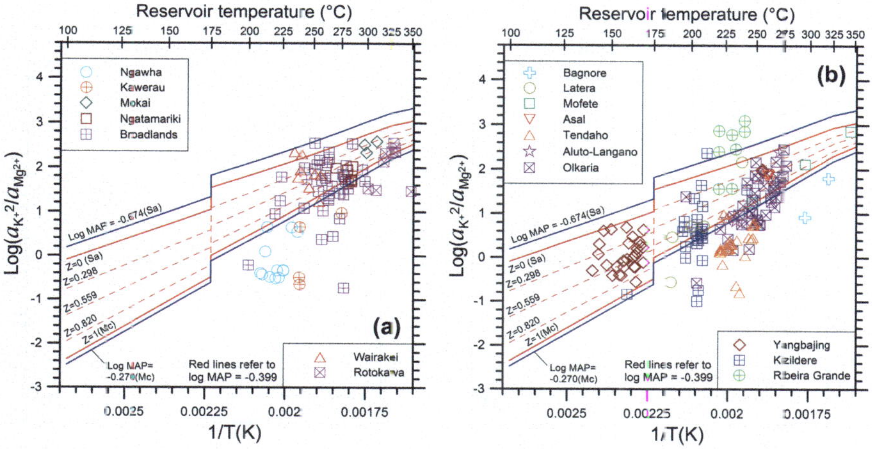

Fig. 7.3 K^2/Mg-log activity ratios of the selected reservoir liquids from the geothermal systems of **a** New Zealand and **b** miscellaneous sites as a function of aquifer temperature. Also shown are the K^2/Mg-log activity ratios fixed by equilibrium coexistence of clinochlore, muscovite, either chalcedony (below 175 °C) or quartz/chalcedony (above 175 °C), and adularia with ordering parameter Z = 1 and 0 (red solid lines), Z = 0.820, 0.559, and 0.298 (red dashed lines), all for log MAP = −0.399, as well as for log MAP = −0.674 and Z = 0 and for log MAP = −0.270 and Z = 1 (blue solid lines)

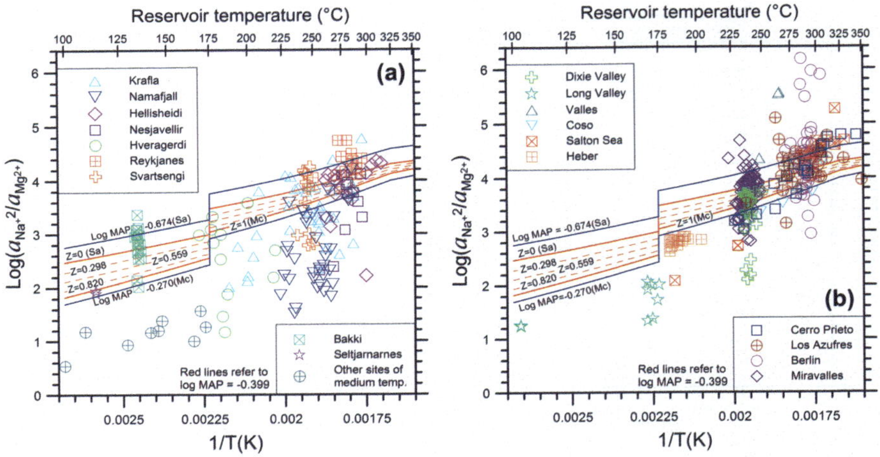

Fig. 7.4 Na^2/Mg-log activity ratios of the selected reservoir liquids from the geothermal systems of **a** Iceland and **b** North-Central America as a function of aquifer temperature. Also shown are the Na^2/Mg-log activity ratios fixed by equilibrium coexistence of clinochlore, muscovite, either chalcedony (below 175 °C) or quartz/chalcedony (above 175 °C), albite, and adularia with ordering parameter Z = 1 and 0 (red solid lines), Z = 0.820, 0.559, and 0.298 (red dashed lines), all for log MAP = −0.399, as well as for log MAP = −0.674 and Z = 0 and for log MAP = −0.270 and Z = 1 (blue solid lines)

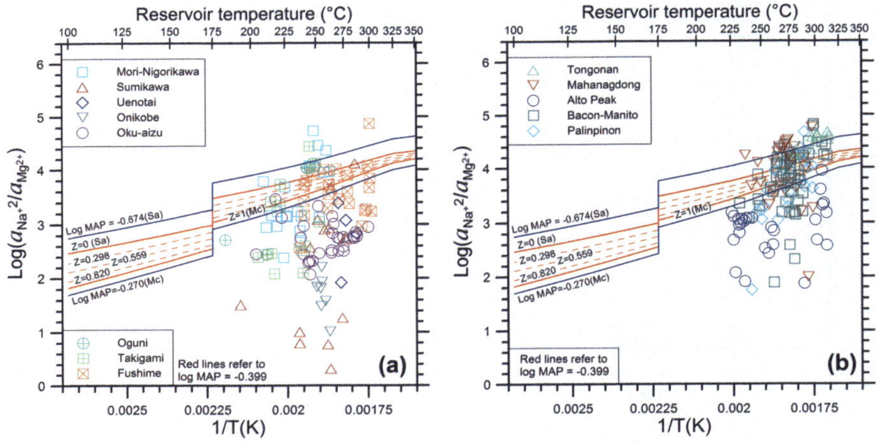

Fig. 7.5 Na^2/Mg-log activity ratios of the selected reservoir liquids from the geothermal systems of **a** Japan and **b** The Philippines as a function of aquifer temperature. Also shown are the Na^2/Mg-log activity ratios fixed by equilibrium coexistence of clinochlore, muscovite, either chalcedony (below 175 °C) or quartz/chalcedony (above 175 °C), albite, and adularia with ordering parameter Z = 1 and 0 (red solid lines), Z = 0.820, 0.559, and 0.298 (red dashed lines), all for log MAP = −0.399, as well as for log MAP = −0.674 and Z = 0 and for log MAP = −0.270 and Z = 1 (blue solid lines)

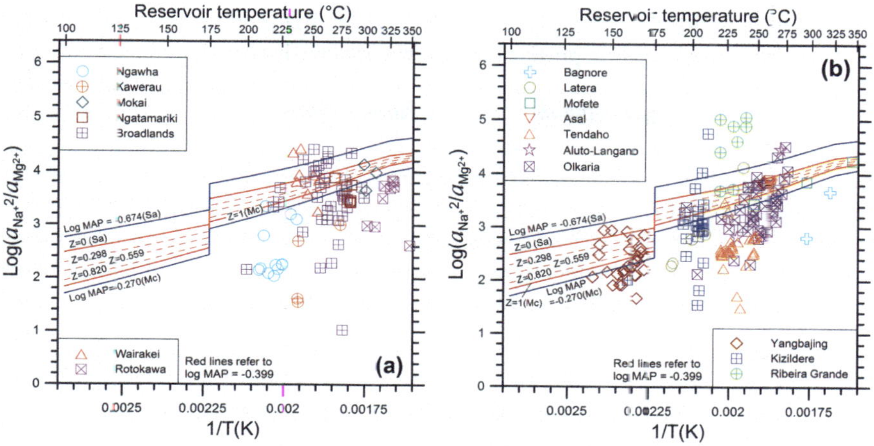

Fig. 7.6 Na^2/Mg-log activity ratios of the selected reservoir liquids from the geothermal systems of **a** New Zealand and **b** miscellaneous sites as a function of aquifer temperature. Also shown are the Na^2/Mg-log activity ratios fixed by equilibrium coexistence of clinochlore, muscovite, either chalcedony (below 175 °C) or quartz/chalcedony (above 175 °C), albite, and adularia with ordering parameter Z = 1 and 0 (red solid lines), Z = 0.820, 0.559, and 0.298 (red dashed lines), all for log MAP = −0.399, as well as for log MAP = −0.674 and Z = 0 and for log MAP = −0.270 and Z = 1 (blue solid lines)

1. The impact of the adularia order-disorder is more important on the K^2/Mg log-activity ratio than on the Na^2/Mg log-activity ratio. This fact is not surprising due to the distinct stoichiometric coefficients of adularia, 2.8 and 0.8, in reactions (7.1) and (7.2) controlling the K^2/Mg and Na^2/Mg log-activity ratios, respectively.
2. The effect of adularia order-disorder on the K^2/Mg log-activity ratio is slightly lower than that caused by the compositional changes of hydrothermal chlorites and illites, considering a range of $\pm 1\sigma$ around the mean a_{Clc} and a_{Ms} values. In contrast, the influence of adularia order-disorder on the Na^2/Mg log-activity ratio is appreciably lower than that brought about by the chemical variations of hydrothermal illites and chlorites, referring again to an interval of $\pm 1\sigma$ around the average a_{Clc} and a_{Ms} values. Also this fact is not surprising because adularia has different stoichiometric coefficients in reactions (7.1) and (7.2) whereas the stoichiometric coefficient of clinochlore is the same in both reactions and the same applies to muscovite.

Also shown in Figs. 7.1, 7.2, and 7.3 and in Figs. 7.4, 7.5, and 7.6 are the K^2/Mg- and Na^2/Mg-log activity ratios, respectively, of the selected geothermal liquids against the reciprocal of their aquifer temperatures (in K). Several reservoir liquids have K^2/Mg- and Na^2/Mg-log activity ratios within the range expected based on the different chemistry of hydrothermal chlorites and illites and the varying ordering parameter of adularia. Nevertheless, many reservoir liquids have K^2/Mg- and especially Na^2/Mg-log activity ratios lower than theoretical values. The Mg concentration of these geothermal liquids is probably not representative of mineral-solution equilibrium due to presence of excess magnesium which is ascribable to different causes, as already pointed out in Sect. 5.6.2, including: (i) addition of Mg-rich fluids either external to the geothermal system or proceeding from shallower and colder parts of the geothermal system itself and (ii) cooling of the reservoir liquids and Mg acquisition, for instance during their travel from the upflow zone towards the outflow zone or due to breakthrough of re-injected brines (e.g., Fournier 1991). A few reservoir liquids have K^2/Mg- and especially Na^2/Mg-log activity ratios higher than theoretical values, possibly due to incorporation of Mg in the lattice of precipitating calcite or other solid phases, as already underscored in Sect. 5.6.2.

7.3 The Log MAP Values of the Selected Reservoir Liquids and Related Implications

The binary diagrams of Figs. 7.1, 7.2, 7.3, 7.4, 7.5, and 7.6 allows a comparison between the theoretical K^2/Mg- and Na^2/Mg-log activity ratios controlled by reactions (7.1) and (7.2) and the corresponding ratios of all the selected geothermal liquids in a single view. However, these binary diagrams do not allow a precise assessment of the consistency or inconsistency between theoretical and observed values for each aqueous solution. To gain more information on this subject, the log MAP values of the 973 geothermal liquids with Mg concentration higher than detection limit were computed and processed as described below.

First, the log $K_{K\text{–}Mg}$ and log $K_{Na\text{–}Mg}$ values were obtained for each geothermal liquid by inserting the reservoir temperature and the ordering parameter Z of adularia apparently in equilibrium with it into Eqs. (7.7) or (7.8) and Eqs. (7.9) or (7.10), respectively. The Z values >1 were assumed equal to one, whereas the Z values <0 were set to zero. Second, knowing the log $K_{K\text{–}Mg}$ and log $K_{Na\text{–}Mg}$ values as well as the K^2/Mg- and Na^2/Mg-log activity ratios of each aqueous solution, their log MAP values were computed by means of Eqs. (7.5) and (7.6), respectively. For each geothermal liquid, the log MAP values given by Eqs. (7.5) and (7.6) agree within less than 0.01 log-units[1]. Among the 973 reservoir liquids with Mg concentration higher than detection limit, 587 (60% of the cases) have negative log MAP values, ranging from −0.000639 to −2.59, while 386 (40% of the entries) have positive MAP values, varying from +0.00233 to +3.18.

The log MAP values computed from the K^2/Mg-log activity ratios are contrasted with these activity ratios in the binary diagrams of Figs. 7.7, 7.8, and 7.9, whereas the log MAP values calculated on the basis of the Na^2/Mg-log activity ratios are not considered to avoid unnecessary repetitions. In the diagrams of Figs. 7.7, 7.8, and 7.9, the reservoir liquids of a given geothermal system are distributed close or relatively

[1]These negligible differences are due to the approximations in the calculations of the ordering parameter Z of adularia by means of Eq. (6.2), log $K_{K\text{–}Mg}$ by using Eqs. (7.7) or (7.8), and log $K_{Na\text{–}Mg}$ by utilizing Eqs. (7.9) or (7.10).

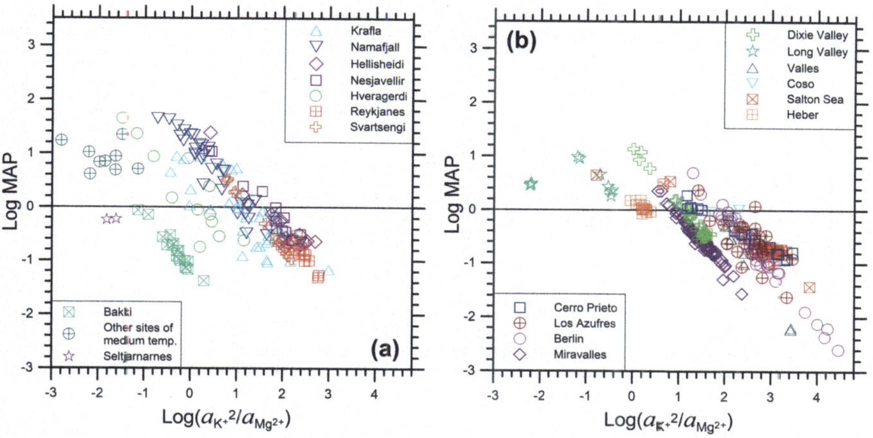

Fig. 7.7 Correlation diagrams between the log MAP values and K^2/Mg-log activity ratio for the selected reservoir liquids from the geothermal systems of **a** Iceland and **b** North-Central America

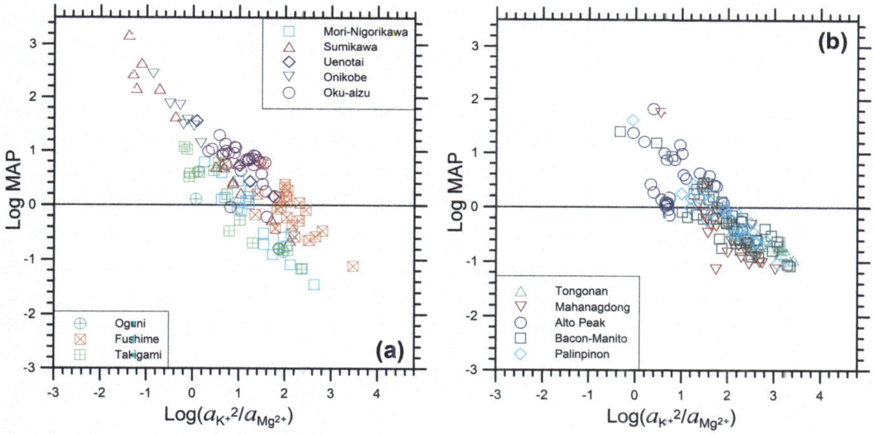

Fig. 7.8 Correlation diagrams between the log MAP values and K^2/Mg-log activity ratio for the selected reservoir liquids from the geothermal systems of **a** Japan and **b** The Philippines

close to a straight line of slope −1 and intercept depending on the aquifer temperature and the order/disorder degree of adularia. Tight trends are observed for the reservoir liquids of geothermal systems with little variations in both aquifer temperature and ordering parameter of adularia, as those coming from Bakki (Fig. 7.7a), Miravalles and Dixie Valley (Fig. 7.7b), and the shallow reservoir of Yangbajing (Fig. 7.9b). In contrast, the reservoir liquids of geothermal systems with significant changes in both aquifer temperature and ordering parameter of adularia, as is the case of Los Azufres (Fig. 7.7b), Fushime and Oku-aizu (Fig. 7.8a), Alto Peak (Fig. 7.8b), Broadlands (Fig. 7.9a) exhibit a considerable spread of data points. It should not be forgotten,

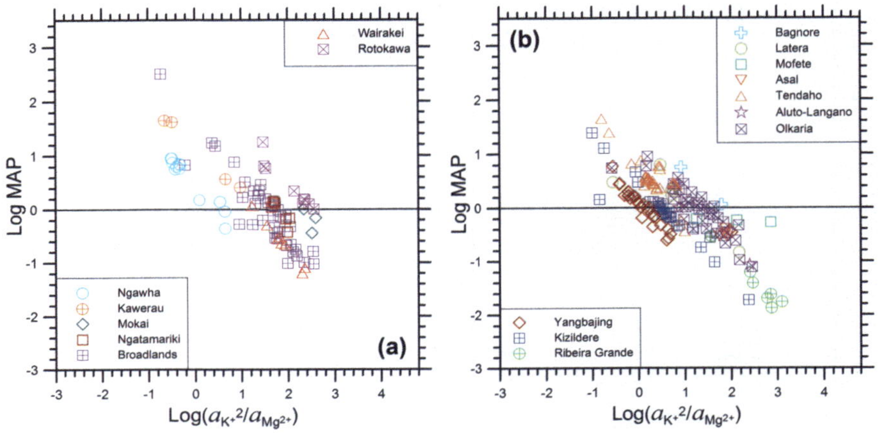

Fig. 7.9 Correlation diagrams between the log MAP values and K^2/Mg-log activity ratio for the selected reservoir liquids from the geothermal systems of **a** New Zealand and **b** miscellaneous sites

however, that the relation between log MAP values and K^2/Mg-log activity ratios may be influenced also by the number of samples available for each geothermal system.

Recalling that MAP is equal to $a_{\mathrm{Clc}}^{0.2} \cdot a_{\mathrm{Ms}}^{0.8}$ and that both a_{Clc} and a_{Ms} cannot be higher than one, it follows that the maximum permissible value of log MAP is zero for an aqueous solution in equilibrium with the hydrothermal minerals involved in reactions (7.1) and (7.2). Consequently, only the geothermal liquids with negative log MAP values may be representative of reservoir conditions, although possible effects of Mg loss cannot be excluded as already underscored above and in Sect. 5.6.2. In contrast, the geothermal liquids with positive log MAP values are certainly affected by presence of excess (non-equilibrium) Mg, for the reasons given above and in Sect. 5.6.2.

Minimum muscovite activities of the illites in apparent equilibrium with the 587 reservoir liquids with negative log MAP values were computed for $a_{\mathrm{Clc}} = 1$ and minimum clinochlore activities of the chlorites in apparent equilibrium with these aqueous solutions were calculated for $a_{\mathrm{Ms}} = 1$. Computed minimum values of muscovite activity range between 5.79×10^{-4} and 0.998, with a mean of 0.292, a median of 0.209 and a standard deviation of 0.248. Calculated minimum values of clinochlore activity distribute in a much wider interval, from 1.13×10^{-13} to 0.991, with an average of 0.0720, a median of 0.00191 and a standard deviation of 0.0720. These computed minimum values of a_{Ms} and a_{Clc} partly overlap the ranges of muscovite and clinochlore activities of the hydrothermal illites and chlorites (see Sect. 7.1).

Furthermore, computed minimum values of a_{Ms} and a_{Clc} could be used to speculate, to some extent, on both the lithology of reservoir rocks and the possible seawater recharge, as already recalled in Sect. 4.4.2. In fact, on the one hand, chlorite chemistry depends upon the composition of both reservoir rock and other hydrothermal minerals (e.g., Albee 1962; De Caritat et al. 1993). On the other hand, the composition of chlorites can be controlled by fluid chemistry, at least in part. For instance, the Mg-rich chlorites of Reykjanes and Bouillante might be due to the inflow of Mg-rich seawater into these geothermal systems, as recognized by Arnórsson (1978a) for Reykjanes and Sanjuan et al. (2001) for Bouillante. Nevertheless, the computed minimum values of a_{Ms} and a_{Clc} have limited importance for practical purposes.

Let us now consider the 386 geothermal liquids with positive MAP values. For these geothermal liquids, it is possible to compute the maximum molality of free Mg^{2+} ion fixed by equilibrium with the hydrothermal minerals participating to reactions (7.1) and (7.2), through the following two relations:

$$\log m_{Mg^{2+},max.eq} = 2 \cdot \log a_{K^+} - \log \gamma_{Mg^{2+}} - \log K_{K-Mg} \tag{7.11}$$

$$\log m_{Mg^{2+},max.eq} = 2 \cdot \log a_{Na^+} - \log \gamma_{Mg^{2+}} - \log K_{Na-Mg} \tag{7.12}$$

which are obtained rearranging Eqs. (7.5) and (7.6) and assuming unit values for both a_{Ms} and a_{Clc}. Activities of muscovite and clinochlore <1 would lead to lower molalities of free Mg^{2+} ion. For the reservoir liquids with $0 \leq Z \leq 1$, the $m_{Mg^{2+},max.eq}$ given by Eq. (7.11) matches closely the $m_{Mg^{2+},max.eq}$ obtained by Eq. (7.12), with an average deviation of 0.16 ± 0.25 (1σ) %. For the reservoir liquids with $Z > 1$ or $Z < 0$, the adularia ordering parameter was set to 1 and 0, respectively, leading to higher discrepancies between the two $m_{Mg^{2+},max.eq}$ values computed by means of Eqs. (7.11) and (7.12).

For the 386 geothermal liquids with positive MAP values, the maximum molality of free Mg^{2+} ion, computed for mineral-solution equilibrium and $a_{Ms} = a_{Clc} = 1$ is contrasted with the analytical molality of free Mg^{2+} ion in the correlation plots of Figs. 7.10, 7.11, and 7.12. The shifts to the right of the equality line allow one to evaluate the minimum concentration of excess (non-equilibrium) Mg present in each reservoir liquid. In turn, these data can be used to recognize and quantify the processes controlling excess Mg, such as cooling and mixing (see above). To be recalled that mixing of Mg-rich, low-temperature waters from shallow levels with SiO_2-rich, high temperature waters from deep levels may lead to smectite scaling as recognized at Tongonan, Palinpinon, and Bacon-Manito in The Philippines (Reyes and Cardile 1989), in several geothermal district heating systems in Iceland (Gunnlaugsson and Einarsson 1989; Kristmannsdóttir et al. 1989; Hauksson et al. 1995), at Onikobe, Japan (Ajima et al. 1998), and in the Mori geothermal power plant (Kasai et al. 2000). Cooling by boiling was instead invoked to explain smectite scaling in the surface facilities of the Mindanao geothermal field (Dulce et al. 2010).

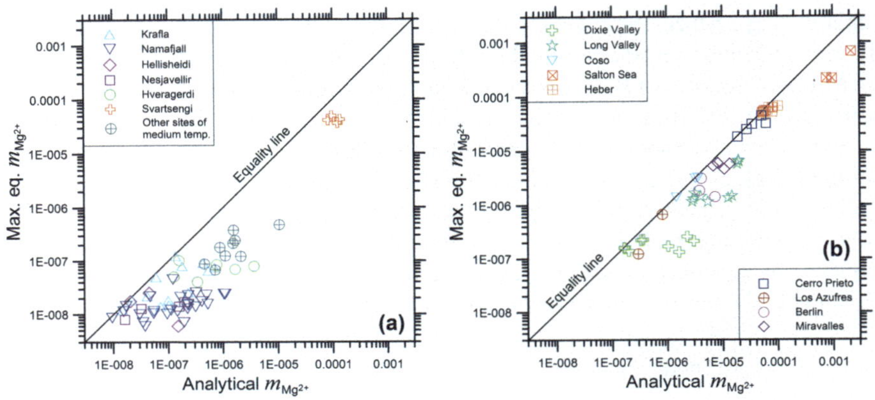

Fig. 7.10 Correlation plot between the maximum equilibrium molality of free Mg^{2+} ion and the analytical molality of free Mg^{2+} ion for the selected reservoir liquids with positive MAP values from the geothermal systems of **a** Iceland and **b** North-Central America

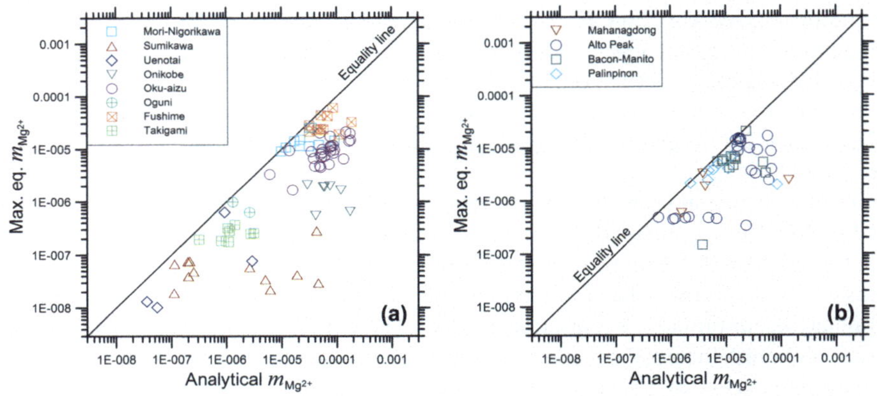

Fig. 7.11 Correlation plot between the maximum equilibrium molality of free Mg^{2+} ion and the analytical molality of free Mg^{2+} ion for the selected reservoir liquids with positive MAP values from the geothermal systems of **a** Japan and **b** The Philippines

7.4 Final Considerations on the Use of the K^2/Mg- and Na^2/Mg-Activity Ratios

Summing up, the K^2/Mg- and Na^2/Mg-activity ratios of reservoir liquids are controlled by the aquifer temperature, the order-disorder degree of hydrothermal adularia, and the mineral activity product MAP $= a_{Clc}^{0.2} \cdot a_{Ms}^{0.8}$, involving the activity of clinochlore in hydrothermal chlorite and the activity of muscovite in hydrothermal illite.

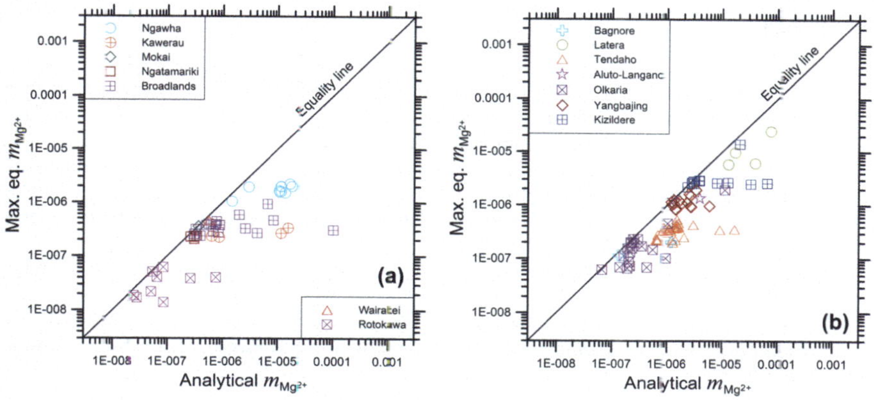

Fig. 7.12 Correlation plot between the maximum equilibrium molality of free Mg^{2+} ion and the analytical molality of free Mg^{2+} ion for the selected reservoir liquids with positive MAP values from the geothermal systems of **a** New Zealand and **b** miscellaneous sites

For the reservoir liquids with negative MAP values, the K^2/Mg- and Na^2/Mg-activity ratios can be used to obtain information on the minimum activities of clinochlore in chlorites and of muscovite in illites, which are in apparent equilibrium with the aqueous solutions of interest. For the reservoir liquids with positive MAP values, the K^2/Mg- and Na^2/Mg-activity ratios can be utilized to assess the minimum concentrations of excess (non-equilibrium) Mg and, consequently, to identify and quantify the processes controlling excess Mg, such as cooling and mixing, which may lead to Mg-silicate (e.g., smectite) scaling. This is probably the most valuable information provided by the activity-based theoretical K–Mg and Na–Mg geoindicators.

The K–Mg and Na–Mg exchange reactions other than (7.1) and (7.2), discussed in Sect. 5.6.3, are not explored here to develop alternative K–Mg and Na–Mg geoindicators for two main reasons, namely the restricted temperature range in which they are expected to be effective and the limitations caused by the varying characteristics of the relevant Mg-bearing minerals. For what concerns the first reason, the exchange reactions involving phlogopite, diopside, tremolite, and talc should work above ~280 °C, whereas the exchange reactions including Mg-montmorillonite and dolomite should be active below ~180 °C (e.g., Henley and Ellis 1983; Reyes 1990). For what concerns the second reason, phlogopite is a component of biotite, diopside is a component of clinopyroxene, tremolite is a component of amphibole, and Mg-montmorillonite is a component of smectites, whereas the thermodynamic properties of dolomite are strongly influenced by order-disorder. Therefore, alternative K–Mg and Na–Mg geoindicators are expected to be affected by limitations similar to or even worse than those of the K–Mg and Na–Mg geoindicators derived above for the exchange reactions involving clinochlore as Mg-bearing solid phase.

References

Ajima S, Todaka N, Murakake H (1998) An interpretation of smectite precipitation in production wells caused by the mixing of different geothermal fluids. In: Proceedings of 23rd Workshop on Geothermal Reservoir Engineering. Stanford University, pp 264–269

Albee AL (1962) Relationships between the mineral association, chemical composition and physical properties of the chlorite series. Am Mineral 47:851–870

Arnórsson S (1978) Major element chemistry of the geothermal sea-water at Reykjanes and Svartsengi. Iceland Mineral Mag 42:209–220

De Caritat P, Hutcheon I, Walshe JL (1993) Chlorite geothermometry: a review. Clay Clay Miner 41:219–239

Dulce RG, Aragon GM, Sambrano BG, Bayrante LF (2010) Genesis of smectite scales in Mindanao geothermal production field, Philippines. In: Proceedings World Geothermal Congress. Bali, Indonesia, p 9

Fournier RO (1991) Water geothermometers applied to geothermal energy. In: Application of Geochemistry in Geothermal Reservoir Development. F. D'Amore, co-ordinator, UNITAR, pp 37–69

Giggenbach WF (1988) Geothermal solute equilibria. derivation of Na–K–Mg–Ca geoindicators. Geochim Cosmochim Ac 52:2749–2765

Gunnlaugsson E, Einarsson A (1989) Magnesium-silicate scaling in mixture of geothermal water and deaerated fresh water in a district heating system. Geothermics 18:113–120

Hauksson T, Þórhallsson S, Gunnlaugsson E, Albertsson A (1995) Control of magnesium silicate scaling in district heating systems. World Geotherm Congr, Florence, Italy, pp 18–31

Helgeson HC, Delany JM, Nesbitt HW, Bird DK (1978) Summary and critique of the thermodynamic properties of rock-forming minerals. Am J Sci 278A:229

Henley RW, Ellis AJ (1983) Geothermal systems ancient and modern: a geochemical review. Earth-Sci Rev 19:1–50

Kasai K, Sato K, Kimura SI, Shakunaga N, Obara K (2000) Characterization of smectite scale and scale inhibition test by pH control at the Mori geothermal power plant, Japan. Proc World Geothermal Congr 1331–1336

Kristmannsdóttir H, Ólafsson M, Thórhallsson S (1989) Magnesium silicate scaling in district heating systems in Iceland. Geothermics 18:191–198

Reyes AG (1990) Petrology of Philippine geothermal systems and the application of alteration mineralogy to their assessment. J Volcanol Geoth Res 43:279–309

Reyes AG, Cardile CM (1989) Characterization of clay scales forming in Philippine geothermal wells. Geothermics 18:429–446

Sanjuan B, Brach M, Lasne E (June, 2001) Bouillante geothermal fluid: mixing and water/rock interaction processes at 250 °C. In: Cidu R (ed.), Water-rock interaction, WRI-10, international symposium on water-rock interaction, vol 2. Villasimius, Italy. A.A. Balkema, pp 911–914

Chapter 8
The Activity-Based Theoretical K–Ca and Na–Ca Geoindicators

Abstract Our activity-based theoretical K-Ca and Na-Ca geoindicators were tested with about one thousand reservoir liquids. The theoretical K-Ca geothermometers reproduce aquifer temperature with an average error of 3.8 °C and the error is lower than 10 °C in 97.3% of the cases. The theoretical Na-Ca geothermometers reproduce aquifer temperature with an average error of 6.9 °C and the error is less than 15 °C in 93.5% of the cases. The CO_2 fugacities given by our theoretical activity-based K-Ca and Na-Ca f_{CO_2}-indicators match the CO_2 fugacities given by the K-Ca f_{CO_2}-indicator of Giggenbach (1984), with an absolute deviation of 0.33 log-units on average, and reproduce the CO_2 fugacities given by speciation calculations, with an absolute deviation of 0.78 log-units on average. Furthermore, these theoretical K-Ca and Na-Ca geothermometers and f_{CO_2}-indicators allow one to identify the Ca-bearing solid phase in equilibrium with each reservoir liquid, either a Ca-Al-silicate (laumontite or clinozoisite or prehnite or wairakite) or calcite. This indication represents a substantial improvement with respect to the traditional K-Ca and Na-Ca geoindicators, is probably more reliable than the results of multicomponent chemical geothermometry, being slightly affected by pH and Al concentration, and may give qualitative clues on well permeability.

8.1 The K–Ca and Na–Ca Exchange Reactions and the Univariant Reactions Involving Calcite and a Ca–Al-Silicate

Under relatively low f_{CO_2} values, the K^2/Ca-activity ratio of reservoir liquids is probably controlled by equilibrium coexistence of hydrothermal minerals comprising a Ca–Al-silicate, adularia, a silica mineral and, in some cases, muscovite as well, as already recalled in Sect. 5.4.5. Plausible Ca–Al-silicates are laumontite [Lmt; $CaAl_2Si_4O_{12}·4H_2O$], clinozoisite [Czo; $Ca_2Al_3Si_3O_{12}(OH)$], prehnite [Prh; $Ca_2Al_2Si_3O_{10}(OH)_2$], and wairakite [Wrk; $CaAl_2Si_4O_{12}·2H_2O$]. They are involved in the K–Ca exchange reactions (5.131)–(5.134), which are rewritten and renumbered here for convenience:

R. Cioni and L. Marini, *A Thermodynamic Approach to Water Geothermometry*, Springer Geochemistry, https://doi.org/10.1007/978-3-030-54318-1_8

$$2KAlSi_3O_8 + 4H_2O + Ca^{2+} = CaAl_2Si_4O_{12} \cdot 4H_2O + 2SiO_2 + 2K^+ \quad (8.1)$$

$$2.25KAlSi_3O_8 + 0.5H_2O + Ca^{2+} = 0.25KAl_3Si_3O_{10}(OH)_2 + 0.5Ca_2Al_3Si_3O_{12}(OH) + 4.5SiO_2 + 2K^+ \quad (8.2)$$

$$2.5KAlSi_3O_8 + H_2O + Ca^{2+} = 0.5KAl_3Si_3O_{10}(OH)_2 + 0.5Ca_2Al_2Si_3O_{10}(OH)_2 + 4.5SiO_2 + 2K^+ \quad (8.3)$$

$$2KAlSi_3O_8 + 2H_2O + Ca^{2+} = CaAl_2Si_4O_{12} \cdot 2H_2O + 2SiO_2 + 2K^+. \quad (8.4)$$

Likewise, under relatively low f_{CO_2} values, the Na^2/Ca-activity ratio is probably constrained by equilibrium coexistence of the same hydrothermal minerals either with albite *in lieu* of adularia, if muscovite is absent, or with albite *in lieu* of a part of adularia, if muscovite is present, as described by the Na–Ca exchange reactions (5.127)–(5.130), which are written again and renumbered here below:

$$2NaAlSi_3O_8 + 4H_2O + Ca^{2+} = CaAl_2Si_4O_{12} \cdot 4H_2O + 2SiO_2 + 2Na^+ \quad (8.5)$$

$$2NaAlSi_3O_8 + 0.25KAlSi_3O_8 + 0.5H_2O + Ca^{2+} = 0.25KAl_3Si_3O_{12}(OH)_2 + 0.5Ca_2Al_3Si_3O_{12}(OH) + 4.5SiO_2 + 2Na^+ \quad (8.6)$$

$$2NaAlSi_3O_8 + 0.5KAlSi_3O_8 + H_2O + Ca^{2+} = 0.5KAl_3Si_3O_{12}(OH)_2 + 0.5Ca_2Al_2Si_3O_{12}(OH)_2 + 4.5SiO_2 + 2Na^+ \quad (8.7)$$

$$2NaAlSi_3O_8 + 2H_2O + Ca^{2+} = CaAl_2Si_4O_{12} \cdot 2H_2O + 2SiO_2 + 2Na^+ \quad (8.8)$$

As already noted in Sect. 5.5, under relatively high f_{CO_2} values, the Ca–Al-silicates are not stable, and the K^2/Ca-activity ratio is probably governed by equilibrium coexistence of adularia, muscovite, a silica mineral, and calcite, according to reaction (5.139), which is rewritten and renumbered here under:

$$3KAlSi_3O_8 + CO_{2(g)} + Ca^{2+} + H_2O = CaCO_3 + KAl_3Si_3O_{12}(OH)_2 + 6SiO_2 + 2K^+. \quad (8.9)$$

Reaction (8.9) represents the basis of the K–Ca f_{CO_2}-indicator of Giggenbach (1988). The Na^2/Ca activity ratio is probably controlled by a similar reaction, with albite *in lieu* of a part of adularia:

$$2NaAlSi_3O_8 + KAlSi_3O_8 + CO_{2(g)} + Ca^{2+} + H_2O = CaCO_3 + KAl_3Si_3O_{12}(OH)_2 + 6SiO_2 + 2Na^+. \quad (8.10)$$

Reaction (8.10) represents the basis for a Na–Ca f_{CO_2}-indicator, which is derived in this book, following early suggestions of Ellis (1970) and Chiodini et al. (1991).

As already underscored in Sect. 5.5, equilibrium coexistence of calcite and a Ca–Al-silicate (either laumontite or clinozoisite or prehnite or wairakite), together with adularia, muscovite, and a silica mineral, acts as a CO_2 buffer (Giggenbach 1981, 1984, 1988; Arnórsson et al. 1983; Arnórsson 1985; Arnórsson and Gunnlaugsson 1985). In other terms, these mineral assemblages constrain f_{CO_2}, at any specified temperature, as described by the following univariant reactions:

$$\begin{aligned} KAl_3Si_3O_{10}(OH)_2 + 4SiO_2 + CaCO_3 + 3H_2O &= KAlSi_3O_8 \\ + CaAl_2Si_4O_{12} \cdot 4H_2O + CO_{2(g)} \end{aligned} \tag{8.11}$$

$$\begin{aligned} 0.75KAl_3Si_3O_{10}(OH)_2 + 1.5SiO_2 + CaCO_3 &= 0.75KAlSi_3O_8 \\ + 0.5Ca_2Al_3Si_3O_{12}(OH) + CO_{2(g)} + 0.5H_2O \end{aligned} \tag{8.12}$$

$$\begin{aligned} 0.5KAl_3Si_3O_{10}(OH)_2 + 1.5SiO_2 + CaCO_3 &= 0.5KAlSi_3O_8 \\ + 0.5Ca_2Al_2Si_3O_{10}(OH)_2 + CO_{2(g)} \end{aligned} \tag{8.13}$$

$$\begin{aligned} KAl_3Si_3O_{10}(OH)_2 + 4SiO_2 + CaCO_3 + H_2O &= KAlSi_3O_8 \\ + CaAl_2Si_4O_{12} \cdot 2H_2O + CO_{2(g)}, \end{aligned} \tag{8.14}$$

To be noted that reactions (8.11)–(8.14) are obtained by summation of previous reactions. In detail:

- reaction (8.11) is the algebraic sum of either reactions (8.1) and (8.9) or reactions (8.5) and (8.10),
- reaction (8.12) is the algebraic sum of either reactions (8.2) and (8.9) or reactions (8.6) and (8.10),
- reaction (8.13) is the algebraic sum of either reactions (8.3) and (8.9) or reactions (8.7) and (8.10),
- reaction (8.14) is the algebraic sum of either reactions (8.4) and (8.9) or reactions (8.8) and (8.10).

The f_{CO_2}-temperature relationships corresponding to reactions (8.11)–(8.14) define the upper f_{CO_2} value of Ca–Al-silicate stability. At any given temperature, Ca–Al-silicates are stable at lower f_{CO_2} values, whereas calcite is stable at higher f_{CO_2} values. Moreover, the mineral assemblages involved in reactions (8.11)–(8.14) represent the more or less hypothetical condition of full mineral-solution equilibrium. However, none of these CO_2 buffers is ubiquitously efficacious in natural geothermal systems (Grant 1982; Giggenbach 1982). The probable reason for this is that a continuous flux of CO_2 occurs through the geothermal systems in the same manner as the heat flux, as suggested both by field evidence (Mahon et al. 1980) and by water/rock hydrothermal experiments (Kacandes and Grandstaff 1989). Thus, the univariant equilibrium conditions involving the minerals acting as CO_2 buffers may be established only in some parts of the geothermal systems, whereas in general the f_{CO_2} is an externally fixed parameter. If f_{CO_2} is externally controlled, one of the minerals participating in the univariant reactions (8.11)–(8.14) is no longer part of the equilibrium assemblage, and CO_2 is involved in bivariant reactions, such as reactions (8.9) and (8.10).

8.2 The Activities of Ca-Endmembers in Hydrothermal Ca–Al-Silicates and Calcite

Before gaining further insights from the reactions presented in the previous section, it is useful to recall the available information on the activity of Ca-endmembers in Ca–Al-silicates and calcite from active geothermal systems (Table 8.1), which was discussed in Sects. 4.5.3 for clinozoisite, 4.6.2 for prehnite, 4.7.4 for wairakite, 4.7.5 for laumontite, and 4.9.1 for calcite.

Since adularia, albite and silica minerals occur as pure or relatively pure solid phases in geothermal systems, their activities do not depart significantly from unity, as already noted many times. The same holds true for calcite and, even if to a lesser extent, for laumontite, whereas the activities of wairakite and especially of prehnite and clinozoisite exhibit higher deviations from one, but still relatively limited considering the mean and median values (Table 8.1). The activity of muscovite diverges even more from unity compared to prehnite and clinozoisite. In fact, it varies between 0.007 and 0.875, with an average of 0.654, a median of 0.688 and a standard deviation of 0.156, in 75 white micas from several geothermal systems (see Sect. 4.3.2 for further details).

Table 8.1 Main statistical parameters for the activity of the Ca-endmembers in Ca–Al-silicates and calcite from active geothermal systems

Ca-endmember	Solid solution	N	Mean	Median	Std dev	Min	Max
Clinozoisite	Clinozoisite/epidote	436	0.703	0.706	0.057	0.523	0.888
Prehnite	Prehnite/ferri-prehnite	127	0.763	0.779	0.154	0.373	0.998
Wairakite	Wairakite/analcime	136	0.891	0.920	0.090	0.573	0.999
Laumontite	Laumontite/alkali-laumontite	45	0.946	0.951	0.031	0.873	0.992
Calcite	Calcite-rich trigonal carbonate	313	0.972	0.988	0.043	0.705	1.000

8.3 The Log K of the K–Ca and Na–Ca Exchange Reactions and of the Univariant Reactions Involving Calcite and a Ca–Al-Silicate

Assuming that the activity of water does not differ significantly from 1, if the ionic strength of the aqueous phase is less than ~1 mol/kg, the decimal logarithm of the thermodynamic equilibrium constants of the reactions presented in Sect. 8.1 can be generally written in the following simplified forms:

$$\log K_{K-Ca,Lmt} \cong \log\left(\frac{a^2_{K^+}}{a_{Ca^{2+}}}\right) + \log a_{Lmt} \tag{8.15}$$

$$\log K_{K-Ca,Czo} \cong \log\left(\frac{a^2_{K^+}}{a_{Ca^{2+}}}\right) + 0.25 \cdot \log a_{Ms} + 0.5 \cdot \log a_{Czo} \tag{8.16}$$

$$\log K_{K-Ca,Prh} \cong \log\left(\frac{a^2_{K^+}}{a_{Ca^{2+}}}\right) + 0.5 \cdot \log a_{Ms} + 0.5 \cdot \log a_{Prh} \tag{8.17}$$

$$\log K_{K-Ca,Wrk} \cong \log\left(\frac{a^2_{K^+}}{a_{Ca^{2+}}}\right) + \log a_{Wrk} \tag{8.18}$$

$$\log K_{Na-Ca,Lmt} \cong \log\left(\frac{a^2_{Na^+}}{a_{Ca^{2+}}}\right) + \log a_{Lmt} \tag{8.19}$$

$$\log K_{Na-Ca,Czo} \cong \log\left(\frac{a^2_{Na^+}}{a_{Ca^{2+}}}\right) + 0.25 \cdot \log a_{Ms} + 0.5 \cdot \log a_{Czo} \tag{8.20}$$

$$\log K_{Na-Ca,Prh} \cong \log\left(\frac{a^2_{Na^+}}{a_{Ca^{2+}}}\right) + 0.5 \cdot \log a_{Ms} + 0.5 \cdot \log a_{Prh} \tag{8.21}$$

$$\log K_{Na-Ca,Wrk} \cong \log\left(\frac{a^2_{Na^+}}{a_{Ca^{2+}}}\right) + \log a_{Wrk} \tag{8.22}$$

$$\log K_{K-Ca,Cal} \cong \log\left(\frac{a^2_{K^+}}{a_{Ca^{2+}}}\right) - \log f_{CO_2} + \log a_{Ms} \tag{8.23}$$

$$\log K_{Na-Ca,Cal} \cong \log\left(\frac{a^2_{Na^+}}{a_{Ca^{2+}}}\right) - \log f_{CO_2} + \log a_{Ms} \tag{8.24}$$

$$\log K_{Lmt-Cal} \cong \log f_{CO_2} + \log a_{Lmt} - \log a_{Ms} \tag{8.25}$$

$$\log K_{Czo-Cal} \cong \log f_{CO_2} + 0.5 \cdot \log a_{Czo} - 0.75 \cdot \log a_{Ms} \quad (8.26)$$

$$\log K_{Prh-Cal} \cong \log f_{CO_2} + 0.5 \cdot \log a_{Prh} - 0.5 \cdot \log a_{Ms} \quad (8.27)$$

$$\log K_{Wrk-Cal} \cong \log f_{CO_2} + \log a_{Wrk} - \log a_{Ms} \quad (8.28)$$

As discussed in Sects. 4.2.3 and 5.3.3, Al–Si order-disorder on the tetrahedral sites of adularia causes further complications. Therefore, adopting the same approach of Sects. 6.1 and 7.1, adularias with ordering parameter Z varying from 0 (i.e., high-sanidine) to 1 (i.e., maximum-microcline) at steps of 0.1 units were alternatively considered to take part to reactions (8.1)–(8.4), (8.6), (8.7), and (8.9)–(8.14). The silica mineral involved in these reactions was assumed to be chalcedony at $T < 175$ °C or quartz/chalcedony at $T > 175$ °C. The thermodynamic properties of these reactions were computed as a function of temperature, at pressure of 1 bar for $T < 100$ °C and at water saturation pressure for $T \geq 100$ °C, using the SUPCRT92 code.

8.4 Derivation of the Activity-Based Theoretical K–Ca and Na–Ca Geothermometers and f_{CO_2}-Indicators

The logarithm of the thermodynamic equilibrium constants of reactions (8.1)–(8.10) for pure solid phases were fitted against both the inverse of the absolute temperature and the ordering parameter of adularia (if present in the reaction), considering as two distinct datasets the data below 175 °C, with chalcedony controlling undissociated SiO_2, and those above 175 °C, with quartz/chalcedony governing undissociated SiO_2. Each regression equation and the average activities of pertinent solid phases were then inserted into the corresponding relation linking the log K to the K^2/Ca- and Na^2/Ca-log activity ratios, i.e., Eqs. (8.15)–(8.24).

The relations derived from Eqs. (8.15) to (8.22) were solved with respect to temperature, thus obtaining four K–Ca geothermometers and four Na–Ca geothermometers. The relationships resulting from Eqs. (8.23) and (8.24) were solved with respect to the logarithm of CO_2 fugacity, thus obtaining a K–Ca f_{CO_2}-indicator and a Na–Ca f_{CO_2}-indicator.

The uncertainties caused by variations in the activities of relevant components in the solid solutions of interest, that is muscovite in illite, laumontite in laumontite/alkali-laumontite, clinozoisite in clinozoisite/epidote, prehnite in prehnite/ferri-prehnite, and wairakite in wairakite/analcime were evaluated considering the average activities plus one standard deviation and the average activities minus one standard deviation (see below).

8.4.1 The Theoretical K–Ca and Na–Ca Laumontite Geothermometers

For reactions (8.1) and (8.5), comprising laumontite, the following four geothermometric equations were obtained from Eqs. (8.15) and (8.19) and the log $K_{K\text{–}Ca,Lmt}$ and log $K_{Na\text{–}Ca,Lmt}$ values listed in Tables 8.2 and 8.3, respectively:

$$T_{K-Ca,Lmt}\,(^{\circ}C) = \frac{2481.5 + 1158.1 \cdot Z}{4.0237 + 1.4861 \cdot Z - \log\left(\frac{a^2_{K+}}{a_{Ca^{2+}}}\right)} - 273.15, \quad \text{for } T < 175\,^{\circ}C \tag{8.29}$$

$$T_{K-Ca,Lmt}\,(^{\circ}C) = \frac{2100.3 + 1157.4 \cdot Z}{3.3178 + 1.4845 \cdot Z - \log\left(\frac{a^2_{K+}}{a_{Ca^{2+}}}\right)} - 273.15, \quad \text{for } 175 < T < 325\,^{\circ}C \tag{8.30}$$

$$T_{Na-Ca,Lmt}\,(^{\circ}C) = \frac{940.13}{2.4914 - \log\left(\frac{a^2_{Na+}}{a_{Ca^{2+}}}\right)} - 273.15, \quad \text{for } T < 175\,^{\circ}C \tag{8.31}$$

$$T_{Na-Ca,Lmt}\,(^{\circ}C) = \frac{1050.2}{2.9090 - \log\left(\frac{a^2_{Na+}}{a_{Ca^{2+}}}\right)} - 273.15, \quad \text{for } 175 < T < 325\,^{\circ}C \tag{8.32}$$

It must be noted that the log K values of reaction (8.5), which are reported in Table 8.3, depend on temperature only because adularia is not involved in this reaction. The average value of log a_{Lmt}, −0.0241, is considered in Eqs. (8.29)–(8.32). Due to deviations of laumontite activity of $+1\sigma$ and -1σ from the mean value, log a_{Lmt} assumes the values −0.0101 and −0.0386, respectively. These variations in log a_{Lmt}, determine differences of ±0.3 to ±1.1 °C, ±0.9 to ±1.8 °C, ±1.1 to ±3.0 °C, and ±2.7 to ±5.0 °C in the temperatures calculated by means of Eqs. (8.29), (8.30), (8.31), and (8.32), respectively, representing their nominal errors.

8.4.2 The Theoretical K–Ca and Na–Ca Clinozoisite Geothermometers

For reactions (8.2) and (8.6), involving clinozoisite, the following four geothermometric functions were obtained, based on Eqs. (8.16) and (8.20) and the log $K_{K\text{–}Ca,Czo}$ and log $K_{Na\text{–}Ca,Czo}$ values reported in Tables 8.4 and 8.5, respectively:

Table 8.2 Logarithm of the thermodynamic equilibrium constant of reaction (8.1), log $K_{K\text{–}Ca,Lmt}$, for pure laumontite, chalcedony (for T < 175 °C) or quartz/chalcedony (for T > 175 °C), and adularia, as a function of the temperature and the ordering parameter Z of adularia (P = 1 bar for T < 100 °C; P = P_{sat} for T ≥ 100 °C)

T (°C)	Z = 1	Z = 0.9	Z = 0.8	Z = 0.7	Z = 0.6	Z = 0.5	Z = 0.4	Z = 0.3	Z = 0.2	Z = 0.1	Z = 0
0.01	−7.9665	−7.6912	−7.4158	−7.1404	−6.8650	−6.5897	−6.3143	−6.0389	−5.7635	−5.4882	−5.2128
25	−6.7611	−6.5213	−6.2815	−6.0416	−5.8018	−5.5619	−5.3221	−5.0823	−4.8424	−4.6026	−4.3627
50	−5.7679	−5.5581	−5.3483	−5.1386	−4.9288	−4.7190	−4.5092	−4.2994	−4.0896	−3.8798	−3.6700
75	−4.9389	−4.7548	−4.5708	−4.3867	−4.2027	−4.0186	−3.8346	−3.6505	−3.4665	−3.2824	−3.0984
100	−4.2385	−4.0768	−3.9150	−3.7532	−3.5915	−3.4297	−3.2680	−3.1062	−2.9444	−2.7827	−2.6209
125	−3.6403	−3.4980	−3.3557	−3.2135	−3.0712	−2.9289	−2.7866	−2.6444	−2.5021	−2.3598	−2.2175
150	−3.1238	−2.9987	−2.8736	−2.7485	−2.6235	−2.4984	−2.3733	−2.2482	−2.1231	−1.9980	−1.8729
175	−2.6734	−2.5635	−2.4537	−2.3439	−2.2341	−2.1242	−2.0144	−1.9046	−1.7948	−1.6849	−1.5751
175	−2.4930	−2.3831	−2.2733	−2.1635	−2.0536	−1.9438	−1.8340	−1.7242	−1.6143	−1.5045	−1.3947
200	−2.1058	−2.0096	−1.9134	−1.8173	−1.7211	−1.6249	−1.5287	−1.4326	−1.3364	−1.2402	−1.1441
225	−1.7611	−1.6771	−1.5932	−1.5093	−1.4254	−1.3416	−1.2577	−1.1738	−1.0899	−1.0060	−0.9221
250	−1.4495	−1.3767	−1.3039	−1.2311	−1.1583	−1.0855	−1.0127	−0.9400	−0.8672	−0.7944	−0.7216
275	−1.1640	−1.1012	−1.0385	−0.9758	−0.9131	−0.8504	−0.7877	−0.7250	−0.6623	−0.5996	−0.5369
300	−0.9023	−0.8486	−0.7951	−0.7416	−0.6882	−0.6347	−0.5812	−0.5277	−0.4742	−0.4207	−0.3673
325	−0.6723	−0.6270	−0.5820	−0.5369	−0.4919	−0.4469	−0.4018	−0.3568	−0.3117	−0.2667	−0.2217
350	−0.5027	−0.4650	−0.4278	−0.3905	−0.3532	−0.3159	−0.2786	−0.2413	−0.2041	−0.1668	−0.1295

Table 8.3 Logarithm of the thermodynamic equilibrium constant of reaction (8.5), log $K_{Na\text{-}Ca,Lmt}$, for pure laumontite, chalcedony (for T < 175 °C) or quartz/chalcedony (for T > 175 °C), and albite, as a function of temperature (P = 1 bar for T < 100 °C; P = P_{sat} for T ≥ 100 °C)

T (°C)	0.01	25	50	75	100	125	150	175
log K	−0.9783	−0.6816	−0.4386	−0.2330	−0.0548	0.1025	0.2440	0.3731
T (°C)	175	200	225	250	275	300	325	350
log K	0.5535	0.6634	0.7672	0.8671	0.9647	1.0581	1.1375	1.1749

$$T_{K-Ca,Czo}\,(^{\circ}C) = \frac{1302.9 \cdot Z + 3697.9}{6.3540 + 1.6719 \cdot Z - \log\left(\frac{a_{K+}^2}{a_{Ca2+}}\right)} - 273.15, \quad \text{for } T < 175\,^{\circ}C \tag{8.33}$$

$$T_{K-Ca,Czo}\,(^{\circ}C) = \frac{1302.1 \cdot Z + 4121.3}{7.6909 + 1.6700 \cdot Z - \log\left(\frac{a_{K+}^2}{a_{Ca2+}}\right)} - 273.15, \quad \text{for } 175 < T < 325\,^{\circ}C \tag{8.34}$$

$$T_{Na-Ca,Czo}\,(^{\circ}C) = \frac{144.81 \cdot Z + 2170.4}{4.8576 + 0.1859 \cdot Z - \log\left(\frac{a_{Na+}^2}{a_{Ca2+}}\right)} - 273.15, \quad \text{for } T < 175\,^{\circ}C \tag{8.35}$$

$$T_{Na-Ca,Czo}\,(^{\circ}C) = \frac{144.72 \cdot Z + 3042.6}{7.2239 + 0.1856 \cdot Z - \log\left(\frac{a_{Na+}^2}{a_{Ca2+}}\right)} - 273.15, \quad \text{for } 175 < T < 325\,^{\circ}C \tag{8.36}$$

The average value of the term $0.25 \cdot \log a_{Ms} + 0.5 \cdot \log a_{Czo}$, −0.1227, is considered in Eqs. (8.33)–(8.36). Due to deviations of clinozoisite and muscovite activity of +1σ and −1σ from the mean values, the term $0.25 \cdot \log a_{Ms} + 0.5 \cdot \log a_{Czo}$ assumes the values −0.0825 and −0.1706, respectively. These variations in the term $0.25 \cdot \log a_{Ms} + 0.5 \cdot \log a_{Czo}$, cause differences of ±0.7 to ±2.3 °C, ±1.8 to ±3.8 °C, ±1.3 to ±4.1 °C, and ±2.5 to ±5.3 °C in the temperatures computed by means of Eqs. (8.33), (8.34), (8.35), and (8.36), respectively, representing their nominal uncertainties.

8.4.3 *The Theoretical K–Ca and Na–Ca Prehnite Geothermometers*

For reactions (8.3) and (8.7), including prehnite, the following four geothermometric equations were obtained on the basis of Eqs. (8.17) and (8.21) and the log $K_{K\text{-}Ca,Prh}$ and log $K_{Na\text{-}Ca,Prh}$ values listed in Tables 8.6 and 8.7, respectively:

Table 8.4 Logarithm of the thermodynamic equilibrium constant of reaction (8.2), log $K_{K\text{–}Ca,Czo}$, for pure clinozoisite, muscovite, chalcedony (for T < 175 °C) or quartz/chalcedony (for T > 175 °C), and adularia, as a function of the temperature and the ordering parameter Z of adularia (P = 1 bar for T < 100 °C; P = P_{sat} for T ≥ 100 °C)

T (°C)	Z = 1	Z = 0.9	Z = 0.8	Z = 0.7	Z = 0.6	Z = 0.5	Z = 0.4	Z = 0.3	Z = 0.2	Z = 0.1	Z = 0
0.01	−10.3855	−10.0757	−9.7659	−9.4561	−9.1463	−8.8365	−8.5267	−8.2169	−7.9071	−7.5973	−7.2875
25	−8.8680	−8.5982	−8.3284	−8.0585	−7.7887	−7.5189	−7.2491	−6.9793	−6.7094	−6.4396	−6.1698
50	−7.5821	−7.3461	−7.1101	−6.8741	−6.6381	−6.4021	−6.1661	−5.9300	−5.6940	−5.4580	−5.2220
75	−6.4769	−6.2699	−6.0628	−5.8557	−5.6487	−5.4416	−5.2346	−5.0275	−4.8205	−4.6134	−4.4063
100	−5.5145	−5.3325	−5.1505	−4.9685	−4.7866	−4.6046	−4.4226	−4.2406	−4.0586	−3.8766	−3.6946
125	−4.6662	−4.5061	−4.3460	−4.1860	−4.0259	−3.8659	−3.7058	−3.5457	−3.3857	−3.2256	−3.0656
150	−3.9100	−3.7693	−3.6286	−3.4878	−3.3471	−3.2064	−3.0656	−2.9249	−2.7842	−2.6435	−2.5027
175	−3.2289	−3.1053	−2.9818	−2.8582	−2.7347	−2.6111	−2.4876	−2.3640	−2.2405	−2.1169	−1.9934
175	−2.8229	−2.6994	−2.5758	−2.4523	−2.3287	−2.2052	−2.0816	−1.9581	−1.8345	−1.7110	−1.5874
200	−2.2249	−2.1167	−2.0085	−1.9003	−1.7921	−1.6839	−1.5757	−1.4675	−1.3593	−1.2511	−1.1429
225	−1.6748	−1.5804	−1.4860	−1.3916	−1.2973	−1.2029	−1.1085	−1.0141	−0.9198	−0.8254	−0.7310
250	−1.1627	−1.0807	−0.9988	−0.9169	−0.8350	−0.7532	−0.6713	−0.5894	−0.5075	−0.4256	−0.3438
275	−0.6804	−0.6097	−0.5392	−0.4687	−0.3982	−0.3276	−0.2571	−0.1866	−0.1160	−0.0455	0.0250
300	−0.2251	−0.1647	−0.1045	−0.0444	0.0158	0.0760	0.1361	0.1963	0.2565	0.3166	0.3768
325	0.1963	0.2473	0.2979	0.3486	0.3993	0.4499	0.5006	0.5513	0.6020	0.6526	0.7033
350	0.5562	0.5985	0.6405	0.6824	0.7243	0.7663	0.8082	0.8502	0.8921	0.9341	0.9760

Table 8.5 Logarithm of the thermodynamic equilibrium constant of reaction (8.6), log $K_{Na\text{-}Ca,Czo}$, for pure clinozoisite, muscovite, chalcedony (for $T < 175$ °C) or quartz/chalcedony (for $T > 175$ °C), adularia, and albite, as a function of the temperature and the ordering parameter Z of adularia ($P = 1$ bar for $T < 100$ °C; $P = P_{sat}$ for $T \geq 100$ °C)

T (°C)	Z = 1	Z = 0.9	Z = 0.8	Z = 0.7	Z = 0.6	Z = 0.5	Z = 0.4	Z = 0.3	Z = 0.2	Z = 0.1	Z = 0
0.01	−3.3972	−3.3628	−3.3284	−3.2940	−3.2595	−3.2251	−3.1907	−3.1563	−3.1219	−3.0874	−3.0530
25	−2.7884	−2.7585	−2.7285	−2.6985	−2.6685	−2.6385	−2.6086	−2.5786	−2.5486	−2.5186	−2.4886
50	2.2528	2.2266	−2.2004	−2.1741	−2.1479	−2.1217	−2.0955	−2.0692	−2.0430	−2.0168	−1.9906
75	−1.7710	−1.7480	−1.7250	−1.7020	−1.6790	−1.6560	−1.6330	−1.6100	−1.5869	−1.5639	−1.5409
100	−1.3308	−1.3106	−1.2903	−1.2701	−1.2499	−1.2297	−1.2095	−1.1892	−1.1690	−1.1488	−1.1286
125	−0.9233	−0.9055	−0.8878	−0.8700	−0.8522	−0.8344	−0.8166	−0.7988	−0.7811	−0.7633	−0.7455
150	−0.5422	−0.5266	−0.5109	−0.4953	−0.4797	−0.4640	−0.4484	−0.4328	−0.4171	−0.4015	−0.3859
175	−0.1825	−0.1687	−0.1550	−0.1413	−0.1275	−0.1138	−0.1001	−0.0864	−0.0726	−0.0589	−0.0452
175	0.2235	0.2372	0.2510	0.2647	0.2784	0.2922	0.3059	0.3196	0.3333	0.3471	0.3608
200	0.5443	0.5564	0.5684	0.5804	0.5924	0.6045	0.6165	0.6285	0.6405	0.6525	0.6646
225	0.8534	0.8639	0.8744	0.8849	0.8954	0.9058	0.9163	0.9268	0.9373	0.9478	0.9583
250	1.1539	1.1630	1.1721	1.1812	1.1903	1.1994	1.2085	1.2176	1.2267	1.2358	1.2449
275	1.4483	1.4561	1.4640	1.4718	1.4796	1.4875	1.4953	1.5031	1.5110	1.5188	1.5266
300	1.7353	1.7420	1.7487	1.7554	1.7621	1.7688	1.7754	1.7821	1.7888	1.7955	1.8022
325	2.0061	2.0118	2.0174	2.0231	2.0287	2.0343	2.0400	2.0456	2.0512	2.0568	2.0625
350	2.2337	2.2384	2.2431	2.2477	2.2524	2.2571	2.2617	2.2664	2.2710	2.2757	2.2804

Table 8.6 Logarithm of the thermodynamic equilibrium constant of reaction (8.3), log $K_{K\text{–}Ca,Prh}$, for pure prehnite, muscovite, chalcedony (for T < 175 °C) or quartz/chalcedony (for T > 175 °C), and adularia, as a function of the temperature and the ordering parameter Z of adularia (P = 1 bar for T < 100 °C; P = P_{sat} for T ≥ 100 °C)

T (°C)	Z = 1	Z = 0.9	Z = 0.8	Z = 0.7	Z = 0.6	Z = 0.5	Z = 0.4	Z = 0.3	Z = 0.2	Z = 0.1	Z = 0
0.01	−8.3997	−8.0555	−7.7113	−7.3671	−7.0228	−6.6786	−6.3344	−5.9902	−5.6459	−5.3017	−4.9575
25	−7.1701	−6.8703	−6.5705	−6.2707	−5.9709	−5.6711	−5.3713	−5.0715	−4.7717	−4.4719	−4.1721
50	−6.1320	−5.8697	−5.6075	−5.3452	−5.0830	−4.8208	−4.5585	−4.2963	−4.0341	−3.7718	−3.5096
75	−5.2427	−5.0126	−4.7825	−4.5525	−4.3224	−4.0924	−3.8623	−3.6322	−3.4022	−3.1721	−2.9420
100	−4.4705	−4.2683	−4.0661	−3.8639	−3.6617	−3.4595	−3.2573	−3.0551	−2.8529	−2.6507	−2.4485
125	−3.7914	−3.6136	−3.4357	−3.2579	−3.0800	−2.9022	−2.7243	−2.5465	−2.3686	−2.1908	−2.0130
150	−3.1870	−3.0306	−2.8742	−2.7179	−2.5615	−2.4052	−2.2488	−2.0924	−1.9361	−1.7797	−1.6233
175	−2.6429	−2.5056	−2.3683	−2.2310	−2.0938	−1.9565	−1.8192	−1.6819	−1.5446	−1.4074	−1.2701
175	−2.2369	−2.0996	−1.9623	−1.8251	−1.6878	−1.5505	−1.4132	−1.2759	−1.1387	−1.0014	−0.8641
200	−1.7634	−1.6431	−1.5229	−1.4027	−1.2825	−1.1623	−1.0420	−0.9218	−0.8016	−0.6814	−0.5612
225	−1.3271	−1.2221	−1.1172	−1.0124	−0.9075	−0.8027	−0.6978	−0.5929	−0.4881	−0.3832	−0.2784
250	−0.9194	−0.8283	−0.7373	−0.6463	−0.5554	−0.4644	−0.3734	−0.2824	−0.1914	−0.1005	−0.0095
275	−0.5336	−0.4550	−0.3767	−0.2983	−0.2200	−0.1416	−0.0632	0.0151	0.0935	0.1719	0.2502
300	−0.1677	−0.1006	−0.0337	0.0331	0.1000	0.1668	0.2337	0.3005	0.3674	0.4342	0.5011
325	0.1704	0.2270	0.2833	0.3396	0.3959	0.4522	0.5085	0.5648	0.6211	0.6774	0.7337
350	0.4523	0.4993	0.5459	0.5925	0.6391	0.6857	0.7323	0.7789	0.8255	0.8722	0.9188

Table 8.7 Logarithm of the thermodynamic equilibrium constant of reaction (8.7), log $K_{Na\text{–}Ca,Prh}$, for pure prehnite, muscovite, chalcedony (for T < 175 °C) or quartz/chalcedony (for T > 175 °C), adularia, and albite, as a function of the temperature and the ordering parameter Z of adularia (P = 1 bar for T < 100 °C; $P = P_{sat}$ for T ≥ 100 °C)

T (°C)	Z = 1	Z = 0.9	Z = 0.8	Z = 0.7	Z = 0.6	Z = 0.5	Z = 0.4	Z = 0.3	Z = 0.2	Z = 0.1	Z = 0
0.01	−1.4115	−1.3426	−1.2738	−1.2049	−1.1361	−1.0672	−0.9984	−0.9295	−0.8607	−0.7919	−0.7230
25	−1.0905	−1.0305	−0.9706	−0.9106	−0.8507	−0.7907	−0.7307	−0.6708	−0.6108	−0.5509	−0.4909
50	−0.8026	−0.7502	−0.6977	−0.6453	−0.5928	−0.5404	−0.4879	−0.4355	−0.3831	−0.3306	−0.2782
75	−0.5368	−0.4907	−0.4447	−0.3987	−0.3527	−0.3067	−0.2607	−0.2147	−0.1687	−0.1226	−0.0766
100	−0.2868	−0.2464	−0.2059	−0.1655	−0.1250	−0.0846	−0.0442	−0.0037	0.0367	0.0772	0.1176
125	−0.0486	−0.0130	0.0226	0.0581	0.0937	0.1293	0.1648	0.2004	0.2360	0.2716	0.3071
150	0.1808	0.2121	0.2434	0.2746	0.3059	0.3372	0.3684	0.3997	0.4310	0.4623	0.4935
175	0.4036	0.4310	0.4585	0.4859	0.5134	0.5409	0.5683	0.5958	0.6232	0.6507	0.6781
175	0.8095	0.8370	0.8645	0.8919	0.9194	0.9468	0.9743	1.0017	1.0292	1.0567	1.0841
200	1.0059	1.0299	1.0540	1.0780	1.1021	1.1261	1.1501	1.1742	1.1982	1.2223	1.2463
225	1.2012	1.2222	1.2431	1.2641	1.2851	1.3061	1.3270	1.3480	1.3690	1.3899	1.4109
250	1.3972	1.4154	1.4336	1.4518	1.4700	1.4882	1.5064	1.5246	1.5428	1.5610	1.5792
275	1.5951	1.6108	1.6265	1.6422	1.6578	1.6735	1.6892	1.7048	1.7205	1.7362	1.7519
300	1.7927	1.8061	1.8195	1.8329	1.8462	1.8596	1.8730	1.8863	1.8997	1.9131	1.9265
325	1.9802	1.9915	2.0028	2.0141	2.0253	2.0366	2.0479	2.0591	2.0704	2.0816	2.0929
350	2.1298	2.1392	2.1485	2.1579	2.1672	2.1765	2.1858	2.1952	2.2045	2.2138	2.2231

$$T_{K-Ca,Prh}\,(^{\circ}C) = \frac{1447.7 \cdot Z + 2573.9}{4.6093 + 1.8577 \cdot Z - \log\left(\frac{a^2_{K^+}}{a_{Ca^{2+}}}\right)} - 273.15, \quad \text{for } T < 175\,^{\circ}C \tag{8.37}$$

$$T_{K-Ca,Prh}\,(^{\circ}C) = \frac{1446.7 \cdot Z + 2876.0}{5.6691 + 1.8555 \cdot Z - \log\left(\frac{a^2_{K^+}}{a_{Ca^{2+}}}\right)} - 273.15, \quad \text{for } 175 < T < 325\,^{\circ}C \tag{8.38}$$

$$T_{Na-Ca,Prh}\,(^{\circ}C) = \frac{289.51 \cdot Z + 1035.4}{3.0844 + 0.3715 \cdot Z - \log\left(\frac{a^2_{Na^+}}{a_{Ca^{2+}}}\right)} - 273.15, \quad \text{for } T < 175\,^{\circ}C \tag{8.39}$$

$$T_{Na-Ca,Prh}\,(^{\circ}C) = \frac{289.33 \cdot Z + 1804.7}{5.3085 + 0.3711 \cdot Z - \log\left(\frac{a^2_{Na^+}}{a_{Ca^{2+}}}\right)} - 273.15, \quad \text{for } 175 < T < 325\,^{\circ}C \tag{8.40}$$

The mean value of the term $0.50 \cdot \log a_{Ms} + 0.5 \cdot \log a_{Prh}$, -0.1509, is considered in Eqs. (8.37)–(8.40). Owing to departures of prehnite and muscovite activity of $+1\sigma$ and -1σ from the mean values, the term $0.50 \cdot \log a_{Ms} + 0.5 \cdot \log a_{Prh}$ becomes equal to -0.0646 and -0.2591, respectively. These changes in the term $0.50 \cdot \log a_{Ms} + 0.5 \cdot \log a_{Prh}$, determine differences of ±1.6 to ±5.3 °C, ±4.0 to ±9.6 °C, ±4.5 to ±14.4 °C, and ±8.4 to ±17.9 °C in the temperatures computed by means of Eqs. (8.37), (8.38), (8.39), and (8.40), respectively, representing their nominal errors.

8.4.4 The Theoretical K–Ca and Na–Ca Wairakite Geothermometers

For reactions (8.4) and (8.8), involving wairakite, the following four geothermometric functions were obtained based on Eqs. (8.18) and (8.22) and the log $K_{K-Ca,Wrk}$ and log $K_{Na-Ca,Wrk}$ values reported in Tables 8.8 and 8.9, respectively:

$$T_{K-Ca,Wrk}\,(^{\circ}C) = \frac{1158.2 \cdot Z + 5457.0}{9.5776 + 1.4861 \cdot Z - \log\left(\frac{a^2_{K^+}}{a_{Ca^{2+}}}\right)} - 273.15, \quad \text{for } T < 175\,^{\circ}C \tag{8.41}$$

$$T_{K-Ca,Wrk}\,(^{\circ}C) = \frac{1157.4 \cdot Z + 5520.6}{9.8775 + 1.4844 \cdot Z - \log\left(\frac{a^2_{K^+}}{a_{Ca^{2+}}}\right)} - 273.15, \quad \text{for } 175 < T < 325\,^{\circ}C \tag{8.42}$$

Table 8.8 Logarithm of the thermodynamic equilibrium constant of reaction (8.4), log $K_{K\text{–}Ca,Wrk}$, for pure wairakite, chalcedony (for T < 175 °C) or quartz/chalcedony (for T > 175 °C), and adularia, as a function of the temperature and the ordering parameter Z of adularia (P = 1 bar for T < 100 °C; $P = P_{sat}$ for T ≥ 100 °C)

T (°C)	Z = 1	Z = 0.9	Z = 0.8	Z = 0.7	Z = 0.6	Z = 0.5	Z = 0.4	Z = 0.3	Z = 0.2	Z = 0.1	Z = 0
0.01	−13.2163	−12.9410	−12.6656	−12.3902	−12.1148	−11.8394	−11.5641	−11.2887	−11.0133	−10.7379	−10.4626
25	−11.1706	−10.9308	−10.6909	−10.4511	−10.2112	−9.9714	−9.7316	−9.4917	−9.2519	−9.0120	−8.7722
50	−9.4480	−9.2382	−9.0284	−8.8186	−8.6088	−8.3991	−8.1893	−7.9795	−7.7697	−7.5599	−7.3501
75	−7.9783	−7.7942	−7.6101	−7.4261	−7.2420	−7.0580	−6.8739	−6.6899	−6.5058	−6.3218	−6.1377
100	−6.7091	−6.5473	−6.3855	−6.2238	−6.0620	−5.9002	−5.7385	−5.5767	−5.4149	−5.2532	−5.0914
125	−5.6013	−5.4590	−5.3167	−5.1744	−5.0322	−4.8899	−4.7476	−4.6053	−4.4631	−4.3208	−4.1785
150	−4.6245	−4.4994	−4.3743	−4.2492	−4.1241	−3.9990	−3.8739	−3.7488	−3.6238	−3.4987	−3.3736
175	−3.7552	−3.6454	−3.5356	−3.4257	−3.3159	−3.2061	−3.0963	−2.9864	−2.8766	−2.7668	−2.6570
175	−3.5748	−3.4649	−3.3551	−3.2453	−3.1355	−3.0256	−2.9158	−2.8060	−2.6962	−2.5863	−2.4765
200	−2.8041	−2.7079	−2.6117	−2.5155	−2.4194	−2.3232	−2.2270	−2.1309	−2.0347	−1.9385	−1.8424
225	−2.1061	−2.0221	−1.9382	−1.8544	−1.7705	−1.6866	−1.6027	−1.5188	−1.4349	−1.3510	−1.2671
250	−1.4674	−1.3945	−1.3218	−1.2490	−1.1762	−1.1034	−1.0306	−0.9579	−0.8851	−0.8123	−0.7395
275	−0.8775	−0.8146	−0.7520	−0.6893	−0.6266	−0.5639	−0.5012	−0.4385	−0.3758	−0.3131	−0.2504
300	−0.3312	−0.2775	−0.2240	−0.1705	−0.1171	−0.0636	−0.0101	0.0434	0.0969	0.1503	0.2038
325	0.1660	0.2113	0.2563	0.3014	0.3464	0.3915	0.4365	0.4816	0.5266	0.5716	0.6167
350	0.5877	0.6253	0.6626	0.6999	0.7372	0.7745	0.8118	0.8490	0.8863	0.9236	0.9609

Table 8.9 Logarithm of the thermodynamic equilibrium constant of reaction (8.8), log $K_{Na-Ca,Wrk}$, for pure wairakite, chalcedony (for T < 175 °C) or quartz/chalcedony (for T > 175 °C), and albite, as a function of temperature (P = 1 bar for T < 100 °C; P = P_{sat} for T ≥ 100 °C)

T (°C)	0.01	25	50	75	100	125	150	175
log K	−6.2281	−5.0910	−4.1187	−3.2723	−2.5253	−1.8584	−1.2567	−0.7088
T (°C)	175	200	225	250	275	300	325	350
log K	−0.5283	−0.0348	0.4221	0.8492	1.2512	1.6292	1.9758	2.2653

$$T_{Na-Ca,Wrk}\,(^{\circ}C) = 0.1490 \cdot \chi^3 + 3.9720 \cdot \chi^2 + 52.5016 \cdot \chi + 207.866, \quad \text{for } T < 175\,^{\circ}C \tag{8.43}$$

$$T_{Na-Ca,Wrk}\,(^{\circ}C) = 0.5157 \cdot \chi^3 + 3.6160 \cdot \chi^2 + 52.5514 \cdot \chi + 199.312, \quad \text{for } 175 < T < 325\,^{\circ}C \tag{8.44}$$

In Eqs. (8.43) and (8.44), $\chi = \log\left(a^2_{Na^+}/a_{Ca^{2+}}\right)$.

To be noted that the log K values of reaction (8.8), which are shown in Table 8.9, are function of temperature only because adularia does not take part to this reaction. The average value of log a_{Wrk}, −0.0501, is considered in Eqs. (8.41)–(8.44). Due to deviations of wairakite activity of $+1\sigma$ and -1σ from the average value, log a_{Wrk} assumes the values −0.0083 and −0.0964, respectively. These variations in log a_{Wrk}, determine differences of ±0.5 to ±1.4 °C, ±1.3 to ±2.7 °C, ±0.9 to ±2.2 °C, and ±2.1 to ±3.4 °C in the temperatures calculated by means of Eqs. (8.41), (8.42), (8.43), and (8.44), respectively, representing their nominal uncertainties.

8.4.5 *The Theoretical K–Ca and Na–Ca Calcite f_{CO_2}-Indicators*

For reactions (8.9) and (8.10), comprising calcite, the following four equations were obtained on the basis of Eqs. (8.23) and (8.24) and the log $K_{K-Ca,Cal}$ and log $K_{Na-Ca,Cal}$ values given in Tables 8.10 and 8.11, respectively

$$\log f_{CO_2,K-Ca} = \log\left(\frac{a^2_{K^+}}{a_{Ca^{2+}}}\right) + \left(\frac{1737.2}{T} - 2.2291\right) \cdot Z - \frac{49{,}305}{T^2} - \frac{1506.0}{T} + 3.5345 \tag{8.45}$$

$$\log f_{CO_2,K-Ca} = \log\left(\frac{a^2_{K^+}}{a_{Ca^{2+}}}\right) + \left(\frac{1736.3}{T} - 2.2270\right) \cdot Z$$

Table 8.10 Logarithm of the thermodynamic equilibrium constant of reaction (8.9), log $K_{K-Ca,Cal}$, for pure calcite, adularia, muscovite, and chalcedony (for T < 175 °C) or quartz/chalcedony (for T > 175 °C), as a function of the temperature and the ordering parameter Z of adularia (P = 1 bar for T < 100 °C; $P = P_{sat}$ for T ≥ 100 °C)

T (°C)	Z = 1	Z = 0.9	Z = 0.8	Z = 0.7	Z = 0.6	Z = 0.5	Z = 0.4	Z = 0.3	Z = 0.2	Z = 0.1	Z = 0
0.01	−1.6797	−1.2667	−0.8536	−0.4405	−0.0275	0.3856	0.7987	1.2117	1.6248	2.0379	2.4509
25	−1.7055	−1.3457	−0.9860	−0.6262	−0.2665	0.0933	0.4531	0.8128	1.1726	1.5323	1.8921
50	−1.7285	−1.4138	−1.0992	−0.7845	−0.4698	−0.1551	0.1596	0.4742	0.7889	1.1036	1.4183
75	−1.7469	−1.4709	−1.1948	−0.9187	−0.6426	−0.3666	−0.0905	0.1856	0.4617	0.7378	1.0138
100	−1.7595	−1.5168	−1.2742	−1.0315	−0.7889	−0.5462	−0.3036	−0.0609	0.1817	0.4243	0.6670
125	−1.7653	−1.5519	−1.3384	−1.1250	−0.9116	−0.6982	−0.4848	−0.2714	−0.0580	0.1555	0.3689
150	−1.7635	−1.5758	−1.3882	−1.2005	−1.0129	−0.8253	−0.6376	−0.4500	−0.2624	−0.0747	0.1129
175	−1.7535	−1.5887	−1.4240	−1.2592	−1.0945	−0.9298	−0.7650	−0.6003	−0.4356	−0.2708	−0.1061
175	−1.2122	−1.0474	−0.8827	−0.7179	−0.5532	−0.3885	−0.2237	−0.0590	0.1057	0.2705	0.4352
200	−1.2223	−1.0780	−0.9337	−0.7895	−0.6452	−0.5010	−0.3567	−0.2124	−0.0682	0.0761	0.2203
225	−1.2201	−1.0942	−0.9684	−0.8425	−0.7167	−0.5909	−0.4650	−0.3392	−0.2134	−0.0875	0.0383
250	−1.2043	−1.0950	−0.9858	−0.8767	−0.7675	−0.6583	−0.5491	−0.4400	−0.3308	−0.2216	−0.1124
275	−1.1740	−1.0798	−0.9858	−0.8917	−0.7977	−0.7036	−0.6096	−0.5156	−0.4215	−0.3275	−0.2335
300	−1.1323	−1.0517	−0.9715	−0.8913	−0.8111	−0.7309	−0.6506	−0.5704	−0.4902	−0.4100	−0.3298
325	−1.0911	−1.0232	−0.9556	−0.8880	−0.8205	−0.7529	−0.6853	−0.6178	−0.5502	−0.4826	−0.4151
350	−1.0824	−1.0259	−0.9700	−0.9141	−0.8582	−0.8022	−0.7463	0.6904	0.6345	0.5785	0.5226

Table 8.11 Logarithm of the thermodynamic equilibrium constant of reaction (8.10), log $K_{Na–Ca,Cal}$, for pure calcite, adularia, albite, muscovite, and chalcedony (for T < 175 °C) or quartz/chalcedony (for T > 175 °C), as a function of the temperature and the ordering parameter Z of adularia (P = 1 bar for T < 100 °C; P = P_{sat} for T ≥ 100 °C)

T (°C)	Z = 1	Z = 0.9	Z = 0.8	Z = 0.7	Z = 0.6	Z = 0.5	Z = 0.4	Z = 0.3	Z = 0.2	Z = 0.1	Z = 0
0.01	5.3085	5.4462	5.5839	5.7216	5.8593	5.9970	6.1347	6.2724	6.4101	6.5477	6.6854
25	4.3741	4.4940	4.6139	4.7338	4.8537	4.9737	5.0936	5.2135	5.3334	5.4533	5.5733
50	3.6008	3.7057	3.8106	3.9155	4.0204	4.1253	4.2302	4.3350	4.4399	4.5448	4.6497
75	2.9590	3.0510	3.1430	3.2351	3.3271	3.4191	3.5111	3.6032	3.6952	3.7872	3.8792
100	2.4242	2.5051	2.5860	2.6669	2.7478	2.8287	2.9095	2.9904	3.0713	3.1522	3.2331
125	1.9776	2.0487	2.1198	2.1910	2.2621	2.3333	2.4044	2.4755	2.5467	2.6178	2.6889
150	1.6043	1.6669	1.7294	1.7920	1.8545	1.9171	1.9796	2.0422	2.1047	2.1672	2.2298
175	1.2930	1.3479	1.4028	1.4577	1.5126	1.5676	1.6225	1.6774	1.7323	1.7872	1.8421
175	1.8343	1.8892	1.9441	1.9990	2.0539	2.1089	2.1638	2.2187	2.2736	2.3285	2.3834
200	1.5470	1.5951	1.6431	1.6912	1.7393	1.7874	1.8355	1.8836	1.9317	1.9797	2.0278
225	1.3081	1.3501	1.3920	1.4340	1.4759	1.5179	1.5598	1.6018	1.6437	1.6856	1.7276
250	1.1123	1.1487	1.1851	1.2215	1.2579	1.2943	1.3307	1.3671	1.4035	1.4398	1.4762
275	0.9546	0.9861	1.0174	1.0487	1.0801	1.1114	1.1428	1.1741	1.2055	1.2368	1.2682
300	0.8281	0.8550	0.8817	0.9084	0.9352	0.9619	0.9887	1.0154	1.0421	1.0689	1.0956
325	0.7187	0.7414	0.7639	0.7864	0.8090	0.8315	0.8540	0.8765	0.8990	0.9216	0.9441
350	0.5951	0.6140	0.6326	0.6512	0.6699	0.6885	0.7072	0.7258	0.7445	0.7631	0.7817

$$-\frac{780{,}426}{T^2}+\frac{1523.2}{T}-0.1345 \qquad (8.46)$$

$$\log f_{CO_2,Na-Ca}=\log\left(\frac{a^2_{Na^+}}{a_{Ca^{2+}}}\right)+\left(\frac{579.07}{T}-0.7430\right)\cdot Z$$
$$-\frac{258{,}609}{T^2}-\frac{1879.8}{T}+3.4685 \qquad (8.47)$$

$$\log f_{CO_2,Na-Ca}=\log\left(\frac{a^2_{Na^+}}{a_{Ca^{2+}}}\right)+\left(\frac{578.72}{T}-0.7423\right)\cdot Z$$
$$-\frac{1{,}081{,}929}{T^2}+\frac{1648.6}{T}-0.8608 \qquad (8.48)$$

Equations (8.45) and (8.47) are valid below 175 °C, whereas Eqs. (8.46) and (8.48) hold true from 175 to 350 °C. Shifts of muscovite activity of $\pm 1\sigma$ from the mean value determine differences of 0.09–0.12 log-units in the log f_{CO_2} values calculated by means of Eqs. (8.45)–(8.48), irrespective of temperature.

8.5 Derivation of the Theoretical f_{CO_2}-Temperature Functions Controlled by Equilibrium Coexistence of a Ca–Al-Silicate and Calcite

Adopting the same approach described in Sect. 8.4, the log K of reactions (8.11)–(8.14) for pure solid phases were fitted against both the absolute temperature reciprocal and the ordering parameter of adularia, considering separately the data below 175 °C, with chalcedony controlling undissociated SiO_2, and those above 175 °C, with quartz/chalcedony governing undissociated SiO_2. Each regression equation and the average activities of pertinent solid phases were then inserted into Eqs. (8.25)–(8.28). These equations were solved with respect to the logarithm of CO_2 fugacity, thus obtaining four f_{CO_2}-temperature functions, one for each Ca–Al-silicate. Again, the uncertainties brought about by variations in the activities of relevant components in the solid solutions of interest were assessed taking into account the average activities plus one standard deviation and the average activities minus one standard deviation.

8.5.1 The f_{CO_2}-Temperature Functions Fixed by Equilibrium Coexistence of Laumontite and Calcite

The log f_{CO_2} values controlled by laumontite/calcite equilibrium coexistence [see reaction (8.11) and Eq. (8.25)] depend on the absolute temperature reciprocal and the ordering parameter Z of hydrothermal adularia as described by the following relations:

$$\log f_{CO_2} = \left(\frac{579.063}{T} - 0.7430\right) \cdot Z - \frac{260{,}888}{T^2} - \frac{2806.6}{T} + 5.9407 \quad \text{for } T < 175\,^{\circ}C \tag{8.49}$$

$$\log f_{CO_2} = \left(\frac{578.708}{T} - 0.7422\right) \cdot Z - \frac{818{,}992}{T^2} - \frac{426.78}{T} + 3.0381 \quad \text{for } 175 < T < 325\,^{\circ}C. \tag{8.50}$$

Equations (8.49) and (8.50) are based on the log $K_{Lmt\text{-}Cal}$ values shown in Table 8.12. Deviations of muscovite and laumontite activities of $\pm 1\sigma$ from the mean values bring about differences of 0.079–0.104 log-units in the log f_{CO_2} values calculated by means of Eqs. (8.49) and (8.50), irrespective of temperature.

8.5.2 The f_{CO_2}-Temperature Functions Fixed by Equilibrium Coexistence of Clinozoisite and Calcite

The log f_{CO_2} values governed by clinozoisite/calcite equilibrium coexistence [see reaction (8.12) and Eq. (8.26)] vary with the absolute temperature inverse and the ordering parameter Z of hydrothermal adularia as defined by the following equations:

$$\log f_{CO_2} = \left(\frac{434.321}{T} - 0.5573\right) \cdot Z - \frac{5493.1}{T} + 10.3015 \quad \text{for } T < 175\,^{\circ}C \tag{8.51}$$

$$\log f_{CO_2} = \left(\frac{434.061}{T} - 0.5568\right) \cdot Z - \frac{5612.4}{T} + 10.4368 \quad \text{for } 175 < T < 325\,^{\circ}C. \tag{8.52}$$

Equations (8.51) and (8.52) were derived from the log $K_{Czo\text{-}Cal}$ values listed in Table 8.13. Shifts of clinozoisite and muscovite activity of $\pm 1\sigma$ from the mean values cause differences of 0.053–0.070 log-units in the log f_{CO_2} values calculated using Eqs. (8.51) and (8.52), irrespective of temperature.

Table 8.12 Logarithm of the thermodynamic equilibrium constant of reaction (8.11), log $K_{Lmt\text{-}Cal}$, for pure laumontite, calcite, adularia, muscovite, and chalcedony (for T < 175 °C) or quartz/chalcedony (for T > 175 °C), as a function of the temperature and the ordering parameter Z of adularia (P = 1 bar for T < 100 °C; P = P_{sat} for T ≥ 100 °C)

T (°C)	Z = 1	Z = 0.9	Z = 0.8	Z = 0.7	Z = 0.6	Z = 0.5	Z = 0.4	Z = 0.3	Z = 0.2	Z = 0.1	Z = 0
0.01	−6.2868	−6.4245	−6.5622	−6.6999	−6.8376	−6.9753	−7.1130	−7.2506	−7.3883	−7.5260	−7.6637
25	−5.0556	−5.1756	−5.2955	−5.4154	−5.5353	−5.6552	−5.7752	−5.8951	−6.0150	−6.1349	−6.2548
50	−4.0394	−4.1443	−4.2492	−4.3541	−4.4590	−4.5639	−4.6688	−4.7737	−4.8785	−4.9834	−5.0883
75	−3.1919	−3.2840	−3.3760	−3.4680	−3.5601	−3.6521	−3.7441	−3.8361	−3.9282	−4.0202	−4.1122
100	−2.4791	−2.5599	−2.6408	−2.7217	−2.8026	−2.8835	−2.9644	−3.0452	3.1261	3.2070	−3.2879
125	−1.8750	−1.9462	−2.0173	−2.0884	−2.1596	−2.2307	−2.3019	−2.3730	−2.4441	−2.5153	−2.5864
150	−1.3604	−1.4229	−1.4855	−1.5480	−1.6105	−1.6731	−1.7356	−1.7982	−1.8607	−1.9233	−1.9858
175	−0.9199	−0.9748	−1.0297	−1.0846	−1.1396	−1.1945	−1.2494	−1.3043	−1.3592	−1.4141	−1.4690
175	−1.2808	−1.3357	−1.3906	−1.4455	−1.5004	−1.5553	−1.6103	−1.6652	−1.7201	−1.7750	−1.8299
200	−0.8835	−0.9316	−0.9797	−1.0278	−1.0759	−1.1240	−1.1720	−1.2201	−1.2682	−1.3163	−1.3644
225	−0.5409	−0.5829	−0.6249	−0.6668	−0.7088	−0.7507	−0.7926	−0.8346	−0.8765	−0.9185	−0.9604
250	−0.2452	−0.2816	−0.3180	−0.3544	−0.3908	−0.4272	−0.4636	−0.5000	−0.5364	−0.5728	−0.6092
275	0.0101	−0.0214	−0.0527	−0.0840	−0.1154	−0.1467	−0.1781	−0.2094	−0.2408	−0.2721	−0.3035
300	0.2300	0.2032	0.1764	0.1497	0.1229	0.0962	0.0695	0.0427	0.0160	−0.0108	−0.0375
325	0.4188	0.3961	0.3736	0.3511	0.3286	0.3060	0.2835	0.2610	0.2385	0.2159	0.1934
350	0.5797	0.5609	0.5423	0.5236	0.5050	0.4863	0.4677	0.4491	0.4304	0.4118	0.3931

Table 8.13 Logarithm of the thermodynamic equilibrium constant of reaction (8.12), log $K_{Czo\text{-}Cal}$, for pure clinozoisite, calcite, adularia, muscovite, and chalcedony (for T < 175 °C) or quartz/chalcedony (for T > 175 °C), as a function of the temperature and the ordering parameter Z of adularia (P = 1 bar for T < 100 °C; $P = P_{sat}$ for T ≥ 100 °C)

T (°C)	Z = 1	Z = 0.9	Z = 0.8	Z = 0.7	Z = 0.6	Z = 0.5	Z = 0.4	Z = 0.3	Z = 0.2	Z = 0.1	Z = 0
0.01	−8.7058	−8.8090	−8.9123	−9.0156	−9.1188	−9.2221	−9.3254	−9.4286	−9.5319	−9.6352	−9.7384
25	−7.1625	−7.2524	−7.3424	−7.4323	−7.5223	−7.6122	−7.7021	−7.7921	−7.8820	−7.9720	−8.0619
50	−5.8536	−5.9323	−6.0109	−6.0896	−6.1683	−6.2470	−6.3256	−6.4043	−6.4830	−6.5616	−6.6403
75	−4.7300	−4.7990	−4.8680	−4.9370	−5.0061	−5.0751	−5.1441	−5.2131	−5.2821	−5.3512	−5.4202
100	−3.7550	−3.8157	−3.8763	−3.9370	−3.9977	−4.0583	−4.1190	−4.1797	−4.2403	−4.3010	−4.3616
125	−2.9009	−2.9542	−3.0076	−3.0610	−3.1143	−3.1677	−3.2210	−3.2744	−3.3277	−3.3811	−3.4344
150	−2.1466	−2.1935	−2.2404	−2.2873	−2.3342	−2.3811	−2.4280	−2.4749	−2.5218	−2.5687	−2.6156
175	−1.4754	−1.5166	−1.5578	−1.5990	−1.6402	−1.6814	−1.7226	−1.7637	−1.8049	−1.8461	−1.8873
175	−1.6108	−1.6520	−1.6931	−1.7343	−1.7755	−1.8167	−1.8579	−1.8991	−1.9402	−1.9814	−2.0226
200	−1.0026	−1.0387	−1.0748	−1.1108	−1.1469	−1.1829	−1.2190	−1.2551	−1.2911	−1.3272	−1.3633
225	−0.4547	−0.4862	−0.5176	−0.5491	−0.5806	−0.6120	−0.6435	−0.6749	−0.7064	−0.7379	−0.7693
250	0.0417	0.0143	−0.0130	−0.0403	−0.0676	−0.0949	−0.1221	−0.1494	−0.1767	−0.2040	−0.2313
275	0.4936	0.4701	0.4466	0.4230	0.3995	0.3760	0.3525	0.3290	0.3055	0.2820	0.2585
300	0.9072	0.8871	0.8670	0.8470	0.8269	0.8068	0.7868	0.7667	0.7467	0.7266	0.7066
325	1.2874	1.2704	1.2535	1.2366	1.2197	1.2028	1.1859	1.1691	1.1522	1.1353	1.1184
350	1.6386	1.6245	1.6105	1.5965	1.5825	1.5685	1.5545	1.5406	1.5266	1.5126	1.4986

8.5.3 The f_{CO_2}-Temperature Functions Fixed by Equilibrium Coexistence of Prehnite and Calcite

The log f_{CO_2} values fixed by prehnite/calcite equilibrium coexistence [see reaction (8.13) and Eq. (8.27)] depend on the absolute temperature reciprocal and the ordering parameter Z of hydrothermal adularia according to the following relations:

$$\log f_{CO_2} = \left(\frac{289.504}{T} - 0.3714\right) \cdot Z - \frac{4369.1}{T} + 8.5567 \quad \text{for T} < 175\,^{\circ}\text{C} \tag{8.53}$$

$$\log f_{CO_2} = \left(\frac{289.321}{T} - 0.3710\right) \cdot Z - \frac{4372.6}{T} + 8.4261 \quad \text{for } 175 < \text{T} < 350\,^{\circ}\text{C}. \tag{8.54}$$

Equations (8.53) and (8.54) were obtained from the log $K_{Prh\text{-}Cal}$ values shown in Table 8.14. Deviations of prehnite and muscovite activity of $\pm 1\sigma$ from the average values cause differences of 0.007–0.010 log-units in the log f_{CO_2} values computed using Eqs. (8.53) and (8.54), irrespective of temperature.

8.5.4 The f_{CO_2}-Temperature Functions Fixed by Equilibrium Coexistence of Wairakite and Calcite

The log f_{CO_2} values constrained by wairakite/calcite equilibrium coexistence [see reaction (8.14) and Eq. (8.28)] depend on the absolute temperature inverse and the ordering parameter Z of hydrothermal adularia as described by the following equations:

$$\log f_{CO_2} = \left(\frac{579.066}{T} - 0.7430\right) \cdot Z - \frac{87{,}786}{T^2} - \frac{6737.3}{T} + 12.7898 \quad \text{for T} < 175\,^{\circ}\text{C} \tag{8.55}$$

$$\log f_{CO_2} = \left(\frac{578.696}{T} - 0.7422\right) \cdot Z - \frac{416{,}559}{T^2} - \frac{5422.6}{T} + 11.1262 \quad \text{for } 175 < \text{T} < 350\,^{\circ}\text{C}. \tag{8.56}$$

Equations (8.55) and (8.56) were derived from the log $K_{Wrk\text{-}Cal}$ values listed in Table 8.15. Shifts of wairakite and muscovite activity of $\pm 1\sigma$ from the average values cause differences of 0.051–0.072 log-units in the log f_{CO_2} values computed using Eqs. (8.55) and (8.56), independent of temperature.

Table 8.14 Logarithm of the thermodynamic equilibrium constant of reaction (8.13), log $K_{Prh\text{-}Cal}$, for pure prehnite, calcite, adularia, muscovite, and chalcedony (for T < 175 °C) or quartz/chalcedony (for T > 175 °C), as a function of the temperature and the ordering parameter Z of adularia (P = 1 bar for T < 100 °C; P = P_{sat} for T ≥ 100 °C)

T (°C)	Z = 1	Z = 0.9	Z = 0.8	Z = 0.7	Z = 0.6	Z = 0.5	Z = 0.4	Z = 0.3	Z = 0.2	Z = 0.1	Z = 0
0.01	−6.7200	−6.7888	−6.8577	−6.9265	−6.9954	−7.0642	−7.1331	−7.2019	−7.2707	−7.3396	−7.4084
25	−5.4646	−5.5245	−5.5845	−5.6444	−5.7044	−5.7644	−5.8243	−5.8843	−5.9442	−6.0042	−6.0642
50	−4.4034	−4.4559	−4.5083	−4.5608	−4.6132	−4.6657	−4.7181	−4.7706	−4.8230	−4.8754	−4.9279
75	−3.4957	−3.5418	−3.5878	−3.6338	−3.6798	−3.7258	−3.7718	−3.8178	−3.8638	−3.9099	−3.9559
100	−2.7110	−2.7515	−2.7919	−2.8324	−2.8728	−2.9132	−2.9537	−2.9941	−3.0346	−3.0750	−3.1155
125	−2.0261	−2.0617	−2.0973	−2.1328	−2.1684	−2.2040	−2.2396	−2.2751	−2.3107	−2.3463	−2.3818
150	−1.4235	−1.4548	−1.4861	−1.5173	−1.5486	−1.5799	−1.6112	−1.6424	−1.6737	−1.7050	−1.7363
175	−0.8894	−0.9169	−0.9443	−0.9718	−0.9992	−1.0267	−1.0542	−1.0816	−1.1091	−1.1365	−1.1640
175	−1.0247	−1.0522	−1.0797	−1.1071	−1.1346	−1.1620	−1.1895	−1.2169	−1.2444	−1.2719	−1.2993
200	−0.5411	−0.5651	−0.5892	−0.6132	−0.6373	−0.6613	−0.6853	−0.7094	−0.7334	−0.7575	−0.7815
225	−0.1069	−0.1279	−0.1489	−0.1699	−0.1908	−0.2118	−0.2328	−0.2537	−0.2747	−0.2957	−0.3167
250	0.2849	0.2667	0.2485	0.2303	0.2121	0.1939	0.1757	0.1575	0.1393	0.1212	0.1030
275	0.6405	0.6248	0.6091	0.5934	0.5777	0.5621	0.5464	0.5307	0.5150	0.4994	0.4837
300	0.9646	0.9512	0.9378	0.9244	0.9111	0.8977	0.8843	0.8710	0.8576	0.8442	0.8308
325	1.2615	1.2502	1.2389	1.2276	1.2164	1.2051	1.1938	1.1826	1.1713	1.1601	1.1488
350	1.5347	1.5253	1.5159	1.5066	1.4973	1.4880	1.4787	1.4693	1.4600	1.4507	1.4414

Table 8.15 Logarithm of the thermodynamic equilibrium constant of reaction (8.14), log $K_{Wrk\text{-}Cal}$, for pure wairakite, calcite, adularia, muscovite, and chalcedony (for T < 175 °C) or quartz/chalcedony (for T > 175 °C), as a function of the temperature and the ordering parameter Z of adularia (P = 1 bar for T < 100 °C; P = P_{sat} for T ≥ 100 °C)

T (°C)	Z = 1	Z = 0.9	Z = 0.8	Z = 0.7	Z = 0.6	Z = 0.5	Z = 0.4	Z = 0.3	Z = 0.2	Z = 0.1	Z = 0
0.01	−11.5366	−11.6743	−11.8120	−11.9497	−12.0874	−12.2250	−12.3627	−12.5004	−12.6381	−12.7758	−12.9135
25	−9.4651	−9.5850	−9.7049	−9.8249	−9.9448	−10.0647	−10.1846	−10.3045	−10.4245	−10.5444	−10.6643
50	−7.7195	−7.8244	−7.9292	−8.0341	−8.1390	−8.2439	−8.3488	−8.4537	−8.5586	−8.6635	−8.7684
75	−6.2313	−6.3233	−6.4154	−6.5074	−6.5994	−6.6914	−6.7835	−6.8755	−6.9675	−7.0595	−7.1516
100	−4.9496	−5.0305	−5.1113	−5.1922	−5.2731	−5.3540	−5.4349	−5.5158	−5.5966	−5.6775	−5.7584
125	−3.8360	−3.9071	−3.9783	−4.0494	−4.1206	−4.1917	−4.2628	−4.3340	−4.4051	−4.4762	−4.5474
150	−2.8610	−2.9236	−2.9861	−3.0487	−3.1112	−3.1738	−3.2363	−3.2989	−3.3614	−3.4239	−3.4865
175	−2.0017	−2.0567	−2.1116	−2.1665	−2.2214	−2.2763	−2.3312	−2.3861	−2.4410	−2.4960	−2.5509
175	−2.3626	−2.4175	−2.4724	−2.5274	−2.5823	−2.6372	−2.6921	−2.7470	−2.8019	−2.8568	−2.9117
200	−1.5818	−1.6299	−1.6780	−1.7261	−1.7742	−1.8223	−1.8703	−1.9184	−1.9665	−2.0146	−2.0627
225	−0.8860	−0.9279	−0.9699	−1.0118	−1.0538	−1.0957	−1.1377	−1.1796	−1.2216	−1.2635	−1.3054
250	−0.2631	−0.2995	−0.3359	−0.3723	−0.4087	−0.4451	−0.4815	−0.5179	−0.5543	−0.5907	−0.6271
275	0.2966	0.2652	0.2338	0.2025	0.1711	0.1398	0.1084	0.0771	0.0457	0.0144	−0.0170
300	0.8011	0.7742	0.7475	0.7208	0.6940	0.6673	0.6405	0.6138	0.5871	0.5603	0.5336
325	1.2571	1.2344	1.2119	1.1894	1.1669	1.1444	1.1218	1.0993	1.0768	1.0543	1.0318
350	1.6701	1.6513	1.6327	1.6140	1.5954	1.5767	1.5581	1.5394	1.5208	1.5022	1.4835

8.6 Plots of the K^2/Ca- and Na^2/Ca-Log Activity Ratios Versus the Absolute Temperature Inverse

The reservoir liquids of interest as well as the theoretical K–Ca and Na–Ca geothermometers and f_{CO_2}-indicators which were derived in Sect. 8.4 are shown in the diagrams of Figs. 8.1, 8.2, 8.3, 8.4, 8.5 and 8.6. The reservoir liquids are indicated by the usual symbols, whereas the K–Ca and Na–Ca theoretical geoindicators are represented by lines of different color, namely moss-green for the laumontite geothermometers, green for the clinozoisite geothermometers, orange for the prehnite geothermometers, olive for the wairakite geothermometers, and grey for the calcite f_{CO_2}-indicators, whose lines refer to different f_{CO_2} values, as specified.

For the geoindicators involving the ordering parameter of adularia, Z was alternatively set at 0.30 (mean value -1σ, dashed lines) and 0.82 (mean value $+1\sigma$, solid lines) to bracket the range of Z values of most hydrothermal adularias in hypothetical equilibrium with the considered reservoir liquids (see Sect. 6.1). The activities of Ca-endmembers in Ca–Al-silicates and of muscovite in illite were assumed equal to the average values for the hydrothermal solid solutions from active geothermal systems (see Sect. 8.2), like in the derivation of the geoindicators (Sect. 8.4).

For the calcite f_{CO_2}-indicator, at the same f_{CO_2} value, the theoretical K^2/Ca-log activity ratios of the aqueous solution for $Z = 0.30$ and $Z = 0.82$ differ by 1.26 log-units at 100 °C, 0.75 log-units at 200 °C, and 0.42 log-units at 300 °C, whereas the differences in the theoretical Na^2/Ca-log activity ratios are 1/3 of previous figures,

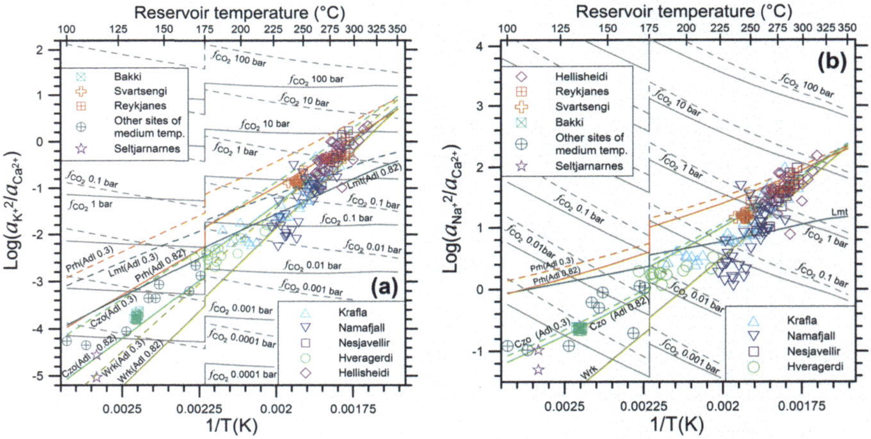

Fig. 8.1 Logarithm of the **a** K^2/Ca and **b** Na^2/Ca activity ratios of the selected reservoir liquids from the geothermal fields of Iceland as a function of the aquifer temperature reciprocal. Also shown are the theoretical log activity ratios fixed by mineral-solution equilibria, for average activities of relevant solid phases. Different f_{CO_2} values, as indicated, were considered for the reactions involving calcite (gray lines). The ordering parameter of adularia was set at 0.30 (dashed lines) and 0.82 (solid lines), to bracket the range of Z values of adularia apparently in equilibrium with the reservoir liquids of interest

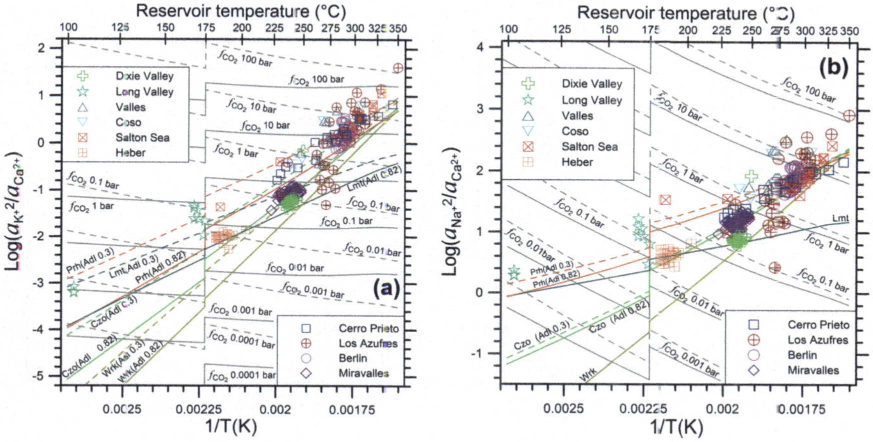

Fig. 8.2 Logarithm of the **a** K^2/Ca and **b** Na^2/Ca activity ratios of the selected reservoir liquids from the geothermal fields of Northern and Central America as a function of the aquifer temperature reciprocal. Also shown are the theoretical log activity ratios fixed by mineral-solution equilibria, for average activities of relevant solid phases. Different f_{CO_2} values, as indicated, were considered for the reactions involving calcite (gray lines). The ordering parameter of adularia was set at 0.30 (dashed lines) and 0.82 (solid lines), to bracket the range of Z values of adularia apparently in equilibrium with the reservoir liquids of interest

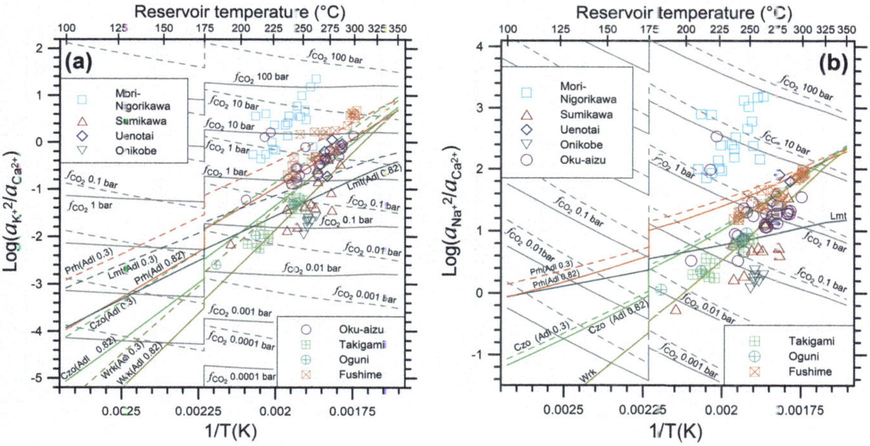

Fig. 8.3 Logarithm of the **a** K^2/Ca and **b** Na^2/Ca activity ratios of the selected reservoir liquids from the geothermal fields of Japan as a function of the aquifer temperature reciprocal. Also shown are the theoretical log activity ratios fixed by mineral-solution equilibria, for average activities of relevant solid phases. Different f_{CO_2} values, as indicated, were considered for the reactions involving calcite (gray lines). The ordering parameter of adularia was set at 0.30 (dashed lines) and 0.82 (solid lines), to bracket the range of Z values of adularia apparently in equilibrium with the reservoir liquids of interest

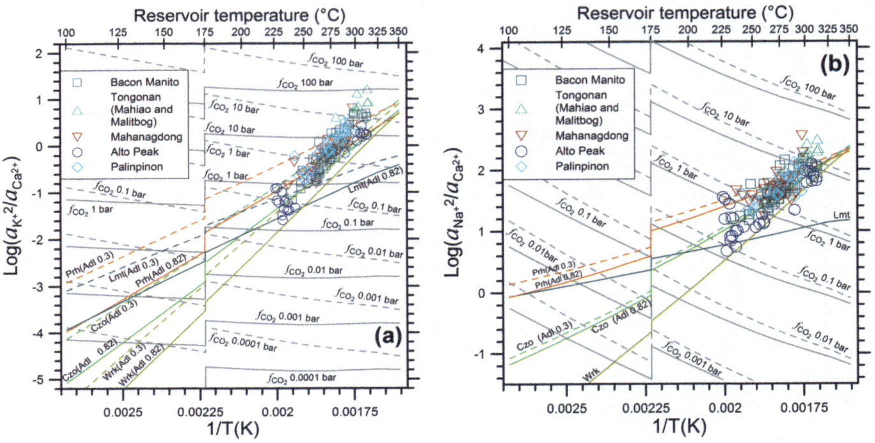

Fig. 8.4 Logarithm of the **a** K^2/Ca and **b** Na^2/Ca activity ratios of the selected reservoir liquids from the geothermal fields of the Philippines as a function of the aquifer temperature reciprocal. Also shown are the theoretical log activity ratios fixed by mineral-solution equilibria, for average activities of relevant solid phases. Different f_{CO_2} values, as indicated, were considered for the reactions involving calcite (gray lines). The ordering parameter of adularia was set at 0.30 (dashed lines) and 0.82 (solid lines), to bracket the range of Z values of adularia apparently in equilibrium with the reservoir liquids of interest

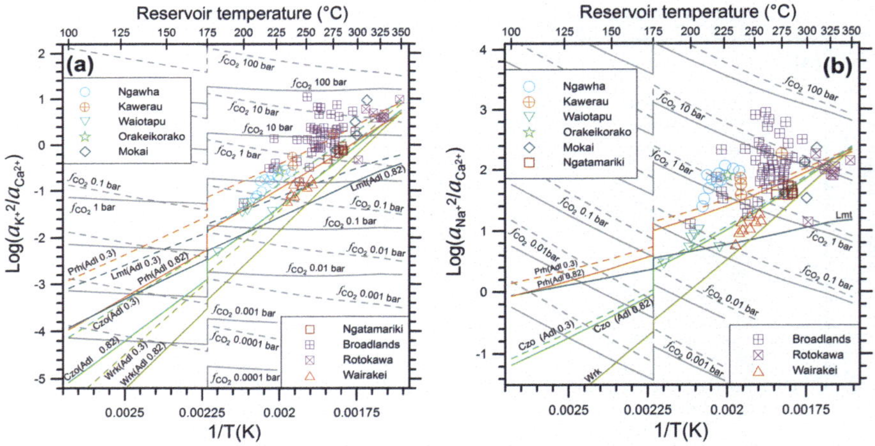

Fig. 8.5 Logarithm of the **a** K^2/Ca and **b** Na^2/Ca activity ratios of the selected reservoir liquids from the geothermal fields of New Zealand as a function of the aquifer temperature reciprocal. Also shown are the theoretical log activity ratios fixed by mineral-solution equilibria, for average activities of relevant solid phases. Different f_{CO_2} values, as indicated, were considered for the reactions involving calcite (gray lines). The ordering parameter of adularia was set at 0.30 (dashed lines) and 0.82 (solid lines), to bracket the range of Z values of adularia apparently in equilibrium with the reservoir liquids of interest

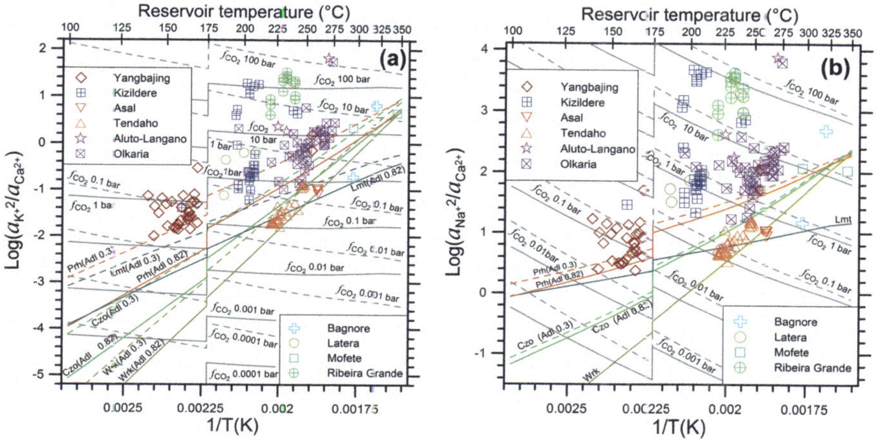

Fig. 8.6 Logarithm of the **a** K^2/Ca and **b** Na^2/Ca activity ratios of the selected reservoir liquids from miscellaneous geothermal fields as a function of the aquifer temperature reciprocal. Also shown are the theoretical log activity ratios fixed by mineral-solution equilibria, for average activities of relevant solid phases. Different f_{CO_2} values, as indicated, were considered for the reactions involving calcite (gray lines). The ordering parameter of adularia was set at 0.30 (dashed lines) and 0.82 (solid lines), to bracket the range of Z values of adularia apparently in equilibrium with the reservoir liquids of interest

being 0.42 log-units at 100 °C, 0.25 log-units at 200 °C, and 0.14 log-units at 300 °C. This different impact of adularia order-disorder is due to its different stoichiometric coefficient, which is 3 in the K–Ca exchange reaction (8.9) controlling the K^2/Ca activity ratio, but is 1 in Na–Ca exchange reaction (8.10), governing the Na^2/Ca activity ratio.

Likewise, for the reactions comprising Ca–Al silicates, adularia order-disorder has a greater impact on the theoretical K^2/Ca-log activity ratios than on the corresponding Na^2/Ca-log activity ratios. Again, these differences are related to the higher stoichiometric coefficient of adularia in the K–Ca exchange reactions (8.1), (8.2), (8.3), and (8.4), compared to the Na–Ca exchange reactions (8.5), (8.6), (8.7), and (8.8), respectively. In particular, the theoretical Na^2/Ca-log activity ratios governed by the Na–Ca exchange reactions (8.5) and (8.8), involving laumontite and wairakite, respectively, do not depend on Z because adularia does not participate to these reactions.

In the plots of the K^2/Ca log-activity ratio versus the aquifer temperature reciprocal as well as in the diagrams of the Na^2/Ca log-activity ratio versus the aquifer temperature inverse, the lines of the wairakite, clinozoisite and prehnite geothermometers converge gradually with increasing temperatures and diverge progressively with decreasing temperatures, whereas the lines of the laumontite geothermometers have lower slope and cross the lines of the wairakite, clinozoisite and prehnite geothermometers. Consequently, for the reservoir liquids of medium-low temperature (in the range 100–200 °C approximately), the attainment of equilibrium with either prehnite or clinozoisite or wairakite results in substantially different K^2/Ca and Na^2/Ca log-activity ratios, at any given temperature and Z value of adularia. In other words, in principle, it is possible to establish if a given low-temperature reservoir liquid is in equilibrium with either prehnite or clinozoisite or wairakite. However, there are some ambiguities owing to the possible attainment of equilibrium with laumontite instead of prehnite or clinozoisite.

In contrast, the high-temperature reservoir liquids may appear to be in equilibrium (or close to it) with two or even three Ca–Al-silicates at the same time, which is obviously an impossible condition because it would violate the Gibbs' phase rule. This apparent multiple-equilibrium condition is due to the small differences in the Gibbs free energies and log K values of the K–Ca and Na–Ca exchange reactions involving prehnite, clinozoisite, and wairakite above 200 °C approximately. Furthermore, the differences between the Gibbs free energies of these reactions become smaller and smaller with increasing temperature.

The reservoir liquids of interest can be separated into three different groups based on their position with respect to the lines of the Ca–Al-silicate geothermometers in the diagrams of Figs. 8.1, 8.2, 8.3, 8.4, 8.5 and 8.6, namely those situated along these lines, those that are found above these lines and those that are located below these lines.

Most reservoir liquids from the geothermal fields of Iceland (Fig. 8.1), Northern and Central America (Fig. 8.2), and the Philippines (Fig. 8.4), as well as the Japanese reservoir liquids from Uenotai, Takigami, Oguni, and Fushime (Fig. 8.3), the New Zealand reservoir liquids from Waiotapu, Ngatamariki, Rotokawa, and Wairakei (Fig. 8.5), and those from Asal and Tendaho (Fig. 8.6) are positioned in the area below the dashed line of the prehnite geothermometers and above the solid lines of the wairakite and laumontite geothermometers (intersecting at ~250 °C), indicating that these reservoir liquids are in equilibrium with a Ca–Al-silicate or close to this condition. Therefore, the K^2/Ca and Na^2/Ca log-activity ratios of these reservoir liquids can be inserted into the geothermometric functions controlled by laumontite, Eqs. (8.29)–(8.32), clinozoisite, Eqs. (8.33)–(8.36), prehnite, Eqs. (8.37)–(8.40), and wairakite, Eqs. (8.41)–(8.44), to compute the temperatures possibly occurring in the geothermal reservoir.

Several reservoir liquids are situated above the dashed line of the prehnite geothermometers and are in apparent equilibrium with calcite, under the CO_2 fugacity values indicated by the grey lines. This is the case of all the reservoir liquids from Mori-Nigorikawa (Fig. 8.3), Ngawha, Orakeikorako, and Kawerau (Fig. 8.5), Kizildere, Ribeira Grande, Aluto-Langano, and Latera (Fig. 8.6). Moreover, most reservoir liquids from Long Valley and some from Valles, Coso, and Los Azufres (Fig. 8.2), a few from The Philippines geothermal fields (Fig. 8.4), several from Broadlands (Fig. 8.5), Yangbajing and Olkaria (Fig. 8.6) are also in apparent equilibrium with calcite. The word apparent is necessary, because the K^2/Ca and Na^2/Ca log-activity ratios of these reservoir liquids could be affected by Ca loss due to precipitation of calcite or other Ca-bearing solid phases. If this is not the case, the K^2/Ca and Na^2/Ca log-activity ratios of these reservoir liquids can be inserted into Eqs. (8.45)–(8.48) to calculate the f_{CO_2} presumably present in the geothermal aquifer.

Below the solid lines of the wairakite and laumontite geothermometers (intersecting at ~250 °C) there are only few sample points, including all the reservoir liquids from Onikobe, some from Sumikawa (Fig. 8.3), some from Namafjall (Fig. 8.1), and one from Los Azufres (Fig. 8.2). The low K^2/Ca and Na^2/Ca log-activity ratios of these aqueous solutions might be due to rock dissolution promoted by their initial acidity, causing preferential acquisition of Ca, as proposed by Truesdell and Nakanishi (2005) for Onikobe. Accepting this interpretation, these reservoir liquids are probably in disequilibrium with both Ca–Al-silicates and calcite and cannot be used to compute neither the temperature nor the CO_2 fugacity of the geothermal aquifer.

8.7 Plots of CO_2 Fugacity Versus the Absolute Temperature Inverse

The theoretical K–Ca and Na–Ca calcite f_{CO_2}-indicators, that is Eqs. (8.45)–(8.48) provide the same CO_2 fugacities, with deviations varying from 0.0001% to 1.7% with respect to the average f_{CO_2} values. These deviations increase with decreasing f_{CO_2} and are ascribable to numerical approximations. Therefore, it is permissible to average the results of the K–Ca and Na–Ca calcite f_{CO_2}-indicators. Strictly speaking, the use of these f_{CO_2}-indicators is not correct for the reservoir liquids in equilibrium with Ca–Al-silicates, because these reservoir liquids are not in equilibrium with calcite. Nevertheless, this action is tolerable, since it is likely that these reservoir liquids are not too far from calcite saturation.

The f_{CO_2} values of the selected reservoir liquids, obtained by averaging the results of the K–Ca and Na–Ca calcite f_{CO_2}-indicators, as well as the four theoretical f_{CO_2}-temperature functions controlled by equilibrium coexistence of a Ca–Al-silicate and calcite (which were derived in Sect. 8.5) are reported in the diagrams of CO_2 fugacity (on a logarithmic scale) versus the reservoir temperature inverse of Figs 8.7, 8.8 and 8.9. Also in these diagrams, the reservoir liquids are represented by the usual symbols, whereas the f_{CO_2}-temperature functions are indicated by lines of the same colors adopted in Figs. 8.1, 8.2, 8.3, 8.4, 8.5 and 8.6, namely moss-green for the laumontite-calcite functions, green for the clinozoisite-calcite functions, orange for the prehnite-calcite functions, and olive for the wairakite-calcite functions.

Again, the ordering parameter of adularia was assumed equal to 0.30 (dashed lines) or 0.82 (solid lines) to bracket the range of Z values of hydrothermal adularias apparently in equilibrium with the considered reservoir liquids (see Sect. 6.1). The activities of relevant endmembers in the solid solutions of interest, that is muscovite in illite, laumontite in laumontite/alkali-laumontite, clinozoisite in clinozoisite/epidote, prehnite in prehnite/ferri-prehnite, and wairakite in wairakite/analcime, were assumed equal to the average values for the hydrothermal minerals from active geothermal systems (see Sect. 8.2), like in the derivation of the theoretical f_{CO_2}-temperature functions controlled by equilibrium coexistence of a Ca-Al-silicate and calcite (Sect. 8.5).

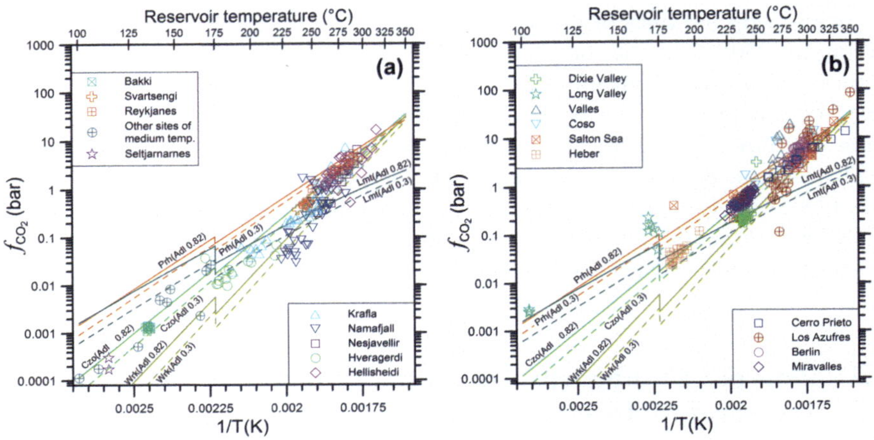

Fig. 8.7 Diagram of CO_2 fugacity (on a logarithmic scale) computed by averaging the results of the K–Ca and Na–Ca calcite f_{CO_2}-indicators versus the reservoir temperature inverse for the selected reservoir liquids from the geothermal fields of **a** Iceland and **b** Northern and Central America. Also shown are the lines of equilibrium coexistence of calcite and a Ca–Al-silicate, drawn for the ordering parameter of adularia Z = 0.82 (solid lines) and Z = 0.30 (dashed lines), assuming average activities of relevant solid phases

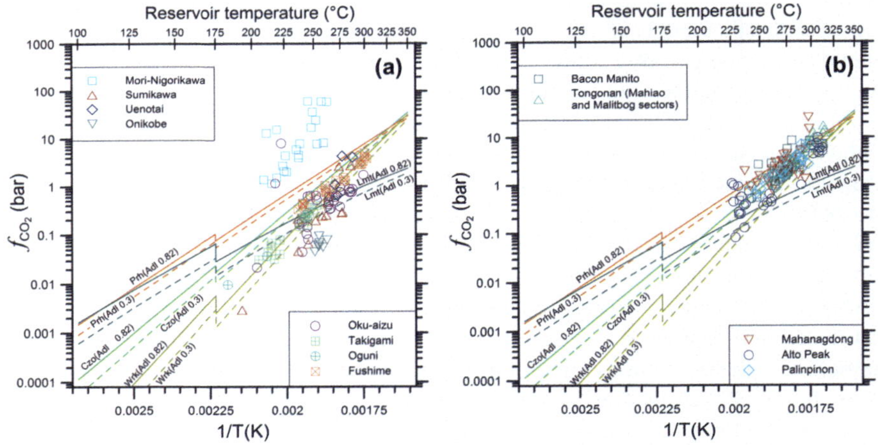

Fig. 8.8 Diagram of CO_2 fugacity (on a logarithmic scale) computed by averaging the results of the K–Ca and Na–Ca calcite f_{CO_2}-indicators versus the reservoir temperature inverse for the selected reservoir liquids from the geothermal fields of **a** Japan and **b** The Philippines. Also shown are the lines of equilibrium coexistence of calcite and a Ca–Al-silicate, drawn for the ordering parameter of adularia Z = 0.82 (solid lines) and Z = 0.30 (dashed lines), assuming average activities of relevant solid phases

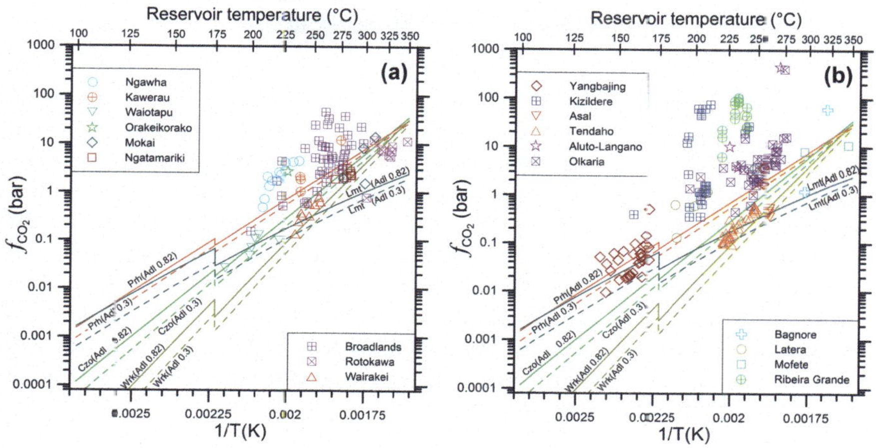

Fig. 8.9 Diagram of CO_2 fugacity (on a logarithmic scale) computed by averaging the results of the K–Ca and Na–Ca calcite f_{CO_2}-indicators versus the reservoir temperature inverse for the selected reservoir liquids from the geothermal fields of **a** New Zealand and **b** miscellaneous sites. Also shown are the lines of equilibrium coexistence of calcite and a Ca–Al-silicate, drawn for the ordering parameter of adularia Z = 0.82 (solid lines) and Z = 0.30 (dashed lines), assuming average activities of relevant solid phases

Similar to what was observed in the plots of Figs. 8.1, 8.2, 8.3, 8.4, 8.5 and 8.6, also in the diagrams of Figs. 8.7, 8.8 and 8.9, the log f_{CO_2}-temperature functions constrained by equilibrium coexistence of calcite and wairakite, calcite and clinozoisite, and calcite and prehnite converge progressively with increasing temperatures and diverge gradually with decreasing temperatures, whereas the two functions fixed by equilibrium coexistence of calcite and laumontite have lower slope and intersect the other calcite/Ca–Al-silicates functions.

Not surprisingly, sample points are distributed in the plots of Figs. 8.7, 8.8 and 8.9 similar to what is observed in Figs. 8.1, 8.2, 8.3, 8.4, 8.5 and 8.6. In fact:

1. The reservoir liquids situated along the lines of the Ca–Al-silicate geothermometers in the plots of Figs. 8.1, 8.2, 8.3, 8.4, 8.5 and 8.6 are found along the lines constrained by equilibrium coexistence of calcite and a Ca–Al-silicate in the diagrams of Figs. 8.7, 8.8 and 8.9, confirming their condition of equilibrium with one of the considered Ca–Al-silicates.
2. The reservoir liquids positioned above the lines of the Ca–Al-silicate geothermometers in the plots of Figs. 8.1, 8.2, 8.3, 8.4, 8.5 and 8.6 are encountered above the lines constrained by equilibrium coexistence of calcite and a Ca–Al-silicate in the diagrams of Figs. 8.7, 8.8 and 8.9, confirming their condition of equilibrium with calcite.
3. The few reservoir liquids located below the lines of the Ca–Al-silicate geothermometers, in the plots of Figs. 8.1, 8.2, 8.3, 8.4, 8.5 and 8.6 are found below the lines constrained by equilibrium coexistence of calcite and a Ca–Al-silicate

in the diagrams of Figs. 8.7, 8.8 and 8.9, being in disequilibrium with both Ca–Al-silicates and calcite.

All in all, the similarity in the distribution of sample points in the two series of diagrams is expected, but the additional information provided by the diagrams of Figs. 8.7, 8.8 and 8.9 is the distinction among: (a) The reservoir liquids in equilibrium with a Ca–Al-silicate, to which it is permissible to apply the theoretical K–Ca and Na–Ca geothermometers derived in Sect. 8.4, that is Eqs. (8.29)–(8.44), to compute the temperatures possibly occurring in the geothermal reservoir. (b) The reservoir liquids in equilibrium with calcite, which are suitable for using the theoretical K–Ca and Na–Ca calcite f_{CO_2}-indicators, i.e., Eqs. (8.45)–(8.48), to calculate the f_{CO_2} presumably present in the geothermal aquifer. (c) The reservoir liquids in disequilibrium with both Ca–Al-silicates and calcite, which cannot be used to compute neither the temperature nor the CO_2 fugacity of the geothermal aquifer.

8.8 Use of the K–Ca and Na–Ca Activity-Based Theoretical Geoindicators

The diagrams examined in Sects. 8.6 and 8.7 provide an useful overview. However, it is necessary to decide for each individual reservoir liquid if its K^2/Ca and Na^2/Ca log-activity ratios can be used to estimate the temperature or the CO_2 fugacity of the geothermal aquifer. To this purpose, it is advisable to compare the CO_2 fugacity computed by averaging the results of the K–Ca and Na–Ca calcite f_{CO_2}-indicators, that is Eqs. (8.45)–(8.48), with the maximum CO_2 fugacity given by the f_{CO_2}-temperature functions controlled by equilibrium coexistence of a Ca–Al-silicate and calcite, that is Eqs. (8.49)–(8.56). If the average CO_2 fugacity given by the K–Ca and Na–Ca calcite f_{CO_2}-indicators is higher than the maximum CO_2 fugacity of Ca–Al-silicate/calcite equilibrium coexistence, then the reservoir liquid can be assumed to be in saturation with calcite, and the computed CO_2 fugacity can be considered reliable and representative of the geothermal aquifer. If the opposite is true, then the reservoir liquid can be assumed to be in equilibrium with a Ca–Al silicate and the different theoretical K–Ca and Na–Ca geothermometers can be applied to it.

We tested this approach using the 1013 reservoir liquids of interest. Results are presented and discussed here below. First, the 23 reservoir liquids presumably in disequilibrium with both Ca–Al-silicates and calcite, being situated below the lines of the Ca–Al-silicate geothermometers, and the Sumikawa sample SM-2_78, with Ca concentration lower than detection limit, were excluded from further processing. The 24 excluded reservoir liquids comprise 8 entries of Namafjall, 1 of Los Azufres, 8 of Sumikawa, and 7 of Onikobe.

Among the remaining 990 reservoir liquids, 706 are suitable for using the theoretical K–Ca and Na–Ca geothermometers, whereas 283 are appropriate for utilizing the theoretical K–Ca and Na–Ca f_{CO_2}-indicators. The computed temperatures and related errors are discussed in Sect. 8.8.1, whereas the calculated CO_2 fugacities and associated uncertainties are presented in Sect. 8.8.2.

8.8.1 Temperatures Given by the Theoretical, Activity-Based K–Ca and Na–Ca Geothermometers and Related Uncertainties

For each reservoir liquid suitable for geothermometry, four K–Ca temperatures and four Na–Ca temperatures were computed. Among them, the K–Ca and Na–Ca temperatures closest to each aquifer temperature were adopted, maintaining the information on the Ca–Al-silicate mineral controlling the geothermometer(s) used to calculate these adopted temperatures. The adopted K–Ca and Na–Ca temperatures are contrasted with the aquifer temperature in the diagrams of Figs. 8.10, 8.11, 8.12, and 8.13, referring to the clinozoisite, wairakite, laumontite, and prehnite geothermometers.

As a whole, the error on the adopted K-Ca temperatures (i.e., the absolute value of their deviation from the aquifer temperature) ranges between 0.0 and 16.9 °C, with an average of 3.8 °C, a median of 3.0 °C and a standard deviation of 3.0 °C. The errors on the adopted K–Ca temperatures are ≤5 °C in 519 cases (73.5% of the total), >5 and ≤10 °C in 168 cases (23.8% of the total), and >10 °C in 19 cases (2.7% of the total).

The error on the adopted Na–Ca temperatures varies between 0 and 36.9 °C, with a mean of 6.9 °C, a median of 6.0 °C and a standard deviation of 5.3 °C. The errors on the adopted Na–Ca temperatures are ≤5 °C in 332 cases (47.0% of the total),

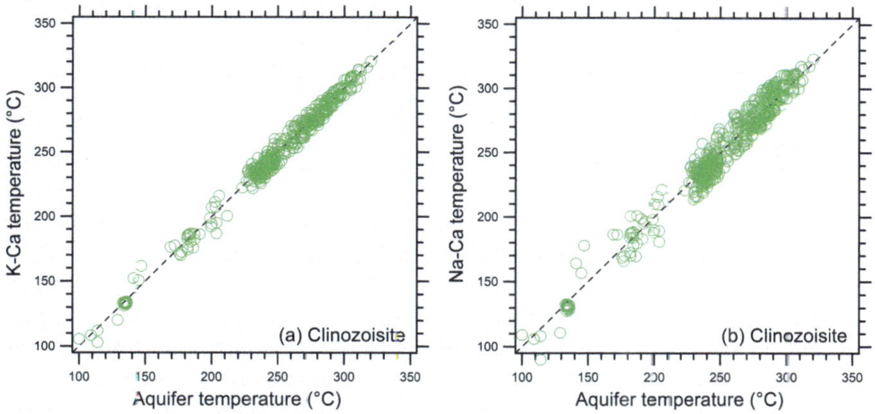

Fig. 8.10 Diagram of the aquifer temperature versus the adopted **a** K–Ca and **b** Na–Ca clinozoisite temperatures

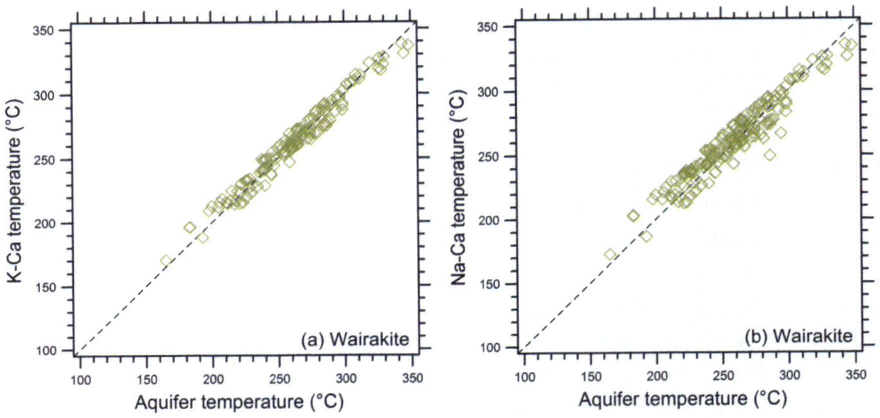

Fig. 8.11 Diagram of the aquifer temperature versus the adopted **a** K–Ca and **b** Na–Ca wairakite temperatures

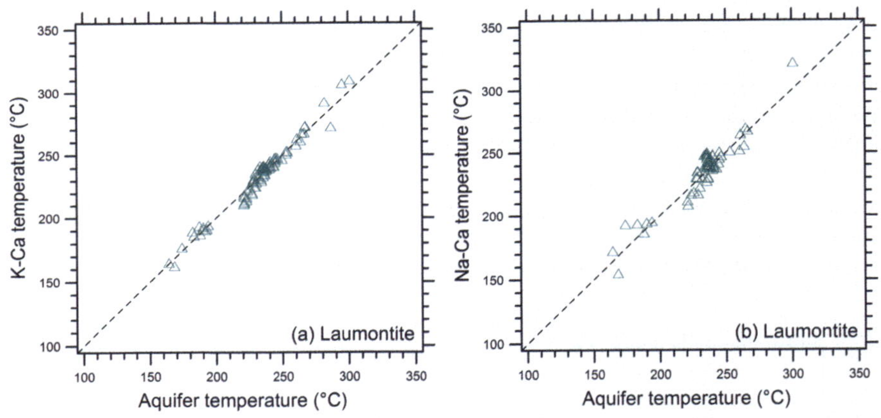

Fig. 8.12 Diagram of the aquifer temperature versus the adopted **a** K–Ca and **b** Na–Ca laumontite temperatures

>5 °C and ≤10 °C in 201 cases (28.5% of the total), >10 °C and ≤15 °C in 127 cases (18.0% of the total), and >15 °C in 46 cases (6.5% of the total).

The adopted K–Ca temperatures include 421 clinozoisite temperatures, 141 wairakite temperatures, 93 laumontite temperatures, and 51 prehnite temperatures. The adopted Na–Ca temperatures include 459 clinozoisite temperatures, 169 wairakite temperatures, 63 laumontite temperatures, and 15 prehnite temperatures. The concordant cases (i.e., the adopted K–Ca and Na–Ca temperatures refer to the same Ca–Al-silicate mineral) are 634, corresponding to 90% of the total, whereas the discordant cases (i.e., the adopted K–Ca and Na–Ca temperatures refer to distinct

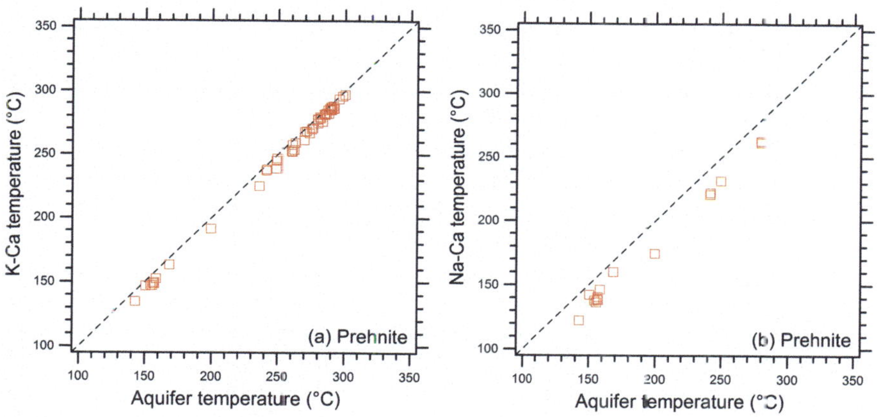

Fig. 8.13 Diagram of the aquifer temperature versus the adopted **a** K–Ca and **b** Na–Ca prehnite temperatures

Ca–Al-silicate minerals) are 72, equivalent to 10% of the total, which is an acceptably low value.

The 421 K–Ca clinozoisite temperatures vary from 103 to 321 °C, with an average of 248 °C, a median of 258 °C, and a standard deviation of 81 °C. The error on the K–Ca clinozoisite temperatures ranges between 0 and 17 °C, with a mean of 3.4 °C, a median of 6.0 °C, and a standard deviation of 7.1 °C.

The 141 K–Ca wairakite temperatures range from 170 to 338 °C with a mean of 262 °C, a median of 263 °C, and a standard deviation of 33 °C. The error on the K–Ca wairakite temperatures varies between 0 and 14 °C, with an average of 5.0 °C, a median of 5.0 °C, and a standard deviation of 3.5 °C.

The 93 K–Ca laumontite temperatures vary from 163 to 310 °C, with an average of 234 °C, a median of 239 °C, and a standard deviation of 25 °C. The error on the K–Ca laumontite temperatures ranges between 0 and 14 °C, with a mean of 3.2 °C, a median of 2.0 °C, and a standard deviation of 2.9 °C.

The 51 K–Ca prehnite temperatures range from 135 to 297 °C, with a mean of 249 °C, a median of 239 °C, and a standard deviation of 67 °C. The error on the K–Ca prehnite temperatures varies between 0 and 11 °C, with an average of 4.5 °C, a median of 6.5 °C, and a standard deviation of 2.1 °C.

The 459 Na–Ca clinozoisite temperatures vary from 90 to 323 °C, with an average of 250 °C, a median of 252 °C, and a standard deviation of 44 °C. The error on the Na–Ca clinozoisite temperatures ranges between 0 and 30 °C, with a mean of 6.5 °C, a median of 5.0 °C, and a standard deviation of 5.0 °C.

The 169 Na–Ca wairakite temperatures range from 173 to 335 °C, with mean and median of 261 °C and standard deviation of 30 °C. The error on Na–Ca wairakite temperatures varies between 0 and 37 °C, with an average of 7.8 °C, a median of 8.0 °C, and a standard deviation of 5.6 °C.

The 63 Na–Ca laumontite temperatures vary from 155 to 322 °C, with an average of 235 °C, a median of 239 °C, and a standard deviation of 24 °C. The error on Na–Ca laumontite temperatures ranges between 0 and 21 °C, with a mean of 5.8 °C, a median of 5.0 °C, and a standard deviation of 4.7 °C.

The 15 Na–Ca prehnite temperatures range from 123 to 263 °C, with a mean of 176 °C, a median of 147 °C, and a standard deviation of 50 °C. The error on Na–Ca prehnite temperatures varies between 8 and 25 °C, with an average of 16.5 °C, a median of 17.0 °C, and a standard deviation of 4.4 °C.

The average and median values of the K–Ca clinozoisite temperatures agree with the average and median values of the Na–Ca clinozoisite temperatures within a few degrees. Similarly, there is a good agreement between the average and median values of the K–Ca and Na–Ca temperatures for wairakite and laumontite as well. In contrast, the average and median values of the K–Ca prehnite temperatures are at variance with the mean and median values of the Na–Ca prehnite temperatures. This discrepancy is due, at least partly, to the different number of cases, 51 for the K–Ca prehnite temperatures versus 15 for the Na–Ca prehnite temperatures.

The higher number of adopted K–Ca and Na–Ca clinozoisite temperatures compared to the other Ca–Al silicate temperatures is in accordance with the widespread occurrence of hydrothermal epidote in active geothermal systems. Moreover, the K–Ca and Na–Ca laumontite, clinozoisite, and wairakite temperatures are in satisfactory agreement with the distinct distribution of these three hydrothermal minerals in active geothermal systems, where the stable Ca–Al-silicate is wairakite or epidote or prehnite at high temperatures, typically 200–300 °C, and laumontite at lower temperatures, as already recalled in Chaps. 4 and 5.

The errors on K–Ca and Na–Ca temperatures are partly explained by the deviation of the activities of relevant minerals from the average values, considering that deviations of $+1\sigma$ and -1σ from the mean value of the pertinent activity term cause uncertainties of 3.8, 2.7, 1.8, and 9.6 °C in the K–Ca temperatures of clinozoisite, wairakite, laumontite, and prehnite, respectively, whereas the uncertainties of this type on the corresponding Na–Ca temperatures are 5.3, 3.4, 5.0, and 17.9 °C, respectively.

Finally, the identification of the Ca–Al-silicate in equilibrium with each reservoir liquid may provide a qualitative indication on well permeability. In fact, according to Reyes (1990), wairakite is an indicator of high permeability, whereas prehnite and laumontite (if abundant) are indicators of poor permeability. Prehnite apparently requires little flow of geothermal fluids through the rocks to form as it appears to recrystallize easily, with increase in temperature, from the primary Fe–Mg minerals, such as pyroxene (Reyes 1990).

8.8.2 CO_2 Fugacities Given by the Theoretical, Activity-Based K–Ca and Na–Ca Calcite f_{CO_2}-Indicators and Related Uncertainties

The CO_2 fugacity was computed by means of the theoretical K–Ca and Na–Ca calcite f_{CO_2}-indicators for 283 reservoir liquids coming from the geothermal fields of Krafla, Namafjall, Nesjavellir, Hellisheidi, Dixie Valley, Long Valley, Valles, Coso, Salton Sea, Los Azufres, Berlin, Mori-Nigorikawa, Uenotai, Oku-aizu, Bacon Manito, Tongonan (Mahiao and Malitbog sectors), Mahanagdong, Alto Peak, Palinpinon, Ngawha, Kawerau, Orakeikorako, Mokai, Broadlands, Rotokawa, Yangbajing, Kizildere, Aluto-Langano, Olkaria, Bagnore, Latera, Mofete, and Ribeira Grande. All these reservoir liquids are presumably in equilibrium with calcite. Since the CO_2 fugacities given by the theoretical K–Ca and Na–Ca calcite f_{CO_2}-indicators are practically equal (see Sect. 8.7), the average of the two values was taken and plotted against both:

1. the CO_2 fugacities calculated by means of speciation calculations (Figs. 8.14a, 8.15a, 8.16a, and 8.17a), that is combining the analyses of the liquid and vapor phases separated at known pressure, temperature conditions, usually using the computer program WATCH, and
2. the CO_2 fugacities computed using the K–Ca f_{CO_2}-indicator of Giggenbach (1984; Figs. 8.14b, 8.15b, 8.16b, and 8.17b), which is discussed in Sect. 5.5.

The reasons of the deviations observed in these diagrams are discussed in Sect. 5.5.2 and are disregarded here to avoid unnecessary repetitions. The absolute value of the difference between the CO_2 fugacities given by the theoretical K–Ca

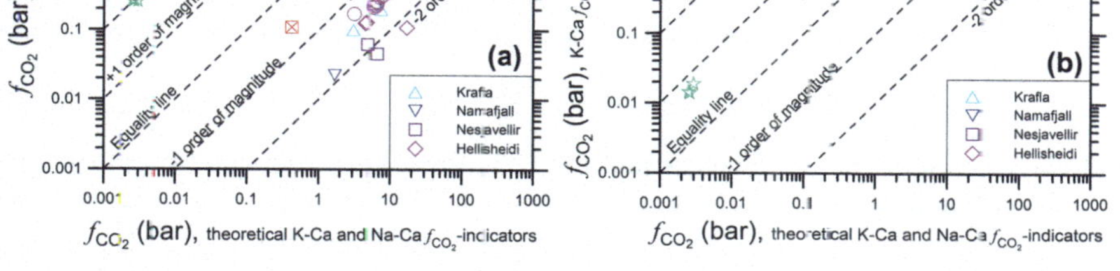

Fig. 8.14 Log-log diagram of the average CO_2 fugacity given by the theoretical K–Ca and Na–Ca calcite f_{CO_2}-indicators versus **a** the CO_2 fugacity obtained from speciation calculations and **b** the CO_2 fugacity computed using the K–Ca f_{CO_2}-indicator of Giggenbach (1984) for the geothermal fields of Iceland and Northern-Central America

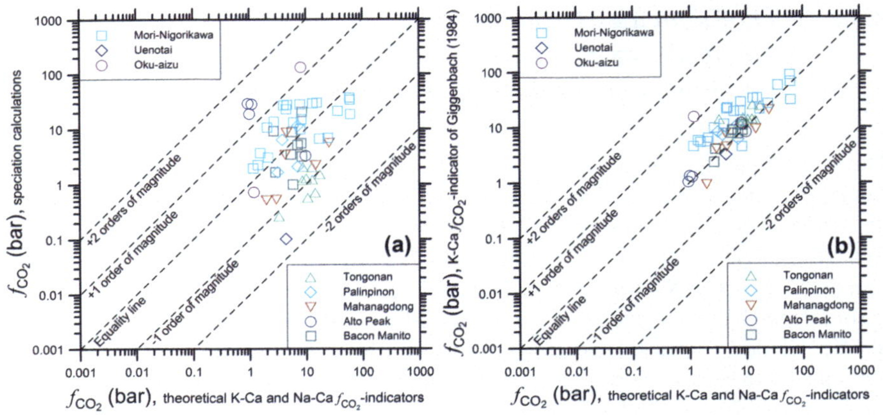

Fig. 8.15 Log-log diagram of the average CO_2 fugacity given by the theoretical K–Ca and Na–Ca calcite f_{CO_2}-indicators versus **a** the CO_2 fugacity obtained from speciation calculations and **b** the CO_2 fugacity computed using the K–Ca f_{CO_2}-indicator of Giggenbach (1984) for the geothermal fields of Japan and the Philippines

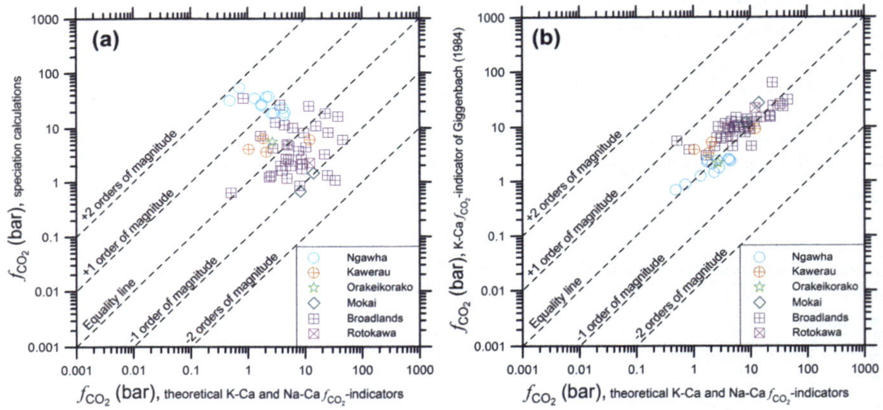

Fig. 8.16 Log-log diagram of the average CO_2 fugacity given by the theoretical K–Ca and Na–Ca calcite f_{CO_2}-indicators versus **a** the CO_2 fugacity obtained from speciation calculations and **b** the CO_2 fugacity computed using the K–Ca f_{CO_2}-indicator of Giggenbach (1984) for the geothermal fields of New Zealand

and Na–Ca calcite f_{CO_2}-indicators and the CO_2 fugacities calculated by means of speciation calculations ranges between 0.0093 and 2.19 log-units, with an average of 0.78 log-units, a median of 0.66 log-units, and a standard deviation of 0.54 log-units. The absolute value of these differences is ≤0.50 log-units in 110 cases (38.9% of the total), >0.50 and ≤1.00 log-units in 78 cases (27.6% of the total), >1.00 and ≤1.50 log-units in 63 cases (22.3% of the total) >1.50 and ≤2.00 log-units in 28 cases (9.9% of the total), and >2.00 log-units in 4 cases (1.4% of the total).

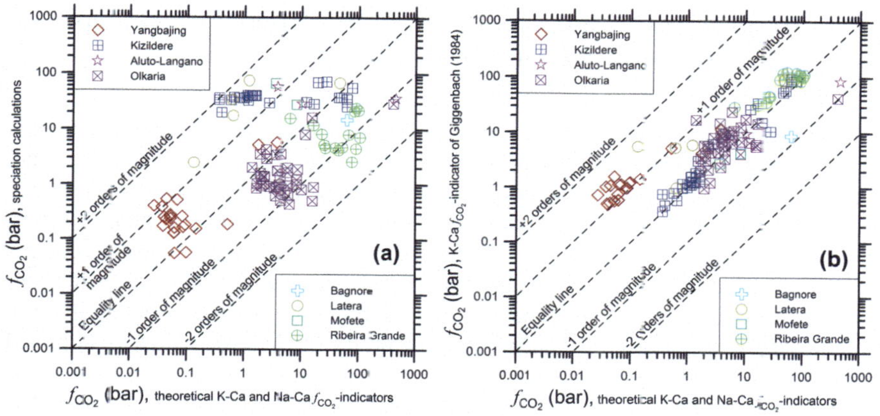

Fig. 8.17 Log-log diagram of the average CO_2 fugacity given by the theoretical K–Ca and Na–Ca calcite f_{CO_2}-indicators versus **a** the CO_2 fugacity obtained from speciation calculations and **b** the CO_2 fugacity computed using the K–Ca f_{CO_2}-indicator of Giggenbach (1984) for the miscellaneous geothermal fields

The absolute value of the difference between the CO_2 fugacities given by the theoretical K–Ca and Na–Ca calcite f_{CO_2}-indicators and the CO_2 fugacities computed using the K–Ca f_{CO_2}-indicator of Giggenbach (1984) varies between 0.00029 and 1.62 log-units, with a mean of 0.33 log-units, a median of 0.24 log-units, and a standard deviation of 0.32 log-units. The absolute value of these differences is ≤0.50 log-units in 228 cases (80.6% of the total), >0.50 and ≤1.00 log-units in 34 cases (12.0% of the total), >1.00 and ≤1.50 log-units in 20 cases (7.1% of the total), and >1.50 log-units in 1 case (0.4% of the total).

The satisfactory agreement between the CO_2 fugacities calculated by means of the theoretical K–Ca and Na–Ca calcite f_{CO_2}-indicators and the CO_2 fugacities given by the K–Ca f_{CO_2}-indicator of Giggenbach (1984) is not surprising. In fact, the theoretical K–Ca calcite f_{CO_2}-indicator developed in this work is an improved version of the K–Ca f_{CO_2}-indicator of Giggenbach (1984). This improvement is mainly due to the use of the activities of K^+ and Ca^{2+} ions instead of the total concentrations of K and Ca, respectively.

Following the rigorous approach developed here, the CO_2 fugacities given by the theoretical K–Ca and Na–Ca calcite f_{CO_2}-indicators were considered reliable if higher than the maximum CO_2 fugacity of Ca–Al-silicate/calcite coexistence. However, the theoretical K–Ca and Na–Ca calcite f_{CO_2}-indicators can be probably used even if the CO_2 fugacities obtained by means of the theoretical K–Ca and Na–Ca calcite f_{CO_2}-indicators are somewhat lower than the maximum CO_2 fugacity of Ca–Al-silicate/calcite coexistence, provided that differences are not too large.

8.9 Final Considerations on the Activity-Based Theoretical K–Ca and Na–Ca Geoindicators

The activity-based theoretical K–Ca and Na–Ca geothermometers developed in this work are based on exchange reactions involving laumontite, clinozoisite, prehnite, and wairakite, whereas the activity-based theoretical K–Ca and Na–Ca f_{CO_2}-indicators implemented here are built on reactions including calcite. Adularia participates to the exchange reactions controlling all the theoretical K–Ca geothermometers and f_{CO_2}-indicators and some Na–Ca geothermometers and f_{CO_2}-indicators. Consequently, these functions involve the ordering parameter of hydrothermal adularia in hypothetical equilibrium with each reservoir liquid of interest, which is obtained from the Na–K activity-ratio and the reservoir temperature (see Chap. 6).

The theoretical K–Ca and Na–Ca geoindicators were tested with about one thousand reservoir liquids. The theoretical K–Ca geothermometers reproduce aquifer temperature with an average error of 3.8 °C and the error is lower than 10 °C in 97.3% of the cases. The theoretical Na–Ca geothermometers reproduce aquifer temperature with an average error of 6.9 °C and the error is less than 15 °C in 93.5% of the cases.

The CO_2 fugacities given by theoretical K–Ca and Na–Ca f_{CO_2}-indicators match satisfactorily the CO_2 fugacities given by the K–Ca f_{CO_2}-indicator of Giggenbach (1984), with an absolute deviation of 0.33 log-units on average, whereas the agreement with the CO_2 fugacities computed by means of speciation calculations is less good, with an absolute deviation of 0.78 log-units on average.

These good performances of the activity-based theoretical K–Ca and Na–Ca geothermometers, and to a lower extent of the f_{CO_2}-indicators as well, are expected because the activities of the Ca-bearing endmembers in relevant hydrothermal minerals from active geothermal systems (i.e., laumontite in laumontite/alkali-laumontite, clinozoisite in clinozoisite/epidote, prehnite in prehnite/ferri-prehnite, wairakite in wairakite/analcime, and calcite in calcite-rich trigonal carbonates) and the activity of muscovite in hydrothermal illites from active geothermal systems do not deviate too much from the average values and average activities do not depart too much from unity (see Chap. 4 and Sect. 8.2).[1]

In addition to these good performances in terms of computed aquifer temperatures and CO_2 fugacities, the theoretical K–Ca and Na–Ca geothermometers and f_{CO_2}-indicators allow one to identify the Ca-bearing solid phase in equilibrium with each reservoir liquid, either a Ca–Al-silicate (laumontite or clinozoisite or prehnite or wairakite) or calcite. This indication represents a significant step forward with respect to the results of the traditional K–Ca and Na–Ca geoindicators and is probably more reliable than the outcomes of multicomponent chemical geothermometry (see Sect. 5.9), being marginally affected by pH and Al concentration. Finally, the identification of the Ca–Al-silicate in equilibrium with each reservoir liquid may

[1]Following the same line of reasoning, K–Ca and Na–Ca theoretical geoindicators based on the exchange reactions involving grossular were not developed because grossular activity in 190 garnet solid solutions from active geothermal systems ranges from $<2.27 \cdot 10^{-7}$ (in 22 cases) to 0.674, with average of 0.127, median of 0.0142, and standard deviation of 0.179.

provide a qualitative indication on well permeability. In fact, wairakite is an indicator of high permeability, whereas prehnite and laumontite (if abundant) are indicators of poor permeability (Reyes 1990).

It is advisable, not to say it is mandatory, to complement the application of the theoretical K–Ca and Na–Ca geothermometers and f_{CO_2}-indicators derived in this book with the plots of the K^2/Ca log-activity ratio versus the absolute temperature reciprocal, the Na^2/Ca log-activity ratio versus the absolute temperature inverse (see Sect. 8.6) and the CO_2 fugacity versus the absolute temperature reciprocal (see Sect. 8.7), all providing useful overviews through the eyeball comparison of all the aqueous solutions of interest with the theoretical geoindicators.

References

Arnórsson S (1985) The use of mixing models and chemical geothermometers for estimating underground temperatures in geothermal systems. J Volcanol Geotherm Res 23:299–335

Arnórsson S, Gunnlaugsson E (1985) New gas geothermometers for geothermal exploration—calibration and application. Geochim Cosmochim Acta 49:1307–1325

Arnórsson S, Gunnlaugsson E, Svavarsson H (1983) The chemistry of geothermal waters in Iceland. III. Chemical geothermometry in geothermal investigations. Geochim Cosmochim Acta 47:567–577

Chiodini G, Cioni R, Guidi M, Marini L (1991) Chemical geothermometry and geobarometry in hydrothermal aqueous solutions: a theoretical investigation based on a mineral-solution equilibrium model. Geochim Cosmochim Acta 55:2709–2727

Ellis AJ (1970) Quantitative interpretation of chemical characteristics of hydrothermal systems. Geothermics 2:516–528

Giggenbach WF (1981) Geothermal mineral equilibria. Geochim Cosmochim Acta 45:393–410

Giggenbach WF (1982) "Geothermal mineral equilibria". Reply to a comment by M.A. Grant. Geochim Cosmochim Acta 46:2681–2683

Giggenbach WF (1984) Mass transfer in hydrothermal alterations systems. Geochim Cosmochim Acta 48:2693–2711

Giggenbach WF (1988) Geothermal solute equilibria. Derivation of Na-K-Mg-Ca geoindicators. Geochim Cosmochim Acta 52:2749–2765

Grant MA (1982) On the lack of a unique relation between CO_2 partial pressure and temperature in geothermal system. Comment on "Geothermal mineral equilibria" by W F. Giggenbach. Geochim Cosmochim Acta 46:2677–2680

Kacandes GH, Grandstaff DE (1989) Differences between geothermal and experimentally derived fluids: how well do hydrothermal experiments model the composition of geothermal reservoir fluids? Geochim Cosmochim Acta 53:343–358

Mahon WAJ, McDowell GD, Finlayson JB (1980) Carbon dioxide: its role in geothermal systems. New Zeal J Sci 23:133–148

Reyes AG (1990) Petrology of Philippine geothermal systems and the application of alteration mineralogy to their assessment. J Volcanol Geotherm Res 43:279–309

Truesdell AH, Nakanishi S (2005) Chemistry of neutral and acid production fluids from the Onikobe geothermal field, Miyagi Prefecture, Honshu, Japan. In: Use of isotope techniques to trace the origin of acidic fluids in geothermal systems. IAEA-TECDOC-1448, pp 169–193

Chapter 9
Conclusions and Way Forwards

Abstract In this chapter, the main characteristics of the theoretical, activity-based Na-K, K-Ca and Na-Ca geoindicators and the suggested procedure for their use are summarized. Possible improvements related to the choice of the thermodynamic database, the implementation of further theoretical, activity-based K-Ca and Na-Ca geoindicators, and the prediction of the ordering parameter of adularia on the basis of water chemistry are briefly discussed.

The main results of our research are the theoretical, activity-based Na–K, K–Ca and Na–Ca geoindicators, which can be used to compute aquifer temperature and CO_2 fugacity through the procedure summarized in Sect. 9.3. These geoindicators express the temperature dependence of the thermodynamic equilibrium constant of the exchange reactions involving pertinent hydrothermal minerals and take into account the average activities of relevant endmembers in solid solutions as well as Al–Si order-disorder on the tetrahedral sites of adularia if needed.

The same approach was applied also to derive K–Mg and Na–Mg geoindicators, but the usefulness of these geoindicators is limited by the extremely variable activity of clinochlore in chlorites. At best, the K–Mg and Na–Mg geoindicators can be used to assess the minimum concentrations of excess (non-equilibrium) Mg and, consequently, to identify and quantify related cooling and/or mixing processes, which may lead to Mg-silicate (e.g., smectite) scaling, as shown in Chap. 7.

9.1 The Theoretical, Activity-Based Na–K Geoindicators

The theoretical Na–K geoindicators were implemented building on previous findings of Bird and Norton (1981), who showed that Na–K geothermometry is complicated by the different structural state of alkali feldspars and Al–Si order-disorder on their tetrahedral sites. Consequently, a single Na–K geothermometer valid everywhere does not exist. Quite surprisingly, these important findings of Bird and Norton (1981)

R. Cioni and L. Marini, *A Thermodynamic Approach to Water Geothermometry*, Springer Geochemistry, https://doi.org/10.1007/978-3-030-54318-1_9

have received little attention in the scientific community and several geochemists have continued to propose different calibrations for the Na–K geothermometer.

Further evidence of interest for the elaboration of theoretical Na–K geoindicators is provided by the data available for both the authigenic alkali feldspars present in sedimentary rocks (e.g., Kastner and Waldbaum 1968; Morad 1978; Kastner and Siever 1979; Fishman et al. 1995) and the highly pure K-feldspars and albites synthesized at low temperatures by using hydroxide gels (Flehmig 1977). These data consistently indicate that K-feldspar crystals have various degrees of Al–Si ordering, whereas albite crystals are well-ordered. Assuming that hydrothermal alkali feldspars present in active geothermal systems have order-disorder characteristics similar to those of authigenic alkali feldspars present in sedimentary rocks and synthesized at low temperatures by means of hydroxide gels, the Na/K activity ratio of reservoir liquids is expected to be a function of both temperature and the ordering parameter of hydrothermal adularia, Z, which is equal to 1 for maximum-microcline, the proxy of totally ordered adularia, and assumes the value of 0 for high-sanidine, the proxy of completely disordered adularia.

In agreement with these expectations, 950 of the 1013 selected geothermal liquids (corresponding to 93.8% of the total) have Na/K activity ratio intermediate between the values fixed by low-albite/maximum microcline and low-albite/high-sanidine equilibrium coexistence, providing a strong support to our hypothesis. Interestingly and not surprisingly, also the Na–K geothermometers calibrated by different authors are limited by low-albite/maximum microcline and low-albite/high-sanidine equilibrium co-occurrence.

The two limiting theoretical Na–K geothermometers derived in this work, involving low-albite and either high-sanidine or maximum-microcline have little usefulness because they give equilibrium temperatures differing by 125 ± 3 °C on average, in the range 100–350 °C. Therefore, it is advisable to use the Na/K log activity ratio for computing the ordering parameter Z of hydrothermal adularia in hypothetical equilibrium with the aqueous solution of interest and to use this information in the theoretical K–Ca and Na–Ca geoindicators implemented in this work, representing the subject of the next section.

9.2 The Theoretical, Activity-Based K–Ca and Na–Ca Geoindicators

The theoretical, activity-based K–Ca and Na–Ca geothermometers implemented in this work are controlled by the exchange reactions involving different Ca–Al silicates, namely laumontite, clinozoisite, prehnite, and wairakite, whereas the K–Ca and Na–Ca theoretical, activity-based f_{CO_2}-indicators obtained here are governed by reactions comprising calcite. In addition to Ca–Al silicates or calcite, adularia is also included in the exchange reactions controlling all the theoretical K–Ca geoindicators and some Na–Ca geoindicators. Consequently, these relations are function of the ordering parameter of hydrothermal adularia which, in turn, depends on the Na–K activity-ratio and aquifer temperature, as recalled in previous section.

A total of about one thousand reservoir liquids were utilized to test the theoretical, activity-based K–Ca and Na–Ca geoindicators. The temperature given by the K–Ca geothermometers deviates from aquifer temperature by 3.8 °C on average, and the absolute value of the difference between K–Ca temperature and aquifer temperature is lower than 10 °C in 97.3% of the cases. The temperature calculated by means of the Na–Ca geothermometers differs from aquifer temperature by 6.9 °C on average and the absolute value of the difference between Na–Ca temperature and aquifer temperature is lower than 15 °C in 93.5% of the cases.

The K–Ca and Na–Ca f_{CO_2}-indicators give CO_2 fugacities showing absolute deviation of 0.78 log-units, on average, from the CO_2 fugacities calculated by means of speciation calculations. However, the CO_2 fugacities computed by means of the K–Ca and Na–Ca f_{CO_2}-indicators are in satisfactory agreement with the CO_2 fugacities given by the K–Ca f_{CO_2}-indicator of Giggenbach (1984) having an absolute deviation of 0.33 log-units on average.

The good performance of the theoretical, activity-based K–Ca and Na–Ca geothermometers, and to a lower extent of the f_{CO_2}-indicators as well, are not surprising because the controlling exchange reactions involve well-behaved hydrothermal minerals. In fact, the activities of the Ca-bearing endmembers (i.e., laumontite, clinozoisite, prehnite, wairakite, and calcite) in relevant hydrothermal minerals from active geothermal systems (i.e., the solid solutions made up of laumontite/alkali-laumontite, clinozoisite/epidote, prehnite/ferri-prehnite, wairakite/analcime, and trigonal carbonate) exhibit limited variations from the average values and these average activities are not too different from unity.[1]

Moreover, the Ca-bearing mineral in equilibrium with each reservoir liquid, either a Ca–Al-silicate (laumontite or clinozoisite or prehnite or wairakite) or calcite can be identified by using the K–Ca and Na–Ca theoretical geothermometers and f_{CO_2}-indicators. This indication is a considerable improvement with respect to the outcomes of the traditional K–Ca and Na–Ca geoindicators, and is probably more reliable than the results of multicomponent chemical geothermometry (see Sect. 5.9), being slightly influenced by pH and Al concentration. Finally, the identification of the Ca–Al-silicate in equilibrium with each reservoir liquid may give a qualitative indication on well permeability. In fact, wairakite is an indicator of high permeability, whereas prehnite and laumontite (if abundant) are indicators of poor permeability (Reyes 1990).

[1]Incidentally, the exchange reactions involving grossular were disregarded because grossular activity in 190 garnet solid solutions from active geothermal systems spans a very large range, from less than 2.27×10^{-7} (for 22 cases) to 0.674, with an average of 0.127 and a standard deviation of 0.179.

9.3 Suggested Procedure for the Use of the Theoretical, Activity-Based Na–K, K–Ca and Na–Ca Geoindicators

The suggested procedure for using the theoretical, activity-based Na–K, K–Ca and Na–Ca geoindicators includes a series of steps as outlined below and schematically presented in the workflow of Fig. 9.1.

Step 1. Data organization, including the calculation/assessment of the aquifer temperature for the aqueous solution of interest based on silica geothermometers or other available data, such as measured aquifer temperature and discharge enthalpy for wells.

Step 2. Speciation calculations to compute the activities of relevant species by means of WATCH, PHREEQC Interactive, EQ3/6, or similar computer programs. If the charge unbalance is lower than 10%, it is possible to proceed to the next step. Otherwise the procedure should be stopped.

Step 3. Calculation of the pH fixed by mineral-solution equilibrium, pH_{mse}, by means of Eq. (3.10), involving aquifer temperature, total ionic salinity, and the CO_2 fugacity given by speciation-saturation calculations. If the difference between pH_{mse} and the pH given by speciation calculations is less than 1 pH unit, it is possible to proceed to the next step. Otherwise, it is advisable to change the reconstruction of reservoir fluids, adopting a different approach, as discussed in Sect. 3.1.

Step 4. Calculation of the ordering parameter Z of hydrothermal adularia in hypothetical equilibrium with the aqueous solution of interest on the basis of aquifer temperature and the Na/K activity ratio.

Step 5. Calculation of CO_2 fugacity by means of the K–Ca and Na–Ca calcite f_{CO_2}-indicators, that is Eqs. (8.45)–(8.48) and the f_{CO_2}—temperature functions controlled by equilibrium coexistence of a Ca–Al-silicate and calcite, that is Eqs. (8.49)–(8.56), based on aquifer temperature and the ordering parameter Z of adularia. If the CO_2 fugacity given by the K–Ca and Na–Ca calcite f_{CO_2}-indicators is higher than the maximum CO_2 fugacity of Ca–Al-silicate/calcite coexistence, then the aqueous solution of interest is in equilibrium with calcite, and the computed CO_2 fugacity is representative of the geothermal aquifer. Otherwise, the K–Ca and Na–Ca geothermometers should be used (see next step) instead of the K–Ca and Na–Ca calcite f_{CO_2}-indicators.

Step 6. Calculation of the K–Ca and Na–Ca temperatures using the different theoretical, activity-based K–Ca and Na–Ca geothermometers, that is Eqs. (8.29)–(8.44), considering the ordering parameter Z of adularia where needed.

Step 7. Comparison of the different K–Ca and Na–Ca temperatures with the aquifer temperature established in step 1, selection of the K–Ca and Na–Ca temperatures closest to the aquifer temperature, and identification of the Ca–Al silicate(s) in equilibrium with the aqueous solution of interest or close to this condition.

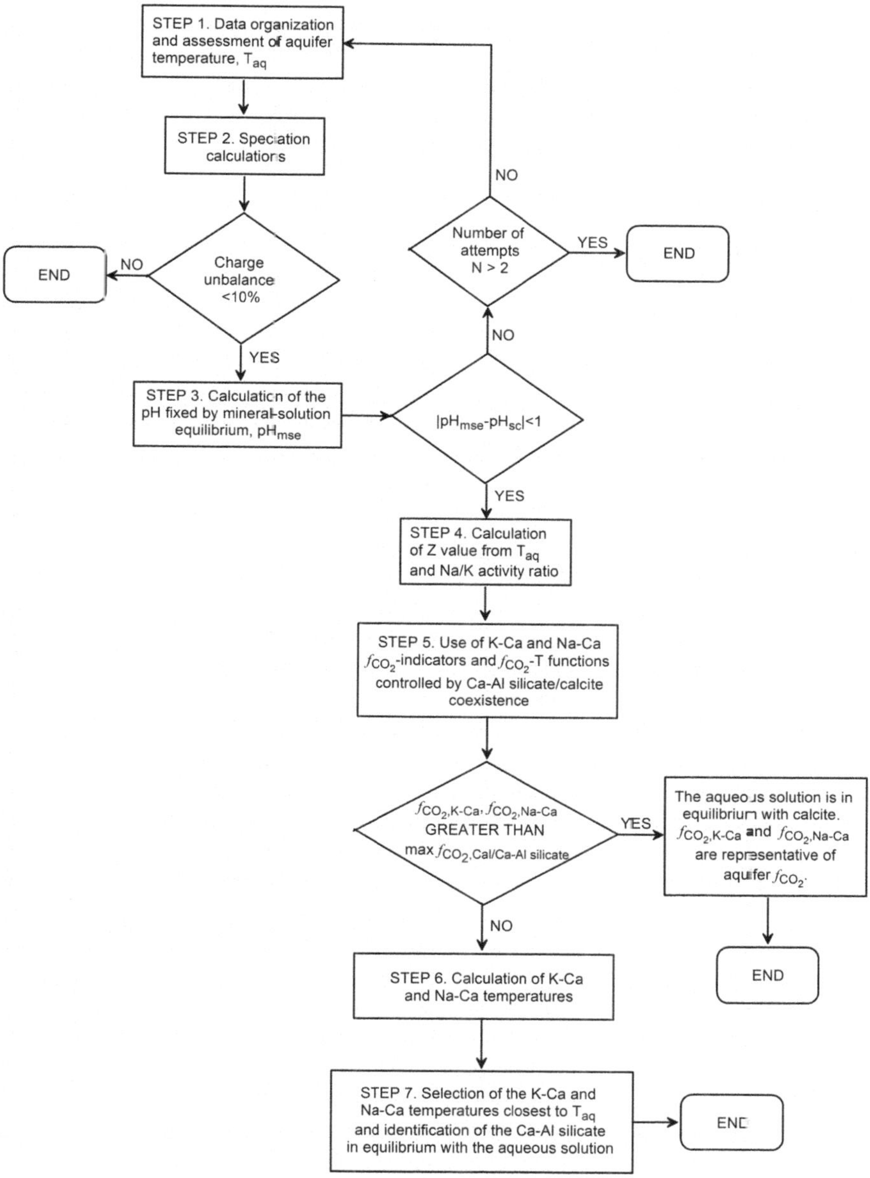

Fig. 9.1 Workflow of the suggested procedure for the use of the theoretical, activity-based Na–K, K–Ca and Na–Ca geoindicators

In addition to this procedure, it is advisable, not to say it is mandatory, to use the diagrams of K^2/Ca log-activity ratio vs. the absolute temperature reciprocal, Na^2/Ca log-activity ratio vs. the absolute temperature inverse and CO_2 fugacity vs. the absolute temperature reciprocal, allowing an eyeball comparison of all the aqueous solutions of interest with the theoretical geoindicators.

9.4 The Way Forward

The approach to water geothermometry and f_{CO_2} evaluation proposed in this book was thoroughly tested with over one thousand reservoir liquids coming from different geothermal fields worldwide. Therefore, we are confident that it can be applied elsewhere with a high probability of success. Nevertheless, we are aware that the work we did is far from complete for at least two reasons.

The first reason relates to our choice of the thermodynamic database, that will certainly make many people turn up their noses. As long as a standard, internationally accepted thermodynamic database does not exist, it could be interesting to use different thermodynamic databases for calculating the thermodynamic equilibrium constants of the Na–K, K–Ca, Na–Ca, K–Mg, and Na–Mg exchange reactions of interest. This is a fairly simple exercise, which we have deliberately chosen not to perform, apart from the test of alkali feldspars transition temperatures in Sects. 4.2.4 and 6.1, because we trust the thermodynamic database implemented by Helgeson and coworkers and we did not want to complicate too much this book.

A second reason is the limited number of Ca–Al silicates taken into account in this work to implement the theoretical, activity–based K–Ca and Na–Ca geoindicators. In addition to the Ca–Al silicates considered here, that is laumontite, clinozoisite, prehnite, and wairakite, other Ca–Al silicates and Ca-silicates occur as hydrothermal minerals in active geothermal systems and, therefore, are of interest for the elaboration of water geothermometers and f_{CO_2}-indicators. It is worth recalling the Ca-bearing zeolites (e.g., heulandite, stilbite, chabazite, thomsonite, scolecite, mordenite, yugawaralite, levyne, gismondine, and gmelinite), which are stable at low temperatures, usually below 200 °C, as well as actinolitic amphibole, Ca clinopyroxene, andradite-grossular garnet, and wollastonite, which are stable at high temperatures, generally $\geq$300 °C (e.g., Bird et al. 1984). Some of these minerals, however, exhibit large ranges of compositions that might complicate their use for the derivation of suitable geoindicators, similar to what we noted for garnet. Titanite (sphene) might be another Ca-bearing mineral of interest, together with other Ti-bearing solid phases, for the elaboration of geoindicators. The main problem, however, is the lack of reliable thermodynamic data as discussed in Sect. 9.4 and, again, we return to the previous issue.

Furthermore, there is room for improvement related to the ordering parameter Z of adularia, which is a subject of major importance. In fact, in the suggested procedure, Z is computed as a function of the Na/K activity ratio and aquifer temperature, but this approach introduces a sort of vicious circle in geothermometric calculations. Since previous researches (Martin 1969; Flehmig 1977) have shown that the structural state and the degree of ordering of alkali feldspars depends on pH and Na and K concentrations, it should be possible to predict Z on the basis of water chemistry. We have tried to investigate this possibility but our attempts were unsuccessful. Experiments of alkali feldspar synthesis using hydroxide gels might provide further evidence on the parameters possibly controlling their degree of ordering. Other indications might results from the chemistry of hydrothermal alkali feldspars in geothermal aquifers hosting reservoir liquids rich in boron and ammonium, such as Ngawha and Monte Amiata, because both B^{3+} and NH_4^+ ions are incorporated in the alkali feldspars lattice.

Moreover, other activity ratios might be of interest for geothermometry, e.g., the $SO_4/(F)^2$ ratio, and for the implementation of f_{CO_2}-indicators, e.g. the HCO_3/F and $(HCO_3)^2/SO_4$ ratios, as pointed out by Chiodini et al. (1991). We intentionally avoided to consider these potential geo-indicators in order to conclude this book (or perhaps its first edition) and submit our results, albeit partial, to the attention of the scientific community.

References

Bird DK, Norton DL (1981) Theoretical prediction of phase relations among aqueous solutions and minerals: Salton Sea geothermal system. Geochim Cosmochim Acta 45:1479–1494

Bird DK, Schiffman P, Elders WA, Williams AE, McDowell SD (1984) Calc-silicate mineralization in active geothermal systems. Econ Geol 79:671–695

Chiodini G, Cioni R, Guidi M, Marini L (1991) Chemical geothermometry and geobarometry in hydrothermal aqueous solutions: a theoretical investigation based on a mineral-solution equilibrium model. Geochim Cosmochim Acta 55:2709–2727

Fishman NS, Turner CE, Brownfield IK (1995) Authigenic albite in a Jurassic alkaline, saline lake deposit, Colorado Plateau. Evidence for early diagenetic origin. U.S. Geological Survey Bulletin 1808, Evolution of Sedimentary Basins, San Juan Basin, pp 1–13

Flehmig W (1977) The synthesis of feldspars at temperatures between 0–80 °C, their ordering behaviour and twinning. Contrib Mineral Petr 65:1–9

Giggenbach WF (1984) Mass transfer in hydrothermal alterations systems. Geochim Cosmochim Acta 48:2693–2711

Kastner M, Waldbaum DR (1968) Authigenic albite from Rhodes. Am Mineral 53:1579–1602

Kastner M, Siever R (1979) Low temperature feldspars in sedimentary rocks. Am J Sci 279:435–479

Martin RF (1969) The hydrothermal synthesis of low albite. Contrib Mineral Petr 23:323–339

Morad S (1978) Feldspars in sedimentary rocks. In: Sedimentology. Encyclopedia of earth science. Springer, Berlin, Heidelberg

Reyes AG (1990) Petrology of Philippine geothermal systems and the application of alteration mineralogy to their assessment. J Volcanol Geother Res 43:279–309

The manufacturer's authorised representative in the EU is Springer Nature Customer Service Centre GmbH, Europaplatz 3, 69115 Heidelberg, Germany. If you have any concerns regarding our products, please contact ProductSafety@springernature.com

Printed and bound by CPI Group (UK) Ltd, Croydon, CR0 4YY
30/09/2026
02225507-0017